Edited by
Bruno Pignataro

Discovering the Future of Molecular Sciences

Related Titles

Pignataro, B. (ed.)

Molecules at Work

Selfassembly, Nanomaterials, Molecular Machinery

2012

ISBN: 978-3-527-33093-5

Pignataro, B. (ed.)

New Strategies in Chemical Synthesis and Catalysis

2012

ISBN: 978-3-527-33090-4

Pignataro, B. (ed.)

Ideas in Chemistry and Molecular Sciences

Advances in Synthetic Chemistry

2010

ISBN: 978-3-527-32539-9

Pignataro, B. (ed.)

Ideas in Chemistry and Molecular Sciences

Where Chemistry Meets Life

2010

ISBN: 978-3-527-32541-2

Pignataro, B. (ed.)

Ideas in Chemistry and Molecular Sciences

Advances in Nanotechnology, Materials and Devices

2010

ISBN: 978-3-527-32543-6

Pignataro, B. (ed.)

Tomorrow's Chemistry Today

Concepts in Nanoscience, Organic Materials and Environmental Chemistry Second edition

2009

ISBN: 978-3-527-32623-5

Edited by
Bruno Pignataro

Discovering the Future of Molecular Sciences

WILEY-VCH
Verlag GmbH & Co. KGaA

Editor

Prof. Bruno Pignataro
Università di Palermo
Dipartimento di Fisica e Chimica
Viale delle Scienze ed. 17
90128 Palermo
Italy

Library of Congress Card No.: applied for

British Library Cataloguing-in-Publication Data
A catalogue record for this book is available from the British Library.

Bibliographic information published by the Deutsche Nationalbibliothek
The Deutsche Nationalbibliothek lists this publication in the Deutsche Nationalbibliografie; detailed bibliographic data are available on the Internet at <http://dnb.d-nb.de>.

Print ISBN: 978-3-527-33544-2
ePDF ISBN: 978-3-527-67325-4
ePub ISBN: 978-3-527-67324-7
Mobi ISBN: 978-3-527-67323-0
oBook ISBN: 978-3-527-67322-3

Cover-Design Adam-Design, Weinheim, Germany
Typesetting Laserwords Private Limited, Chennai, India
Printing and Binding Markono Print Media Pte Ltd, Singapore

Printed on acid-free paper

Contents

Preface

This book is the last of the series based on The European Young Chemist Award (EYCA) competition and it reports on some of the latest hits of chemistry by young excellence.

The EYCA is indeed aimed to showcase and recognize the excellent research being carried out by young scientists (less than 35 years old) working in the chemical sciences. In particular, it is intended to honour and encourage younger chemists whose current research displays a high level of excellence and distinction. It seeks to recognize and reward younger chemists of exceptional ability who show promise for substantial future achievements in chemistry-related research fields.

The inaugural award was bestowed during the first European Chemistry Congress, which took place at the ELTE Convention Centre in Budapest in 2006, while the second and the third were in 2008 and 2010 during the same conferences in Torino (Italy) and Nürnberg (Germany), respectively.

The quality of the young chemists competitors was so high that I decided in all these cases to edit books collecting their contributions. Thus always with Wiley-VCH as Publisher and under the patronage of the major European Chemical Societies and the European Association for Chemical and Molecular Sciences (EuCheMS) and of the Italian Chemical Society (SCI) as sponsors I edited the following books: Tomorrow's Chemistry Today-Concepts in Nanoscience, Organic Materials and Environmental Chemistry (2nd Ed. 2009); Ideas in Chemistry and Molecular Sciences-Advances in Synthetic Chemistry (2010); Ideas in Chemistry and Molecular Sciences-Where Chemistry Meets Life (2010); Ideas in Chemistry and Molecular Sciences-Advances in Nanotechnology, Materials and Devices (2010); Molecules at Work-Self-assembly, Nanomaterials and Molecular Machinery (2012); New Strategies for Chemical Synthesis and Catalysis (2012).

The fourth European Young Chemist Award was presented in Prague (Czech Republic) during the fourth EuCheMS Chemistry Congress (2012).

As it occurred for all the previous awards, the scientific quality of the young chemists competitors was again outstanding.

Just to give an idea of their scientific level and therefore of the expected quality of the chapters in the book, I am delighted and proud to report some very short statements extracted from the supporting letters of some of the competitors of the awards invited by me to contribute to this book.

"In my experience, it will be very difficult to find a scientist of this age with better personality and higher capacities than him"; "He has done stellar work"; "She is a superb scientist with the skills to perform incredibly difficult experiments and to model results based on theory. She has shown the ability to imagine innovative ideas for new research directions"; "I consider him among the most brilliant European chemists of his generation"; "The best way to define to him is as truly exceptional"; "I believe he is one of the leaders of the actual generation of European Chemists"; "I can qualify him without hesitation as the best PhD student I had so far in my career"; "He is a rising star in the field of chemistry"; "He is rapidly being recognized worldwide as one of the leading young European chemists". "He has pioneered a number of new research strands. I consider the candidate to be one of the top, if not the top, person I have mentored".

Two among the authors of the chapters have got the ERC starting grant and some of them got different awards. Much of the scientific production of all the authors is in high-quality Journals with some of the competitors having papers in Nature, Science, Chem. Rev., Angew. Chem., JACS and other important Journals.

After the brief genesis of the book and the above points on the scientific quality of the authors, let me spend some words about its content.

The book is divided into two parts: "Advanced methodologies" and "Materials, Nanoscience and Nanotechnologies".

In the first part there are various collected contributions ranging from analytical methodologies involving recognition issues or mass spectrometry to the area of studies involving electronic energy transfer and pump and probe methodologies as well as micro flow chemistry or advanced calculation methodologies.

The first chapter, entitled "Supramolecular receptors for the recognition of bioanalytes" by Amilan Jose Devadoss (in collaboration with Prof Alexander Schiller and Dr Amrita Ghosh), reports on fluorogenic and chromogenic supramolecular sensors for the recognition of important bioanalytes and their applications in various biological studies. Studies conducted by the author and examples from other researchers are considered. Thus, promising examples for the recognition of bioanalytes like pyrophosphate, nucleoside triphosphates, carbohydrates, lipopolysaccharides and nucleic acids are described. Metal complexes with chromogenic or luminescent motif (mainly of the Zn(II) type), new color- and fluorescence-based polydiacetylene vesicle systems and boronic acids have been the considered receptors. Potential application in biological cell staining, drug delivery, and molecular logic functions has also been summarized. In agreement with the authors I believe that this chapter will inspire new advancement in the research area of bioanalytes recognition and in the discovery of molecular sciences in the future.

To the same broad area of research than that by Devadoss *et al.* belongs the next contribution by Olalla Vázquez. The title is "Methods for DNA recognition". Owing to the paramount importance of DNA for life, the focus is however here on the molecular bases of double stranded DNA (dsDNA) recognition. Special emphasis is placed on recognizing the most relevant conformation under physiological conditions: the so- called B-form of dsDNA. The interaction of natural transcription factors (TFs) with the DNA, gene expression, and the current developments in the design and preparation of synthetic dsDNA binders are considered. As to this last items, within the discussion on the Metallo-DNA and Polypyrroles and bis(benzamidine) binders, I like to mention that a schematic representation of the cytotoxic pathway of the famous cisplatin and the simple explanation of the cell death is reported. In conclusion, I feel that the chapter, in some aspect, tries to provide a contribution to yet incompletely answered important questions in the field, like those pushed by the author: "How do the large and diverse number of DA-binding proteins recognize their specific binding sites? Which are the rules that govern how proteins bind to DNA sequences?"

The next chapter by Yury Tsybin is dedicated to the astonishing advances in high resolution and tandem MS applied to structure analysis of complex molecular systems. In this chapter, following the presentation of the basic principles in mass spectrometry (MS), the Fourier Transform Mass Spectrometer that gives superior resolving power and mass accuracy among all types of mass spectrometers is introduced. Then the configuration and working principles of some modern MS variants, namely, Orbitrap Fourier Transform MS (Orbitrap FTMS), Ion Cyclotron Resonance FTMS (ICR FTMS) and Time of Flight FTMS (TOF FTMS) are described with particular emphasis on the first two because of their wider spread and commercial availability compared to TOF FTMS. This part of the chapter is followed by two sections with a discussion on the applications of high resolution MS and tandem mass spectrometry (MS/MS) in the analysis of complex mixtures or biological samples. The study of peptides and proteins with the emerging field of native mass spectrometry (which aims at preserving the solution phase protein–ligand interactions) and petroleomics (comprehensive molecular structure analysis of crude oils and complex petroleum fractions by high-resolution FTMS) are, for example, research areas that should benefit greatly from these methodologies. Great effort is made by the author to give suggestions on how to improve the actual performance of the available instrumentation in order to cope with the always increasing demand for analytical chemistry.

The next contribution by Elisabetta Collini is entitled "Coherent electronic energy transfer in biological and artificial multichromophoric systems" and deals with electronic energy transfer (EET), a phenomenon that is important for efficient light-harvesting in photosynthesis, the development of fluorescence-based sensor technologies, and improvements in solar cell design. In particular the chapter, well balanced between introductory theorethical problems and experimental studies, focuses on the involvement of quantum-coherence in this type of phenomenon and provides some basis to allow to answer the two following fundamental questions outlined by the author: "To what extent such coherences are really relevant for the efficiency and the mechanism of biological and artificial EET processes? Would it be possible to implement quantum interference effects to control and optimize energy transfer pathways?" After an introductory part in which the author briefly talks of the EET phenomenon, the meaning of electronic coherence in energy transfer, the theorethical interpretation of the energy migration, what mentioned above is done by first presenting the developments of new ultrafast spectroscopy experiments and then describing and discussing some experimental studies on coherent electronic energy transfer in two multichromophoric systems: a light-harvesting antenna isolated from a marine cryptophyte alga and the conjugated polymer MEH-PPV (poly[2-methoxy,5-(2′-ethyl-hexoxy)-1,4-phenylenevinylene.

The next chapter is provided by Danielle Buckley and is entitled "Ultrafast Studies of Carrier Dynamics in Quantum Dots for Next Generation Photovoltaics". It is pointed out here that first generation devices suffer from losses in efficiency because of different causes, while second generation devices make them more appealing because of the lower material and manufacturing costs. Third generation photovoltaics (PVs), also referred to as next generation PVs, aims to correct one or more efficiency losses found in first and second generation devices as well as to lower the costs. Next generation approaches to achieve these improvements include utilizing multi-junction cells, intermediate band cells, hot-carriers, multiple exciton generation (MEG), and spectrum conversion. After some introductory sections talking of concepts that are needed to understand carrier dynamics in quantum dots, this chapter focuses on ultrafast studies of quantum dots that have the potential to contribute to the development of hot carrier and MEG cells. These include transient absorption (TA), time-resolved terahertz spectroscopy (TRTS), and time-resolved photoluminescence (TRPL). In each case ultrafast pulses are used to excite or 'pump' a sample with energy at or above the band gap and the subsequent probe or resulting emission provides information about carrier dynamics. Some issues on the chemistry of the quantum dots used in the third generation PVs are also reported. The overall situation described in the chapter suggests a rapid advancement of quantum dot PV devices.

In the next chapter by Timothy Noël entitled "Micro Flow Chemistry: New Possibilities For Synthetic Chemists" the new possibility for synthetic chemists offered by micro flow chemistry are presented. Starting from a introduction of the basic engineering principles of micro flow, this chapter gives an overview of the most important advantages of micro flow chemistry for the organic synthetic chemist with respect to traditional batch techniques. Thus it is stressed that unusual reaction conditions far from the common laboratory practices such as high temperatures and high pressures or the use of hazardous intermediates, are enabled by microreactor technology. Also, scale-up problems that have to be considered to go from laboratory scale to production scale and the reaction screening and the optimization protocols in microreactors are issues considered in this contribution. The chapter ends with a section where the author says how he sees the field evolving in the near future.

On the basis of recent contributions from the author's laboratories and selected highlights from the Houk and Bickelhaupt research groups, the next chapter by Israel Fernández López is entitled "Understanding trends in reactions barriers" and contributes to an old challenge for chemists: the need to control the reactivity of molecules.

In the chapter, the author demonstrates the good performance of the combined activation strain (ASM) model/ energy decomposition analysis (EDA) method to explore and understand trends in reactivity in various fundamental types of reactions in organic chemistry such as Pericyclic Reactions (Double Group Transfer Reactions, Alder-ene Reactions, 1,3-Dipolar Cycloaddition Reactions, Diels-Alder Reactions) Nucleophilic Substitutions and Additions, S_N2 Reactions, Nucleophilic Additions to Arynes, as well as Unimolecular Processes.

The second Part of the book provides contributions on a series of materials going from polyoxometalates (POMs) to other metal complexes. Nanoparticle assemblies and porous molecular solids are two other considered themes. The two last chapters deal with molecular machines and motors. Nanoscience and nanotechnology issues are often reported in most of these chapters.

The first chapter in this Part is provided by Haralampos N. Miras and is dedicated to the science of molecular metal oxides or POMs. These molecular systems have attracted the attention of research groups over the years, because of their plethora of unique archetypes with applications ranging from catalysis and medicine to molecular electronics, magnetism, energy, and so on. The chapter shows that after a period in which the discovery of new architectures was connected to serendipity it is now possible to design and control to an important extent both the structure as well as the function of the systems. This is achieved essentially by combining the use of new techniques like ESI/MS and the new synthetic approaches discussed in the chapter. The new discoveries and developments in the area has led to a variety of unprecedented architectures and the emergence of intriguing properties and new phenomena, paving the route for the engineering of materials with innovative functionalities. On the other hand, the capability of a real control over the self-assembly processes of these complex chemical systems opens the door for further discoveries towards a well-established and directed functional future as it is written in the title of this contribution.

Again, the second chapter in this Part, by Andrey Seliverstov, Johannes Forster, Johannes Tucher, Katharina Kastner and Carsten Streb, deals with POMs. Let me start the comments on this contribution stressing that, as outlined by the authors, the POMs possess, among others, a great capacity to incorporate a wide range of heterometals into the cluster shell, thus giving access to a large number of cluster derivatives with tunable physicochemical properties.

In this chapter the focus is on the immense potential of these systems for the development of new energy conversion and storage systems. The authors outline first the electrochemical and photochemical activity of POMs and then the applications are considered. Thus treated themes are: the POM photocatalysis and the conversion of light into chemical reactivity; the energy conversion and the splitting of water into oxygen and hydrogen; the oxidation of water to molecular oxygen and protons by using POMs; the photoreductive H_2-generation or the photoreductive CO_2-activation always exploiting POMs. In the second part of the chapter the authors describe the important role of POM ionic liquids (POM-ILs) in the area and after that they report a section on POM-based photovoltaics where the discussion is centered on the fact that POM anions have been employed as redox active components for the assembly of photoelectrical cells for sunlight to electricity conversion. A final section is dedicated to POM-based molecular cluster batteries.

The next chapter is provided by Shigeyosh Inoue and is entitled "The next generation of silylene ligands for better catalysts".

In this chapter after a brief general introduction on silylene (that can be considered as the heavier analog of carbene), bis(silylene), and silylene transition metal complexes, the author reports on the synthesis and catalytic applications of silylene transition metal complexes. Ti, Ni, Pd, Ir, Rh as well as Fe containing complexes have been considered in these respects. The key of the game is that the ligand is always used to modulate the electronic properties of the transition metal. Also, steric effect may be obviously operative when bulky ligands are considered. In agreement with the author I believe that "although a broad range of fascinating achievements have been recently disclosed, this research area is still unexplored, and more fascinating advances will be made in the near future".

The next chapter is provided by Alicia Casitas and is entitled "Halide Exchange Reactions Mediated by Transition Metals". Here the author, after having outlined the practical importance of the halide exchange reactions in various fields, gives an overview of the history and developments of these types of reactions with particular emphasis to the nickel-, palladium-, and copper-mediated reactions. The need to improve the actual situation in order to have milder and more environmentally benign type of reactions and the need to have more efficient and practical synthetic methods are underlined.

The next chapter by Marie-Alexandra Neouze Gauthey is entitled "Nanoparticle assemblies from molecular mediator" and is dedicated to the synthesis and applications of nanoparticle assembly. As to the synthesis, the following methods are reviewed: (i) inter-ligand bonding, where a molecule is introduced between the nanoparticles and will remain in the final material; (ii) template-assisted method, where the template molecules will force the organization of the nanoparticles; (iii) deposition of 2D assemblies, where the interaction with a surface helps to organize the nanoparticle assembly; and (iv) pressure driven assemblies. Then the chapter deals with some applications of such materials. For this reason, plasmonic nanostructures for sensing, communication or signal enhancement, magnetic nanostructures, metamaterials, as well as catalysis are considered.

The next chapter is provided by Shan Jiang in collaboration with Andy Cooper and Abbie Trewin and is entitled "Porous molecular solids". This contribution deals with microporous materials that have pore sizes smaller than 2 nm and are of strong interest as they have potential applications in separations, gas storage, catalysis, sensors, and drug delivery. Porous organic molecular crystals and Porous amorphous molecular materials are both considered. For the first type of systems, porous organic molecules like the well-known calixarenes or other chemical systems are first reviewed. Then an overview is done on the porous organic cage molecules developed by the Cooper's research group and prepared by cycloimination condensation reactions. The work done in other groups is also reported. This is followed by a section dedicated to simulation issues in order to show how useful molecular modeling and simulation tools to design and rationalize the properties of these systems are. A further section deals with applications. As to the amorphous systems, the problems of synthesis and simulation are again taken into account underlining the fact that obviously here they are more challenging with respect to the crystalline systems. In all cases, the structure activity connections and the success since now obtained on the synthetic control of the structures of these systems are highlighted and discussed.

The next contribution is provided by Gabriel Loget and Alexander Kuhn and is entitled "Electrochemical Motors." Here, some examples of moving objects are first presented. Thus, examples of biomotors, chemical motors such as self-electrophoretic swimmers and bubble-propelled swimmers or externally powered motors (which do not need a fuel molecule for the movement like the magnetically-propelled swimmers) are briefly discussed. It is then noted that, because of morphological or chemical reasons as well as being introduced by an electric or magnetic field, some form of asymmetry is always present in all the reported cases. Thus the authors state and show that asymmetry is crucial for the generation of controlled motion; the key concept for the propulsion of particles is asymmetry. Because bipolar electrochemistry, a phenomenon known for a long time and originally used in industrial application for electrolysis or batteries, intrinsically provides a break of symmetry, which can be induced on any kind of conducting object, it is an appealing alternative to the existing mechanisms for motion generation. The chapter is then dedicated to show the potentiality of this methodology and describe different strategies that, by using bipolar electrochemistry, can trigger different types of motion.

The last chapter by Massimo Baroncini and Giacomo Bergamini is entitled "Azobenzene in Molecular and Supramolecular Devices and Machines" and gives a contribution to the design of synthetic nanomachines able to carry out movements at the molecular and supramolecular scale triggered by external stimuli. In the reported examples, azobenzene moieties are part of molecular and supra-molecular architectures in which photoisomerization controls molecular movements and nanoscale interactions.

According to the authors the results described show that "molecular and supramolecular systems capable of performing large-amplitude controlled mechanical movements upon light stimulation can be obtained by careful incremental design strategies, the tools of modern synthetic chemistry, and the paradigms of supramolecular chemistry, together with inspiration from natural systems."

The book is aimed at advanced and specialist researchers. It should be relevant for both readers from academia and industry as it will deal with fundamental contributions as well as possible applications. The contributions come essentially from academia researchers. The audience I feel need this book is Chemists in Advanced Methodologies, Materials, Nanoscience, Nanotechnologies, and Chemical Synthesis areas. The audience with an occasional need for this book should be that of Physicists and Engineers.

I am not aware of books that can compete with the proposed one for the peculiarity of being a book written with the contributions of top-level young chemists. All the chapters are written in a clear and simple way and all try to give perspectives for the future.

Going to the conclusions and in connection with these crucial times I would like to say what one of the fourth EuCheMS Congress attendees told me at the end of the event: Future is done! And one can probably be more optimistic by looking at the creativity shown by this generation of scientists and their ability to develop interdisciplinary and collaborative projects with such a high degree of innovation. Putting everything together I really thing that the book helps in discovering at least a part of the future of the Molecular Science.

I cannot finish this preface without acknowledging the various institutions and people that supported the EYCA rendering possible this new book: the Italian Consiglio Nazionale dei Chimici (CNC) and the Italian Chemical Society (SCI) and their Presidents, Roberto Zingales and Vincenzo Barone, for sponsoring the Award; the Symposia Chairs and Experts involved in the selection of finalists; the Jury for their availability for this hard task; my coworkers for their continuous help; Francesco De Angelis, Sergio Facchetti and Nineta Majcen for the help and encouragement; the local organizers with Pavel Drasar for the support; the EYCN, EuCheMS and the fourth EuCheMS Chemistry Congress for their patronage.

Università di Palermo
Palermo, Italy

Bruno Pignataro

List of Contributors

Massimo Baroncini
Università di Bologna
Dipartimento di Chimica
"G. Ciamician"
via Selmi 2
I-40126 Bologna
Italy

Giacomo Bergamini
Università di Bologna
Dipartimento di Chimica
"G. Ciamician"
via Selmi 2
I-40126 Bologna
Italy

Danielle Buckley
University of Colorado Boulder
Department of Chemistry and
Biochemistry
Boulder, CO 80309
USA

Alicia Casitas Montero
Max-Planck-Institut für
Kohlenforschung
Department of Organometallic
Chemistry
Kaiser-Wilhelm-Platz 1
45470 Mülheim an der Ruhr
Germany

and

Max-Planck-Institut für
Kohlenforschung
Kaiser-Wilhelm-Platz 1
45470 Mülheim an der Ruhr
Germany

Elisabetta Collini
Università di Padova
Dipartimento di Scienze
Chimiche
via Marzolo 1
35131 Padova
Italy

Andrew I. Cooper
The University of Liverpool
Department of Chemistry
Crown Street
Liverpool L69 7ZD
UK

Israel Fernández López
Universidad Complutense de
Madrid
Departamento de Química
Orgánica
Facultad de Ciencias Químicas
Avda. Complutense s/n
28040 Madrid
Spain

Johannes Forster
Friedrich-Alexander-University
Erlangen-Nuremberg
Department of Chemistry and
Pharmacy
Inorganic Chemistry II
Egerlandstr. 1
91058 Erlangen
Germany

Amrita Ghosh
University of Bielefeld
Department of Inorganic
Chemistry
Universitätsstraße 25
Fakultät für Chemie
D-33501 Bielefeld
Germany

Shigeyoshi Inoue
Institut für Chemie
Anorganische Chemie
Technische Universität Berlin
Straße des 17. Juni 135
Sekr. C2
D-10623 Berlin
Germany

Shan Jiang
The University of Liverpool
Department of Chemistry
Crown Street
Liverpool L69 7ZD
UK

D. Amilan Jose
Friedrich Schiller University Jena
Faculty of Chemistry and Earth
Sciences
Institute for Inorganic and
Analytical Chemistry
Humboldtstrasse 8
D-07743 Jena
Germany

and

Department of Chemistry
National Institutes of Technology
Kurukshetra
Haryana-136119
Thanesar
India

Katharina Kastner
Friedrich-Alexander-University
Erlangen-Nuremberg
Department of Chemistry and
Pharmacy
Inorganic Chemistry II
Egerlandstr. 1
91058 Erlangen
Germany

and

University of Ulm
Institute of Inorganic Chemistry I
Albert-Einstein-Allee 11
89081 Ulm
Germany

Alexander Kuhn
Université de Bordeaux
ISM, ENSCBP
UMR 5255
16 Avenue Pey Berland
33607 Pessac
France

Gabriel Loget
University of California-Irvine
Department of Chemistry
Irvine
California 92697
United States

Haralampos N. Miras
The University of Glasgow
School of Chemistry
Glasgow G12 8QQ
UK

Marie-Alexandra Neouze Gauthey
Institute of Materials Chemistry
Vienna University of Technology
Getreidemarkt 9/165
1060 Vienna
Austria

and

Interdisciplinary Laboratory on
Nanometric and Supramolecular
Organization (LIONS)
CEA Saclay
DSM, IRAMIS
NiMBE 91191
Gif-sur-Yvette Cedex
Note de Palaiseau
France

Timothy Noël
Eindhoven University of
Technology
Micro Flow Chemistry and
Process Technology
Department of Chemistry and
Chemical Engineering
Den Dolech 2 (STW 1.48)
5612 AZ, Eindhoven
The Netherlands

Alexander Schiller
Friedrich Schiller University Jena
Faculty of Chemistry and Earth
Sciences
Institute for Inorganic and
Analytical Chemistry
Humboldtstrasse 8
D-07743 Jena
Germany

Andrey Seliverstov
Friedrich-Alexander-University
Erlangen-Nuremberg
Department of Chemistry and
Pharmacy
Inorganic Chemistry II
Egerlandstr. 1
91058 Erlangen
Germany

and

University of Ulm
Institute of Inorganic Chemistry I
Albert-Einstein-Allee 11
89081 Ulm
Germany

Carsten Streb
Friedrich-Alexander-University
Erlangen-Nuremberg
Department of Chemistry and
Pharmacy
Inorganic Chemistry II
Egerlandstr. 1
91058 Erlangen
Germany

and

University of Ulm
Institute of Inorganic Chemistry I
Albert-Einstein-Allee 11
89081 Ulm
Germany

Abbie Trewin
The University of Liverpool
Department of Chemistry
Crown Street
Liverpool L69 7ZD
UK

Yury O. Tsybin
Biomolecular Mass Spectrometry Laboratory
Institute of Chemical Sciences and Engineering
Ecole Polytechnique Fédérale de Lausanne
av. Forel 1015 Lausanne
Switzerland

Johannes Tucher
Friedrich-Alexander-University Erlangen-Nuremberg
Department of Chemistry and Pharmacy
Inorganic Chemistry II
Egerlandstr. 1
91058 Erlangen
Germany

and

University of Ulm
Institute of Inorganic Chemistry I
Albert-Einstein-Allee 11
89081 Ulm
Germany

Olalla Vázquez
Universidade de Santiago de Compostela
Department of Organic Chemistry
Cruz Gallastegui 15-4A
36001 Pontevedra
Spain

Part I
Advanced Methodologies

Discovering the Future of Molecular Sciences, First Edition. Edited by Bruno Pignataro.

1
Supramolecular Receptors for the Recognition of Bioanalytes

D. Amilan Jose, Amrita Ghosh, and Alexander Schiller

Detection, identification, and imaging of specific analytes are of broad interest in chemical as well as in biological science. In this regard, molecular sensors play innumerable roles such as in the detection of biological molecules, hazardous materials, and warfare agents, in high-throughput screenings, monitoring biochemical processes, intelligent drug delivery, and molecular logic devices. This chapter focuses on fluorogenic and chromogenic supramolecular sensors for the recognition of important bioanalytes and their applications in various biological studies. A significant amount of literature is available related to this research area [1]. However, our aim is to review the research work carried out by us and selected important examples by others.

1.1
Introduction

Molecular recognition is a basic phenomenon in biological processes. The principle of molecular recognition is the specific interaction between a chemical entity and a target molecule. They are often complementary in their geometric and electronic features [2]. The idea of molecular recognition was first described by Emil Fischer in 1894, who proposed that enzyme and substrate fit together like "lock-and-key" [3]. The recognition mechanism is mediated mainly by supramolecular interactions such as hydrogen bonding, ion-pairing, hydrophobic interactions, and dipolar associations [4]. Several examples for these mechanisms exist in nature, for example, deoxyribonucleic acid (DNA) protein, ribonucleic acid (RNA) ribosome, and antigen antibody recognition. Researchers have shown great interest in the design of artificial systems to mimic these biological recognition processes. In this regard, the concept of supramolecular chemistry provides a route to design such sensor materials according to the technical needs [2]. In fact, supramolecular methods have already been proven to be very successful for biomolecule detection. However, developing new methods capable of detecting trace amounts of biologically relevant analytes, such as anions, nucleic acid, enzymes, microorganisms, and proteins in water, is still a demanding task. Apart from detecting methods, the biggest obstacle

Discovering the Future of Molecular Sciences, First Edition. Edited by Bruno Pignataro.

is identifying suitable receptor systems that are sensitive to specific analytes or families of analytes under physiological conditions.

Great advances have been made in the signaling of small target molecules, such as inorganic anions and metal ions [5, 6]. However, it is still difficult to design highly selective and sensitive receptors for complex bioanalytes, such as nucleoside polyphosphates, proteins, nucleic acids, and complex carbohydrates in water. A large number of active research groups around the world, including those of A. D. Hamilton, A. Das, A. Schiller, B. D. Smith, B. Koenig, B. Singaram, C. M. Niemeyer, C. Schmuck, E. V. Anslyn, I. Hamachi, J. L. Sessler, J. -L. Reymond, J. Yoon, J. W. Steed, K. Severin, P. A. Gale, P. Jr. Anzenbacher, R. Jelinek, S. Matile, S. Shinkai, T. D. James, T. Schrader, W. Nau, and many more contributed toward the development of novel supramolecular receptors for the recognition of important bioanalytes.

Fluorescent and colorimetric receptors for binding to bioanalytes are of enormous importance [7]. Fluorescent sensors are crucial as they generally allow detection of the analyte present in (ultra)trace amounts and offer possibilities for the use as a biological cell imaging reagent. In contrast, chromogenic sensors with visual detection have an edge over others as they allow naked eye detection without the use of any sophisticated instrumentation.

1.2 Bioanalytes

It is essential to know the important functions of the target analytes, so that one can design a suitable receptor for them. Our interest and main focus of this chapter lies in pyrophosphate (PPi), nucleoside triphosphates (NTPs), phosphorylated proteins, and peptides, nucleic acids (DNA and RNA), lipopolysaccharides (LPSs), and carbohydrates. These analytes are ubiquitous in nature; phosphates and its derivatives dominate the living world. Most of the coenzymes are esters of phosphoric or pyrophosphoric acid; the principal reservoirs of biochemical energy are phosphates. Many intermediary metabolites are phosphate esters.

PPi ($P_2O_7^{4-}$ (Figure 1.1) is an essential intermediate in biochemical syntheses and degradation reactions [8]. PPi is one of the important products of adenosine-5′-triphosphate (ATP) hydrolysis under cellular conditions, and the detection of PPi has been investigated as a real-time DNA sequencing method [9]. Recently, signaling of PPi has become an important issue in cancer research. Patients with calcium pyrophosphate dihydrate disease (CPPD) have also been shown to have a high synovial fluid PPi level [10].

NTPs (Figure 1.1), such as ATP, cytidine triphosphate (CTP), uridine triphosphate (UTP), are widespread in living systems and crucial for various cellular functions [11]. Among all NTPs, recognition studies of ATP are well known. ATP is produced mainly in mitochondria and used as an universal energy source for various cellular events. It is also involved in enzymatic processes as a reactive substrate. For example, ATP serves as a phosphate donor in kinase catalyzed protein phosphorylation and also acts as an extracellular signaling mediator [12]. Adenosine-5′-diphosphate

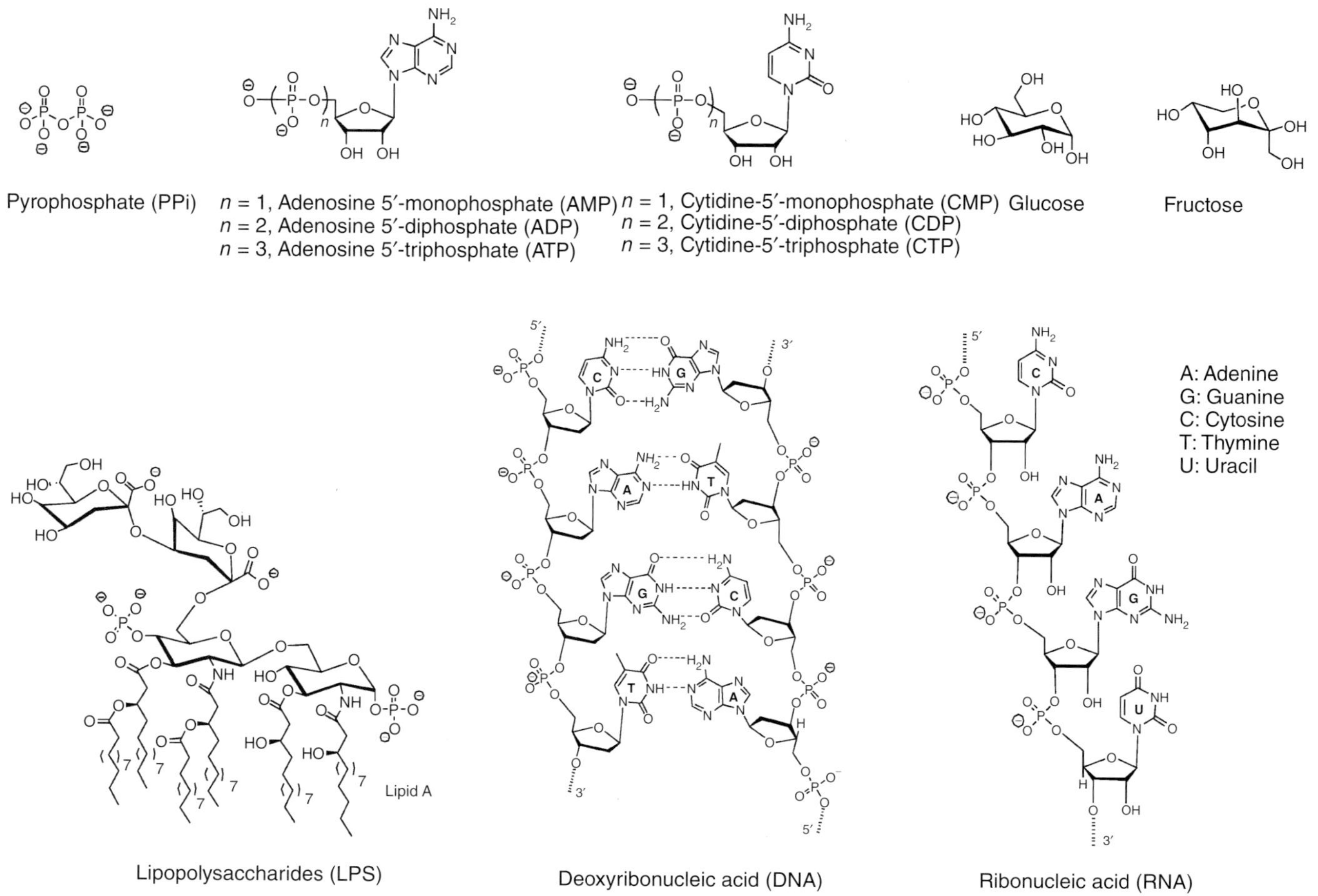

Figure 1.1 Chemical structures of important bioanalytes such as pyrophosphate, nucleoside phosphates (AMP, ADP, ATP, CMP, CDP, and CTP), carbohydrates (fructose and glucose), lipopolysaccharides (LPSs), and nucleic acids (DNA and RNA).

(ADP) and adenosine 5′-monophosphate (AMP) are important for their roles in bioenergetics, metabolism, and the transfer of genetic information.

The genetic materials DNA and RNA are phosphodiesters and they are essential for all known forms of life [11]. DNA is the molecular store of genetic information. The key biological role of RNA is as a messenger; it reads out the genetic code in DNA (transcription) and transports it to the ribosome, where it is decoded into the sequence of a protein (translation) [13]. Single stranded DNAs or RNAs can interact with their complementary strands with high specificity and are useful for nucleic acid detection. Sensor systems for binding nucleic acids have various applications in DNA diagnostics, gene analysis, biological warfare agent detection, forensic investigations, identification of microorganisms in food and environmental samples, and identification of infectious organisms in humans. Currently, the demand for the detection of RNA and DNA sequences for identifying the genetic cause of diseases is rising in medicine [14, 15].

Carbohydrates (saccharides) are known to mediate a large number of biological and pathological events [16]. They are involved in many key biological functions. In the form of glycoproteins, they can alter protein structure and function. As major components of glycolipids, they can play pivotal roles in cell–cell recognition and signaling [17]. They donate nature with structural rigidity, in the form of cellulose, and in the forms of starch and glycogen, they function as the energy store. The simplest biologically important carbohydrates are monosaccharides such as glucose, galactose, and fructose. From a medicinal point of view, the monitoring of glucose has proved of particular importance [18, 19]. In humans, a breakdown in the transport pathways of glucose has been linked to conditions such as cancer and cystic fibrosis [20].

LPSs (Figure 1.1) are amphiphilic molecules present on the outer leaflet of *Gram −ve* bacteria [21]. Despite a great compositional variation depending on their particular bacterial origin, they all consist of a hydrophobic domain known as lipid A (or endotoxin), a nonrepeating "core" oligosaccharide, and a distal polysaccharide. LPSs are one of the most potent microbial inducers of inflammation and of a cascade of physiological events that may lead to toxic shock and death. Sensors that are capable of detecting and identifying different types of LPS can be used to develop devices for bacterial diagnostics [22, 23].

1.3 Metal Complexes as Receptors for Biological Phosphates

Molecular recognition for the application in biology must occur at physiological conditions. Thus, receptors should be able to detect phosphates in aqueous or aqueous buffer solution. Mostly, two types of water soluble receptors are found in the literature: (i) positively charged or neutral nonmetallic receptors and (ii) metal complex based receptors [24]. The first type interacts through weak binding forces, such as hydrogen bonding and stacking interactions. The second type interacts mainly with the analyte through charge–charge interactions. The analyte coordinates to the metal center mimicking many metalloenzymes [25]. Our research interest focuses on both types of receptors. A number of metal ions have been used as receptors for the recognition of biological phosphates (e.g., PPi, NTP), including those of the main group, transition metals, and lanthanides [26]. Researchers have adopted various approaches in the metal-anion coordination to compete with the high hydration energy of phosphates at physiological conditions. However, Zn^{2+} is among the most commonly employed metal center [27]. In addition, coordinatively unsaturated metal complexes as a receptor provide binding sites with high affinity to Lewis bases. Many important bioanalytes (anions, thioles, nucleobases, esters and amides, ureas, etc.) are Lewis bases. They retain a significant affinity even in protic solvents including water.

1.3.1 Fluorescent Zn(II) Based Metal Complexes and Their Applications in Live Cell Imaging

The dipicolylamine (DPA) ligand is often used in zinc complex based sensor systems [28]. It provides a tridentate coordination environment with three nitrogen donors, shows good selectivity for Zn^{2+}, and leaves coordination sites free for anion binding. Two Zn(II)–DPA moieties in a compound exhibit strong binding with biological phosphates in water [28]. Fluorescent chemical receptors based on Zn(II)–DPA complexes for biological phosphates were pioneered by Hamachi and coworkers [29]. They have reported many Zn(II) complexes based on the traditional receptor–linker–fluorophore concept with one or two Zn(II)–DPA moieties as a binding unit, the fluorophore as a signaling unit, and a linker moiety [30].

Xanthene type chemosensor **1** (Figure 1.2a), reported by Hamachi *et al.* [31], was highly selective to nucleoside polyphosphate detection in water. Binding of **1** to ATP leads to a fluorescence turn-on with significant enhancement in the emission intensity (>30-fold). This is actually the recovery of the fluorescence intensity for

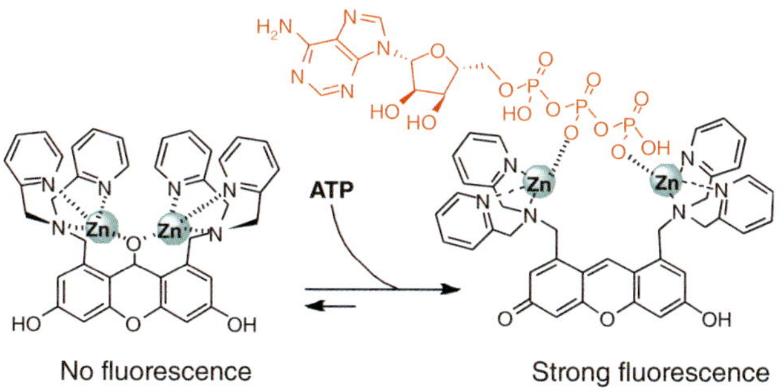

1
(a)

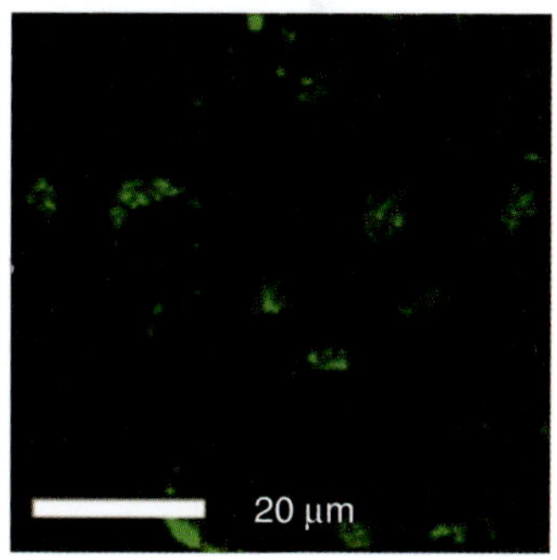

(b)

Figure 1.2 (a) Schematic representation of the turn-on fluorescence sensing mechanism of **1** after binding with ATP. (b) Confocal fluorescence images show fluorescence staining of the intracellular ATP stores in Jurkat obtained by **1**. (Adapted and reproduced with permission from Ref. [31]. Reproduced with permission of American Chemical Society (United States).)

the conjugated xanthene structure, which was quenched after coordination with Zn(II). The association constant for ATP derived from fluorescence titration is $1.3 \times 10^6\ M^{-1}$. The application of **1** as a bioanalytical molecular tool was demonstrated by fluorescence imaging of stored ATP in living Jurkat cells (a human T-cell lymphoblast-like cell line) (Figure 1.2b).

The group of Hamachi has also developed fluorescence resonance energy transfer (FRET) based ratiometric sensors **2** and **3** (Figure 1.3a) for the detection of ATP [32]. It was shown that, for receptors **2** and **3**, the same xanthene backbone of **1** acts as FRET acceptor along with a coumarin FRET donor. After binding with ATP, FRET from the coumarin to the xanthenes was observed. The affinity constant of these receptors toward ATP was calculated to 2.9×10^6 and $7.3 \times 10^6\ M^{-1}$, respectively, in aqueous solution. However, no detectable emission change was observed with monophosphates and phosphodiester species. The significant ratiometric sensing of ATP was used for the real-time visualization of the ATP level inside HeLa cells (cell lines HEK293 and NIH3T3) and the monitoring of two enzyme reactions involving nucleoside polyphosphates. Though these two chemosensors show relatively low selectivity among the polyphosphates, these are applicable to image the ATP level, inside the living cells (Figure 1.3b).

Very recently, the same group also achieved the imaging of nucleoside polyphosphates on plasma membrane surfaces with a lipid modified receptor **4**, which has a same xanthene core as receptor **1** (Figure 1.4). Receptor **4** was able to detect

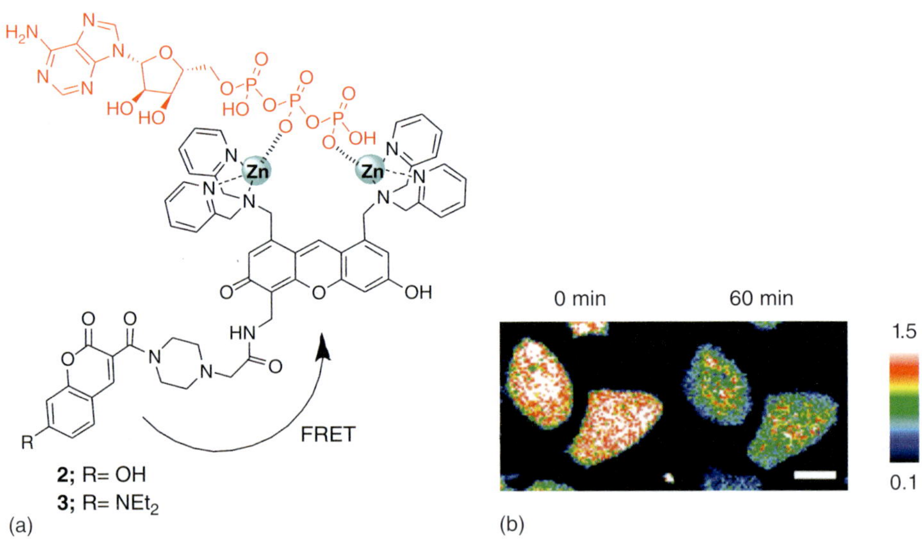

Figure 1.3 (a) Molecular structures for **2** and **3** and the FRET via turn-on fluorescence sensing on binding with ATP. (b) Ratiometric analysis of living cells stained with **2**. Pixel-by-pixel ratio image of a HeLa cell before (0 min) and after (60 min) treatment with 20 mM 2-deoxyglucose and 1 mM potassium cyanide (KCN). Scale bar: 10 µm. (Adapted and reproduced with permission from Ref. [32]. Reproduced with permission of American Chemical Society (United States).)

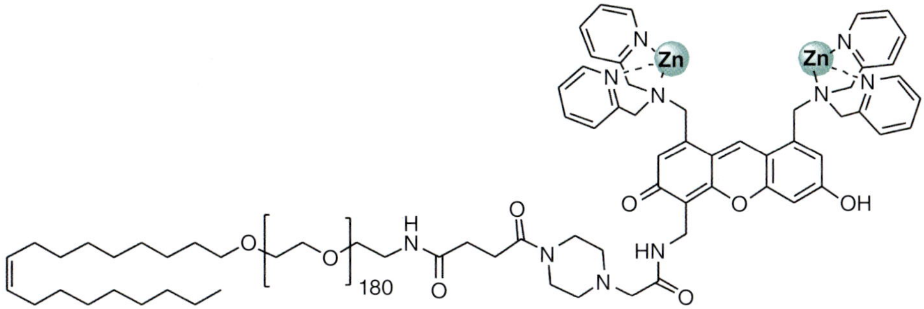

4

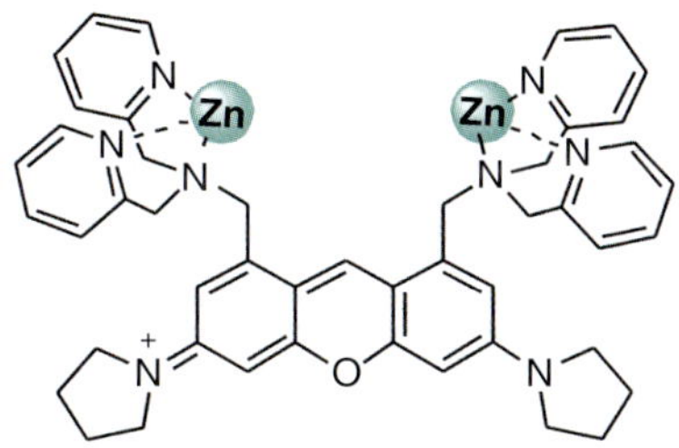

5

Figure 1.4 Receptors used for site-specific imaging of nucleoside polyphosphates.

polyphosphate derivatives XTP (X = A, G, C), XDP (X = A, U), and PPi with a several-fold enhancement in fluorescence intensity.

In a live cell imaging study, sensor **4** with a lipid anchor selectively localized on the plasma membrane surface and detected the extracellular release of nitrophenyl-phosphates (NPPs) during cell necrosis induced by streptolysin. For subcellular imaging of ATP in mitochondria, they have also designed rhodamine-type Zn(II) complex **5**, possessing a cationic pyronin ring instead of xanthenes. Receptor **5** detects the local increase of ATP concentration during apoptosis. Multicolor images were obtained through simultaneous use of **4** and **5** allowing detection of the dynamics of ATP in different cellular compartments at the same time [33].

1.3.2
Chromogenic Zn(II)-Based Metal Receptors and Their Applications in Biological Cell Staining

Most of the existing reports on the recognition of biological phosphates are based on changes in fluorescence properties [34]. Examples for colorimetric detection of biological phosphates in aqueous environments are rare in literature [35]. In an effort to make colorimetric receptors for the recognition of biological phosphates, Amilan Jose *et al.* reported a new chromogenic complex **6**, which can be used to bind ATP in aqueous solutions under biological conditions [36, 37] (Figure 1.5).

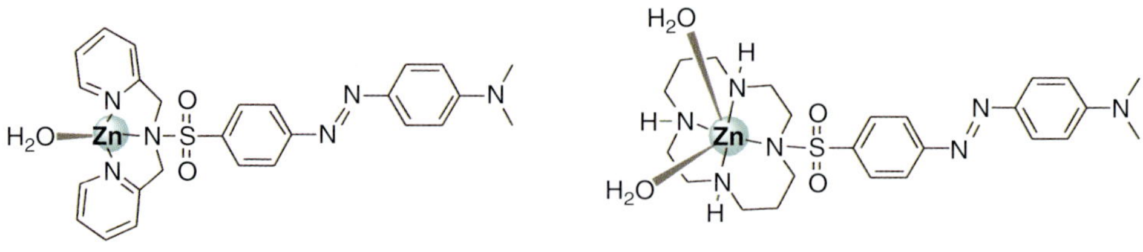

6 7

Figure 1.5 Chemical structures for receptors **6** and **7**.

A Zn(II)–DPA unit in receptor **6** acts as a receptor fragment for ATP recognition, while the dimethylamino phenylazo group acts as the signaling unit for reporting the binding, detectable by a color change.

The selectivity of the receptor **6** toward different biologically important anions was checked in aqueous media. The absorption maximum of **6** was found to be shifted to 484 nm from 463 nm with ATP (Figure 1.6a). However, on addition of ADP, a much smaller red shift (8 nm) in λ_{max} occurred and no distinct change in color could be

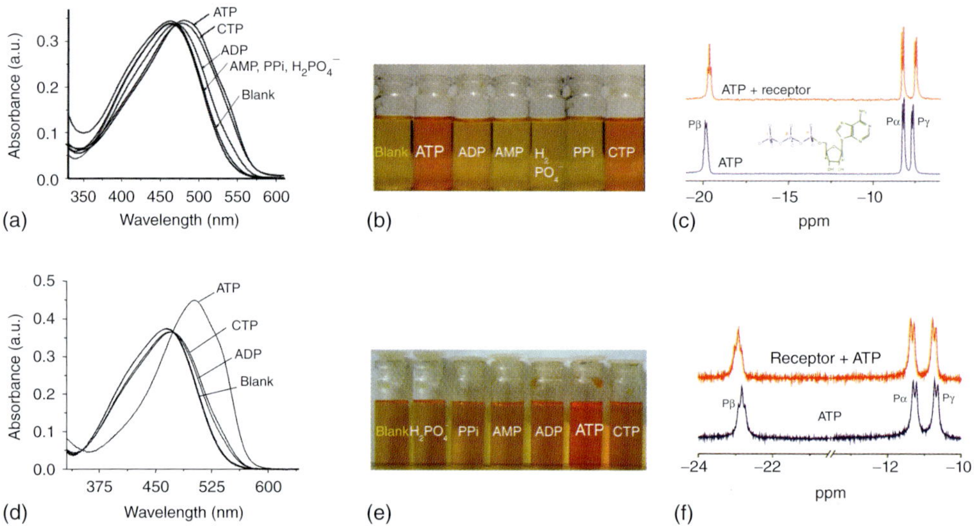

Figure 1.6 (a) Absorbance spectra of **6** (25 μM) in HEPES buffer solution (pH 7.2) at 25 °C in the presence of various anions (250 μM) (b) change in color of **6** in aqueous solution; from left to right: blank, with ATP, ADP, AMP, $H_2PO_4^-$, PPi (anion concentration 100 μM), and CTP (125 μM). (c) Partial ^{31}P-NMR spectrum of ATP in the presence and absence of receptor **6**. (Adapted and reproduced with permission from Ref. [36]. Reproduced with permission of American Chemical Society (United States).) (d) Absorbance spectra of **7** in HEPES buffer solution (pH 7.2) at 25 °C in the presence of various anions (e) change in color of **7** in aqueous solution with ATP, ADP, AMP, $H_2PO_4^-$, PPi, and CTP. (f) Partial ^{31}P-NMR spectrum of ATP in the presence and absence of receptor **7**. (Adapted and reproduced with permission from Ref. [38]. Reproduced with permission of Royal Society of Chemistry (United Kingdom).)

seen by the naked eye (Figure 1.6b). Furthermore, no change in absorption spectra was observed on addition of AMP, PPi, or $H_2PO_4^-$ (Figure 1.6a). For experiments with excess CTP, an almost similar spectral shift was detected as for ATP. Binding constants for ATP and CTP, evaluated from spectrophotometric titration, were found to be 1130 ± 6 and $772 \pm 5\ M^{-1}$, respectively, in aqueous solution (pH ~ 7.2) at 25 °C. The change in color and spectral behavior on binding of ATP to the Zn(II) center in **6** was associated with the perturbation of the energy of the frontier orbitals of the donor amine functionality and the acceptor azo fragment [36]. Binding of ATP was also confirmed by ^{31}P-NMR spectroscopy (Figure 1.6c). Downfield shifts for the ^{31}P signals for γ and β-P atoms signify the binding to Zn-atom of **6** through the oxide of respective phosphate units. An insignificant shift in ^{31}P signals was observed when similar experiments were repeated for ADP, and no shift was observed with AMP. The enhanced electrostatic interaction between triphosphates and **6** is crucial for efficient **6**-O (phosphate) binding compared to other anions. The observed binding preference for ATP or CTP ≫ ADP ≫ AMP could be attributed to the difference in the number of the anionic charges of the phosphate species [36].

Saccharomyces cerevisiae (yeast cells) is a eukaryotic microbe that derives its energy in the form of ATP. The surface of these yeast cells is exposed with negatively charged ATP. The chemosensor **6** could be used for the staining of this type of cells. The colorless yeast cells were viewed under normal light microscopy (Figure 1.7) with and without exposition of **6**. The microscopy images revealed that the treated cells got stained with **6** and the color of the cells changed to reddish yellow (Figure 1.7a).

The change in the color of yeast cells indicates that the negatively charged phosphate groups on the surface of the cells were effectively bound to **6**. The

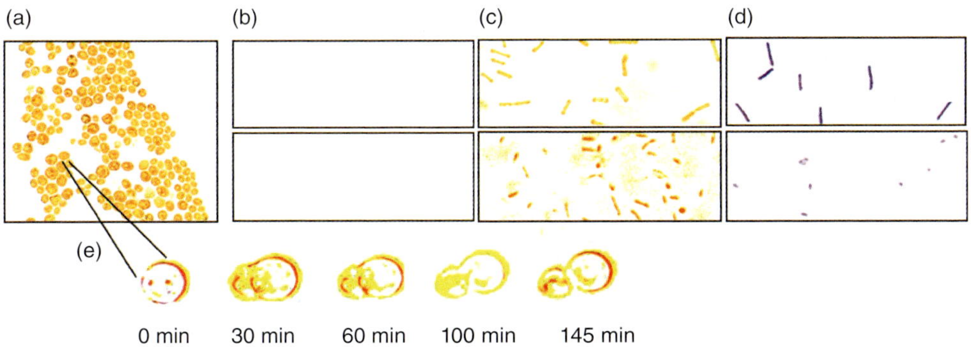

Figure 1.7 Light microscopy images (a) yeast cell with receptor **6**, (b-top) *Gram +ve* Bacillus sp. without any staining agent, (c-top) when treated with **6**, (d-top) when treated with gentian violet dye; *Gram −ve* Pseudomonas sp. (b-bottom) without any staining agents, (c-bottom) when treated with **6**, and (d-bottom) when treated with safranin dye. (e) Light microscopy images of a yeast cell stained with **6** monitored at different time intervals. (Adapted and reproduced with permission from Ref. [36]. Reproduced with permission of American Chemical Society (United States).)

staining was found to be stable as the color of the stained cells remained unchanged even after subsequent washing with water/ethanol. The possibility of using **6** as staining agent for prokaryotes (*Gram +ve* and *Gram −ve* bacterial cells) was also investigated. The experimental results suggested that the single staining agent **6** could distinguish between *Gram +ve* and *Gram −ve* bacteria through distinctly different color intensities and shape of the stained cells. The difference in the staining intensity for two different bacteria can be better understood if one considers the difference in cell structure and cell wall composition of the respective bacteria.

The viability of the cells was observed before and after staining with **6** under the light microscope. Unaffected cell proliferation and growth confirmed that the staining agent **6** was nontoxic and kept cells viable after staining (Figure 1.7e). The growth dynamics was monitored for the same eukaryotic cells and prokaryotic bacteria in an aqueous culture medium in the absence and presence of **6** by recording the change in absorbance spectrum as a function of time. The plot of change in absorbance versus time shows a nice correlation between the cell growth and the ATP generation with progressive growth (lag phase to log phase and then stationary phase and ultimately a decline in the growth curve) for the respective cells during the metabolic processes. The growth profile was also monitored in the absence and presence of two different respiratory inhibitors such as rotenone and cycloheximide (which reduces the ATP generation and cell growth, respectively). Studies with respiratory inhibitors confirm the staining due to the binding of extracellular ATP, produced *in situ*, with **6**.

In a subsequent work by the same group, Zn(II)–cyclam modified chromogenic complex, **7** and its [2]-pseudorotaxane form, **α-CD.7** (CD = cyclodextrin) was reported for preferential binding of ATP, among other biologically important phosphates in aqueous solution (Figures 1.6 and 1.8) [38, 39]. The chemosensor **7** exhibits higher solubility in aqueous medium as well as improved selectivity toward ATP. A visually detectable change in solution color was observed on addition of ATP to **7** (Figure 1.6e). The charge transfer absorption band of **7** at 463 nm was red shifted with a maximum at 503 nm on addition of excess sodium salt of ATP. An insignificant shift of 9 nm was observed when CTP of comparable concentration was added (Figure 1.6d). In contrast, there was almost no change in the spectra and color with other phosphate anions. This confirms that receptor **7** is very selective for ATP among other nucleoside phosphates examined. The relative binding affinity for ATP of $K = 1.9 \times 10^3\ M^{-1}$ was evaluated by systematic titrations in *N′*-2-hydroxyethylpiperazine-*N′*-2 ethanesulfonic acid (HEPES) buffer medium.

The binding affinity of the ATP with **7** is slightly higher than that of **6**. Unlike receptor **6**, ^{31}P–NMR binding studies of **7** with ATP shows upfield shifts for the α-, β-, and γ-phosphorus atoms of the bound ATP (Figure 1.6f). These chemical shifts signals signify the binding to Zn-atom of **7** through oxygen atom bearing the negative charge of the respective phosphate unit. Relatively weaker interaction of the $O_{[\gamma-PO4]}^-$ and $O_{[\alpha-PO4]}^-$ relative to $O_{[\beta-PO4]}^-$ unit accounts for the smaller $\Delta\delta$ shift in ^{31}P–NMR spectra. This observation validates the formation of a hepta coordinated Zn(II)-center in **7** in presence of ATP. The solubility of the **7** (0.045 g l^{-1}) in water was found to be enhanced (0.34 g l^{-1}) in the presence of α-CD. This is

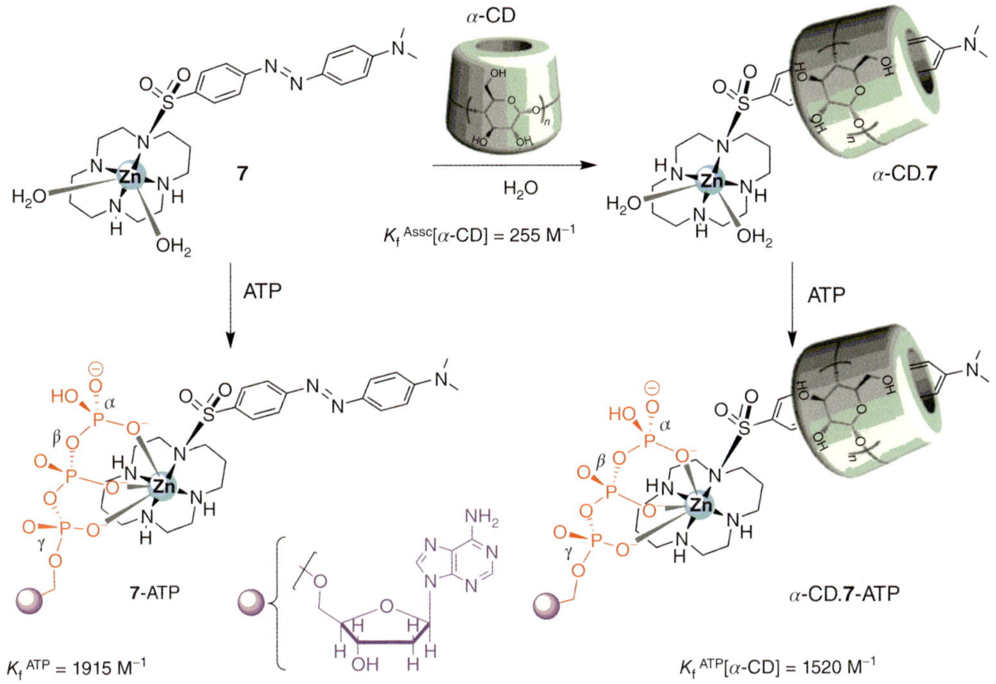

Figure 1.8 Schematic representation of the formation of [2] pseudorotaxane, **α-CD-7**, and the binding of ATP to the Zn(II)-center of **7 or α-CD-7**. (Adapted and reproduced with permission from Ref. [38]. Reproduced with permission of Royal Society of Chemistry (United Kingdom).)

due to the favored nonbonded interactions after inclusion of the hydrophobic azo functionality of **7** into the hydrophobic cavity of the α-CD. High solubility of **7** in water in the presence of α-CD helped to attain a higher effective concentration of **7** and an intensified color change on binding to ATP [38].

The possible use of 7 and **α-CD.7** as staining agents for yeast cells was also studied. The results demonstrated that it could be used as a colorimetric staining agent for eukaryotic yeast cells and can be viewed under a simple light microscope. Staining studies were also conducted with prokaryotic *Bacillus sp.* (*Gram +ve*) and *Pseudomonas sp.* (*Gram −ve*) bacteria (Figure 1.9). The *Gram +ve* bacteria appeared longer in the images as expected, while more intense staining was observed for *Gram −ve* bacteria. After staining, the color of the *Gram +ve* bacteria cells changed from colorless to pink, but in the case of *Gram −ve* bacteria the color change occurred from colorless to violet. The thinner, hydrophilic, and more porous cell walls of the *Gram −ve* bacteria are expected to allow higher excretion of ATP to the cell surface, where it gets bound to the **6**, **7**, and **α-CD.7** thereby causing the efficient staining.

Similarly to **6**, the experimental results again confirmed the retained viability of all the live cells after staining with both **7** and **α-CD.7** and this nontoxic behavior

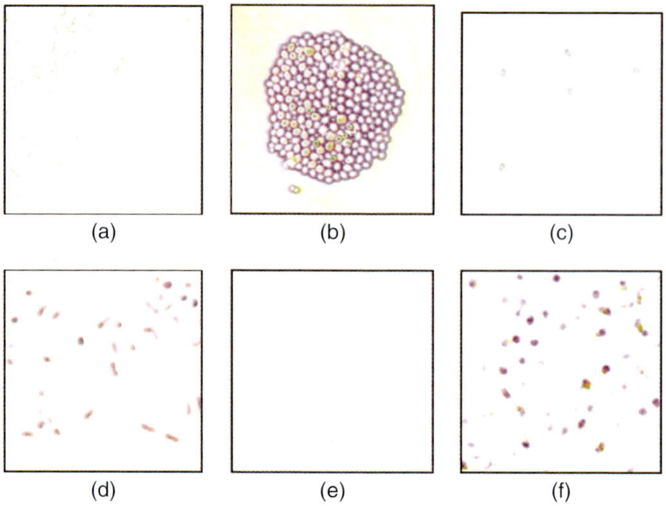

Figure 1.9 Light microscopic images of cells with **α-CD.7** (a) unstained *S. cerevisiae* cells, (b) stained *S. cerevisiae* cells, (c) unstained *Gram +ve* bacteria, (d) stained *Gram +ve* bacteria, (e) unstained *Gram −ve* bacteria, and (f) stained *Gram −ve* bacteria at 25 °C in 10 mM HEPES buffer solution. (Adapted and reproduced with permission from Ref. [38]. Reproduced with permission of Royal Society of Chemistry (United Kingdom).)

could be used for studying the cell growth dynamics of each of these individual microbes [40]. Thereby, colorimetric receptors **6**, **7**, and **α-CD.7** can be useful as efficient viable staining agents for a microorganism through selective recognition of biological phosphate anion. In addition, other interesting Zn(II) complex based receptors for the recognition of PPi and simple phosphate have also appeared in literature [26, 41–44].

1.4
Functionalized Vesicles for the Recognition of Bioanalytes

Vesicular particles are an interesting class of dynamic supramolecular structures and have been employed in diverse applications in biological research, mostly because of their relative ease of preparation and variability in composition [45–47]. Vesicles are often perceived as closely mimicking the cell membrane [48, 49]. These features have promoted the use of vesicles in molecular recognition [50]. Our research interest focused on the development of new color and fluorescence based vesicles for the recognition of bioanalytes [51, 52]. In particular, we were interested in polydiacetylene (PDA) based vesicles as receptors for the detection and analysis of biological analytes. In this section, we describe our research and the ensuing results of PDA vesicles for biomolecular sensing and very interesting examples of other research groups.

1.4.1
Polydiacetylene Based Chromatic Vesicles

Conjugated PDAs are an amazing polymeric system that displays unique chromatic properties [53, 54]. PDA polymers are formed by the 1,4-addition of self-assembled diacetylenic monomers; the reaction is initiated by ultraviolet (UV) irradiation at 254 nm (Scheme 1.1). The resulting polymer is intensely blue in color. Electronic delocalization within the conjugated framework results in an absorption at around 650 nm [55]. The practical use of PDAs arises from their ability to undergo a blue to red visible color transitions in response to different external stimuli, such as temperature, pH, mechanical stress, and chemical and biological species.

The mechanism corresponding to the color change is believed to be an irreversible stress-induced structural transition of the conjugated backbone of the polymer. This direct colorimetric detection strategy bypasses the need for optical reporters and transducers [56, 57]. The lipido-mimetic nature of PDA, that is a hydrophobic tail (long aliphatic chain) and a hydrophilic headgroup (carboxylate), results in the formation of biomimetic assemblies, such as nanoscale vesicular particles in aqueous solutions and monolayers at the air/water interface.

This unique behavior of stimuli-induced blue to red color transition as well as fluorescence enhancement of the PDAs has led to the development of a variety of PDA-based sensing components. Mostly, PDA sensors have been used as thin films or as vesicles in solution. A bottleneck in the development of a PDA sensor assembly for molecular recognition is the preparation of the diacetylene monomers embedded with the suitable recognizing element of choice. Two important approaches have been used to functionalize the surface of the assembly. In the first case, the diacetylenic monomer lipid is covalently modified with the appropriate receptor by synthetic reaction. This allows direct cross-linking of the "receptor-lipid" with the surrounding PDA matrix. In the second case, a receptor molecule is noncovalently incorporated into the PDA matrix analogous to the heterogeneous mixing of molecules in cell membranes [58, 59]. One of the initial demonstrations of PDA sensor for the potential biological application is the colorimetric detection of influenza virus by the sialic acid ligand modified PDA films [60].

1.4.1.1 PDA Based Receptors for Biological Phosphate

Metal ion functionalized vesicles also play an important role for molecular recognition at membrane–water interfaces [49, 61]. The ability to modify the vesicle membranes with metal complex based receptors for biological analytes is an important aspect and immature area in molecular recognition. As described in the previous section, metal complexes, such as Zn(II)–DPA and Zn(II)–cyclen (cyclen = 1,4,7,10-tetraazacyclododecane), can reversibly coordinate anionic analytes of biological origin under physiological conditions with high affinity and selectivity. Many research groups have investigated the application of Zn(II)–DPA receptors in liposomes for binding with biological analytes, molecules transport across membranes, and cell staining [30, 45, 47, 61–63].

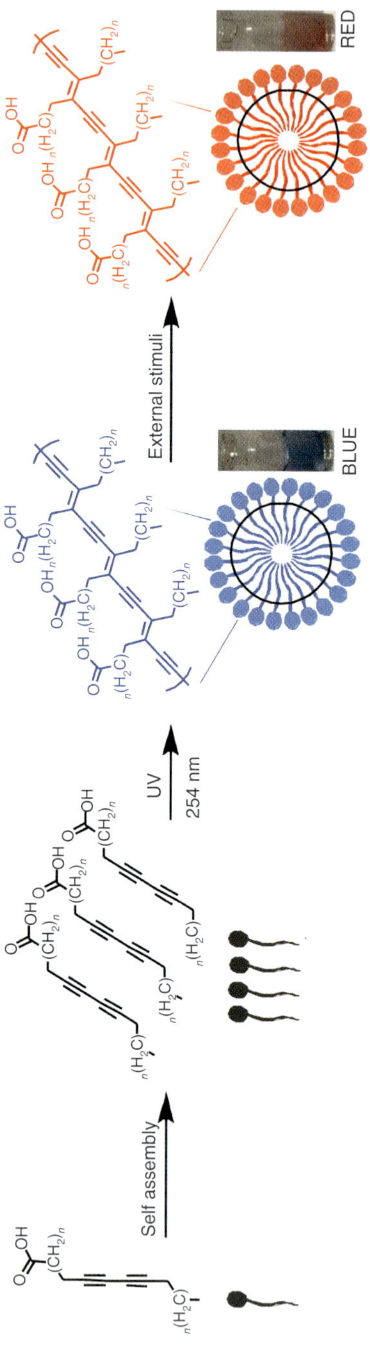

Scheme 1.1 Structural and chromatic properties of polydiacetylene (PDA) based vesicles as receptors.

However, the preparation of self-assembled PDA vesicles with biological phosphate binding amphiphilic Zn(II)–cyclen and Cu(II)–IDA (IDA = iminodiacetato) complexes (Figure 1.10) was reported by Amilan Jose *et al.* for the first time [52]. The receptor functionalized liposomes **LP-8**, **LP-9**, and **LP-10** were prepared by a proper mixture of mono, bis-Zn(II)–cyclen, and Cu(II)–IDA receptor modified diacetylene monomer and the unmodified diacetylene monomer 10,12-tricosadiynoic acid (TCDA) or 10,12-pentacosadiynoic acid (PCDA) in buffered aqueous solution (10 mM, HEPES, pH 7.2). The polymerized self-assembled bilayer vesicles were prepared at room temperature by irradiating the solutions with light at 254 nm, whereby the colorless receptor embedded vesicle solution turned blue. The average size of the liposomes of 160–200 nm was determined by dynamic light scattering (DLS). The absorption spectra of the modified vesicles show distinct absorption bands at 640, 589, and 543 nm. However, on addition of ATP and PPi to the **LP-8** and **LP-9** PDA vesicles, the absorption band at 640 nm disappeared completely and intense absorption bands at 489 and 543 nm were observed (Figure 1.11a). The color of the solution turned red. No changes in the absorption spectra or color were

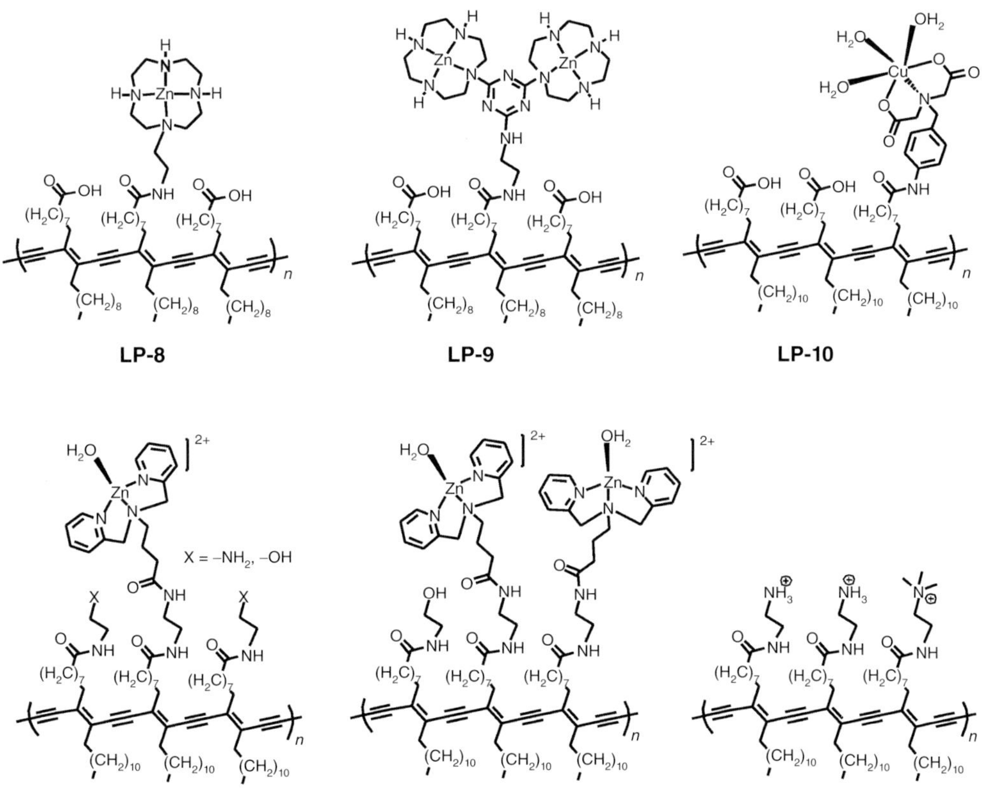

Figure 1.10 Polymeric PDA vesicles prepared from receptor modified diacetylene monomers.

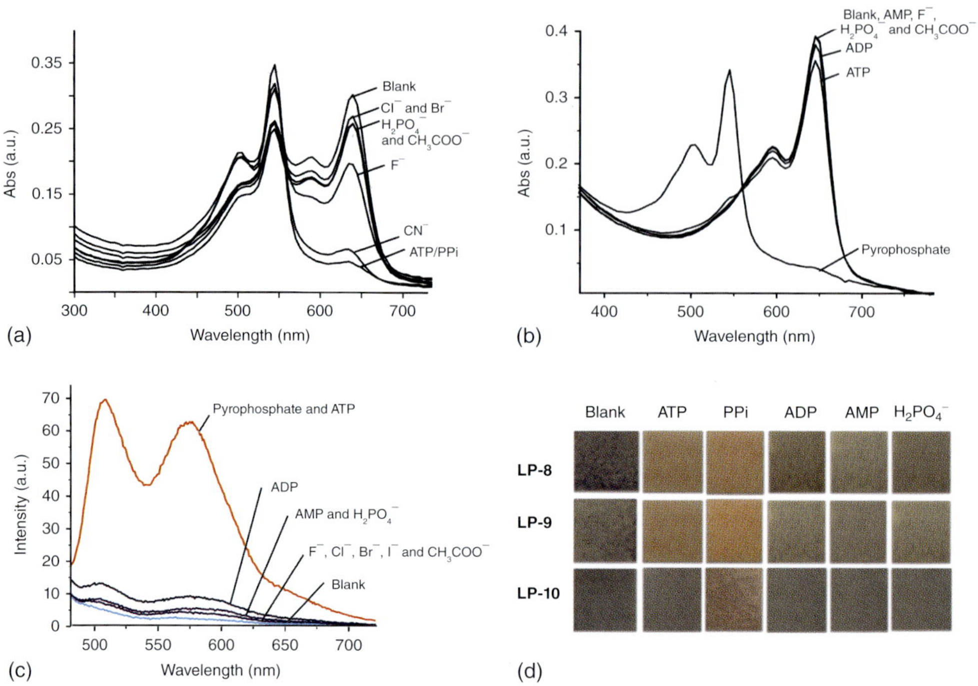

Figure 1.11 (a) UV-visible spectra of **LP-9** in the presence of different anions (aqueous solution, HEPES 10 mM, pH 7.2, 100 equiv. of the anion salt added). (b) UV-visible spectra of **LP-10** in the presence of different anions (aqueous solution, HEPES 10 mM, pH 7.2, 100 equiv. of the anion salt added). (c) Emission spectra of **LP-9** in the presence of different anions (aqueous solution, HEPES 10 mM, pH 7.2, 100 equiv. of the anion salt added). (d) Color change of the receptor embedded test paper with different anions. (Adapted and reproduced with permission from Ref. [52]. Reproduced with permission of Wiley-VCH (Germany).)

observed with other anions such as F^-, Cl^-, Br^-, I^-, $H_2PO_4^-$, CH_3COO^-, AMP, or ADPunder similar conditions.

The color change of the Zn(II)–cyclen modified PDA liposomes from blue to red was quantified via the colorimetric response (CR) by using equation $\%CR = [(A_0 - A_x)/A_x] \times 100$. The absorption ratio before analyte addition is calculated as $A_0 = I_{620}/(I_{620} + I_{490})$ and the absorption ratio after analyte addition followed from $A_x = I_{620}/(I_{620} + I_{490})$, respectively. **LP-9** prepared from dinuclear Zn(II)–cyclen complexes showed an increased affinity to ATP and PPi ions than **LP-8** prepared from mononuclear Zn(II)–cyclen complexes. It was interesting to see that Cu(II)–IDA complex modified vesicles (**LP-10**) behave differently with biological phosphates as compared to Zn(II)–cyclen modified vesicles. Among different phosphates, **LP-10** responded only to PPi and is able to selectively discriminate between ATP and PPi (Figure 1.11b). The binding of ATP and PPi with vesicular receptors were also monitored by emission intensities. The emission spectra of **LP-8**, **LP-9**, and **LP-10** in aqueous buffered solution showed very weak emission bands centered at 625 nm on excitation at 510 nm. The intensity of the emission band increases significantly in the presence of ATP and PPi. Other anions induce only very little or no change in the emission intensities (Figure 1.11c). The intrinsic response of colorimetric vesicles can provide convenient "naked eye" detection and the corresponding analyte affinities are usually in the millimolar (mM) range. For practical applications, test papers of the vesicles were prepared by soaking filter papers in the solution of the vesicles and drying them in air. The blue colored test paper was immersed in the aqueous analyte solution for several seconds and then air-dried. Similarly to solution, the color of the test paper made from **LP-8** and **LP-9** changed with aqueous solutions of ATP and PPi. However, in the case of **LP-10**, only with PPi the color change was obtained (Figure 1.11d).

Ahn *et al.* reported that similar family of PDA vesicles functionalized with Zn(II)-DPA binding unit in solution and solid substrates for PPi [64]. A functionalized liposome (**LP-11**) has been prepared by an 1 : 1 mixture of Zn(II)-DPA functionalized PDA monomer and ethylene diamine (EDA) capped PDA (**PCDA-EDA**) monomer. This solution (0.25 mM HEPES, pH 7.4) was irradiated at 254 nm and treated with zinc nitrate to obtain the corresponding polymerized liposome **LP-11**.

Scanning electron microscopy and DLS analyses confirmed the size distribution range of 40–80 nm. In these cases blue liposome solution (**LP-11**, where $X = NH_2$) became red–purple on interaction with only phosphate or PPi; no color change was observed with the other anions. Even though color changes were observed with ATP and AMP, they caused precipitation, which made them not suitable for further studies. The observed analyte selectivity can be explained by evoking the strong affinity of the Zn(II)-DPA ligands toward phosphate and PPi anions. In contrast, little color change was observed on addition of anions to a solution of **LP-11** (where $X = OH$) suggesting that the presence of amino group also plays a role in anion recognition. The CR values show the high selectivity of the liposome toward phosphate and PPi ions in solution. PPi selective **LP-11** also successfully fabricated into microarray chip. The microarray-chip system selectively responds to PPi and the red fluorescence spot images are clearly visible in the presence of 1 μm to 1 nm range of PPi (Figures 1.12a–j).

Recently, the same group has prepared **LP-12** by mixing 2 : 1 mixture of Zn(II)-DPA functionalized PDA monomer and alcohol-terminated PDA monomer. **LP-12** and **LP-11** differ from PDA monomer ratio used for the liposome preparation. **LP-12** shows a color change from blue to reddish purple and emits fluorescence in the turn-on mode on interaction with phosphatidylserine over other analytes such as phosphatidylcholine, sphingomyelin, and phosphatidylethanolamine [65]. Confocal fluorescence microscopy and fluorescence-activated cell sorting (FACS) analysis demonstrate that liposome responds to apoptotic cells and selectively stains the apoptotic cells in a manner similarly to commercial apoptosis detection kit (Figure 1.12k–n). Cell staining study also demonstrates that liposome can be used to detect apoptotic cells over normal cells.

Quaternary ammonium and primary amine head groups modified PDA sensor system (**LP-13**) for biological phosphates was reported by Juyoung Yoon *et al.* [66]. The sensor displayed a selective colorimetric change and a large fluorescence

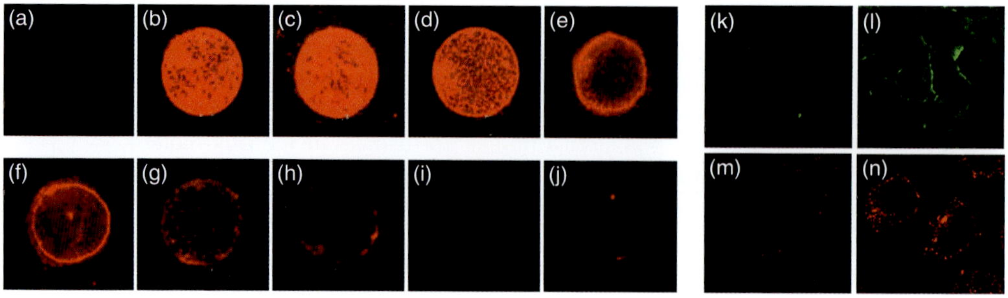

Figure 1.12 Fluorescence images of the liposome chip to pyrophosphate at various concentrations: (a) buffer only, (b) 100 μm, (c) 10 μm, (d) 1 μm, (e) 100 nm, (f) 10 nm, (g) 1 nm, (h) 100 pM, (i) 10 pM, and (j) 1 pM. The images were taken after dipping the chip into each analyte solution (10 mM, pH = 7, HEPES buffer) and incubating for 6 h at room temperature. (Adapted and reproduced with permission from Ref. [64]. Reproduced with permission of Wiley-VCH (Germany).) and confocal fluorescence microscopic images of HeLa cells in HEPES buffer (normal cells: k, l and apoptotic cells: m, n) stained with fluorescein conjugated Annexin V from Aldrich (k,l) and **LP-12** (m,n), respectively. (Adapted and reproduced with permission from Ref. [65]. Reproduced with permission of Wiley-VCH (Germany).)

enhancement in the presence of ATP at pH 7.0 in water among various anions. The ratio between PDA monomers and the control of steric factors were critical points for ATP detection. The possible interaction of phosphate groups in ATP with quaternary ammonium units and ammonium units on the surface of the PDA polymer could be the reason for the selectivity of vesicles toward ATP.

1.4.1.2 **PDA Based Receptors for Lipopolysaccharide**

LPS is highly toxic and biologically active. Owing to its high toxicity, continuous effort has been directed toward the development of specific detection of LPS. PDA liposomes prepared from amino acids functionalized PDA monomers (**14** and **15**) were used for the detection of LPS from five different strains of *Gram −ve* bacteria such as *Escherichia coli O26 : B6*, *Pseudomonas aeruginosa*, *Salmonella minnesota*, *Shigella flexneri*, and *Salmonella enteriditis* [67]. On interaction of liposome with LPS, the color change of the liposomes from blue to red was quantified by calculating the CR value. Four different conditions (i) at RT, (ii) at 35 °C, (iii) with sodium dodecyl sulfate (SDS), and (iv) with ethylenediamine tetraacetic acid (EDTA) were used to get eight CR values for two PDA liposomes prepared from **14** and **15**. The fingerprints obtained from a set of five LPS are unique enough to identify all five LPS unambiguously in a blind test [67].

The group of Schmuck *et al.* reported a peptide-functionalized PDA liposome as a turn-on fluorescent sensor for LPS at micromolar concentrations in water [68]. Inspired by naturally occurring antibiotic polymyxin B (PMB), they designed new fluorescent sensors for LPS. Two diacetylene monomers connected with histidine (**16**) and pentalysine oligopeptide (**17**) have been synthesized by using microwave-assisted peptide synthesis (Figure 1.13). Irradiating a 1 : 9 mixture of highly fluorescent self-assembled PDA monomer (**16** and **17**) led to a complete

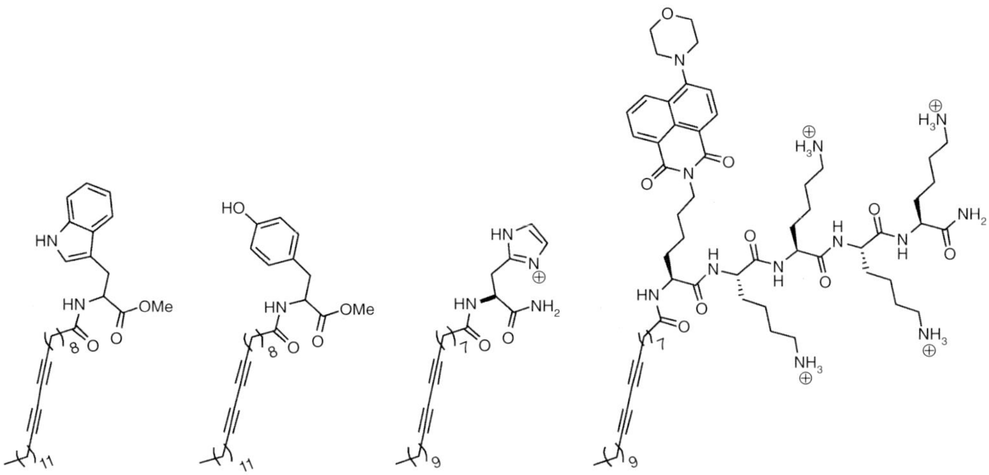

14 **15** **16** **17**

Figure 1.13 Chemical structure of receptor modified PDA monomers for LPS detection.

quenching of the fluorescence. This quenching is caused by an energy transfer from the napthalic acid fluorophore (emission maximum at 540 nm) to the cross-linked polymer (absorption maximum at 536 nm). Binding of submicromolar concentration of LPS to the nonfluorescent polymerized PDA liposomes restored the fluorescence. The fluorescent change is only selective for LPS, compared to other anionic biological relevant species, such as nucleotides, anionic sugars, or ctDNA. Stern-volmer analysis provided a binding constant of $K = 1.5 \times 10^6\ M^{-1}$ (Figure 1.14). Further LPS selective sensor allowed for the fluorescence staining of the membrane of *E. coli* bacteria; control experiment also showed that the LPS selective PDA liposomes are nontoxic to either bacteria or human cells.

1.4.1.3 **PDA Based Receptors for Oligonucleotides and Nucleic Acids**

Sequence-specific DNA detection is important in medical, biological, and biotechnological areas. Techniques for detection of small quantities of DNA find broad potential applications including gene expression monitoring, pharmacogenomic research, drug discovery, viral, bacterial, forensic, and genetic identification [69].

Ma and coworkers developed a colorimetric method for the detection of oligonucleotides by PDA liposomes [70]. PDA vesicles were prepared by the mixture of **TCDA** (70%), dimyristoylphosphatidylcholine (**DMPC**, 29 mol%), and respective oligonucleotides (Probe 1 or Probe 2, 1%). Oligonucleotides were partially complementary to opposite ends of the target DNA. In the presence of the target DNA, 5′-TACGAGTTGAGAATCCTGAATGCG-3′, the PDA vesicles experience a chromic transformation from deep blue to red (Figure 1.15). The force produced on the conjugated backbones of liposomes leads to color transitions that come from the hybridization of two oligonucleotides with target DNA. On the other hand, on addition of mismatched oligonucleotides, no color or absorption spectrum

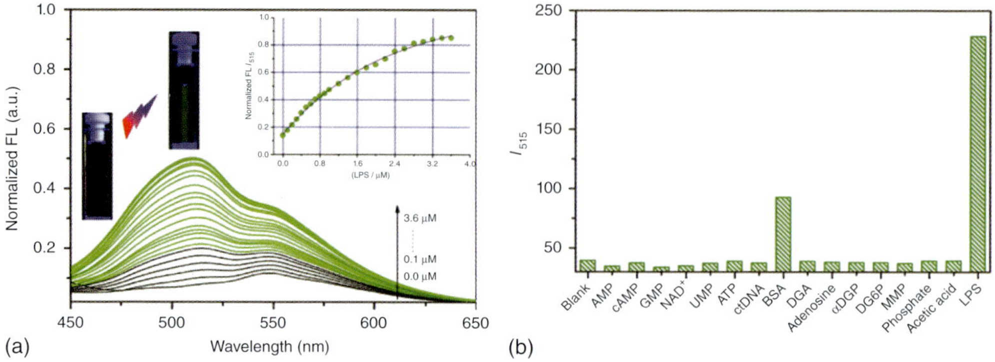

Figure 1.14 (a) Fluorescent emission titration spectra of the PDA liposomes prepared from 10:90 mixture of **16** and **17** with LPS in 10.0 mM DMSO/TBS (v/v = 1/4, pH = 7.4). Inset: Normalized fluorescence intensity at 515 nm versus the concentration of LPS (0–3.6 μM). (b) The selectivity of the increase of fluorescence on addition of various biologically important species. Only LPS (and to a much lesser extent the protein BSA) gives rise to a significant increase in fluorescence. (Adapted and reproduced with permission from Ref. [68]. Reproduced with permission of American Chemical Society (United States).)

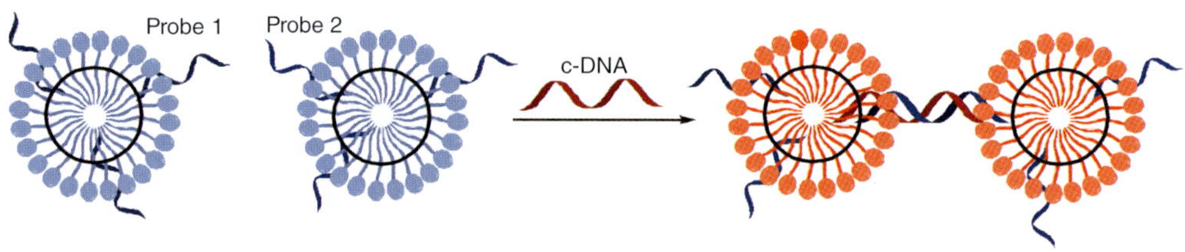

Figure 1.15 Schematic diagrams for the colorimetric detection of DNA using polydiacetylene vesicles functionalized with probe DNA. The sequences of different oligonucleotides used for the study is Probe 1: 5′-TCTCAACTCGTATTTTTT-$(CH_2)_3$-cholesteryl-3′; Probe 2: 5′-cholesteryl-$(CH_2)_3$-TTTTTTCGCATTCAGGAT-3′; target DNA: 5′-TACGAGTTGAGAATCCTGAATGCG-3′; mismatched DNA: 5′-GCGTAACTCCTAAGAGTTGAGCTA-3′. (Figure adapted from Ref. [70].)

changes were observed. These results indicate that the sensing system could be highly specific to target DNA sequences. However, this strategy is tedious and time-consuming because PDA liposomes with different probe oligonucleotides need to be prepared depending on the target DNA.

A novel strategy for the detection of nucleic acids was developed by Kim and Park *et al.* It is based on the positively charged PDA vesicles and negatively charged phosphate backbone of DNA [71]. The PDA liposomes **LP-18** and **LP-19** with positive charges were prepared by using primary and quaternary amine modified diacetylene monomers (Figure 1.16). On addition of the nucleic acids, amplified by common polymerase chain reaction (PCR), the color of polymerized liposomes **LP-18** underwent a transition from blue to red. Interestingly, primary

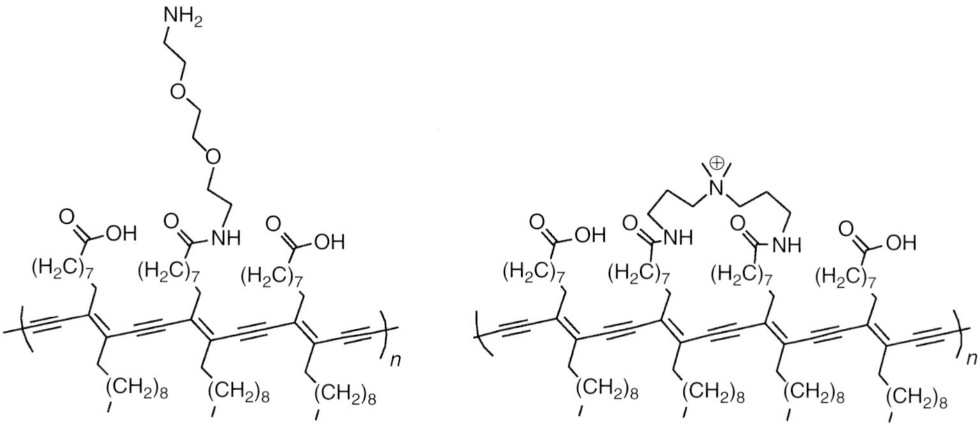

Figure 1.16 Polymeric PDA vesicles prepared from primary and quaternary amine receptor modified diacetylene monomers.

amine-functionalized **LP-18** displayed higher sensitivity than those containing quaternary amine-functionalized **LP-19**.

The main sensing strategy is based on the nonspecific ionic interaction with the positively charged PDA liposome. A simple purification step is required before the detection of DNA; this limits the detecting utility of this system in the aspect of its simplicity. Together with the above example, several other PDA liposome based sensors have been utilized to detect other important bioanalytes, including cations, antibodies, influenza virus, human serum albumin, carbonic anhydrase, *E. coli*, bacterial pore-forming toxin, thrombin, melamine, lectin, heparin, and pathogenic agents [50, 56–58, 72].

1.5 Boronic Acid Receptors for Diol-Containing Bioanalytes

Boronic acids receptors bind with diol units in aqueous solution to form cyclic boronate esters. Receptor design often uses the pK_a drop observed on addition of saccharides to boronic acids; the acidity of the boronic acids is enhanced when 1,2-, 1,3-, or 1,4-diols reversibly react with them to form cyclic boronic esters as five, six, or seven membered rings. This fundamental interaction is still of central importance in the construction of novel sensors for diol-containing (bio)analytes. A number of excellent review articles have been published in the past 2 years on boronic acid sensors [73–76]. Here we describe recent studies of innovative power in this field, such as high glucose selectivity, molecular logic with sugars, saccharide sensing at the few-molecule level of a reporter dye and drug delivery.

Glucose plays a dominant role in metabolic processes. For example, control of blood glucose concentration is of central importance for patients suffering from diabetes mellitus. Thus, there is a strong clinical need for accurate glucose monitoring [75, 76]. Many synthetic boronic acid probes have been developed to selectively detect glucose at physiological concentrations [73, 74]. To develop a successful *in vivo* boronic acid-based glucose sensor, an important criterion that must be met is the preferential binding of glucose over other physiologically significant monosaccharides, such as fructose and ribose derivatives. This remains still a great challenge because most organic boronic acids display higher binding affinities against fructose [73, 75].

The groups of Jiang and James have recently developed a ratiometric fluorescent chemosensor based on an amphiphilic monoboronic acid that is highly selective and sensitive for glucose in aqueous solution [77]. The presence of glucose leads to pyrene excimer emission, while its monomer emission remains unchanged. In contrast, fructose results only in a modest enhancement of the monomer emission (Figure 1.17).

A sensor can also be interpreted as molecular switch using Boolean algebra. Prasanna de Silva *et al.* showed for the first time that molecular fluorescent probes for ions can function as logic gates [78]. Till now, fascinating molecular digital analysis with receptors and carbohydrates, oligonucleotides, oligopeptides, proteins, and metal ions have been shown [79–81]. First applications of these molecular logic gates can be found in the design of smart materials, in the delivery/activation of drugs, and in clinical diagnostics [82]. However, only very few chemical logic studies exist with boronic acids as saccharide receptors [83–85]. In 2012, we described a two-component saccharide probe with logic capability [84]. The combination of a boronic acid-appended viologen and perylene diimide was able to perform a complementary implication/not implication logic function (Figure 1.18). Fluorescence quenching and recovery with fructose was analyzed with fluorescence correlation spectroscopy on the level of a few molecules of the reporting dye. The study was highlighted as JACS Spotlight in 2012 [86].

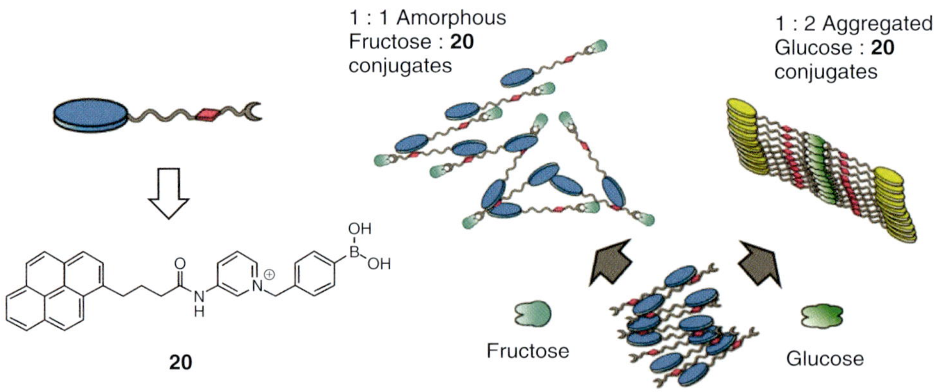

Figure 1.17 Amphiphilic monoboronic acid that is highly selective and sensitive for glucose. (Adapted and reproduced with permission from Ref. [77]. Reproduced with permission of American Chemical Society (United States).)

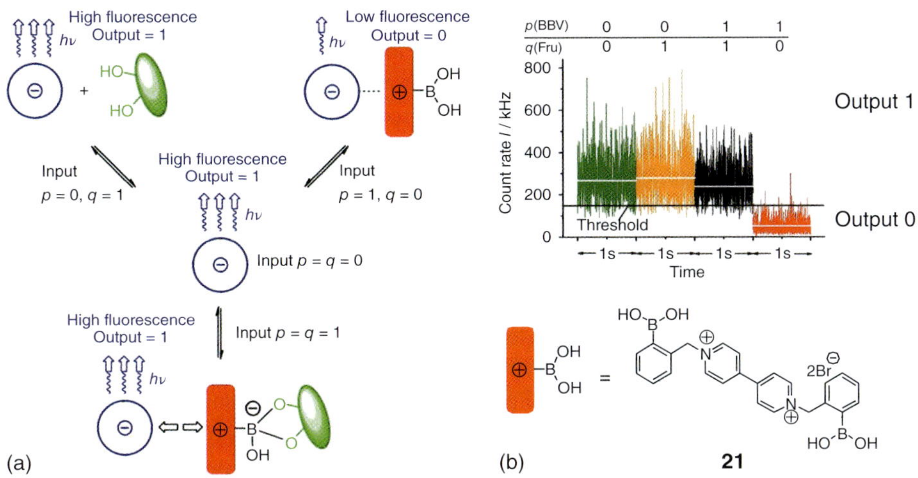

Figure 1.18 (a) A two-component saccharide probe performs IMP logic by an allosteric indicator displacement assay (AIDA). (Blue circle: fluorescent dye 1,6,7,12-tetrakis(4-sulfonylphenoxy)-*N*,*N′*-(2,6-diisopropylphenyl)perylene-3,4:9,10-tetracarboxidiimide (WS-PDI); red rectangle: saccharide receptor and quencher Bis-boronic acid appended Benzyl Viologen **21** (BBV); green ellipse: fructose, Fru.) (b) IMP logic gate via fluorescence transients in a confocal microscope from WS-PDI and the inputs p(BBV) and q(Fru): green (input 0,0), orange (0,1), black (1,1), and red (1,0). Truth table of IMP is also shown. (Adapted and reproduced with permission from Ref. [84]. Reproduced with permission of American Chemical Society (United States).)

Boronic acids can also be used for triggered drug delivery. A remarkable example was recently shown by the group of Kataoka [87]. Therapeutics based on small interfering RNA (siRNA) offer an attractive clinical option because of its ability to silence genes in a highly sequence-specific manner. siRNA was encapsulated by a phenylboronate-functionalized polyion complex (PIC) micelle. It binds to the phenylboronate via ribose of the siRNA thereby stabilizing the complex under conditions equivalent to an extracellular environment. This complex is disrupted in response to the addition of ATP, at a concentration comparable to that inside cells.

1.6
Conclusion and Outlook

To conclude, we have described promising examples for the recognition of important bioanalytes such as PPi, NTP, carbohydrates, LPS, and nucleic acids. Their potential application in biological cell staining, drug delivery, and molecular logic functions has also been mentioned.

Metal complex receptors with chromogenic or luminescent motif provide useful signals to measure the detection process. In addition, metal complex strategies offer great advantages for bioanalytes binding in pure aqueous media with improved solubility of the receptor. The attractive feature of the colorimetric PDA vesicle systems is that they do not require any complex detection method, but it is rather a single step "mix and observe" process [58]. An important challenge in this field is to develop reversible PDA sensors for the detection of bioanalytes. Until now, no reversible PDA sensors for chemical or biological analytes are available.

Boronic acids are the commonly used receptor system for carbohydrate recognition. The combination of a boronic acid-appended viologen receptor and reporter dye was able to perform a logic function with fructose. This work demonstrates ways by which Boolean logic can process information in the field of sugar diagnostics [86]. The ability of boronic acid receptors to function effectively in water for bioanalytes within physiological and environmental scenarios was well studied. In recent years, efforts have moved to a more biological direction with applications in drug delivery, carbohydrate biomarkers, and array analysis.

Future advances in recognition of bioanalytes will involve the development of new intelligent methods to improve the binding properties of receptors in water. Although some examples are available now, discrimination between bioanalytes has been rarely achieved, and thus it is a clear challenge for the future. Recently, fluorine containing materials have been applied for monitoring different biological events such as enzymatic activity, cell viability, and biological reactions [88]. Thus, the development of fluorinated probes and using ^{19}FNMR spectroscopy for the detection of bioanalytes is a challenging and potential task within physiological condition. We believe that this chapter will inspire new advancement in the research area of bioanalytes recognition and discovering the future of molecular sciences.

Acknowledgment

A. S. thanks the Carl-Zeiss foundation for a Junior Professor fellowship and the Fond der chemischen Industrie (FCI). A. S. and D. A. J. thank the EC for financial support through the FP7-project "Novosides" (grant agreement nr. KBBE-4-265854). Research was partially supported by the DFG research unit "Heme and heme degradation products" FOR 1738, TP 7. Thanks also to the Center of Medical Optics and Photonics (CEMOP), the Abbe Center of Photonics (ACP), and the Jena Center of Soft Matter (JCSM) at the Friedrich Schiller University Jena. A. G. thanks the Alexander von Humboldt foundation for a fellowship. A. G. and D. A. J. thanks Dr. A. Das, Prof. Dr. B. König, Prof. Dr. A. Müller, and Prof. Dr. A. Schiller for their constant encouragement and support.

References

1. Gale, P.A. and Steed, J.W. (eds) (2012) *Supramolecular Chemistry: From Molecules to Nanomaterials*, vol. 1–8, John Wiley & Sons, Ltd., Chichester.
2. Lehn, J.-M. (1995) *Supramolecular Chemistry: Concepts and Perspectives*, Wiley-VCH Verlag GmbH, Weinheim, Cambridge.

3. Fischer, E. (1894) Einfluss der Configuration auf die Wirkung der Enzyme. *Ber. Dtsch. Chem. Ges.*, **27**, 2985–2993.
4. Muller-Dethlefs, K. and Hobza, P. (2000) Noncovalent interactions: a challenge for experiment and theory. *Chem. Rev.*, **100**, 143–168.
5. Valeur, B. and Leray, I. (2000) Design principles of fluorescent molecular sensors for cation recognition. *Coord. Chem. Rev.*, **205**, 3–40.
6. Sessler, J.L., Gale, P.A., and Cho, W.-S. (2006) *Anion Receptor Chemistry*, Royal Society of Chemistry, Cambridge.
7. Martinez-Manez, R. and Sancenon, F. (2003) Fluorogenic and chromogenic chemosensors and reagents for anions. *Chem. Rev.*, **103**, 4419–4476.
8. Westheimer, F.H. (1987) Why nature chose phosphates. *Science*, **235**, 1173–1178.
9. Ronaghi, M., Karamohamed, S., Pettersson, B., Uhlén, M., and Nyrén, P. (1996) Real-time DNA sequencing using detection of pyrophosphate release. *Anal. Biochem.*, **242**, 84–89.
10. Kim, S., Lee, D., Hong, J.-I., and Yoon, J. (2009) Chemosensors for pyrophosphate. *Acc. Chem. Res.*, **42**, 23–31.
11. Blackburn, G.M., Gait, M.J., and Williams, D.M. (eds) (2006) *Nucleic Acids in Chemistry and Biology*, 3rd edn, Royal Society of Chemistry.
12. Burnstock, G. (2006) Pathophysiology and therapeutic potential of purinergic signaling. *Pharmacol. Rev.*, **58**, 58–86.
13. Orgel, L. (2000) A simpler nucleic acid. *Science*, **290**, 1306–1307.
14. Liu, J., Cao, Z., and Lu, Y. (2009) Functional nucleic acid sensors. *Chem. Rev.*, **109**, 1948–1998.
15. Sassolas, A., Leca-Bouvier, B.D., and Blum, L.J. (2007) DNA biosensors and microarrays. *Chem. Rev.*, **108**, 109–139.
16. Jin, S., Cheng, Y., Reid, S., Li, M., and Wang, B. (2010) Carbohydrate recognition by boronolectins, small molecules, and lectins. *Med. Res. Rev.*, **30**, 171–257.
17. Sears, P. and Wong, C.-H. (2001) Toward automated synthesis of oligosaccharides and glycoproteins. *Science*, **291**, 2344–2350.
18. Suri, J.T., Cordes, D.B., Cappuccio, F.E., Wessling, R.A., and Singaram, B. (2003) Continuous glucose sensing with a fluorescent thin-film hydrogel. *Angew. Chem. Int. Ed.*, **42**, 5857–5859.
19. Henning, T. (2009) *In Vivo Glucose Sensing*, John Wiley & Sons, Inc., Hoboken, NJ, pp. 113–156.
20. Steiner, M.-S., Duerkop, A., and Wolfbeis, O.S. (2011) Optical methods for sensing glucose. *Chem. Soc. Rev.*, **40**, 4805–4839.
21. Raetz, C.R. and Whitfield, C. (2002) Lipopolysaccharide endotoxins. *Annu. Rev. Biochem.*, **71**, 635–700.
22. Freudenberg, M.A., Tchaptchet, S., Keck, S., Fejer, G., Huber, M., Schütze, N., Beutler, B., and Galanos, C. (2008) Lipopolysaccharide sensing an important factor in the innate immune response to Gram-negative bacterial infections: benefits and hazards of LPS hypersensitivity. *Immunobiology*, **213**, 193–203.
23. Heumann, D. and Roger, T. (2002) Initial responses to endotoxins and Gram-negative bacteria. *Clin. Chim. Acta*, **323**, 59–72.
24. Bianchi, A., Bowman-James, K., and Garcia-Espana, E. (eds) (1997) *Supramolecular Chemistry of Anions*, Wiley-VCH Verlag GmbH, Weinheim.
25. Ambrosi, G., Formica, M., Fusi, V., Giorgi, L., Macedi, E., Micheloni, M., Paoli, P., Pontellini, R., and Rossi, P. (2011) A macrocyclic ligand as receptor and ZnII-complex receptor for anions in water: binding properties and crystal structures. *Chem. Eur. J.*, **17**, 1670–1682.
26. Das, P., Bhattacharya, S., Mishra, S., and Das, A. (2011) Zn(II) and Cd(II)-based complexes for probing the enzymatic hydrolysis of $Na_4P_2O_7$ by alkaline phosphatase in physiological conditions. *Chem. Commun.*, **47**, 8118–8120.
27. Hargrove, A.E., Nieto, S., Zhang, T., Sessler, J.L., and Anslyn, E.V. (2011) Artificial receptors for the recognition of phosphorylated molecules. *Chem. Rev.*, **111**, 6603–6782.
28. Ngo, H.T., Liu, X., and Jolliffe, K.A. (2012) Anion recognition and sensing with Zn(II)-dipicolylamine complexes. *Chem. Soc. Rev.*, **41**, 4928–4965.

29. Ojida, A., Mito-oka, Y., Inoue, M.-a., and Hamachi, I. (2002) First artificial receptors and chemosensors toward phosphorylated peptide in aqueous solution. *J. Am. Chem. Soc.*, **124**, 6256–6258.
30. Sakamoto, T., Ojida, A., and Hamachi, I. (2009) Molecular recognition, fluorescence sensing, and biological assay of phosphate anion derivatives using artificial Zn(II)-Dpa complexes. *Chem. Commun.*, 141–152.
31. Ojida, A., Takashima, I., Kohira, T., Nonaka, H., and Hamachi, I. (2008) Turn-on fluorescence sensing of nucleoside polyphosphates using a xanthene-based Zn(II) complex chemosensor. *J. Am. Chem. Soc.*, **130**, 12095–12101.
32. Kurishita, Y., Kohira, T., Ojida, A., and Hamachi, I. (2010) Rational design of FRET-based ratiometric chemosensors for in vitro and in cell fluorescence analyses of nucleoside polyphosphates. *J. Am. Chem. Soc.*, **132**, 13290–13299.
33. Kurishita, Y., Kohira, T., Ojida, A., and Hamachi, I. (2012) Organelle-localizable fluorescent chemosensors for site-specific multicolor imaging of nucleoside polyphosphate dynamics in living cells. *J. Am. Chem. Soc.*, **134**, 18779–18789.
34. O'Neil, E.J. and Smith, B.D. (2006) Anion recognition using dimetallic coordination complexes. *Coord. Chem. Rev.*, **250**, 3068–3080.
35. Li, C., Numata, M., Takeuchi, M., and Shinkai, S. (2005) A sensitive colorimetric and fluorescent probe based on a polythiophene derivative for the detection of ATP. *Angew. Chem. Int. Ed.*, **44**, 6371–6374.
36. Ghosh, A., Shrivastav, A., Jose, D.A., Mishra, S.K., Chandrakanth, C.K., Mishra, S., and Das, A. (2008) Colorimetric sensor for triphosphates and their application as a viable staining agent for prokaryotes and eukaryotes. *Anal. Chem.*, **80**, 5312–5319.
37. Jose, D.A., Mishra, S., Ghosh, A., Shrivastav, A., Mishra, S.K., and Das, A. (2007) Colorimetric sensor for ATP in aqueous solution. *Org. Lett.*, **9**, 1979–1982.
38. Mahato, P., Ghosh, A., Mishra, S.K., Shrivastav, A., Mishra, S., and Das, A. (2010) Zn(II) based colorimetric sensor for ATP and its use as a viable staining agent in pure aqueous media of pH 7.2. *Chem. Commun.*, **46**, 9134–9136.
39. Mahato, P., Ghosh, A., Mishra, S.K., Shrivastav, A., Mishra, S., and Das, A. (2011) Zn(II)–cyclam based chromogenic sensors for recognition of ATP in aqueous solution under physiological conditions and their application as viable staining agents for microorganism. *Inorg. Chem.*, **50**, 4162–4170.
40. Mahato, P., Ghosh, A., Mishra, S.K., Shrivastava, A., Mishra, S., and Das, A. (2011) Chemical Indexing Equivalent to 157:157207 (IN), Council of Scientific & Industrial Research, India, p. 68.
41. Lee, D.H., Kim, S.Y., and Hong, J.-I. (2004) A fluorescent pyrophosphate sensor with high selectivity over ATP in water. *Angew. Chem. Int. Ed.*, **43**, 4777–4780.
42. Zhang, J.F., Kim, S., Han, J.H., Lee, S.-J., Pradhan, T., Cao, Q.Y., Lee, S.J., Kang, C., and Kim, J.S. (2011) Pyrophosphate-selective fluorescent chemosensor based on 1,8-naphthalimide–DPA–Zn(II) complex and its application for cell imaging. *Org. Lett.*, **13**, 5294–5297.
43. Das, P., Ghosh, A., Kesharwani, M.K., Ramu, V., Ganguly, B., and Das, A. (2011) ZnII–2,2′:6′,2″-Terpyridine-based complex as fluorescent chemosensor for PPi, AMP and ADP. *Eur. J. Inorg. Chem.*, **2011**, 3050–3058.
44. Lee, D.H., Im, J.H., Son, S.U., Chung, Y.K., and Hong, J.-I. (2003) An azophenol-based chromogenic pyrophosphate sensor in water. *J. Am. Chem. Soc.*, **125**, 7752–7753.
45. Gruber, B. and Koenig, B. (2013) Self-assembled vesicles with functionalized membranes. *Chem. Eur. J.*, **19**, 438–448.
46. Gruber, B., Balk, S., Stadlbauer, S., and Koenig, B. (2012) Dynamic interface imprinting: high-affinity peptide binding sites assembled by analyte-induced recruiting of membrane receptors. *Angew. Chem. Int. Ed.*, **51**, 10060–10063.
47. Gruber, B., Stadlbauer, S., Woinaroschy, K., and Koenig, B. (2010) Luminescent vesicular receptors for the recognition of

biologically important phosphate species. *Org. Biomol. Chem.*, **8**, 3704–3714.

48. Tomas, S. and Milanesi, L. (2010) Mutual modulation between membrane-embedded receptor clustering and ligand binding in lipid membranes. *Nat. Chem.*, **2**, 1077–1083.
49. Voskuhl, J. and Ravoo, B.J. (2009) Molecular recognition of bilayer vesicles. *Chem. Soc. Rev.*, **38**, 495–505.
50. Liu, Q. and Boyd, B.J. (2013) Liposomes in biosensors. *Analyst*, **138**, 391–409.
51. Jose, D.A. and Konig, B. (2010) Polydiacetylene vesicles functionalized with N-heterocyclic ligands for metal cation binding. *Org. Biomol. Chem.*, **8**, 655–662.
52. Jose, D.A., Stadlbauer, S., and König, B. (2009) Polydiacetylene-based colorimetric self-assembled vesicular receptors for biological phosphate ion recognition. *Chem. Eur. J.*, **15**, 7404–7412.
53. Ahn, D.J. and Kim, J.-M. (2008) Fluorogenic polydiacetylene supramolecules: immobilization, micropatterning, and application to label-free chemosensors. *Acc. Chem. Res.*, **41**, 805–816.
54. Reppy, M.A. and Pindzola, B.A. (2007) Biosensing with polydiacetylene materials: structures, optical properties and applications. *Chem. Commun.*, **42**, 4317–4338.
55. Okada, S., Peng, S., Spevak, W., and Charych, D. (1998) Color and chromism of polydiacetylene vesicles. *Acc. Chem. Res.*, **31**, 229–239.
56. Yarimaga, O., Jaworski, J., Yoon, B., and Kim, J.-M. (2012) Polydiacetylenes: supramolecular smart materials with a structural hierarchy for sensing, imaging and display applications. *Chem. Commun.*, **48**, 2469–2485.
57. Chen, X., Zhou, G., Peng, X., and Yoon, J. (2012) Biosensors and chemosensors based on the optical responses of polydiacetylenes. *Chem. Soc. Rev.*, **41**, 4610–4630.
58. Jelinek, R. and Kolusheva, S. (2007) Biomolecular sensing with colorimetric vesicles. *Top. Curr. Chem.*, **277**, 155–180.
59. Banerjee, S. and Koenig, B. (2013) Molecular imprinting of luminescent vesicles. *J. Am. Chem. Soc.*, **135**, 2967–2970.
60. Charych, D.H., Nagy, J.O., Spevak, W., and Bednarski, M.D. (1993) Direct colorimetric detection of a receptor-ligand interaction by a polymerized bilayer assembly. *Science*, **261**, 585–588.
61. Lambert, T.N. and Smith, B.D. (2003) Synthetic receptors for phospholipid headgroups. *Coord. Chem. Rev.*, **240**, 129–141.
62. Jiang, H. and Smith, B.D. (2006) Dynamic molecular recognition on the surface of vesicle membranes. *Chem. Commun.*, 1407–1409.
63. DiVittorio, K.M., Leevy, W.M., O'Neil, E.J., Johnson, J.R., Vakulenko, S., Morris, J.D., Rosek, K.D., Serazin, N., Hilkert, S., Hurley, S., Marquez, M., and Smith, B.D. (2008) Zinc(II) coordination complexes as membrane-active fluorescent probes and antibiotics. *ChemBioChem*, **9**, 286–293.
64. Kim, K.M., Oh, D.J., and Ahn, K.H. (2011) Zinc(II)-dipicolylamine-functionalized polydiacetylene-liposome microarray: a selective and sensitive sensing platform for pyrophosphate ions. *Chem. Asian J.*, **6**, 122–127.
65. Cho, Y.-S., Kim, K.M., Lee, D., Kim, W.J., and Ahn, K.H. (2013) Turn-on fluorescence detection of apoptotic cells using a zinc(II)-dipicolylamine-functionalized poly(diacetylene) liposome. *Chem. Asian J.*, **8**, 755–759.
66. Jeon, H., Lee, S., Li, Y., Park, S., and Yoon, J. (2012) Conjugated polydiacetylenes bearing quaternary ammonium groups as a dual colorimetric and fluorescent sensor for ATP. *J. Mater. Chem.*, **22**, 3795–3799.
67. Rangin, M. and Basu, A. (2004) Lipopolysaccharide identification with functionalized polydiacetylene liposome sensors. *J. Am. Chem. Soc.*, **126**, 5038–5039.
68. Wu, J., Zawistowski, A., Ehrmann, M., Yi, T., and Schmuck, C. (2011) Peptide functionalized polydiacetylene liposomes act as a fluorescent turn-on sensor for bacterial lipopolysaccharide. *J. Am. Chem. Soc.*, **133**, 9720–9723.
69. Barone, V., Cacelli, I., Ferretti, A., Monti, S., and Prampolini, G. (2009) Sensors for DNA detection: theoretical investigation of the conformational

properties of immobilized single-strand DNA. *Phys. Chem. Chem. Phys.*, **11**, 10644–10656.

70. Wang, C. and Ma, Z. (2005) Colorimetric detection of oligonucleotides using a polydiacetylene vesicle sensor. *Anal. Bioanal. Chem.*, **382**, 1708–1710.

71. Jung, Y.K., Kim, T.W., Kim, J., Kim, J.-M., and Park, H.G. (2008) Universal colorimetric detection of nucleic acids based on polydiacetylene (PDA) liposomes. *Adv. Funct. Mater.*, **18**, 701–708.

72. Cho, Y.-S. and Ahn, K.H. (2013) Molecular interactions between charged macromolecules: colorimetric detection and quantification of heparin with a polydiacetylene liposome. *J. Mater. Chem. B*, **1**, 1182–1189.

73. Bull, S.D., Davidson, M.G., van den Elsen, J.M.H., Fossey, J.S., Jenkins, A.T.A., Jiang, Y.-B., Kubo, Y., Marken, F., Sakurai, K., Zhao, J., and James, T.D. (2012) Exploiting the reversible covalent bonding of boronic acids: recognition, sensing, and assembly. *Acc. Chem. Res.*, **46**, 312–326.

74. Nishiyabu, R., Kubo, Y., James, T.D., and Fossey, J.S. (2012) Boronic acid building blocks: tools for sensing and separation. *Chem. Commun.*, **47**, 1106–1123.

75. Schiller, A. (2012) in *Molecules at Work. Selfassembly, Nanomaterials, Molecular Machinery* (ed. B. Pignataro), Wiley-VCH Verlag GmbH, Weinheim, pp. 315–338.

76. Schiller, A., Vilozny, B., Wessling, R.A., and Singaram, B. (2011) in *Reviews in Fluorescence*, vol. 2009 (ed. C. Geddes), Springer, New York, pp. 155–191.

77. Huang, Y.-J., Ouyang, W.-J., Wu, X., Li, Z., Fossey, J.S., James, T.D., and Jiang, Y.-B. (2013) Glucose sensing via aggregation and the use of "Knock-Out" binding to improve selectivity. *J. Am. Chem. Soc.*, **135**, 1700–1703.

78. de Silva, A.P., Gunaratne, N.H.Q., and McCoy, C.P. (1993) A molecular photoionic and gate based on fluorescent signalling. *Nature*, **364**, 42–44.

79. de Silva, A.P. and Uchiyama, S. (2007) Molecular logic and computing. *Nat. Nanotechnol.*, **2**, 399–410.

80. Katz, E. and Privman, V. (2010) Enzyme-based logic systems for information processing. *Chem. Soc. Rev.*, **39**, 1835–1857.

81. Pischel, U., Andréasson, J., Gust, D., and Pais, V.F. (2013) Information processing with molecules—Quo vadis? *ChemPhysChem*, **14**, 28–46.

82. Katz, E. (2012) *Biomolecular Information Processing: From Logic Systems to Smart Sensors and Actuators*, Wiley-VCH Verlag GmbH, Weinheim.

83. Pasparakis, G., Vamvakaki, M., Krasnogor, N., and Alexander, C. (2009) Diol-boronic acid complexes integrated by responsive polymers-a route to chemical sensing and logic operations. *Soft Matter*, **5**, 3839–3841.

84. Elstner, M., Weisshart, K., Müllen, K., and Schiller, A. (2012) Molecular logic with a saccharide probe on the few-molecules level. *J. Am. Chem. Soc.*, **134**, 8098–8100.

85. Kikkeri, R., Grünstein, D., and Seeberger, P.H. (2010) Lectin biosensing using digital analysis of Ru(II)-glycodendrimers. *J. Am. Chem. Soc.*, **132**, 10230–10232.

86. Perkel, J.M. (2012) JACS spotlight "A sugar-sensing molecular logic gate? Sweet!". *J. Am. Chem. Soc.*, **134**, 9535–9536.

87. Naito, M., Ishii, T., Matsumoto, A., Miyata, K., Miyahara, Y., and Kataoka, K. (2012) A phenylboronate-functionalized polyion complex micelle for ATP-triggered release of siRNA. *Angew. Chem. Int. Ed.*, **51**, 10751–10755.

88. Srinivas, M., Heerschap, A., Ahrens, E.T., Figdor, C.G., and de Vries, I.J.M. (2010) ^{19}F MRI for quantitative in vivo cell tracking. *Trends Biotechnol.*, **28**, 363–370.

2
Methods of DNA Recognition

Olalla Vázquez

2.1
Introduction

DNA is the biopolymer by which the genetic information of bacteria and all higher organisms is carried. This information is encoded in the DNA sequence, which defines many of the essential features of life. This information is put into practice on specific recognition events, principally through the action of proteins called *transcription factors* (TFs). The binding of the TFs to DNA is normally reversible and noncovalent in nature. Inherently, this process allows the genetic information to respond to the environment. For these reasons, the elucidation of the molecular basis of the DNA recognition is important not only from a fundamental point of view, but also from the perspective of understanding the DNA recognition process, which will provide us the opportunity to design nonnatural agents that can recognize specific sequences on the DNA double helix and introduce additional properties to the recognition process, such as sensing and controllability.

Since the elucidation of the double helix structure of the DNA, the study of DNA has been unwrapped from the initial structural and functional characterization in living organisms to what is now known as DNA technology, encompassing the development of DNA as a tool in biological sciences, and more recently as a structural and nanotechnological scaffold [1]. The deciphering of the human genome [2] and the accessibility of structural data [3] have opened new perspectives in biomedical research, promising improved diagnostic techniques and personalized therapies. However, the development of DNA-based functional and dynamic processes, particularly those employing double stranded DNA (dsDNA), is still in its infancy, although it is increasingly expanding. In this context, the ability to specifically manipulate genetic information processing genome-wide, or to find new ways of codifying physicochemical processes with dsDNA – particularly the development of efficient methods for selective sensing of dsDNA sequences – might offer interesting new approaches to interfere with the DNA information processing at an early stage of gene expression, and to target genomic modifications for research and discovery at the interphase of biomedical and chemical sciences.

Discovering the Future of Molecular Sciences, First Edition. Edited by Bruno Pignataro.

In this chapter the focus is on the molecular basis of dsDNA recognition, the interaction of natural TFs with the DNA, gene expression, and the current developments in the design and preparation of synthetic dsDNA binders. We place special emphasis on the recognition of the most relevant conformation under physiological conditions: the B-form of dsDNA. For those interested in the recognition of higher levels of ordering of the B-DNA and/or supramolecular interactions with other DNA structures, several reviews have recently addressed this subject [4].

2.2
Historical Outline: The Central Dogma

This year, 2013, coincides with the 60th anniversary of the proposal of a double helical structure for DNA by James Watson and Francis Crick, where the X-ray crystallographic data of Rosalind Franklin and Maurice Wilkins were crucial to the discovery [5]. There have been many other milestones that occurred in the middle of the twentieth century that were perhaps equally relevant to the unraveling of the cellular processes, but none had the same repercussions throughout the public and the scientific community. Indeed, the dsDNA model has become one of the most recognized icons of twentieth-century science. The true beauty of the Watson–Crick model was that the structure immediately suggested function: a possible copying mechanism for the genetic material. The fact that this structure could code for and transmit the genetic information aroused more interest than the structure itself.

To a great extent, the characteristics of an organism are determined by its genes, that is, the specific sequences of nucleic acids necessary for the synthesis of a functional polypeptide or RNA molecule [6]. Without a doubt, the flow of information from one generation to the next is a key question for life. In 1957 and after the postulation of the structure of the DNA, Francis Crick proposed the transference of this genetic information in the international meeting of the Society of Experimental Biology, and it was published one year later [7]. He called the basic idea of this hypothesis the *Central Dogma*. This states that the genetic information of the DNA is not directly converted into proteins, but it must first be copied into the RNA. The translation of the RNA into proteins is unidirectional. Until that date, proteins were the best candidates to become the molecules of life, because DNA was generally considered to be chemically, and thus structurally, too simple to contain the presumably complex information for heredity. Nevertheless, this hypothesis gave the DNA a key role in modern biology and it has been a constant point of reference for the proponents of the new molecular biology of late 1950s and 1960s. However, it has also provided a target for many criticisms of the molecular approach as the information is translated into proteins and these cannot modify and affect the genes. The unidirectionality of this formulation has always been one of the most controversial aspects. Indeed, in 1970, Crick himself had to review his initial formulation and explain his own interpretation in a full paper [8].

It is interesting that Francis Crick change the word dogma for naming a scientific theory, but in fact for a long time, his idea did acquire the dimension of *Absolute Truth* [9]. Nevertheless, since his first formulation until now, several corrections have been made to form the original scheme [10]. Likewise, it has been demonstrated that the proteins can have a regulatory effect on the DNA. There are also transmissible proteins (prions) and 98% of the transcriptional production is formed by introns and other RNAs without codified function such as ribozymes, interference RNAs.

Watson and Crick's description of DNA as a double helix and the subsequent deciphering of the genetic code began to move molecular biology to the level of a more quantitative science. Furthermore, two other technological breakthroughs played crucial roles in this development: the ability to rapidly sequence DNA and the emergence of rapid, inexpensive methods of synthesis of shorts strands of DNA. The cell was, and in many ways still remains, a black box, but for the first time there was the possibility to treat biology as a branch of knowledge that at least some day could be understood in mathematical terms. Since the *Central Dogma* provided a directionality to the flow of information in the cells.

2.3 Intermolecular Interaction between the Transcription Factors and the DNA

Gene expression relies on a myriad of carefully orchestrated interactions, which are subtly controlled in time and space. Therefore, gene expression entails many different levels of regulation, any of which cannot be undervalued. However, from a molecular point of view, it has been demonstrated that the regulation mainly takes place in transcription step by high-affinity interactions of the TFs with specific sequences of the DNA. More than 50 years after the structure of the DNA was first proposed by Watson and Crick [11], biologists and chemists are still struggling to fully understand how these proteins interact with the genome. One of the most important questions that remains relates to specificity – how do the large and diverse number of DNA-binding proteins recognize their specific binding sites? Moreover, most of the DNA-binding proteins are part of large families that share DNA-binding domains, but how do they carry out unique functions *in vivo*? Providing answers to these questions is especially timely given the need to accurately annotate the many complete genome sequences, and the answers require a better understanding of the rules that govern how proteins bind to DNA sequences.

As with any other biochemical interaction, the TFs recognize the specific DNA sequences due to chemical and steric complementarity of their molecular surfaces [12]. Therefore, the interaction process between the TFs and the DNA cannot be totally understood without analyzing their main structural features.

2.3.1
The Structure of DNA and Its Role in the Recognition

The cellular genome is arranged on chromosomes where each comprises a dsDNA molecule, packed together with a set of associated proteins. In contrast with the highly variable structures of proteins and RNA, the most frequently observed DNA structure is governed by only a few principles. Nucleobases form hydrogen bonds with other nucleobases. In a simplistic model, DNA secondary structures can be controlled by changing the sequence, altering the protonation state of nucleobases, and/or including metal-based interactions, but the most commonly found conformation under physiological conditions is the B-form of the dsDNA. It is characterized by a fairly uniform right-handed double helical structure, in which the two antiparallel deoxyribonucleotide chains twist around each other. The double helix is a quite rigid and compact molecule. The diameter of the B-DNA measures approximately 20 Å, and each turn of the helix is 34 Å long with 10 bases per turn (Figure 2.1).

The structure is stabilized by intermolecular hydrogen bonds between complementary Watson–Crick base pairs and by hydrophobic interactions that tend to keep the nonpolar surface of the bases away from the surrounding aqueous environments while their polar edges and phosphate groups are exposed to the solvent. Even if it is quite a regular structure, B-DNA shows significant local variability depending on the sequence, which is decisive in the protein interactions. The asymmetry of the nucleotides brings on two grooves with different sizes and geometric features. The major groove is wide and relatively shallow, while the minor one is deeper and narrower. The width of the minor groove is mainly determined by the sequence, and in general, the A/T regions are narrower than G/C ones. Current inspections of the groove hydration properties have revealed a clear qualitative difference in the energetic signature of binding to the major and

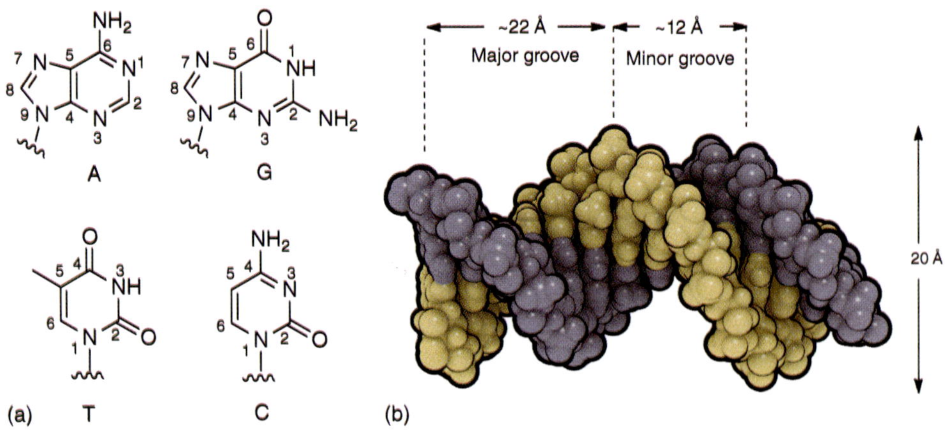

Figure 2.1 (a) Structure of the four nucleobases of the DNA and (b) representation of the ideal B-DNA conformation with the main structural dimensions.

minor grooves; overall, the interaction with the major groove is a process guided predominantly by the enthalpy while the entropic forces drive the binding with the minor groove, despite an unfavorable enthalpy [13].

In most cases, the recognition process takes place through interactions between the exposed functional groups of the bases: the binding of the TF to DNA does not disrupt the DNA packing, although sometimes these interactions can cause local conformational alterations (Figure 2.2).

A simple analysis of the ideal B-DNA conformation makes visible a higher functional variability of the donor–acceptor pattern in the major groove than that in the minor groove. In fact, in the minor groove the A/T and T/A bases pairs are

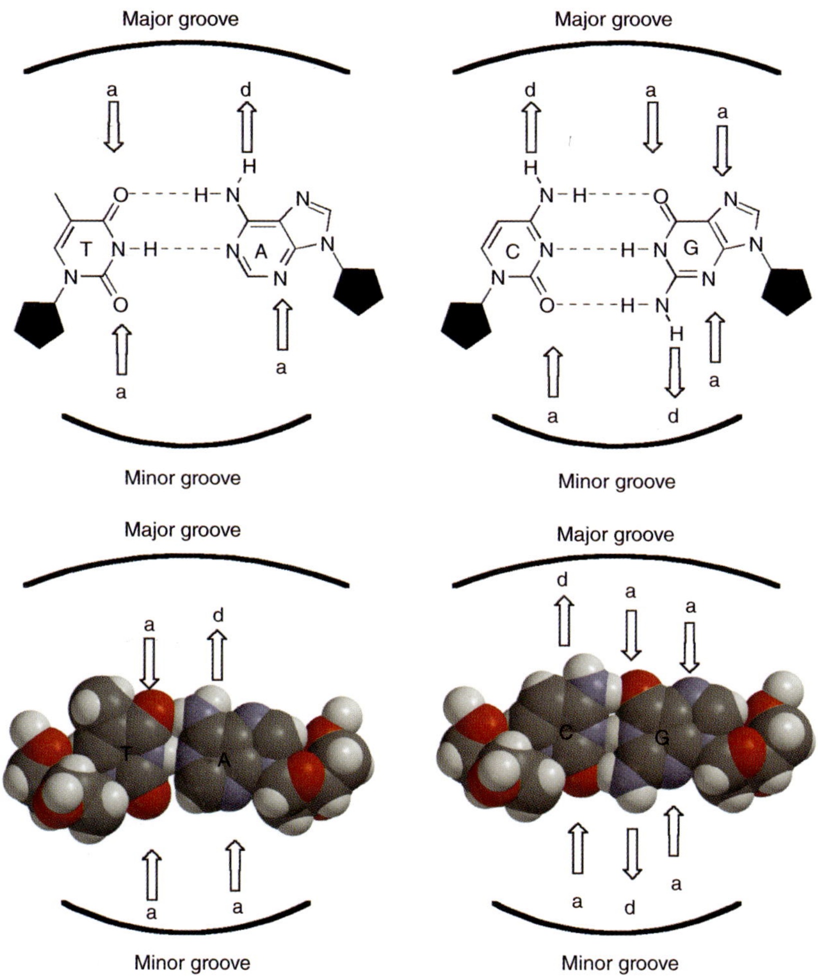

Figure 2.2 Representation of the different pattern of donors (d) and acceptors (a) of the hydrogen bonds of the DNA base pairs in both grooves.

degenerate in their capability to form hydrogen bonds whereas in the major groove all the four possible base pairs can be distinguished by particular donor–acceptor combinations [14]. These aspects, as well as the size and the hydration of the minor groove, justify the fact that the majority of TFs preferably recognize the DNA through the major groove.

2.3.2
DNA Binding Domains of the TF

During the past few years, there has been an exponential increase in the number of DNA-protein structures that have been resolved by NMR and crystallography. There were more than 280 structures of DNA–protein complexes placed in the Protein Data Bank at the end of 2002 [15] and now, 11 years later, there are twice as many. As a consequence, we now have a very good overall picture of the architecture of DNA-binding proteins and how they bind to DNA.

In spite of the fact that the DNA–protein complexes have a high morphologic variability, the study of these complexes at the atomic level has allowed us to formulate general rules of how the interaction occurs. In most of the cases the proteins present an α helix, known as *recognition helix*, in charge of "reading" of the base pairs in the major groove of the DNA. This helix performs most of the specific contacts with the bases in the complex as well as many accessory interactions with the sugar–phosphate backbone. It has been discovered that most of the contacts of the recognition helix in the major groove occur when the axis of this helix is flanked in the DNA backbone [16]. However, the orientation of this helix is not the same in all proteins and there are even helices that can be inserted in the minor groove of the DNA [17], and there are also proteins that use other structural motifs like β sheets or turns in order to recognize DNA. Moreover, these motifs are embedded in a more complex structure essential for the interaction, which is called the *binding domain*.

Several classifications have been established in accordance with the structural motifs the proteins use for the recognition [18]:

1) **Helix-turn-helix (HTH) and homeodomains**: the HTH was the first DNA recognition motif discovered [19]. Traditionally it was defined as a 20 amino acid sequence consisting of two almost perpendicular α helices (~120 °C) connected through a β turn. However, nowadays some authors have extended this definition and included the proteins with longer linkers in this category as long as the orientation of the helices remains the same.
   Apart from the specific contacts of the residue side chains of the recognition helix, there are also other fundamental interactions with the backbone of the DNA in these domains. Thus, in flanking positions of the minor groove the linker and the first helix perform accessory interactions, which serve as a bridge between the major groove and the N-terminus of the recognition helix. Binding to the minor groove usually takes place through short arginine-containing peptide tails residues [11a] such as RQR in Src [20], GRPR in Hin recombinase [21], and RKKR in POU homeodomains [22]. This TF tends to

be called *homeodomain* in eukaryotic cells. It is interesting to comment that there are authors who establish differences between both, beyond the types of organism mentioned. The homeodomain has been defined as the bigger structural motif and is formed by 60 amino acids. This can fold and bind with high affinity with the DNA. Unlike the HTH where the interaction takes places in the form of dimers (or even higher order), the HTH monomers are unable to achieve high-affinity DNA binding (Figure 2.3).

2) **Zinc finger proteins (ZFPs)**: they constitute the largest group of TFs [23]. The structure of a ZFP is characterized by regular sequences of Cys/His residues whose tertiary structure, which is responsible for the recognition, is stabilized by tetrahedral zinc coordination. The DNA-binding domain consists of two antiparallel β sheets and an α helix, which is inserted in the major groove of the DNA. There are different combinations of Cys and His residues (Cys_2His_2, Cys_4, Cys_3His, Cys_6, etc.) but the most common class of these proteins (Cys_2His_2) is about 22 and 30 amino acid long, and contains the sequence X_2-**C**-$X_{2,4}$-**C**-X_{12}-**H**-$X_{3,4,5}$-**H** where X represents hydrophobic residues [24] (Figure 2.4).

Most of the ZFPs interact with DNA as oligomers and it is important to highlight that this association of several zinc finger modules by short linker peptides is essential for the recognition process. To illustrate this we give the example of the ZIF268 protein, which was the first zinc finger discovered in 1982 [25]. This protein consists of the zinc finger modules where, in each one, the key contacts with the bases of the DNA backbone are made through an Arg located immediately before the α helix that interacts with the third base of the primary structure of the DNA (5′-XXG). The third residue of the helix contacts with the second base (5′-XGX) and the sixth one interacts with the first base (5′-GXX). This positioning reflects the importance of cooperation in this matter. In this family the interaction is highly conserved and the basic position of the recognition sites are −1, 2, 3, and 6 (being 1 the first residue of the helix). The manipulation of these key residues has allowed the acquisition of variants that recognize alternative DNA sequences [26] (Figure 2.5).

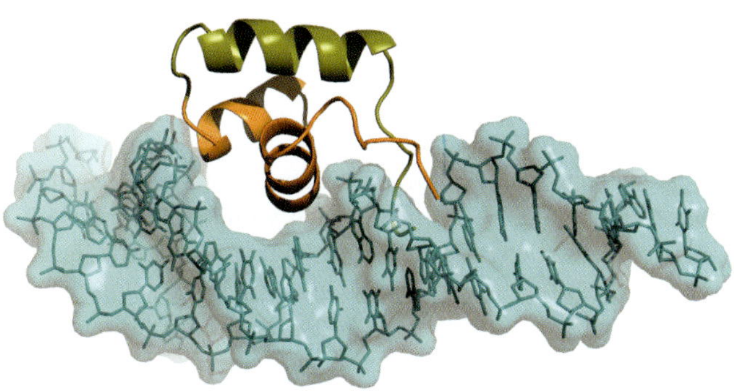

Figure 2.3 Structure of the HTH domain (in orange) of the protein TC3 (PDB: 1TC3).

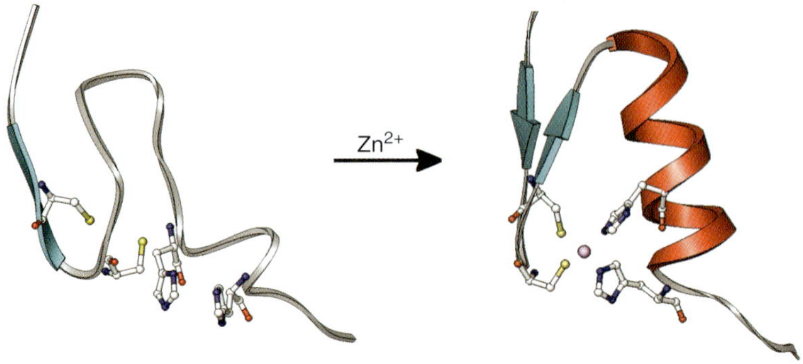

Figure 2.4 Conformational change due to the Zn^{2+} coordination.

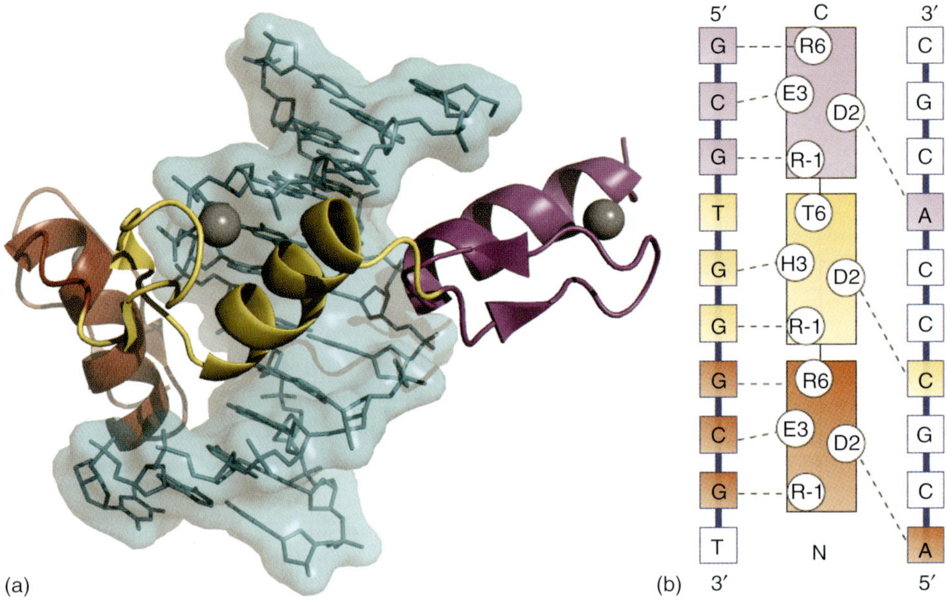

Figure 2.5 (a) Structure of the ZFP ZIF268 (PDB: 1AAY) and (b) diagram of the specific interactions between the DNA and the protein.

The field of zinc finger engineering has progressed remarkably in the past few years. The advances observed suggest that the ZFP design field has reached a stage where it is possible to create novel ZFPs for diverse DNA target sites by design or phage display, or by the combination of both [24a, 27]. Optimized ZFP sequences for target DNA sequences can be designed by using a web-based tool in which the target sequences are automatically converted into ZFP amino acid sequence. The purpose of such designed ZFPs is selective gene targeting by fusing the nucleases, recombinases, and TFs. The length of the recognized nucleotide sequence can be extended simply by connecting zinc finger (ZF) modules in tandem fashion. Through this method, artificial ZFPs are recognized as 9–18 bp DNA sequences and the optimized ZFP amino acid sequences for recognition of base triplets now cover <80% of the total of 64 base triples [28]. In particular, the C_2H_2 domain of Sp1 provided an attractive framework for the design of artificial ZFP with a high DNA specificity. Thus novel DNA-binding proteins using this C_2H_2-type zinc finger domain as a scaffold could be utilized to carry out different activities such as nucleases or TFs. In this context, the group of Sugiura has fueled the development of multiple-zinc fingers that are able to bind to longer regions of DNA [29]. To achieve this ambitious goal they design new peptides constructed by connecting three finger domains of the TF Sp1 by a Kruppel type linker (Thr-Gly-Glu-Lys-Pro) peptide. The affinity of resulting peptide had increased by 30-fold compared to the wild type ZFP. On the basis of this former success, they designed satisfactorily DNA binding fingers in order to regulate gene expression [29a,c].

3) **Basic leucine zipper (bZIP) and helix-loop-helix (HLH)**: these proteins have been exclusively identified in eukaryotic organisms. Their name comes from the shape of the structural motif that forms functional dimers. The bZIP domain is one of the simplest and most effective and it consists only of noncovalent helical heterodimers whose monomeric units consist of about 60 amino acids. Such units show two different subdomains: first the N-terminal *basic region* – so-called because of the presence of positively charged residues like Arg – which consists of approximately 20 amino acids. This region interacts with the DNA by inserting into the major groove of the DNA (Figure 2.6).
The other subdomain is known as *leucine zipper* and contains approximately 30 amino acids in the C-terminus. As its name indicates, it contains Leu residues repeated in each of the seven amino acids that generate the hydrophobic interaction that mediates the dimerization. It has been demonstrated that each monomer alone is unable to interact with the DNA with high affinity, and in the absence of the target DNA, the basic region does not have any defined structure. The HLH motif can be considered as a modification of the bZIP α

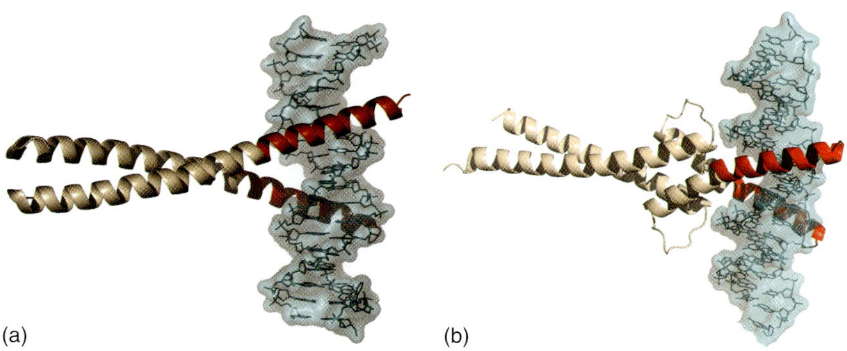

Figure 2.6 (a) Structure of the bZIP domain of the protein GCN4 (PDB: 1YSA) and (b) structure of the domain HLH of the protein SREBP1a (PDB: 1AM9).

helices, where the basic region and the dimerization regions are separated by a turn.

4) **β sheet proteins**: unlike the majority of the TF that rely on α helices for specific DNA recognition, these proteins bind to the DNA through β sheet domains. The most characteristic example is the TATA-binding protein (TBP). This contains 10 antiparallel β sheets, which insert into the minor groove of the DNA and bend it [30].

2.3.3
General Aspects of the Intermolecular Interactions between the TFs and the DNA

As can be deduced from the above discussion, TFs use a large variety of architectural motifs to achieve DNA recognition. It is therefore extremely difficult to extract general rules to explain the selectivity of the binding process. Indeed, many authors have already remarked that there is no simple code that links the binding domains and amino acids with DNA specific sequences [31]. Even in the case of zinc fingers, in which all members share a common structural motif but bind different DNA sequences, it has been extremely difficult to decipher a general recognition code [32]. However, despite the lack of simple sequence recognition patterns, it is possible to formulate some general principles, which provide a general framework to rationalize the molecular basis of the specific DNA recognition and to design new DNA-binding peptides.

Generally speaking, specific binding is defined as a molecular association in which a particular molecule is tightly and exclusively bound by forming an energetically and kinetically stabilized complex structure [33]. The affinity constant of this process is usually in the range $10^8-10^{11}\ M^{-1}$ [34]. In contrast, a nonspecific interaction is defined as a random interaction ($K_a < 10^4-10^6\ M^{-1}$). In connection with the TFs, there are basically five major types of direct interactions between proteins and nucleic acids [35]:

1) **Salt bridges and hydrogen bonds between the DNA phosphodiester backbone and amino acid residues with basic side chains (Lys, Arg, and His)**: these contacts usually do not confer specificity to the binding but are essential for the interaction as they increase the thermodynamic stability of the complex and help to anchor the recognition domain in the correct orientation.
2) **Hydrogen bonds between the sugars or bases in the DNA and polar side chains in the proteins**: they are critical interactions for specificity. An analysis of the available structures of protein–DNA complexes shows that Arg, Lys, Ser, and Thr participate most in this type of hydrogen bonding. Interestingly, Asp and Glu almost never take part, probably owing to the unfavorable interaction with DNA. Within this group bidentate hydrogen bonds between a single side chain and a base or base pair are especially important because they provide an efficient way of increasing the bond energy per amino acid/base pair while conferring a higher degree of specificity (Figure 2.7).

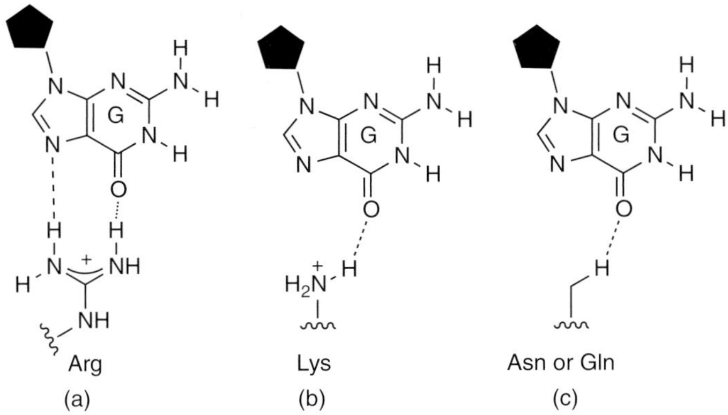

Figure 2.7 (a–c) Examples of the characteristic hydrogen bonds.

3) **Nonpolar contacts between the DNA base pairs and apolar amino acid side chains**: despite their lack of directionality, the hydrophobic interactions are recognized as important components in protein–DNA binding.
4) **Water-mediated hydrogen bonds are relatively common**: in most cases they are established between the DNA backbone and positively charged polar amino acid side chains such as Arg, Lys, Asn, Gln, or even with those that bear a negative charge like Glu and Asp.
5) **van der Waals forces between the bases and the side change of the polar residues**: these interactions are with hydrogen bonds and are the most relevant for specificity.

The thermodynamic factors of the DNA recognition process can be classified as enthalpic factors or entropic ones. Within the first group, the hydrogen bonds, the van der Waals forces, and the electrostatic interactions favor binding while the solvation of the polar groups and the deviation of the ideal geometric parameters affect this interaction unfavorably. Regarding the entropic factors, the hydrophobic interactions and the ionic distribution of the complex contribute favorably, while the loss of freedom, vibration, rotational, and transitional grades are detrimental to binding [33].

From a kinetic point of view, DNA recognition was postulated as a simple two-step process. In the first step, nonspecific long-range electrostatic interactions bring together the TFs and the DNA. The formation of this initial nonspecific complex is followed by the sliding of the protein along the DNA until it finds the target sequence. In this moment the specific interactions are established and the high-affinity sequence-specific complex is thus formed (Figure 2.8).

Finally, it is important to highlight that the linear sequence of base pairs in a binding site is only a small part of the story, and that the three-dimensional structure of both macromolecules must be taken into account to fully understand the protein–DNA recognition. For instance, the appropriate folding of the DNA-binding domain is critical to preorganize, stabilize, and deliver the recognition

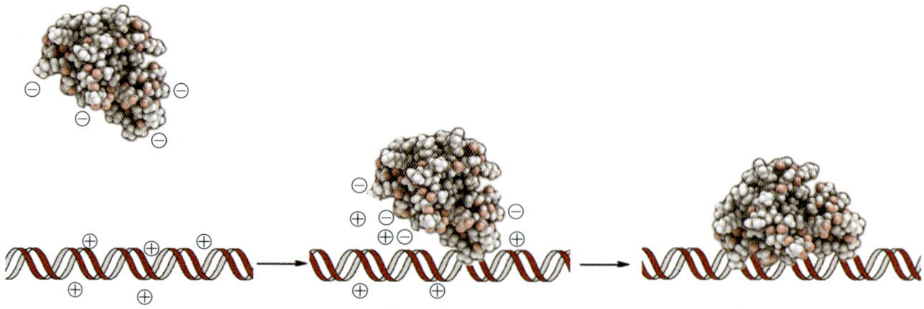

Figure 2.8 Hypothesis of the general mechanism of recognition.

elements in an appropriate orientation. Moreover, the protein regions outside the binding domain can also establish contacts with the main role of promoting homo- or hetero-oligomerization with other proteins or elements of the transcriptional machinery. This ability is very important, not only to allow high DNA affinities, but also because this permits the recognition of long sequences, which ensures site-selective binding within genomes as large as a human's. Furthermore, the ability to multimerize with diverse partners drastically expands the opportunities for recognizing diverse sets of DNA sequences (combinatorial gene regulation) from a few protein partners [36]. Another important aspect derives from the different propensities of certain DNA sequences to adopt unusual or distorted local conformations (indirect readout) [37].

2.4
Miniature Versions of Transcription Factors

The availability of tailored synthetic TFs, which are capable of mimicking their behavior and abilities of DNA recognition, is of high interest not only for basic research but also because these compounds can present relevant therapeutic properties. Unfortunately, the lack of simple rules for DNA recognition, the need to not only consider the structural references of the DNA–TF complexes but also the thermodynamic factors influencing the recognition, and our poor understanding of protein folding make it very difficult to design synthetic alternatives. In fact, it is well established that the DNA-binding domains of the TFs cannot interact with their consensus DNA sites when they are isolated from the rest of the protein. Sometimes even the removal of residues far away from the binding domain can have drastic consequences in the recognition capabilities [24c]. Therefore, in order to design this type of system, the implementation of innovative chemical strategies and the trial and error tests of the various designs based on structural data is inevitable. Despite the challenge, during the past years, several small-scale TFs have been synthesized, which mimic the specific DNA recognition.

2.4.1
Synthetic Modification of bZIP Transcription Factors

The structural simplicity of bZIP proteins makes them the most used TFs family in the design of simplified nonnatural versions. The pioneering work in this field was performed by the Kim group in 1990. He replaced the dimerization element of the homodimeric protein general control nonderepressible (GCN)4, the leucine zipper, with a disulfide bridge without the loss of recognition capability at 4 °C [38]. This approach was also used for other heterodimeric bZIP [39].

After this initial breakthrough, a large number of covalent and noncovalent dimerization units have been used: chiral organic bisphenyl linker [40], supramolecular associations by cyclodextrin/adamantly inclusion complex [41], and octahedral coordination Fe(II) complex [42] (Figure 2.9).

In addition to these static dimerization elements, the group of Mascareñas reported the application of photoswitchable azobenzene units as a switchable linker for controlling the DNA binding. This design took advantage of the large conformational change between the cis and trans isomers to modulate DNA-binding affinity of the synthetic construct [43]. This approach has been subsequently used by other researchers to modulate the interaction of the GCN4 protein and

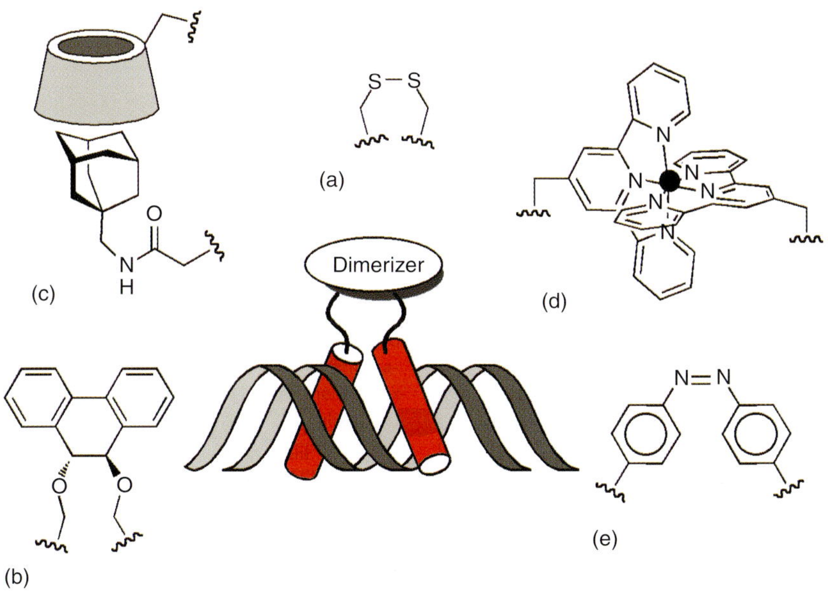

Figure 2.9 Schematic representation of some artificial dimers of bZIP TF that result from substitution of the leucine zipper coiled coil with synthetic linkers. (a) The first example of this strategy by Kim. (b) Chiral organic bisphenyl linker by Morii. (c) Supramolecular associations by cyclodextrin/adamantly inclusion complexes by Morii. (d) Octahedral coordination Fe(II) complex by Schepartz. (e) Azobenzene linker for light controlled DNA binding by Mascareñas.

other proteins through conformational changes in the leucine zipper [44]. In particular, the Woolley group has made spectacular developments in the application of azobenzene units [45] and other azoderivates [46] for the control of DNA recognition. One of the most remarkable examples is the one where the activity of a bZIP protein, the AP-1 TF, can be photocontrolled in living cells through the introduction of a designed azobenzene-cross-linked dominant negative peptide, XAFosW [47] (Figure 2.10).

As an alternative approach for controlling the DNA-binding process of the bZIP proteins by external stimuli, Mascareñas group has recently developed an electrostatic turn-off strategy. They demonstrated that appending a negatively charged Glu_8 tail to a peptide dimer derived from the GCN4 TF leads to an effective suppression of its DNA binding. Moreover, specific DNA recognition could be restored by irradiation with UV light by using a photolabile linker between the acidic tail and the DNA-binding peptide [48]. On the other hand, the Futaki group developed the first example of metal-responsible bZIP proteins by introducing metal-coordinating iminodiacetic acid side chains into the leucine zipper region of the GCN4 DNA-binding domain. Thus, on addition of Co(II) the peptide displayed a significantly reduced DNA affinity [49].

2.4.2
Residue Grafting

Residue grafting is an approach where a stable three-dimensional fold or at least at a particular secondary structure element, it is used as a scaffold into which selected residues are inserted in order to build a particular epitope [50]. The Schepartz

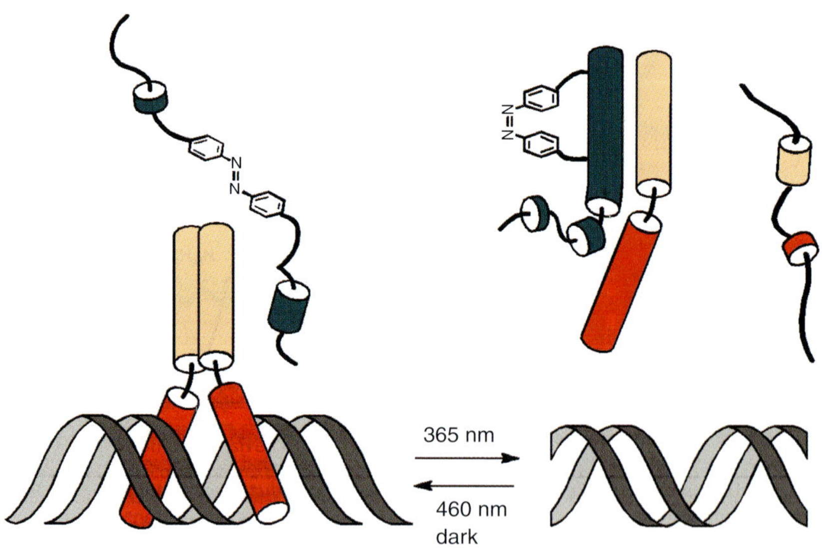

Figure 2.10 Schematic representation of photocontrol of the activity of the AP-1.

group was the first to show a successful application of a residue-grafting strategy to obtain high-affinity monomeric DNA-binding peptides [51]. They inserted the interaction-key residues of the GCN4 protein into the solvent-exposed positions of a small folded structure, the avian pancreatic polypeptide (aPP), in order to obtain the peptide PPrb. The structure of the aPP consists of a short α helix merged through a turn with a polyproline helix, which stabilizes the secondary structure. Following this straightforward approach, they synthesized 42-mer peptides that were able to interact with the target DNA sequence specifically and with high affinity. Subsequent optimizations in the N-terminus gave rise to small-scale peptides with the same DNA recognition capabilities as the natural TF at room temperature without the need for dimerization. Following this successful implementation of the grafting strategy, they also demonstrated the versatility of this approach by obtaining miniature homeodomain proteins and introducing these into the structure of the versatile aPP peptide [52].

A number of other researchers have pursued related strategies to obtain high-affinity and specific DNA recognition peptides. The Morii group employed a structure-based design to build a small domain from the F-actin bundling protein (villin) that features the interaction key residues of GCN4. Unfortunately, the monomeric construct did not show better DNA binding than the basic-region helix by itself. However, the introduction of the leucine zipper domain resulted in dimeric peptides that displayed good DNA affinity [53]. Recently, inspired by the structure of bZIP, the Jon group has designed high-affinity peptides, named *aptoides*. They contain a stabilizing scaffold (tryptophan zipper) and two target-binding regions. In this example, a fluorescently labeled aptide was used successfully for fluorescent imaging of a tumor-specific protein *in vivo* [54].

2.4.3
Conjugation in Order to Develop DNA Binding Peptides

As previously commented, isolated monomeric DNA-binding domains of most TF are unable to recognize DNA. However, several groups have developed strategies to achieve this. The Verdine group was the pioneer in this field. In 1995, Verdine reported that if the basic region of the GCN4 protein is released intramolecularly into the major groove, the entopic cost is greatly diminished making specific recognition possible [55] (Figure 2.11).

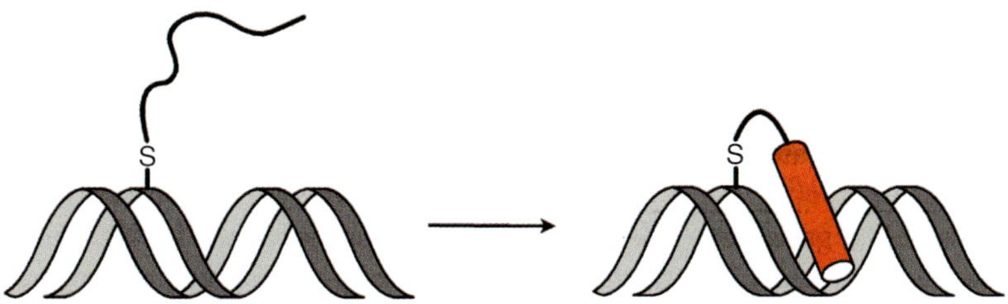

Figure 2.11 Schematic representation of oligopeptide designed by the group of Verdine.

On the basis of the same principle, several groups have restored the DNA recognition of TF fragments by conjugating small peptides from the TF DNA-binding domains (unable to display tight and specific binding by themselves) to fluorogenic dyes [56] or metallic-coordination complex [57] that intercalate into the DNA. It is interesting to highlight that the peptide-metal complex conjugates of Barton have also demonstrated promising nuclease activities. For example, conjugation of the rhodium intercalator $[Rh(phi)_2(phen')]^{3+}$ to a short oligopeptide (13 residues) derived from the α3-helix of the phage P22 repressor displayed sequence-specific DNA photocleavage after irradiation at 313 nm [58]. In a more recent development, the peptide fragment was substituted with a *de novo* zinc-chelating sequence based on the active site of metal-containing hydrolases, thereby allowing DNA cleavage under mild conditions [59].

In this context, a better imitation of the nature would be if both interaction units were sequence-specific. The Mascareñas group initially reported a successful bivalent strategy, where the conjugation of a minor groove binding distamycin analog with the basic region of the GCN4 [60] or with the domain Cys_2His_2 of the TF GAGA [61] generated hybrids with remarkable DNA-binding properties. Further refinements in the minor groove binder led to versatile *aza*-bisbenzamidines conjugates that were able to capture the DNA-binding domain of the Jun bZIP [62]. Most recently, they demonstrated that the DNA binding of monomeric peptides could be restored when conjugated to environment sensitive *aza*-bisbenzamidines. Importantly, the fluorogenic properties of this minor binder allow one to obtain details of the DNA interaction that are eluded in electrophoresis mobility shift assays [63].

2.5 Intermolecular Interaction Between Small Molecules and the DNA

2.5.1 General Concepts

One of the fields more extensively studied within Chemical Biology and Medicinal Chemistry is the synthesis of therapeutic molecules, including those that target dsDNA. Unfortunately, this goal is extremely difficult to attain, even with current technologies. Indeed, in most cases the candidate molecules fail in the first steps of the process. The near future of this field demands an interdisciplinary approach where the development of new compounds exploits the cutting edge of advances in Chemical Biology techniques.

Owing to the toxicity of the initial DNA binders, they were considered to be of no clinical value. However, the recent approval of the DNA minor groove binding agent *trabectidin* (Yondelis®) for the treatment of patients with soft tissue sarcomas and ovarian cancer in Europe has revived the interest in the small molecules that are able to interact with the DNA [64].

In general terms, these molecules interact with DNA through four different modes [65]:

1) **External electrostatic interaction**: the polyanionic surface of the DNA can establish electrostatic interactions with metallic cations (alkali metals: group I and group II in the periodic table). These interactions provide stability for the folded DNA structure. Likewise, cationic organic molecules (polyamines [66], etc.) stabilize the DNA by avoiding the repulsive forces between the DNA backbone phosphates as well as by causing an entropic contribution to the bonding free energy. These bindings show fast kinetics of complex formation ($Ka < 10^8 - 10^9\ M^{-1}\ s^{-1}$) but they are not sequence specific.
   A recent example of a compound that presents such interaction is the trinuclear oligomer of platinum (II). As it does not have reactive Pt–Cl groups, it interacts exclusively with the DNA through electrostatic contacts and hydrogen bonds with the oxygen atoms of the DNA phosphate groups (Figure 2.12).
2) **Intercalation**: intercalators are small molecules that contain an optimal planar aromatic heterocyclic surface of 39 Å^2, which can be inserted and stacked between the bases pair of double helical DNA [67]. A common structural characteristic of all the intercalators is that they are aromatic π-deficient systems of two or more six-membered rings (naphthalene represents the minimal unit). The interaction can be generated in any of the grooves and the forces involved in the processes are π-stacking and electrostatic interactions, charge-transference effects, and hydrogen bonds. This type of recognition entails a distortion of the DNA helix and it is generally considered non-sequence-specific, although it is has been observed that these molecules have certain preference for G/C base pairs. A typical example of an intercalator is ethidium bromide, which is regularly used in the fluorescent labeling of DNA, and it has been used to determine affinity constant by competition assays.
   Nowadays the classic structure of the intercalating agents has been combined with different groups, such as another intercalating unit (bisintercalators) [68], in order to increase the affinity of the interaction, or small fragments of TF as was discussed before [56, 57]. It can also be modified with other moieties [69] to improve its specificity and thermal stability, or metallic complexes of ruthenium and rhodium, which supply the intercalator with luminescent

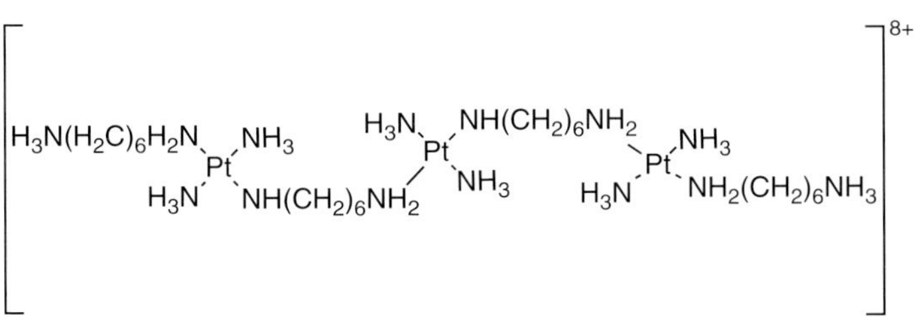

Figure 2.12 Structure of the Pt(II) complex that interacts with the DNA through electrostatic forces and hydrogen bonds.

properties or DNA cleavage activity, respectively. Particularly, over the past few decades, there has been an increasing interest in the application of these metal-coordination complexes for the development of DNA-based probes, reactive agents, and therapeutics [70]. A detailed discussion of metallo-DNA binders can be found in the following pages (Figure 2.13).

3) **Covalent interaction (alkylation agents)**: the N^7 and C^2-NH_2 positions of the guanine and the nitrogen atoms N^3 and N^7 of adenine are the positions most favorable to undergo these modifications. Molecules such as the antitumor agent *cisplatin*, the antibiotic *antramycin*, and the antitumor *mitomycin A* form this group (Figure 2.14).
4) **Insertion in the grooves of the DNA**: some small oligonucleotides (<10–20 bp) can form a triplex helix by the insertion into the major groove of the DNA. Such interaction is established through the formation of Hoogsteen bonds between the base pairs [71]. The interaction is sequence specific and is observed only when the DNA presents a broad and uninterrupted purine sequence. These agents have been used to control gene expression [72] (Figure 2.15).
In contrast to the TFs, small molecules are generally prone to selectively bind the minor groove. The first natural products where such selective interaction was observed were dicationic compounds [73]. The interaction in the minor groove is determined by the formation of hydrogen bonds, van der Waals forces, and electrostatic and hydrophobic interactions. Another common feature in this type of interaction, which is not indispensable, is the complementarity

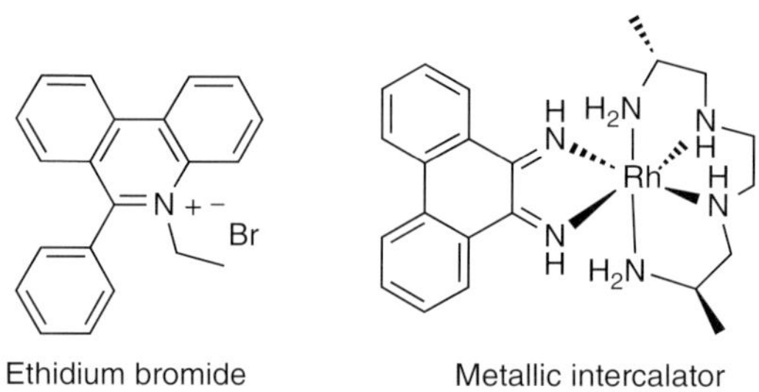

Figure 2.13 Structure of some intercalators.

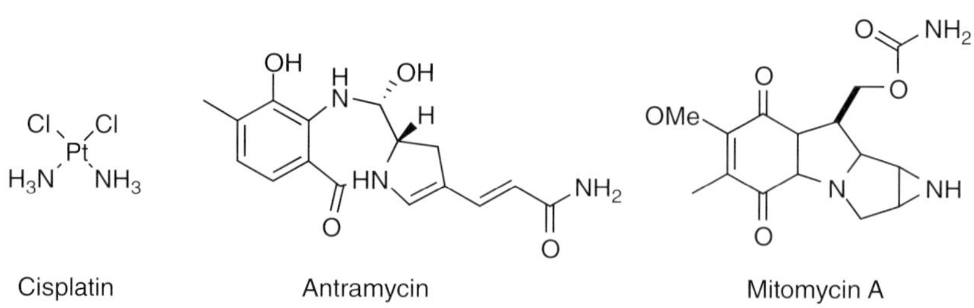

Figure 2.14 Structure of some alkylation agents.

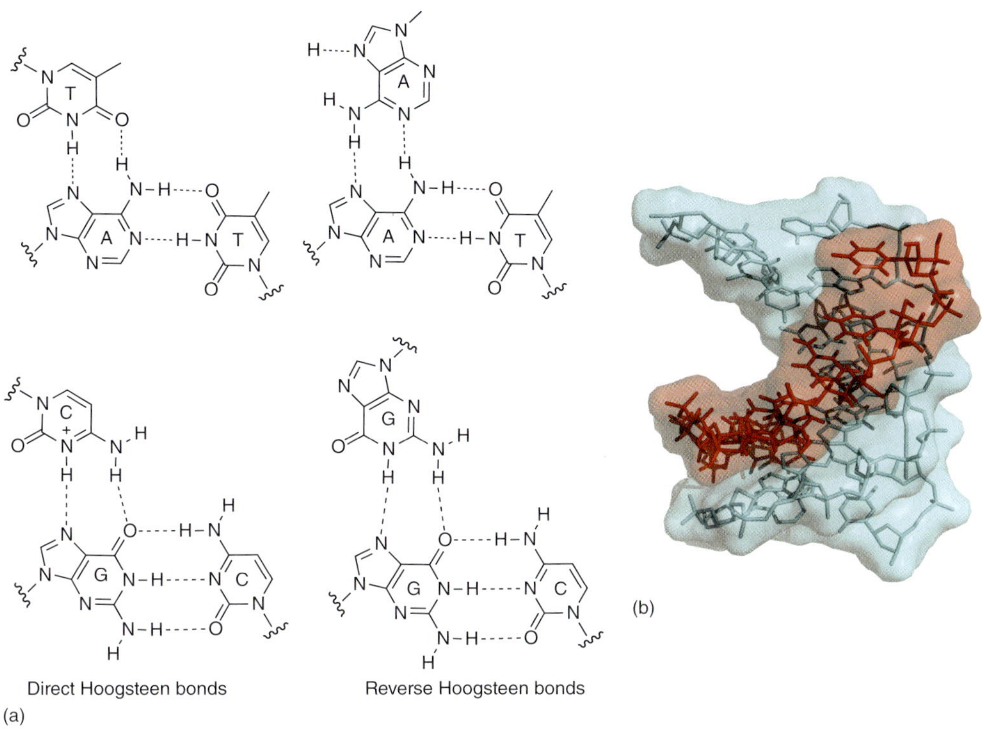

Figure 2.15 (a) Representation of the Hoogsteen bonds and (b) crystal structure of a triple helix (PDB: 1PNN).

of the shape (*isohelicity*) [74]. These compounds often have a concave shape that fits in the minor groove of A/T rich sites because, as it was explained before, these regions are narrower and deeper than the ones rich in pairs G/C. This fact benefits the insertion due to the maximization of the van der Waals forces. On the other hand, the amine group of the guanines protrudes and interferes with the interaction [75]. Minor groove binders include natural products such as *neptrosin* and *distamycin A*, as well as synthetic molecules such as *berenil, hoechst 33258*, and *propamidine*, and nuclear stains such as 4',6-diamidino-2-phenylindole (DAPI) (Figure 2.16).

A main challenge with respect to the application of these compounds as a chemotherapeutic anticancer, antiviral, or antibacterial drug is attaining spacial and temporal control of the drug–DNA interaction in order to enable high selectivity. Recently there has been an increasing interest in the development of DNA binders whose DNA binding can be modulated by external extimuli. Progress in this direction has been extremely slow. However, there are several important examples of photochromic intercalators such as spiropyran [76], a chromene [77], a benzothiazoloquinolinium [78], and dithienylethenes [79], where only one form of the photochromic molecules interacts with the DNA, and the DNA binding can

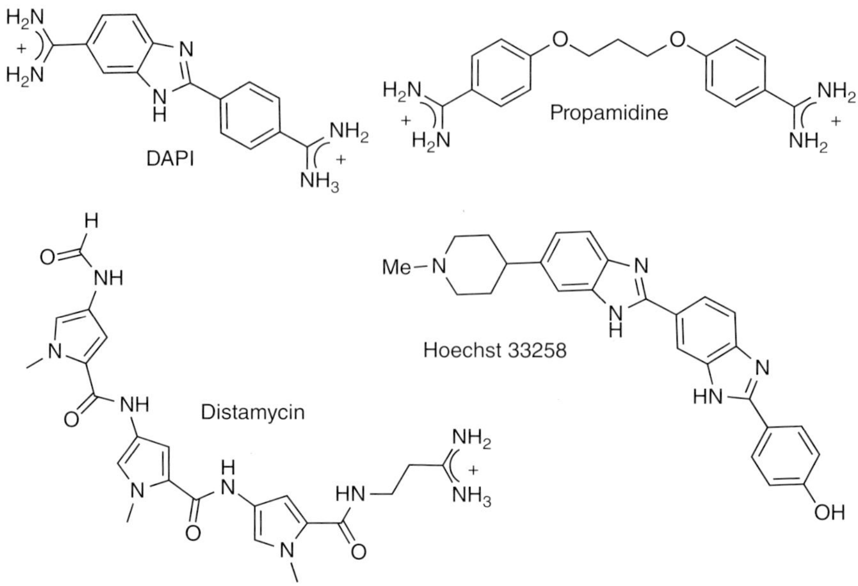

Figure 2.16 Structure of the most representative minor groove binders.

therefore be modulated by irradiation. Franz and Mascareñas groups have reported the application of a temporary caging approach for controlling the interaction of DNA alkylating and minor-groove binding agents, respectively [80]. In addition to these developments, and similar to the case of the mimetic of TFs, the introduction of an azobenzene as a control element was also used with small binders [81].

As metallo-DNA binders and minor groove DNA-binding agents constitute the most relevant examples of DNA recognition with small molecules in terms of specificity and applicability, we focus here on the most representative ones within these groups.

2.5.2
Metallo-DNA Binders: From Cisplatin to Rh Metallo-Insertors

Nature shows very little precedent in the use of transition metal complexes as DNA molecular recognition agents and, consequently, at first glance, this choice could seem odd with regard to DNA recognition. However, since the serendipitous discovery of the anticancer properties of cisplatin in the mid-1960s by Barnett Rosenberg, the field of novel metallo-DNA binders has had an explosive growth and advancement, especially over the past two decades.

Rosenberg observed that the cell division of *Escherichia coli* was inhibited when they were exposed to electric fields produced by platinum electrodes. Further extensive studies led him to the conclusion that the cis form of platinum (IV) complex $[PtCl_4(NH_3)_2]$ was the agent responsible for the hampering of the growth of the bacteria. These platinum ammine complexes were formed by electrolysis at the platinum electrodes. Given these results, it was reasoned that the complexes might be interesting to test for their cancer activity. The cisplatin is indeed a real success story [82]. Today, over 30 years after its approval as a chemotherapeutic agent, it is still one of the world's best-selling cancer drugs. Cisplatin is routinely used for the treatment of testicular cancer, where it is responsible for curing over 90% of cases, and ovarian cancer, and it also plays a vital role in the treatment of cervical, bladder, and head/neck tumors. To date, the only DNA-targeted metal-containing drugs in clinical use are platinum-based drugs.

From Rosengberg's early studies, the implication of the DNA in the cytotoxic properties was clear. Prior to DNA attack, cisplatin undergoes the replacement of its chloride ligands by water as soon as it enters the body. However, the high concentration of chloride ions in blood plasma (about 100 mM) limits the hydrolysis. Once it is inside the cancer cells, where intracellular chloride concentration is relatively low (about 4–20 mM), it undergoes rapid hydrolysis, paving the way to activate cationic species capable of reacting both monofunctionally and bifunctionally (Figure 2.17).

In vitro studies have shown that this monoaquated platinum species is responsible for at least 98% of platinum binding to DNA within the cell nucleus. Although activated cisplatin can interact with various biomolecules, its antitumor activity derives from its capacity to form bifunctional DNA cross-linkers. The predominant adducts are the bifunctional guanine–guanine intrastrand cross-linkers, which trigger a specific distortion of the DNA duplex. One or more DNA-binding proteins

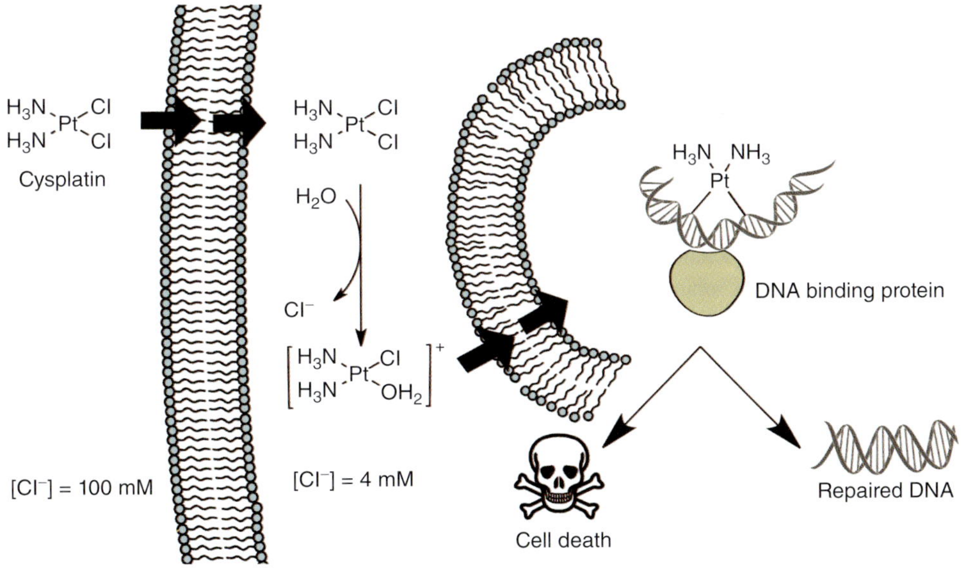

Figure 2.17 Schematic representation of the cytotoxic pathway of cisplatin.

can recognize this bending and they might initiate DNA damage repair or signal for apoptosis.

Despite the success, the effectiveness of cisplatin treatment is limited. Therefore, there is a current search for novel metallo-DNA binders with higher efficacy, improved aqueous solubility, and less severe side effects. One strategy, developed by Sadler group to avoid toxic side effects, consists in using Pt(IV) prodrugs, whose activity can be triggered by showing light directly on the side of the cancer [83]. Among them, *cis,trans,cis*-$[Pt(N_3)_2(OH)_2(NH_3)_2]$, and *cis,trans*-$[Pt(N_3)_2(OH)_2(en)_2]$ can form cross-linkings similarly to the ones with cisplatin. Furthermore, photoactivatable diazido platinum (IV) complexes, which can cross-link distant sites on DNA, appear to produce lesions of DNA that are more difficult to repair than the ones produced by cisplatin, and therefore provide the basis for a novel form of photochemotherapy [84].

Apart from the covalent interactions of the platinum-based chemotherapeutics, other metallo-binders recognize the DNA through noncovalent interactions. These can be classified into two groups: intercalation and insertion. Metallo-intercalators unwind the DNA and insert their planar ligand between two intact base pairs, while metallo-insertors eject the bases of a single base pair and their planar ligand acts as a π-stacking replacement in the DNA base stack. These interactions have been extended into three dimensions mainly by Barton group using octahedral Rh, Ru, or Os complexes containing multi-heterocyclic aromatic ligands. Octahedral metallo-binders are able to target DNA specifically by matching the shape, symmetry, and functionalities of the metal complex to that of the DNA target. The most studied example is probably $[Ru(phen)_2(dppz)]^{2+}$, which luminesces brightly on the addition of duplex DNA. Many analogs have been synthesized and extensively reviewed [85]. While these ruthenium and dppz-based metallo-intercalators have proven to be powerful molecular light switches for the detection of the DNA, rhodium ones are efficient photoactivated DNA cleavage agents. In this case the most well-studied examples are the ones that employ the phi ligand as the intercalator such as $[Rh(bpy)_2(phi)]^{3+}$, $[Rh(phen)_2(phi)]^{3+}$, and $[Rh(phi)_2(bpy)]^{3+}$. This reactivity enables us to directly mark and characterize the site of intercalation [86]. Another remarkable example is the Δ-$[Rh(phi)(DPB)_2]^{3+}$, which represents the best metallo-inhibitor of the activity of *XbaI* restriction endonuclease at the palindromic site until now [87]. Thus, metallo-intercalators have been found to be useful not only as probes for structures, but also as mimics, probes, and perhaps inhibitors of DNA-binding proteins.

Until very recently, no examples of DNA-binding insertors had been reported. However, the research of the group of Barton into mismatch-specific DNA-binding agents has led to the discovery of a family of rhodium complexes that bind DNA via this unique mode. They show important applications; thus, $[Rh(bpy)_2(chrysi)]^{3+}$ and $[Rh(bpy)_2(phzi)]^{3+}$ have been used in diagnostic applications, for the discovery of single nucleotide polymorphisms and for designing new chemotherapeutics [88].

2.5.3 Polypyrroles and Bis(benzamidine) Minor Groove Binders and Their Use as Specific dsDNA Sensors

Distamycin A is the classic example of a minor groove DNA binder. It consists of aromatic units connected by peptide bonds. Distamycin A inserts in the minor groove, spanning from 4 to 6 bp [89] (Figure 2.18).

In 1989, the group of Wemmer observed that distamycin A could form antiparallel dimers with DNA [90]. The tendency to form 1 : 1 or 2 : 1 complexes is highly dependent on the DNA sequence. Circular dichroism studies showed that the tendency of the formation of 2 : 1 complexes decreases in the following order: AAGTT, ATATA ≥ AAACT > AATAA > AAATA > AAAGT > AATAT > TAAAA ≥ AAATT ≥ AAAAA ≥ ATAAA, AAAAT [91]. The discovery of this dimeric interaction, together with the results of Lown and Dickerson – which demonstrated that if some groups of the pyrrole (Py) of the distamycin were exchanged by imidazole (Im), then the interaction with guanine cytosine (GC) pairs were facilitated [92] – induced Dervan group to investigate the system of recognition with both units covalently bound (Figure 2.19).

The effect of the substitution of pyrroles with other heterocycles was also studied. Dervan and coworkers have designed and developed a great variety of hairpin polyamides where it is possible to talk about a code and induce the interaction with a specific sequence by the suitable combination of different heterocycles [93]. The systematic study developed by Dervan has achieved tremendous advances, and his hairpin polyamides have aroused a great interest in the fields of Chemical Biology and Biomedicine.

The discovery of the genetic origin of many diseases and influence of particular genotypes on the response to treatments are fostering the development of new strategies in order to detect specific DNA sequences and visualize the DNA.

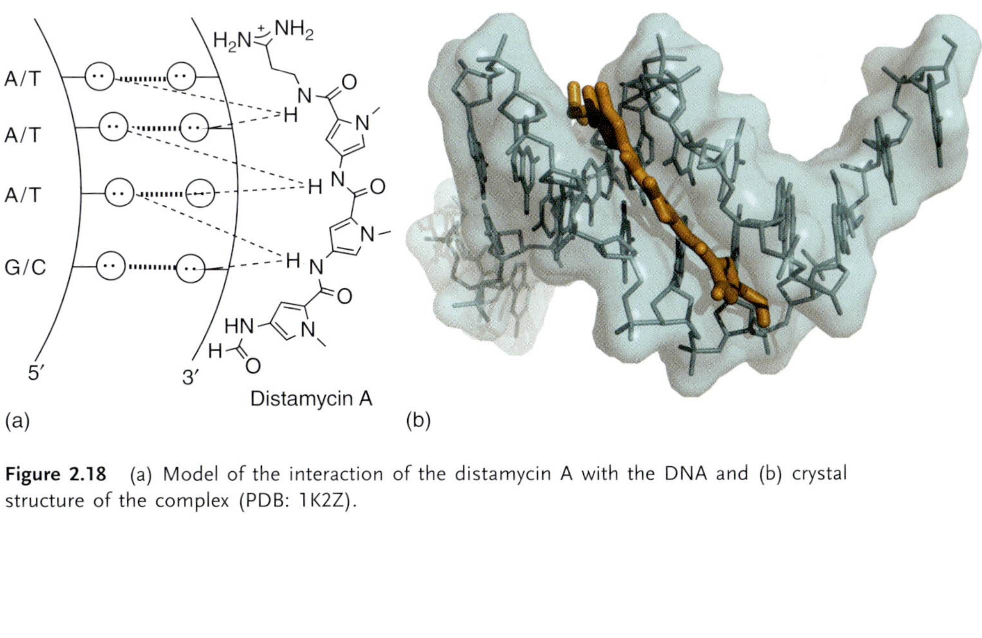

Figure 2.18 (a) Model of the interaction of the distamycin A with the DNA and (b) crystal structure of the complex (PDB: 1K2Z).

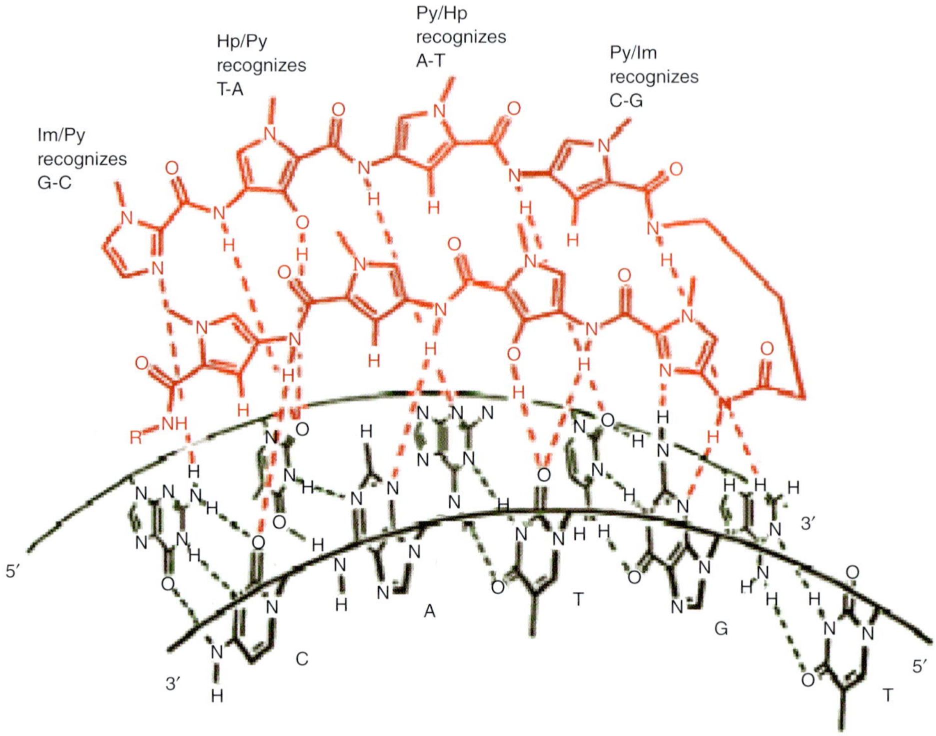

Figure 2.19 Three-dimensional structure of the interactions that take place in the recognition of the DNA by the polyamide hairpins.

Most methods are indirect ones, which are based on hybridization of single-stranded DNA (conjugated to fluorophores) to complementary sequences. Direct probing of dsDNA without denaturation is much less developed [94]. The blue dyes, Hoechst [95] and DAPI, are the most popular and robust DNA probes for live cell imaging, although, unfortunately, they are incapable of sensing specific dsDNA sequences. During the past few years, different color probes such as DRAQ5 [96] or the first standard green DNA dye, C61, that work nicely in live cells at 5 mM with lower toxicity have been reported. However, both fail in terms of specificity [97]. In this context, and taking into account the discussion above, hairpin polyamides can be tailored to recognize specific DNA sequences, and therefore are the most promising candidates as sensors. Indeed, they have already been used in sequence-specific sensing of dsDNA. In 2003, Dervan group prepared the first example of a sequence-specific sensor based on hairpin polyamides. They were conjugated to a tetramethyl rhodamine (TMR) through a very short linker [98]. The resulting probe displayed weak emission in aqueous solution; nevertheless, on the binding to the DNA, fluorescence emission increases 10-fold [99]. Similar probes based on the conjugation of the Cy3 fluorescent dye to a hairpin oligonucleotide or thiazol orange (TO) were also prepared [100]. In 2007, the same group presented new types of modified polyamides that included fluorescent units in their core structures [101]. These hairpins displayed a significant increase in their fluorescence emission on interaction with specific DNA sequences, thus providing a method to selectively detect DNA sequences without denaturing them. Unfortunately, the extreme structural modification in the hairpins compromised the kinetics of recognition.

Despite the importance of hairpin polyamides, this type of binder has some limitations in connection with synthesis as well as a certain lack of specificity in the case of some derivatives, although a novel efficient method to synthesize cyclic polyamides has been reported recently, where the specificity seems to be better than the alternative hairpin polyamide [102]. Another important problem derives from the fact that interaction is exclusively through the minor groove, which involves a limitation in terms of interfering with the action of the TFs [103]. Furthermore, these molecules show difficulties in uptake, particularly in reaching the nuclei. They were just able to reach it in certain occasions and with slow kinetics.

Bis(benzamidine) compounds such as pentamidine or propamidine also recognize the A/T-rich sequences through the minor groove [104]. These molecules are very interesting from a pharmacological point of view, as they are stable and show good uptake properties in a large number of cells [105]. Indeed, the pentamidine has been recently used instead of trypanosomiases, leishmaniasis, and pneumocystis carnii pneumonia, and even in the treatment of cisplatin-resistant cancer [106] (Figure 2.20).

The excellent pharmacological properties together with their structural simplicity have encouraged the search for new analogs with better recognition properties and limited toxicity. Compounds with aromatic linkers such as furamidine or its prodrug DB289 are especially relevant. In fact, DB289 is in the clinical phase of the African trypanosomiasis and it is very active as an antiparasitic drug [107].

The Mascareñas group recently reported the development of new bis(benzamidine) fluorogenic minor groove binders. The structures of these compounds include highly polarized aromatic rings that endow the systems with strong fluorogenicity; on insertion of the binding agents into the apolar DNA minor groove, there is a large emission increase. They developed a simple and

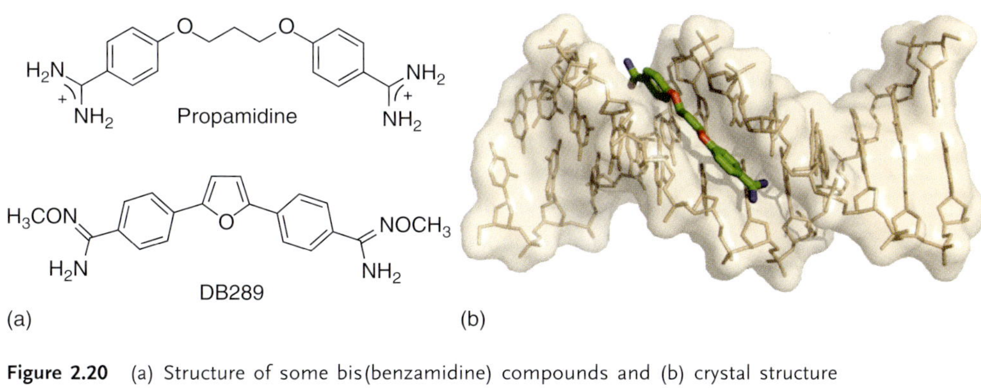

Figure 2.20 (a) Structure of some bis(benzamidine) compounds and (b) crystal structure of the propamidine-DNA complex (PDB 102D).

direct method for rapid monitoring, quantification of the DNA-binding affinity, and the selectivity of the interaction as well as the characterization of the binding of other nonfluorescent minor groove binders [108]. In a later report, the authors demonstrated that direct conjugation of the basic fluorogenic unit with other fluorophores (coumarine or lanthanide chelates, among others) allowed effective energy transfer and the observation of emission at long wavelengths [109].

2.6 Outlook

The deciphering of the human genome and the identification of the genetic components of many diseases have converted the field of development of synthetic DNA-recognition agents into one of the most significant fields within Chemical Biology. However, the advances in this area have been relatively slow, in spite of the availability of a large amount of structural data. Indeed, we are still far from fully understanding the molecular and biophysical basis underlying the selective interactions, especially in the case of TFs and other proteins. Future progress in the area may further combine rational design with combinatorial-selection methods and pursue the concepts of cooperativity and multivalence, similarly to what occurs in nature. Besides, it is also clear that we need to incorporate the three-dimensional organization encoded on the DNA into our drug design models in order to achieve sequence-specific recognition analogous to natural macromolecules. Undoubtedly, the tools of Chemical Biology will play a critical role in the future developments in this area. In addition, we hope that some of these systems can, in the future, demonstrate therapeutic potential.

Acknowledgments

I would like to express my gratitude to Professor Pignataro, the Book Editor, for the invitation to participate in such a great initiative. I would also like to thank Professor Eugenio Vázquez who read the manuscript and provided critical feedback.

References

1. Khakshoor, O. and Kool, E.T. (2011) *Chem. Commun.*, **47**, 7018–7024.
2. (a) Venter, J.C. *et al.* (2001) *Science*, **291**, 1304–1351; (b) Pennisi, E. (2003) *Science*, **300**, 409; (c)International Human Genome Sequencing Consortium (2004) *Nature*, **431**, 931–945.
3. Egli, M. (2004) *Curr. Opin. Chem. Biol.*, **8**, 580–591.
4. (a) Ghosh, I., Stains, C.I., Ooib, A.T., and Segal, D. (2006) *J. Mol. Biosyst.*, **2**, 551–560; (b) Boer, R.D., Canalsa, A., and Coll, M. (2009) *Dalton Trans.*, **3**, 399–414.
5. (a) Bansal, M. (2003) *Curr. Sci.*, **85**, 1556–1563; (b) Macgregor, R.B. Jr., and Poon, G.M.K. (2003) *Comput. Biol. Chem.*, **27**, 461–467.

6. Lodish, H., Berk, A., Zipursky, L.S., Matsudaira, P., Baltimore, D., and Darnell, J. (2000) *Molecular Cell Biology*, Chapter 9, W. H. Freeman, New York, ISBN: 0-7167-3136-3.
7. Crick, F.H.C. (1958) *Symp. Soc. Exp. Biol.*, **12**, 138–163.
8. Crick, F.H.C. (1970) *Nature*, **227**, 561–563.
9. (a) Baltimore, D. (1970) *Nature*, **226**, 1209–1211; (b) Commoner, B. (1964) *Nature*, **203**, 486–491; (c) Commoner, B. (1968) *Nature*, **220**, 334–340.
10. (a) Thieffry, D.C. (1998) *Trends Biochem. Sci.*, **23**, 312–316; (b) Alonso, L. (2005) *INVYCIE*, **342**, 92–96.
11. Watson, J.D. and Crick, F.H. (1953) *Nature*, **171**, 737–738.
12. (a) Rohs, R., West, S.M., Sosinsky, A., Liu, P., Mann, R.S., and Honig, B. (2009) *Nature*, **461**, 1248–1254; (b) Garvie, C.W. and Wolberger, C. (2001) *Mol. Cell*, **8**, 937–946; (c) Rohs, R., Jin, X., West, S.M., Joshi, R., Honig, B., and Mann, R.S. (2010) *Annu. Rev. Biochem.*, **79**, 233–269.
13. Privalov, P.L., Dragan, A.I., Crane-Robinson, C., Breslauer, K.J., Remeta, D.P., and Minetti, C.A.S.A. (2007) *J. Mol. Biol.*, **365**, 1–9.
14. Luscombe, N.M., Laskowski, J., Feng, Z., Gilliland, G., Bhat, T.N., and Weissig, H. (2000) *Nucleic Acids Res.*, **28**, 235–242.
15. Berman, H.M., Westbrook, R.A., and Thornton, J.M. (2001) *Nucleic Acids Res.*, **29**, 2860–2861.
16. Suzuki, M. and Gerstein, M. (1995) *Proteins*, **23**, 525–535.
17. Lewis, M., Chang, G., Horton, N.C., Kercher, M.A., Pace, H.C., Schumacher, M.A., Brennan, R.G., and Lu, P. (1996) *Science*, **271**, 1247–1254.
18. (a) Latchman, D.S. (2004) *Eukaryotic Transcription Factors* Capítulo 4, Elsevier Ltd, Italy, ISBN: 0-12-437178-7; (b) Luscombe, N.M., Austin, S.E., Berman, H.M., and Thornton, J.M. (2000) *Genome Biol.*, **1**, 1–37; (c) Pabo, C.O. (1992) *Annu. Rev. Biochem.*, **61**, 1053–1094.
19. Harrison, S.C. (1991) *Nature*, **353**, 715–719.
20. Joshi, R., Passner, J.M., Rohs, R., Jain, R., Sosinsky, A., Crickmore, M.A., Jacob, V., Aggarwal, A.K., Honig, B., and Mann, R.S. (2007) *Cell*, **131**, 530–543.
21. Feng, J.-A., Johnson, R.C., and Dickerson, R.E. (1994) *Science*, **263**, 348–355.
22. Remènyi, A., Tomilin, A., Pohl, E., Lins, K., Philippsen, A., Reinbold, R., Schçler, H.R., and Wilmanns, M. (2001) *Mol. Cell*, **8**, 569–580.
23. Klug, A. and Schwabe, J.W.R. (1995) *FASEB J.*, **9**, 597–604.
24. (a) Dhanasekaran, D., Negi, S., and Sugiura, Y. (2006) *Acc. Chem. Res.*, **39**, 45–52; (b) Krishna, S., Majumdar, I., and Grishin, N. (2003) *Nucleic Acids Res.*, **31**, 532–550; (c) Pabo, C.O., Peisach, E., and Grant, R.A. (2001) *Annu. Rev. Biochem.*, **70**, 313–340.
25. Iuchi, S. and Kuldell, N. (2005) *Zinc Finger Proteins: From Atomic Contact to Cellular Function*, Chapter 1, Kluwer Academic/Plenum Publishers, New York, ISBN: 0-306-48229-0.
26. Jamieson, A.C., Miller, J.C., and Pabo, C.O. (2003) *Nat. Rev. Drug Discovery*, **2**, 361–368.
27. Jantz, D., Amann, B.T., Gatto, G.J. Jr., and Berg, J.M. (2004) *Chem. Rev.*, **104**, 789–799.
28. (a) Tanaka, K., Tengeiji, A., Kato, T., Toyama, N., and Shionoya, M. (2003) *Science*, **299**, 1212–1213; (b) Tanaka, K., Clever, G.H., Takezawa, Y., Yamada, Y., Kaul, C., Shionoya, M., and Carell, T. (2006) *Nat. Nanotechnol.*, **1**, 190–194.
29. (a) Nagaoka, M. and Sugiura, Y. (1996) *Biochemistry*, **35**, 8761–8768; (b) Kamiuchi, T., Abe, E., Imanishi, M., Kaji, T., Nagaoka, M., and Sugiura, Y. (1998) *Biochemistry*, **37**, 13827–13834; (c) Imanishi, M., Hori, Y., Nagoaka, M., and Sugiura, Y. (2000) *Biochemistry*, **39**, 4383–4390.
30. Allemann, R.K. and Egli, M. (1997) *Chem. Biol.*, **4**, 643–650.
31. (a) Pabo, C.O. and Nekludova, L. (2000) *J. Mol. Biol.*, **301**, 597–624; (b) Pazos, E., Mosquera, J., Vázquez, M.E., and Mascareñas, J.L. (2011) *ChemBioChem*, **12**, 1958–1973.

32. Choo, Y. and Klug, A. (1997) *Curr. Opin. Struct. Biol.*, **7**, 117–125.

33. Masayuki, O. and Nakamura, H. (2000) *Genes Cell*, **5**, 319–326.

34. Vázquez, M.E., Caamaño, A.M., and Mascareñas, J.L. (2003) *Chem. Soc. Rev.*, **32**, 338–349.

35. Pomerantz, J.L., Sharp, P.A., and Pabo, C.O. (1995) *Science*, **267**, 93.

36. Jones, N. (1990) *Cell*, **61**, 9–11.

37. Nekludova, L. and Pabo, C.O. (1994) *Proc. Natl. Acad. Sci. U.S.A.*, **91**, 6948–6952.

38. Talanian, R.V., McKnight, C.J., and Kim, P.S. (1990) *Science*, **249**, 769–771.

39. (a) Park, C., Campbell, J.L., and Goddard, W.A. III, (1992) *Proc. Natl. Acad. Sci. U.S.A.*, **89**, 9094–9096; (b) Park, C., Campbell, J.L., and Goddard, W.A. III, (1993) *Proc. Natl. Acad. Sci. U.S.A.*, **90**, 4892–4896.

40. Morii, T., Simomura, M., Morimoto, S., and Saito, I. (1993) *J. Am. Chem. Soc.*, **115**, 1150–1151.

41. (a) Ueno, M., Murakami, A., Makino, K., and Morii, T. (1993) *J. Am. Chem. Soc.*, **115**, 12575–12576; (b) Morii, T., Yamane, J., Aizawa, Y., Makino, K., and Sugiura, Y. (1996) *J. Am. Chem. Soc.*, **118**, 10011–10017.

42. (a) Cuenoud, B. and Schepartz, A. (1993) *Science*, **259**, 510–513; (b) Palmer, C.R., Sloan, L.S., Adrian, J.C., Cuenod, B. Jr., Paolella, D.N., and Schepartz, A. (1995) *J. Am. Chem. Soc.*, **117**, 8899–8907.

43. Caamaño, A.M., Vázquez, M.E., Martínez-Costas, J., Castedo, L., and Mascareñas, J.L. (2000) *Angew. Chem. Int. Ed.*, **39**, 3104–3107.

44. Kneissl, S., Loveridge, E.J., Williams, C., Crump, M.P., and Allemann, R.K. (2008) *ChemBioChem*, **9**, 3046–3054.

45. (a) Woolley, G.A., Jaikaran, A.S.I., Berezovski, M., Calarco, J.P., Krylov, S.N., Smart, O.S., and Kumita, J.R. (2006) *Biochemistry*, **45**, 6075–6084; (b) Woolley, G.A. (2005) *Acc. Chem. Res.*, **38**, 486–493.

46. Samanta, S., Qin, C., Lough, A.J., and Woolley, G.A. (2012) *Angew. Chem. Int. Ed.*, **26**, 6452–6455.

47. Zhang, F., Timm, K.A., Arndt, K.M., and Woolley, G.A. (2010) *Angew. Chem. Int. Ed.*, **49**, 3943–3946.

48. Jiménez-Balsa, A., Pazos, E., Martínez-Albardonedo, B., Mascareñas, J.L., and Vázquez, M.E. (2012) *Angew. Chem. Int. Ed.*, **51**, 8825–8229.

49. Azuma, Y., Imanishi, M., Yoshimura, T., Kawabata, T., and Futaki, S. (2009) *Angew. Chem. Int. Ed.*, **48**, 6853–6856.

50. (a) Robinson, J.A. (2009) *ChemBioChem*, **10**, 971–973; (b) Nygren, P.-A. and Skerra, A. (2004) *J. Immunol. Methods*, **290**, 3–28; (c) Fernandez-Carneado, J., Grell, D., Durieux, P., Hauert, J., Kovacsovics, T., and Tuchscherer, G. (2000) *Biopolymers*, **55**, 451–458.

51. Zondlo, N.J. and Schepartz, A. (1999) *J. Am. Chem. Soc.*, **121**, 6938–6939.

52. Montclare, J.K. and Schepartz, A. (2003) *J. Am. Chem. Soc.*, **125**, 3416–3417.

53. Morii, T., Sato, S.-I., Hagihara, M., Mori, Y., Imoto, K., and Makino, K. (2002) *Biochemistry*, **41**, 2177–2183.

54. Kim, S., Kim, D., Jung, H.H., Lee, I.-H., Kim, J.I.L., Suh, J.-Y., and Jon, S. (2012) *Angew. Chem. Int. Ed.*, **51**, 1890–1894.

55. Stanojevic, D. and Verdine, G.L. (1995) *Nat. Struct. Biol.*, **2**, 450–457.

56. (a) Thompson, M. and Woodbury, N.W. (2000) *Biochemistry*, **39**, 4327–4338; (b) Thompson, M. and Woodbury, N.W. (2001) *Biophys. J.*, **81**, 1793–1804.

57. (a) Sardesai, N.Y., Lin, S.C., Zimmermann, K., and Barton, J.K. (1995) *Bioconjugate Chem.*, **6**, 302–312; (b) Sardesai, N.Y. and Barton, J.K. (1997) *J. Biol. Inorg. Chem.*, **2**, 762–771.

58. Sardesai, N.Y., Zimmermann, K., and Barton, J.K. (1994) *J. Am. Chem. Soc.*, **116**, 7502–7508.

59. (a) Fitzsimons, M.P. and Barton, J.K. (1997) *J. Am. Chem. Soc.*, **119**, 3379–3380; (b) Copeland, K.D., Fitzsimons, M.P., Houser, R.P., and Barton, J.K. (2002) *Biochemistry*, **41**, 343–356.

60. (a) Vázquez, M.E., Caamaño, A.M., Martínez-Costas, J., Castedo, L., and Mascareñas, J.L. (2001) *Angew. Chem.*

Int. Ed., **40**, 4723–4725; (b) Blanco, J.B., Vázquez, O., Martinez-Costas, J., Castedo, L., and Mascareñas, J.L. (2005) *Chem. Eur. J.*, **11**, 4171–4178; (c) Blanco, J.B., Vázquez, M.E., Martinez-Costas, J., Castedo, L., and Mascareñas, J.L. (2003) *Chem. Biol.*, **10**, 713–722.

61. Vázquez, O., Vázquez, M.E., Blanco, J.B., Castedo, L., and Mascareñas, J.L. (2007) *Angew. Chem. Int. Ed.*, **46**, 6886–6890.
62. Sánchez, M.I., Vázquez, O., Martínez-Costas, J., Vázquez, M.E., and Mascareñas, J.L. (2012) *Chem. Sci.*, **3**, 2383–2387.
63. Sánchez, M.I., Vázquez, O., Martínez-Costas, J., Vázquez, M.E., and Mascareñas, J.L. (2013) *Chem. Eur. J.* doi: 10.1002/chem.201300519, accepted.
64. Cai, X., Gray, P.J. Jr., and Von Hoff, D. (2009) *Cancer Treat. Rev.*, **35**, 437–450.
65. (a) Wilhelmsson, L.M., Lincoln, P., and Nordén, B. (2006) in *Sequence-Specific DNA Binding Agents* (ed. M. Waring), Chapter 4, RSC Publishing, Cambridge, ISBN: 10: 0-85404-370-5; (b) Lorente, A. and Fernández, M.J. (2008) *An. Quím.*, **104**, 280–289; (c) Hannon, M.J. (2007) *Chem. Soc. Rev.*, **36**, 280–295.
66. Thomas, T.J. and Bloomfield, V.A.J. (1984) *Biopolymers*, **23**, 1295–1306.
67. Lerman, L.S. (1961) *J. Mol. Biol.*, **3**, 18.
68. (a) Berthet, N., Michon, J., Lhomme, J., Teulade-Fichou, M.P., Vigneron, J.-P., and Lehn, J.-M. (1999) *Chem. Eur. J.*, **5**, 3625–3630; (b) Biron, E. and Voyer, N. (2008) *Org. Biomol. Chem.*, **6**, 2507–2515.
69. Davies, J., Dodson, E.J., and Moore, M.H. (1995) *Biochemistry*, **34**, 415–425.
70. (a) Richards, A.D. and Rodger, A. (2007) *Chem. Soc. Rev.*, **36**, 471–483; (b) Zeglis, B.M., Pierre, V.C., and Barton, J.K. (2007) *Chem. Commun.*, 4565–4579; (c) Haas, K.L. and Franz, K.J. (2009) *Chem. Rev.*, **109**, 4921–4960; (d) Bruijnincx, P.C.A. and Sadler, P.J. (2008) *Curr. Opin. Chem. Biol.*, **12**, 197–206.
71. Praseuth, D., Guieysse, A.L., and Hélène, C. (1999) *Biochim. Biophys. Acta*, **1489**, 181–206.
72. Da Ros, T., Spalluto, G., Prato, M., Saison-Behmoaras, T., Boutorine, A., and Cacciari, B. (2005) *Curr. Med. Chem.*, **12**, 71–88.
73. (a) Wemmer, D.E. (2000) *Proc. Annu. Rev. Biophys. Biomol. Struct.*, **29**, 439–461; (b) Nelson, S.M., Ferguson, L.R., and Denny, W.A. (2007) *Mutat. Res.*, **623**, 24–40.
74. Nguyen, B., Lee, M.P.H., Hamelberg, D., Joubert, A., Bailly, C., Brun, R., Neidle, S., and Wilson, D.W. (2002) *J. Am. Chem. Soc.*, **124**, 13680–13681.
75. Hamelberg, D., Williams, L.D., and Wilson, W.D. (2001) *J. Am. Chem. Soc.*, **123**, 7745–7755.
76. (a) Andersson, J., Li, S., Lincoln, P., and Andréasson, J. (2008) *J. Am. Chem. Soc.*, **130**, 11836–11837; (b) Hammarson, M., Andersson, J., Li, S., Lincoln, P., and Andréasson, J. (2010) *Chem. Commun.*, **46**, 7130–7132.
77. Paramonov, S.V., Lokshin, V., Ihmels, H., and Fedorova, O.A. (2011) *Photochem. Photobiol. Sci.*, **10**, 1279–1282.
78. Berdnikova, D., Fedorova, O., Gulakova, E., and Ihmels, H. (2012) *Chem. Commun.*, **48**, 4603–4605.
79. Pace, T.C.S., Müller, V., Li, S., Lincoln, P., and Andréasson, J. (2013) *Angew. Chem. Int. Ed.* doi: 10.1002/anie.201209773, in press.
80. (a) Ciesienski, K.L., Hyman, L.M., Yang, D.T., Haas, K.L., Dickens, M.G., Holbrook, R.J., and Franz, K.J. (2010) *Eur. J. Inorg. Chem.*, **15**, 2224–2228; (b) Sánchez, M.I., Vázquez, O., Vázquez, M.E., and Mascareñas, J.L. (2011) *Chem. Commun.*, **47**, 11107–11109.
81. (a) Sissi, C., Rossi, P., Felluga, F., Formaggio, F., Palumbo, M., Tecilla, P., Toniolo, C., and Scrimin, P. (2001) *J. Am. Chem. Soc.*, **123**, 3169–3170; (b) Basak, A., Mitra, D., Kar, M., and Biradha, K. (2008) *Chem. Commun.*, **26**, 3067–3069; (c) Ghosh, S., Usharani, D., De, S., Jemmis, E.D., and Bhattacharya, S. (2008) *Chem. Asian J.*, **3**, 1949–1961.
82. (a) Alderden, R.A., Hall, M.D., and Hambley, T.W. (2006) *J. Chem. Educ.*,

83, 728–734; (b) Pizarro, A.M. and Sadler, P.J. (2009) *Biochemie*, **91**, 1198–1211; (c) Moucheron, C. (2009) *New J. Chem.*, **33**, 235–245; (d) Reedijk, J. (2009) *Eur. J. Inorg. Chem.*, **10**, 1303–1312.

83. (a) Szacilowski, K., Macyk, W., and Drzewiecka-Matuszek, A. (2005) *Chem. Rev.*, **105**, 2647–2694; (b) Ronconi, L. and Sadler, P.J. (2008) *Chem. Commun.*, **2**, 235–237; (c) Hazel, I.A.P., Ronconi, L., and Sadler, P.J. (2009) *Chem. Eur. J.*, **15**, 1588–1596.

84. (a) Kasparkova, J., Mackay, F.S., Brabec, V., and Sadler, P.J. (2003) *J. Biol. Inorg. Chem.*, **8**, 741–745; (b) Vogler, A. and Hlawtsch, J. (2003) *Angew. Chem. Int. Ed.*, **42**, 335–339; (c) Bednarski, P.J., Grünert, R., Zielzki, M., Wellner, A., Mackay, F.S., and Sadler, P.J. (2006) *Chem. Biol.*, **13**, 61–67.

85. (a) Elias, B. and Kirsch-De Mesmaeker, A. (2006) *Coord. Chem. Rev.*, **250**, 1627–1641; (b) Kirsch-De Mesmaeker, A., Moucheron, C., and Boutonnet, N. (1998) *J. Phys. Org. Chem.*, **11**, 566–576; (c) Xiong, Y. and Ji, L.-N. (1999) *Coord. Chem. Rev.*, **711**, 185–186711.

86. Sitlani, A., Long, E.C., Pyle, A.M., and Barton, J.K. (1992) *J. Am. Chem. Soc.*, **114**, 2303–2312.

87. (a) Sitlani, A., Dupureur, C.M., and Barton, J.K. (1993) *J. Am. Chem. Soc.*, **115**, 12589–12590; (b) Sitlani, A. and Barton, J.K. (1994) *Biochemistry*, **33**, 12100–12108.

88. (a) Hart, J.R., Johnson, M.D., and Barton, J.K. (2004) *Proc. Natl. Acad. Sci. U.S.A.*, **101**, 14040–14044; (b) Fink, D., Aebi, S., and Howell, S.D. (1998) *Clin. Cancer Res.*, **4**, 1–6.

89. Lacy, E.R., Madsen, E.M., Lee, M., and Wilson, W.D. (2003) in *Small Molecule DNA and RNA Binders* (eds M. Demeunynck, C. Bailly, and D.E. Wilson), Chapter 15, Wiley-VCH Verlag GmbH, Weinheim, ISBN: 3-527-30595-5.

90. Pelton, J.G. and Wemmer, D.E. (1989) *Proc. Natl. Acad. Sci. U.S.A.*, **86**, 5723–5727.

91. Chen, F. and Sha, F. (1998) *Biochemistry*, **37**, 11144–11151.

92. (a) Kopka, M.L., Yoon, C., Goodsell, D., Pjura, P., and Dickerson, R.E. (1985) *Proc. Natl. Acad. Sci. U.S.A.*, **82**, 1376–1380; (b) Lown, J.W., Krowicki, K., Bhat, U.G., Skorobogaty, A., Ward, B., and Dabrowiak, J.C. (1986) *Biochemistry*, **25**, 7408–7416.

93. (a) Dervan, P.B. (2001) *Bioorg. Med. Chem.*, **9**, 2215–2235; (b) Dervan, P.B. and Edelson, B.S. (2003) *Curr. Opin. Struct. Biol.*, **13**, 284–299; (c) Dervan, P.B., Poulin-Kerstien, A.T., Fechter, E.J., and Edelson, B.S. (2005) *Top. Curr. Chem.*, **253**, 1–31.

94. (a) Ghosh, I., Stains, C.I., Ooi, A.T., and Segal, D.J. (2006) *Mol. Biosyst.*, **2**, 551–560; (b) Pazos, E., Vázquez, O., Mascareñas, J.L., and Vázquez, M.E. (2009) *Chem. Soc. Rev.*, **38**, 3348–3359; (c) Kang, N.-Y., Ha, H.-H., Yun, S.-W., Yu, Y.H., and Chang, Y.-T. (2011) *Chem. Soc. Rev.*, **40**, 3613–3626.

95. Crissman, H.A. and Hirons, G.T. (1994) *Methods Cell Biol.*, **41**, 195–209.

96. Martin, R.M., Leonhardt, H., and Cardoso, M.C. (2005) *Cytometry A*, **67A**, 45–52.

97. Feng, S.H., Kim, Y.K., Yang, S.Q., and Chang, Y.T. (2010) *Chem. Commun.*, **46**, 436–438.

98. Rucker, V.C., Foister, S., Melander, C., and Dervan, P.B. (2003) *J. Am. Chem. Soc.*, **125**, 1195–1202.

99. Rucker, V.C., Dunn, A.R., Sharma, S., Dervan, P.B., and Gray, H.B. (2004) *J. Phys. Chem.*, **108**, 7490–7494.

100. (a) Warren, C.L., Kratochvil, N.C.S., Hauschild, K.E., Foister, S., Brezinski, M.L., Dervan, P.B., Phillips, G.N. Jr., and Ansari, A.Z. (2006) *Proc. Natl. Acad. Sci. U.S.A.*, **103**, 867–872; (b) Fechter, E.J., Olenyuk, B., and Dervan, P.B. (2005) *J. Am. Chem. Soc.*, **127**, 16685–16691.

101. Chenoweth, D.M., Viger, A., and Dervan, P.B. (2007) *J. Am. Chem. Soc.*, **129**, 2216–2217.

102. Moronaga, H., Bando, T., Takagaki, T., Yamamoto, M., Hashiya, K., and Sugiyama, H. (2011) *J. Am. Chem. Soc.*, **133**, 18924–18930.

103. (a) Best, T.P., Edelson, B.S., Nickols, N.G., and Dervan, P.B. (2003) *Proc. Natl. Acad. Sci. U.S.A.*, **100**, 12063–12068; (b) Edelson, B.S., Best, T.P., Olenyuk, B., Nickols, N.G., Doss, R.M., Foister, S., Heckel, A., and Dervan, P.B. (2004) *Nucleic Acid Res.*, **32**, 2802–2818.

104. Nunn, C.M. and Neidle, S. (1995) *J. Med. Chem.*, **38**, 2317–2325.

105. (a) Lansiaux, A., Dassonneville, L., Facompre, M., Kumar, A., Stephens, C.E., Bajic, M., Tanious, F., Wilson, W.D., Boykin, D.W., and Bailly, C. (2002) *J. Med. Chem.*, **45**, 1994–2002; (b) Susbielle, G., Blattes, R., Brevet, V., Monod, C., and Kas, E. (2005) *Curr. Med. Chem. Anticancer Agents*, **5**, 409–420; (c) Reddy, B.S.P., Sondhi, S.M., and Lown, J.W. (1999) *Pharmacol. Ther.*, **84**, 1–111; (d) Neidle, S., Kelland, L.R., Trent, J.O., Simpson, I.J., Boykin, D.W., Kumar, A., and Wilson, W.D. (1997) *Bioorg. Med. Chem. Lett.*, **7**, 1403–1408.

106. (a) Bray, P.G., Barrett, M.P., Ward, S.A., and de Koning, H.P. (2003) *Trends Parasitol.*, **19**, 232–239; (b) Bouteille, B., Oukem, O., Bisser, S., and Dumas, M. (2003) *Fundam. Clin. Pharmacol.*, **17**, 171–181; (c) Rosypal, A.C., Werbovetz, K.A., Salem, M., Stephens, C.E., Kumar, A., Boykin, D.W., Hall, J.E., and Tidwell, R.R. (2008) *J. Parasitol.*, **94**, 743–749.

107. (a) Mathis, A.M., Bridges, A.S., Ismail, M.A., Kumar, A., Francesconi, I., Anbazhagan, M., Hu, Q., Tanious, F.A., Wenzler, T., Saulter, J., Wilson, W.D., Brun, R., Boykin, D.W., Tidwell, R.R., and Hall, J.E. (2007) *Antimicrob. Agents Chemother.*, **51**, 2801–2810; (b) Wilson, W.D., Tanious, F.A., Mathis, A., Tevis, D., Hall, J.E., and Boykin, D.W. (2008) *Biochimie*, **90**, 999–1014; (c) Thuita, J.K., Karanja, S.M., Wenzler, T., Mdachi, R.E., Ngotho, J.M., Kagira, J.M., Tidwell, R., and Brun, R. (2008) *Acta Trop.*, **108**, 6–10.

108. Vázquez, O., Sánchez, M.I., Martínez-Costas, J., Vázquez, M.E., and Mascareñas, J.L. (2010) *Org. Lett.*, **12**, 216–219.

109. Vázquez, O., Sánchez, M.I., Mascareñas, J.L., and Vázquez, M.E. (2010) *Chem. Commun.*, **46**, 5518–5520.

3
Structural Analysis of Complex Molecular Systems by High-Resolution and Tandem Mass Spectrometry

Yury O. Tsybin

3.1
Dissecting Molecular Complexity with Mass Spectrometry

Advances in addressing the grand challenges of the twenty-first century, particularly within health, energy, environmental, and material sciences, rely on the molecular level qualitative and quantitative information. The constantly improving analytical capabilities for molecular structural analysis reveal higher levels of molecular complexity in biological and environmental samples and raise the bar of the analytical requirements.

Diverse applications mentioned above deal with molecules ranging in size from a few atoms, for example, metabolites or small pharmaceuticals, to hundreds of atoms, for example, peptides and proteins, and further to tens of thousands of atoms, for example, covalently and noncovalently bound protein complexes. When complex mixtures of these molecules are analyzed, the level of intermolecular complexity may exceed the limit of the currently employed analytical techniques of molecular structural analysis [1–3] (Figure 3.1).

Mass spectrometry (MS) is the most sensitive and selective technique for molecular structural analysis. MS-based structural analysis begins with ionization – the process of transforming neutral or already charged molecules from gas, solution, or solid state into the gas-phase ions. The charged particles are then transferred to the vacuum environment of a mass spectrometer. Electric and magnetic fields are applied to the charged molecule (an ion) with mass m and the number of elementary charges z for its trajectory manipulation and subsequent ion detection as a function of m/z ratio. As a result, a mass spectrum is obtained with m/z values as the x-axis and ion abundance (typically relative to the most abundant ion in the mass spectrum, or a base peak) as the y-axis. The ability of unambiguous ion detection and identification depends on the analytical characteristics of a mass spectrometer (Figure 3.2).

Resolution or resolving power of a mass spectrometer is a central analytical characteristic responsible for distinguishing the peaks in a mass spectrum [4–7]. The resolving power of a mass spectrometer achieved for a given peak in a mass spectrum is typically defined as the ratio of m/z value of this peak to

Discovering the Future of Molecular Sciences, First Edition. Edited by Bruno Pignataro.

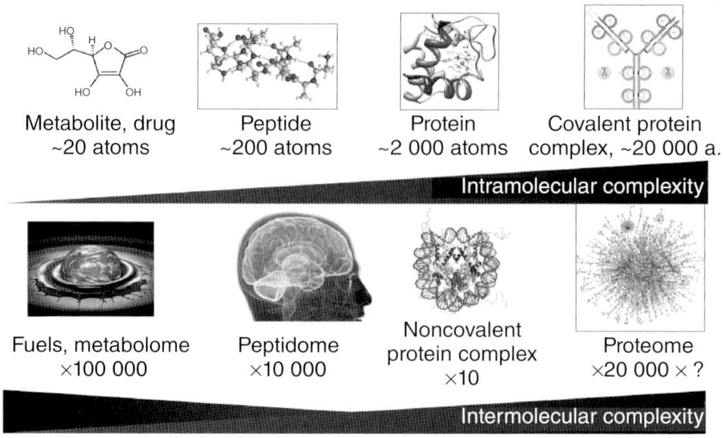

Figure 3.1 A bird's eye view on intra- and intermolecular complexity.

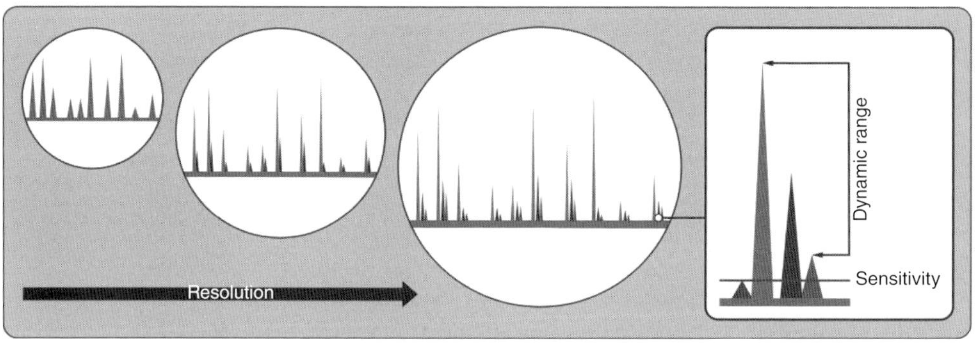

Figure 3.2 Analytical characteristics of a mass spectrometer: resolution (resolving power), sensitivity, and dynamic range. Figure adapted based on the original idea of Jean-Luc Wolfender (University of Geneva, Switzerland).

the peak's full width at half the maximum height, $\Delta_{FWHM}(m/z)$ or $\Delta_{50\%}(m/z)$. For the most MS-based applications, resolving power varies between 1000 and 100 000, though it can reach the level of ~1 000 000 and above for some selected applications. The term resolution is oftentimes, and in this chapter, employed with the same definition in mind. However, strictly speaking [4], the resolution should be defined via the distance between the two peaks, $(m/z)_1$ and $(m/z)_2$, equal to $\Delta(m/z) = (m/z)_2 - (m/z)_1$, with a specified level of valley between the peaks. Typically, the valley between the peaks is chosen as 50% of the least abundant peak. For example, if the two peaks are of the same height and the valley between the peaks is 50% of their height, the resolution is equal to the value of $\Delta_{50\%}(m/z)$, which is used to determine the resolving power. In this case, the resolution is quantitatively equal to the resolving power. At an appropriate level of resolving power, other analytical characteristics of a mass spectrometer can be defined (Figure 3.2). Spectral dynamic range is the ratio between the most abundant ion in the mass spectrum to the least abundant one [7]. This characteristic should not be confused with the dynamic range of concentration of analytes in the sample, for example, protein dynamic range in plasma. Sensitivity is the analytical characteristic indicating the lowest abundance signal that can be seen in a mass spectrum. Finally, mass accuracy shows how close the measured mass of an ion to the true value is. In practice, mass accuracy is defined via the ratio of the difference between the experimental m/z and theoretical m/z values, $(m/z)_{exp} - (m/z)_{theor}$, divided on the theoretical value, $(m/z)_{theor}$, because the error of the theoretical estimate of the true mass is normally much smaller than the error of the experimental mass. As the obtained absolute value of mass accuracy is a small number, it is typically multiplied by 1 000 000 to represent it in ppm (parts per million). Many MS applications require mass accuracy of ~1–5 ppm, whereas for some selected applications, for example, petroleomics, an order of magnitude better mass accuracy is needed.

Most of the chemical elements that make up the molecules are naturally present in a number of isotopes. The isotopes of a given chemical element differ in the number of neutrons although they have the same number of protons. Therefore, molecular mass is an isotopic distribution. For example, the two most abundant isotopes of carbon are ^{12}C (the most abundant carbon isotope) and ^{13}C (its abundance relative to ^{12}C isotope is 1.1%). Owing to the high frequency of occurrence of carbon in biological and organic molecules, primarily ^{13}C isotope is responsible for their characteristic isotopic distributions (Figure 3.3).

Examples shown in Figure 3.3 demonstrate the need for high (ubiquitin) and very high (antibody) resolving power required for visualizing their isotopic distributions. Any isotopic distribution of a peptide or a protein starts with a monoisotopic mass. It is defined as a mass of an ion with the lightest isotope of each chemical element present in a molecule. For example, peptides and proteins are composed only of five chemical elements: C, H, N, O, and S. The lowest mass isotopes of all these five elements are also the most abundant ones. Therefore, oftentimes, the monoisotopic mass of a biomolecule is referred to an ion that is composed of the most abundant isotopes. However, that is not true in the general case. If a molecule contains a chemical element, for example, Pt, which most abundant isotope is not the lightest one, then the appropriate (using the lightest isotopes) definition of the monoisotopic mass should be used. Figure 3.3 also demonstrates that following a certain number of chemical elements present in a molecule, the monoisotopic ions become minor components of isotopic distributions. The corresponding spectral dynamic range increases from ~20 for ubiquitin to $\sim 10^{32}$ for an antibody, whereas the state-of-the-art mass spectrometry provides spectral dynamic range of not more than 10^5. Taken together with the low amounts of biological samples consumed during mass measurements and considering that detector saturation should be avoided, it is obvious that the monoisotopic ions of large proteins, for example, antibodies, will not be present in the mass spectra.

Accurate mass measurements of intact proteins of a medium size, ~10–30 kDa, require high mass resolution for acquiring protein isotopic distributions. Ideally, a protein monoisotopic mass is used for accurate mass assignment (Figure 3.3a).

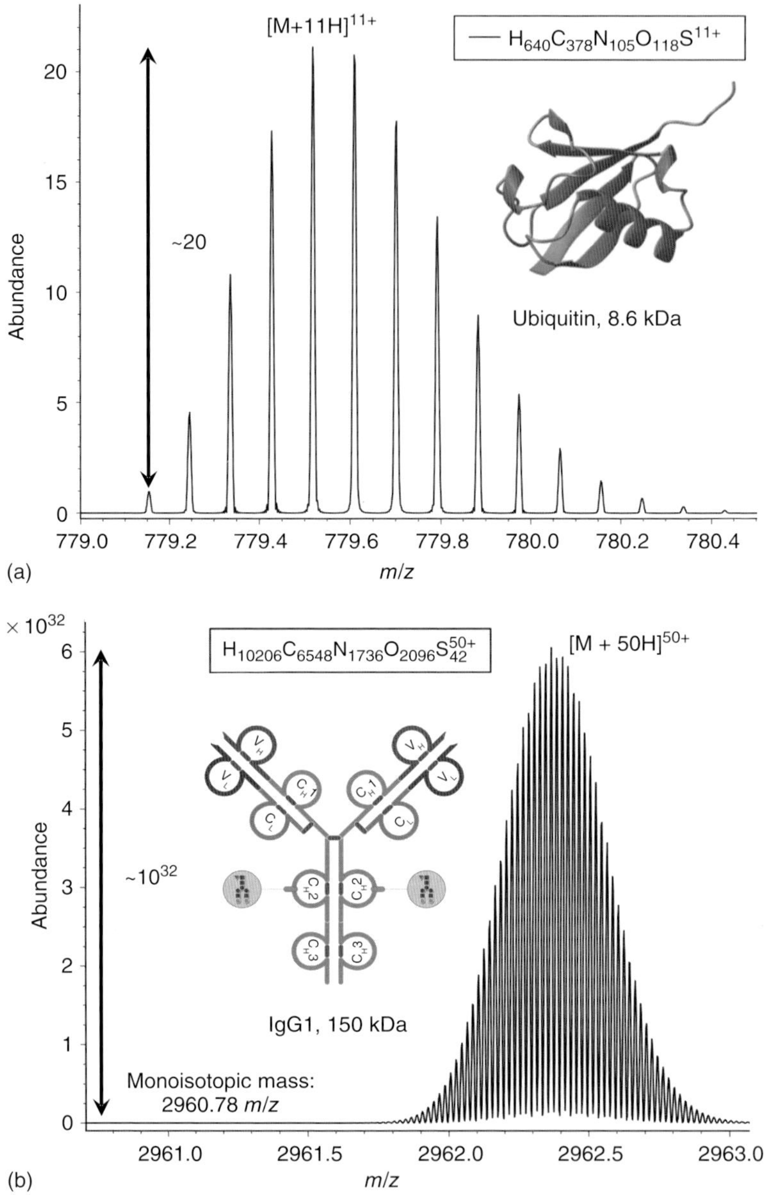

Figure 3.3 Protein mass measurements by mass spectrometry showing (theoretical) baseline-resolved ^{13}C isotopic distributions of (a) $[M+11H]^{11+}$ ions of ubiquitin and (b) $[M+50H]^{50+}$ ions of a monoclonal antibody, IgG. The arrows show the corresponding monoisotopic masses and the dynamic ranges.

With increasing protein size, this is not detectable and typically the most abundant isotopes are used for mass assignment. Further increase of a protein molecular weight, above 30 kDa, requires higher resolving power, but renders protein monoisotopic mass analysis useless (Figure 3.3b). Owing to the symmetric isotopic distributions for large proteins, accurate molecular weight measurements coincide with the measurement of the apex of the isotopic distribution envelope. Therefore, mass measurements of such large proteins performed at low resolution provide comparable degree of mass accuracy as the high-resolution ones.

3.2
Advances in Fourier Transform Mass Spectrometry

Measuring the frequency of an ion oscillating in electric or magnetic fields forms the basis of Fourier transform mass spectrometry (FTMS) that provides superior resolving power and mass accuracy among all types of mass spectrometers (Figure 3.4) [4, 5, 7–9]. Historically, FTMS was introduced in the 1970s by Marshall and Comisarow [7]. FTMS was represented by the Fourier transform ion cyclotron resonance mass spectrometer (FT-ICR MS), which has a magnetic field-based mass analyzer (ICR cell) and employs Fourier transform (FT) for simultaneous detection of all ions with different m/z. Since its early days to the present, FT-ICR MS has provided the highest resolving power and mass accuracy for molecular structural analysis. However, there are technical challenges associated with high magnetic field (up to 21 T as of now) required for high-performance FT-ICR MS. These include the substantial operational costs and variable robustness of the superconducting magnets, as well as difficulties of efficient ion transfer through the extreme magnetic field gradients. Therefore, there always existed a strong incentive toward implementation of FTMS without magnetic field. As a result, at the end of the twentieth century, FTMS principle was successfully extended by Makarov to an electrostatic ion trap, the Orbitrap [8–10]. Orbitrap FTMS made a remarkable entrance into the MS market in 2005 and its popularity grows with every year, as its performance is constantly improving [9]. To a large extent, the rapid progress in Orbitrap FTMS is due to the comprehensive knowledgebase developed by the FT-ICR MS community. In parallel, the idea of FTMS implementation in other types of electrostatic ion traps, specifically in the form of a double reflectron time-of-flight (TOF) MS with an induced current-based ion detector placed axially between the reflectrons, has been entertained by a number of groups, including Zajfman *et al.* [11] and Gonin [12]. Despite a recent progress in the field of TOF FTMS and related instrument development, for example, by groups of McLuckey [13], Antoine [14], and Jarrold [15], the technique remains at the research-grade scale and is available only in a few laboratories in the world. Nevertheless, the envisioned benefits of this technique for molecular structural analysis continue to drive the progress in TOF FTMS domain, which is expected to result in the near future commercial-grade implementation. It is important to note that the principle of ion detection in all FTMS instruments mentioned above is the nondestructive

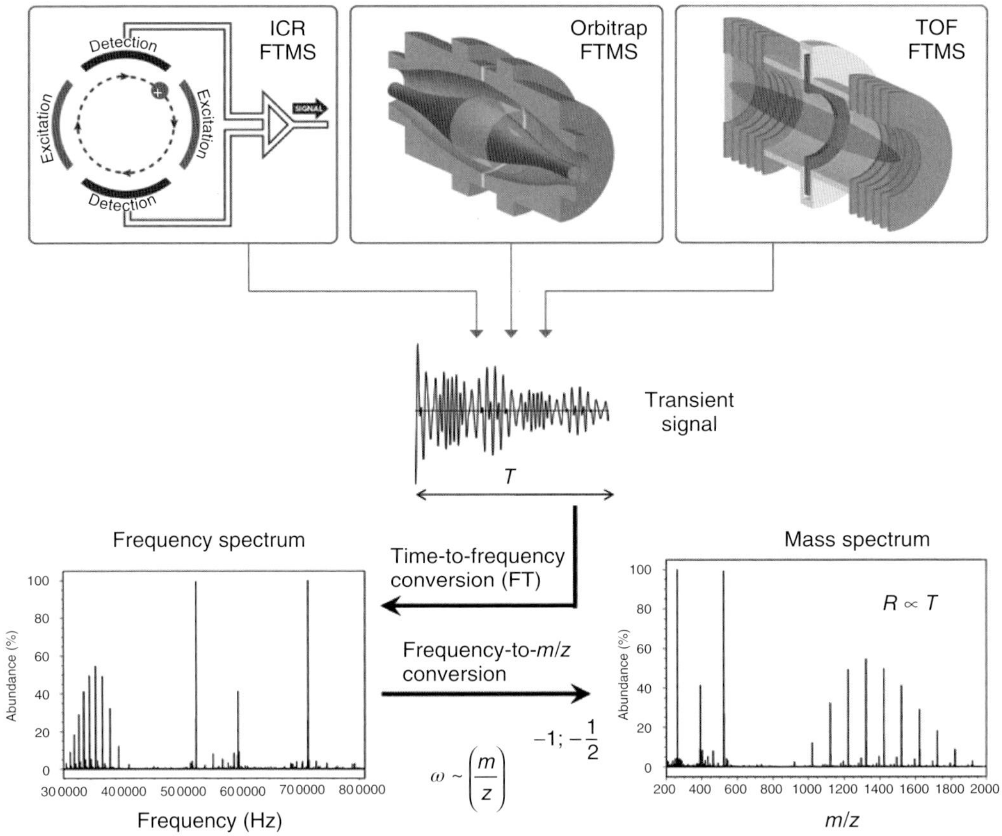

Figure 3.4 The principle of Fourier transform mass spectrometry (FTMS). The frequency of ion motion in magnetic (ICR) or electrostatic (Orbitrap, TOF) fields is measured using induced current ion detection. The measured signal is represented as a "transient signal" – signal voltage as a function of time over a period of user-defined time period T. Fast Fourier transformation (FT) allows time to frequency conversion. The resultant frequency spectrum is then converted into the mass spectrum using the known relations between ion frequency and *m*/*z* values. The resolving power in FTMS is proportional to the transient length T.

induced current ion detection. In part, the difficulties associated with this type of ion detection and the corresponding signal processing is holding up the development of TOF FTMS. Therefore, other research groups, namely Verenchikov [16] and JEOL [17], combined the original method of ion detection by destructive ion collisions with surfaces (e.g., using multichannel plates, conversion dynodes, and electron multipliers) with an extended ion flight pass in a field-free environment using multiple-pass configurations. Nowadays, the multireflectron (MR) TOF MS instruments reach the level of resolving power offered by entry-level Orbitrap and ICR FTMS instruments. In the following, we focus only on Orbitrap and ICR FTMS due to their wider spread and commercial availability compared to TOF FTMS.

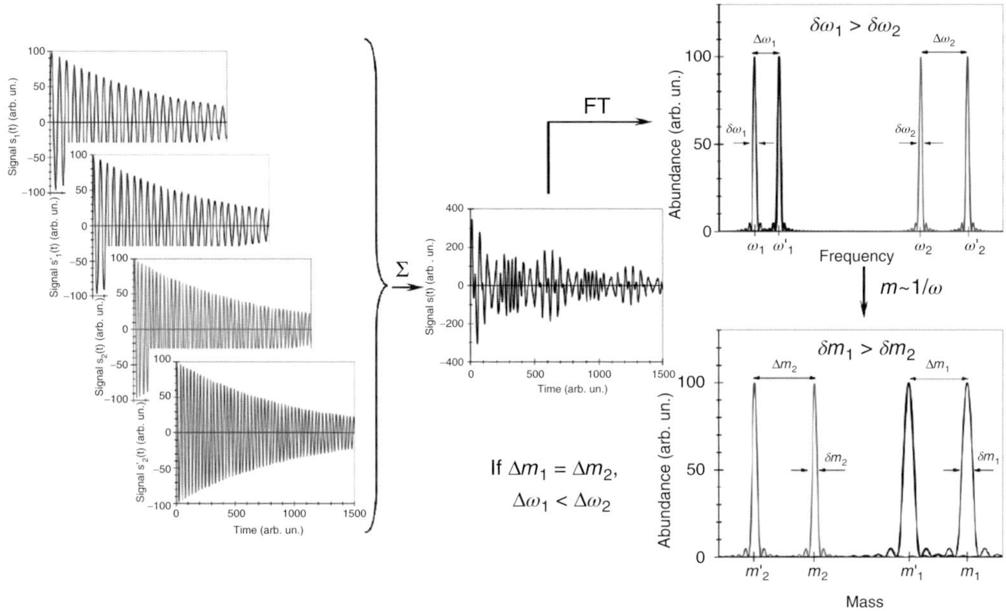

Figure 3.5 The FTMS workflow: (i) harmonic (sinusoidal) transient signals from all ions are mixed together at the ion detector, (ii) time-domain transient signal is converted into the frequency spectrum using Fourier transformation, and (iii) frequency spectrum is further converted into the mass spectrum using the known relation between frequency and *m/z*. Example of a resulting FT-ICR mass spectrum is shown.

The key to the power of FTMS lies in the ability to perform multiple periodical measurements of highly reproducible ion oscillations over a long period of time (milliseconds to seconds). Similarly to time-domain signal processing in NMR spectroscopy, the application of fast FT allows for deciphering complex transient signals with a number of sinusoidal functions (up to hundreds of thousands) and their representation in a frequency domain (Figures 3.4 and 3.5) [7].

In the frequency domain, the peaks corresponding to ions of different *m/z* values all have the same FWHM, as in NMR. In contrast to NMR, on conversion to the *m/z* domain, peaks change their widths. That is due to the nonlinear relationship between the ion frequencies and corresponding *m/z* values. Specifically, in FT-ICR MS, the frequency of periodic ion motion is inversely proportional to *m/z* and in Orbitrap FTMS this frequency is inversely proportional to the square root of *m/z*. Therefore, higher frequency ions are those with lower *m/z* values (and more narrow peaks), whereas lower frequency ions are those with higher *m/z* values (and broader peaks). The distances between the peaks (e.g., in isotopic distributions) in the mass and frequency spectra also differ correspondingly (Figure 3.5).

3.3
Advances in Mass Analyzers for FT-ICR MS

The heart of FT-ICR MS is the ICR cell located in the homogeneous magnetic field and employed to trap the ions injected from external ion sources, isolate the ions of interest, transform them by diverse means of ion activation and dissociation reactions, and detect the products of these reactions as well as the intact species using induced current ion detection principle. The ICR cell acts as a mass analyzer due to the relation between the m/z ratio of the ion and the frequency of ion rotation in a magnetic field in a plane perpendicular to the magnetic field axis [7]. Figure 3.6 shows configurations of the ICR cells currently being employed in FT-ICR MS with a focus on recent developments.

Figure 3.6a shows the standard cylindrical open-ended ICR cell [18]. It traps ions in the axial, along the magnetic field, direction with the cylindrical trapping electrodes, aimed at approaching the harmonic trapping field. The central section of this cell is employed for ion detection and consists of a pair of detection electrodes and a pair of excitation electrodes. For the optimum cell performance conditions, the angular extent of each of these excitation and detection electrodes is 90°. Figure 3.6b shows a modification of the cylindrical open-ended ICR cell where both ion excitation and detection events are carried out with the same pair of 120° electrodes. This cell design has been introduced by Marshall and coworkers in 2013 to optimize the efficiencies of both ion excitation and detection. The ability to employ the same pair of electrodes for both events is provided by the specifically designed

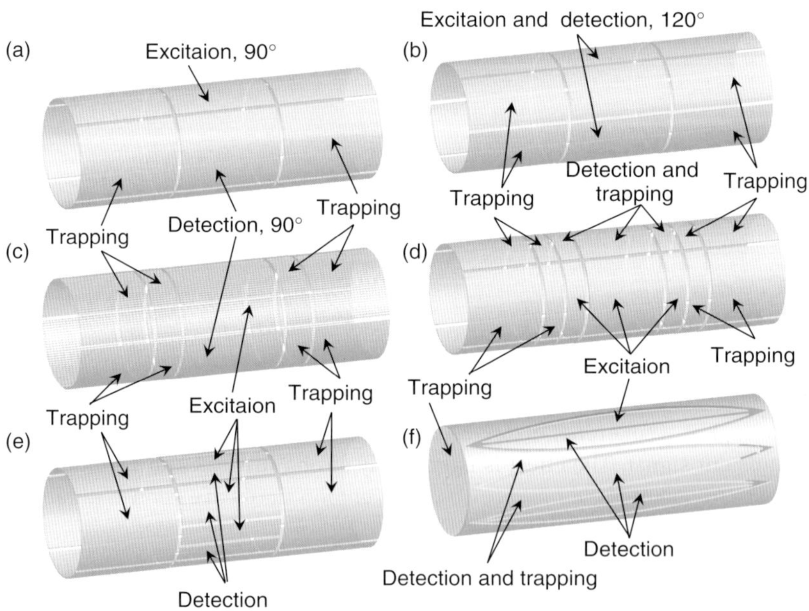

Figure 3.6 (a–f) Configurations of the main types of ICR cells employed in FT-ICR MS.

electric circuit. Figure 3.6c shows another variant of a cylindrical open-ended ICR cell. In this configuration, developed and commercialized by Thermo Scientific, an additional set of trapping electrodes is introduced and a high-transparency grid is added along the whole length of the ICR cell to provide an improved ion RF excitation [8]. Small, <1 V, DC potentials, known as the offset potentials, can be applied to the detection and excitation electrodes to optimize the position of ion cloud inside the ICR cell. Note that the RF excitation voltage is applied only to the grid. The inner diameter of the grid electrode is the same as that of the detection electrodes. Figure 3.6d is a basic model of a compensated harmonized ICR cell [19, 20]. The compensated cells provide improved, closer to parabolic, trapping potential along the z-axis by dividing the trapping and excitation/detection cylinders into a number of sections, typically in three, five, or seven sections. Figure 3.6e shows another variant of an open-ended cylindrical ICR cell that aims at detection of higher order harmonics (frequency multiplies) of ion reduced cyclotron frequencies in FT-ICR MS. As resolution is a function of frequency of ion rotation, recording higher frequency signals is beneficial for FTMS applications. Particularly, to achieve the same resolution in a typical LC-MS experiment, the required acquisition time with the detection of higher harmonics is shorter. For example, if there is a set of 16 electrodes that can be employed for ion excitation and detection compared to the typical set of 4 electrodes as in Figure 3.6a, the detection of the quadruple frequency multiple can be realized (Figure 3.6e). Thus, four times faster data acquisition can be performed, as has been demonstrated by Tsybin and Gorshkov. Finally, Figure 3.6f depicts the cylindrical close-ended "dynamically" harmonized ICR cell, or ParaCell developed and commercialized by Bruker Daltonics in collaboration with Nikolaev in 2012 [21–23]. The particular distinction of this cell is in the configuration of the trapping field. Besides the end-cap trapping electrodes with spherical surfaces, each of the trapping electrodes along the z-axis has a special X-shape with wide ends and a narrow center. The cell is divided into four quadrants as the standard ICR cell, similarly to Figure 3.6a. Each quadrant has two X-shaped trapping electrodes. Ion excitation and detection is provided similarly to the ICR cell shown in Figure 3.6a, where each of the quadrants is employed for either ion detection or excitation. Therefore, the X-shaped electrodes are employed not only for trapping but also for ion excitation and detection. The provided trapping field has a special structure along the z-axis that is defined by the shape of the X-electrodes. When ions move axially, along the z-axis, in both directions, the trapping field they sample is harmonic if averaged over the ion's circular trajectory in the cross-sectional dimension. This ICR cell configuration results in improved coherency of ion motion and significantly increases the lifetime of ion transient signal. The effect is particularly strong in weak and medium, for example, 7 T, magnetic fields. To acquire transient ion signal of extended duration, improved vacuum conditions are needed. The principal benefit of the ParaCell is in the ability to acquire mass spectra with extreme, >10 000 000, resolution for selected applications, and, routinely, with resolution at ~1 000 000. The main drawback is the spectral rate – achieving these ultrahigh levels of resolution requires acquisition times that are not compatible with time-constrained experiments, such as LC-MS.

3.4
Advances in Mass Analyzers for Orbitrap FTMS

In contrast to the ICR cells, Orbitrap mass analyzers are employed only for ion trapping and immediate ion detection [8, 9]. They are composed of the two main electrodes shaped to enable ion trapping and periodic motion in electrostatic fields – the central spindle-shape electrode and the outer split electrode, also employed for ion detection (Figure 3.7).

The ion detection principle is similar to the one in FT-ICR MS (Figure 3.5). The mass analyzer properties of Orbitrap are based on the relation between m/z ratio of an ion and the frequency of ion axial oscillations in the electrostatic field. For a given acquisition time, the achievable resolving power of Orbitrap FTMS is directly proportional to the frequency of ion axial motion in an Orbitrap mass analyzer [24]. The electric field between the central and outer electrodes of an Orbitrap defines this frequency. Higher electric field provides higher frequency, and thus an increased resolution [25]. Figure 3.7 summarizes the main parameters of Orbitrap mass analyzers currently employed in Orbitrap FTMS, including standard, high-field, and ultra-high-field configurations [4, 9].

In a standard Orbitrap mass analyzer, the potential applied to the central electrode is $\pm$3.5 kV and the inner diameter of the outer electrode is 30 mm (central electrode diameter of maximum 12 mm). The high-field Orbitrap mass analyzer provides higher electric field between the central electrode and the outer electrode. The potential applied to the central electrode is also $\pm$3.5 kV, whereas the inner diameter of the outer electrode is decreased to 20 mm, but the maximum outer diameter of the central electrode is reduced only to 10 mm. The ultra-high-field

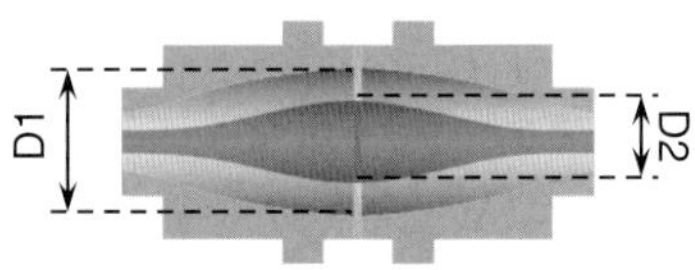

Parameters	Standard orbitrap	High-field orbitrap	Ultra-high-field orbitrap
D1 (mm)	30	20	20
D2 (mm)	12	10	10
Potential* (kV)	3.5	3.5	5
Frequency** (kHz)	300	530	640
Resolution***	15k	25k (50k)	30k (60k)

Figure 3.7 Parameters of Orbitrap mass analyzers. Notes: * – potential of the central electrode during ion detection (negative for positive ion detection and positive for negative ion detection); ** – approximate frequency of ion axial motion for an ion with m/z 524; and *** – approximate resolution (resolving power) achieved for the 524 m/z ion in 192 ms detection period and magnitude (absorption) mode FT processing.

Orbitrap mass analyzer is supplied with an even higher electric field for ion acceleration and periodic motion. The potential applied to the central electrode is ±5 kV, whereas the dimensions of other electrodes are the same as for the high-field Orbitrap mass analyzer. Typical values of ion axial motion frequencies for an ion at m/z 524 demonstrate the corresponding dependence on the electric field (Figure 3.7).

Figure 3.8 shows the main configurations of high-resolution mass spectrometers that employ Orbitrap mass analyzers described above [8, 9]. Figure 3.8a shows the implementation of a standard Orbitrap mass analyzer (as employed in Exactive™ series Orbitrap FTMS) [26], the middle panel of a high-field Orbitrap mass analyzer (as employed in Orbitrap Elite™ FTMS) [25], and the bottom panel of an ultra-high-field Orbitrap mass analyzer (as employed in Orbitrap Fusion™ FTMS). In all these configurations, Orbitrap mass analyzer is employed solely for low-to high-resolution mass measurements without further ion transformation, for example, ion activation and dissociation. Another principal component of these instruments is the C-trap (Figure 3.8). As ion excitation in Orbitrap FTMS is achieved on-the-fly on fast and coherent ion cloud injection, the C-trap is typically used to store the ions before their injection into the Orbitrap [9]. Figure 3.9 details the main steps of an interplay between the C-trap and the Orbitrap for ion detection: (i) ion injection from an external ion source or ion accumulation device and ion capture in the C-trap; (ii) ion cloud squeezing in the C-trap for efficient ion injection into the Orbitrap; (iii) pulsed ion injection in the Orbitrap and formation of ion rings; and (iv) coherent motion of ion rings and ion detection in the Orbitrap.

Efficiency of ion transfer and capture in the C-trap is a function of pressure conditions, which are regulated by the flow of gas (typically nitrogen) into the HCD (higher-energy collision-induced dissociation) trap (Figures 3.7 and 3.8). The HCD trap is connected to the C-trap and serves for ion capture, accumulation, and activation/dissociation [9, 27]. Owing to the additional two turbo pumps in Orbitrap Elite™ compared to the other two configurations, this mass analyzer may be operated at an order of magnitude better vacuum conditions. That translates into improved ion signal stability at longer, >3 s, acquisition times.

3.5 Applications of High-Resolution Mass Spectrometry

High-resolution MS is particularly useful in the performance-demanding analysis of extremely complex mixtures of small molecules (e.g., petroleomics), intact protein mass measurements, and complex mixtures of product ions in tandem mass spectrometry (MS/MS) of proteins (top-down MS and proteomics).

Comprehensive molecular structure analysis of crude oils (petroleum) and complex petroleum fractions by high-resolution FTMS is known as *petroleomics* [28–32]. Thanks, primarily, to the efforts of the group of Marshall, the FT-ICR MS is nowadays the central petroleomics-grade technique. The performance of modern FT-ICR MS combined with the dedicated calibration and data analysis

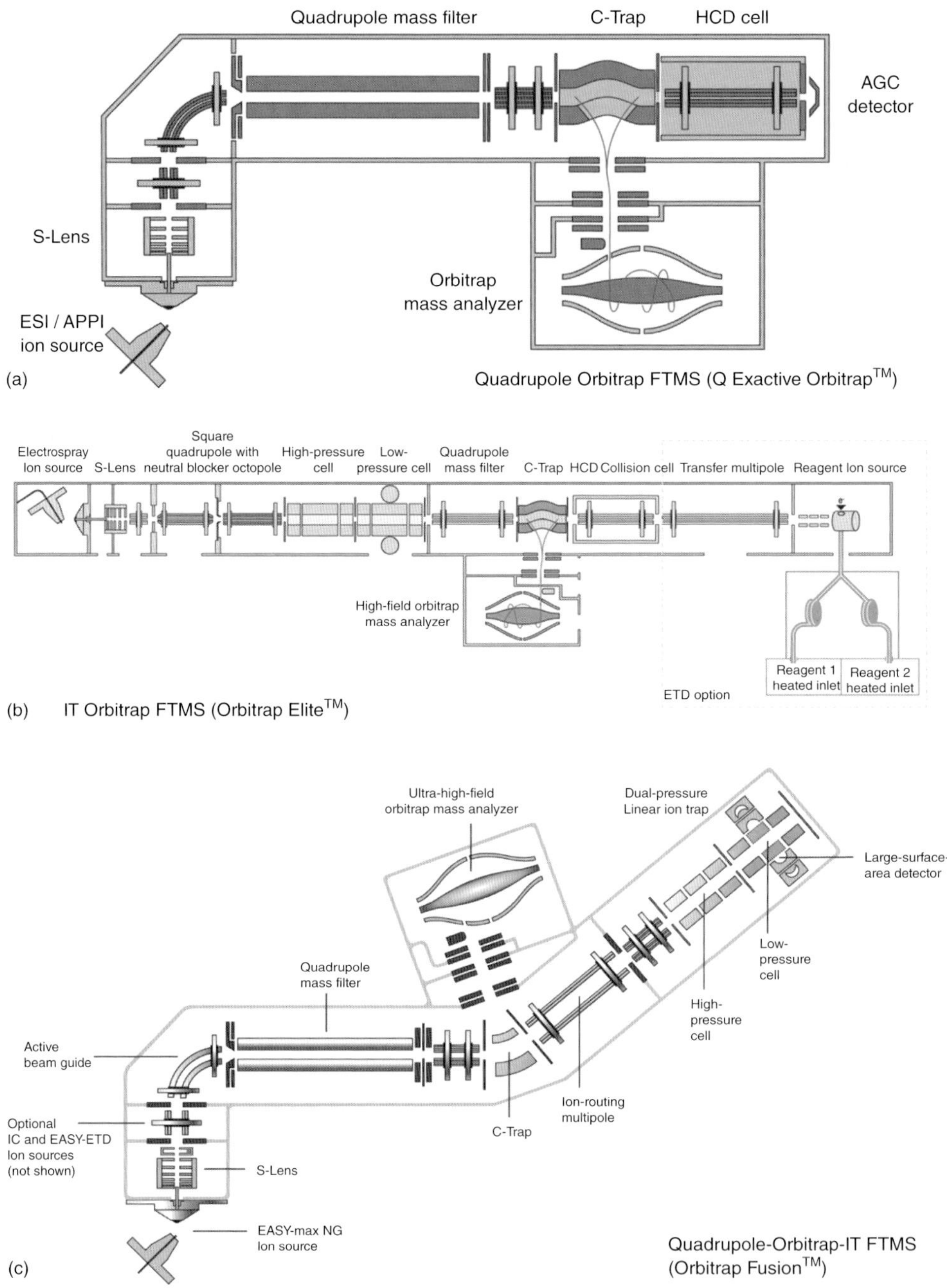

Figure 3.8 (a–c) Configurations of the main types of Orbitrap-based FT mass spectrometers. Figures adapted from *www.planetorbitrap.com*, courtesy of Thermo Scientific.

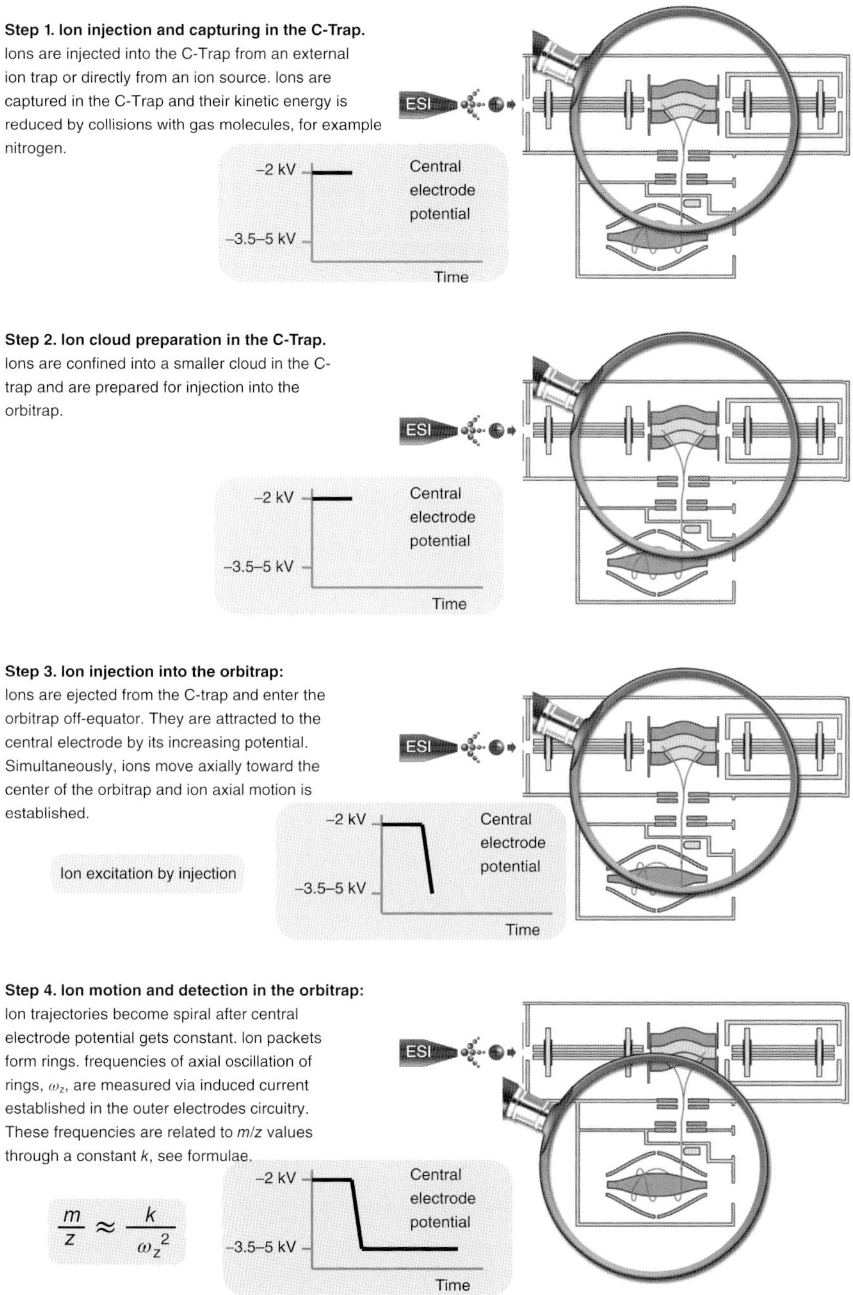

Figure 3.9 (a–d) The working principle of Orbitrap FTMS technology. Case of a positive ion analysis is shown.

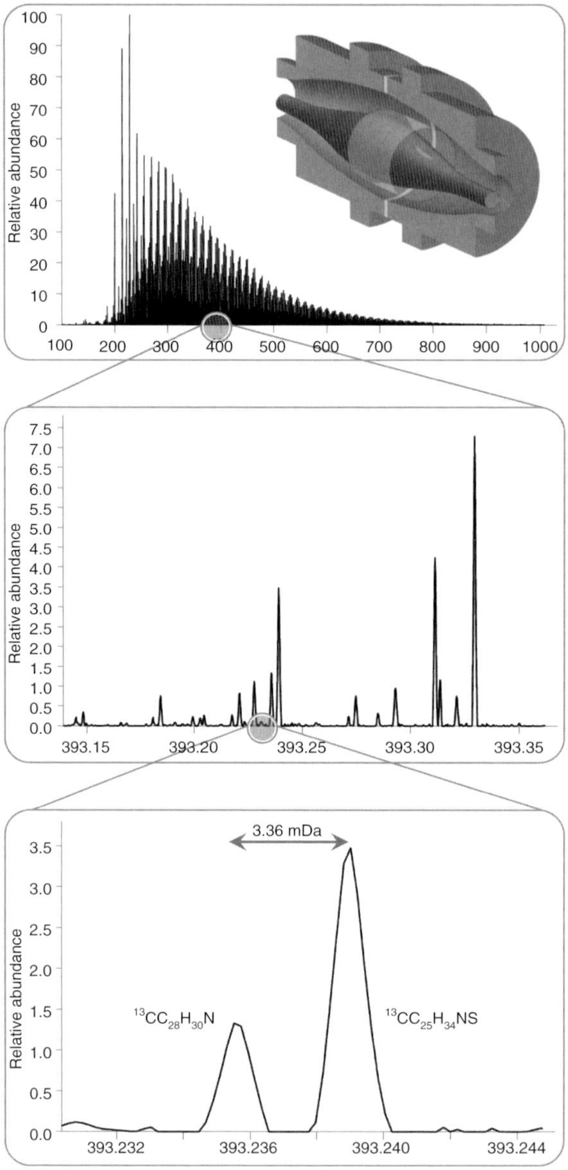

Figure 3.10 Comprehensive composition analysis of complex mixtures of small molecules by high-resolution FTMS. Shown is the mass spectrum of a crude oil fraction obtained with positive mode electrospray ionization high-field Orbitrap FTMS. The achieved level of resolving power is ~480 000 at *m/z* 400 (transient duration of 1.5 s).

procedures is believed to be sufficient to address the qualitative analysis of the most complex mixtures. Recently, high-field Orbitrap FTMS has shown some promise in approaching the capabilities required for petroleomics (Figure 3.10) [30].

The critical enabling factor for implementation of petroleomics on Orbitrap FTMS is the availability of a mass scale calibration method providing the required level of mass accuracy for unambiguous peak assignment. For example, an iterative method of FTMS mass spectra recalibration using an empirical estimation of the mass calibration function demonstrates the required level of performance for both ICR and, importantly, Orbitrap FTMS [33]. Further progress in Orbitrap FTMS-based petroleomics presumably will involve detection of longer transient signals (at least up to 3–6 s). The intrinsic benefit for accurate mass analysis of petroleum-type samples is the regularly repeating patterns of molecular structures. This benefit diminishes when sample nature remains to be very complex, but regular molecular patterns are not present, for example, as observed for biofuels [34]. Therefore, novel approaches to mass spectra calibration are to be investigated.

Figure 3.11 shows a mass spectrum of overlapping but baseline resolved isotopic distributions of a monomer and a dimer of a protein superoxide dismutase (SOD) obtained with Orbitrap FTMS.

The mass spectrum shown in Figure 3.11 was received in the so-called native mass spectrometry conditions, which aims at preserving the solution phase

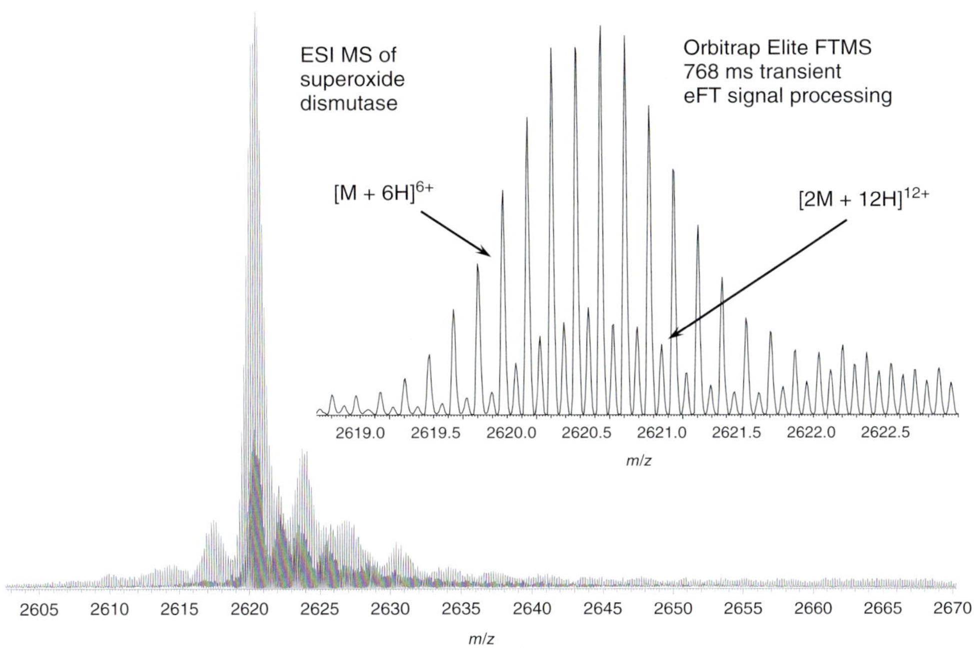

Figure 3.11 Positive ion mode electrospray ionization high-field Orbitrap FTMS of superoxide dismutase (SOD). Both monomer and dimer of SOD are present and their isotopic envelopes are baseline resolved even at relatively high *m/z* values.

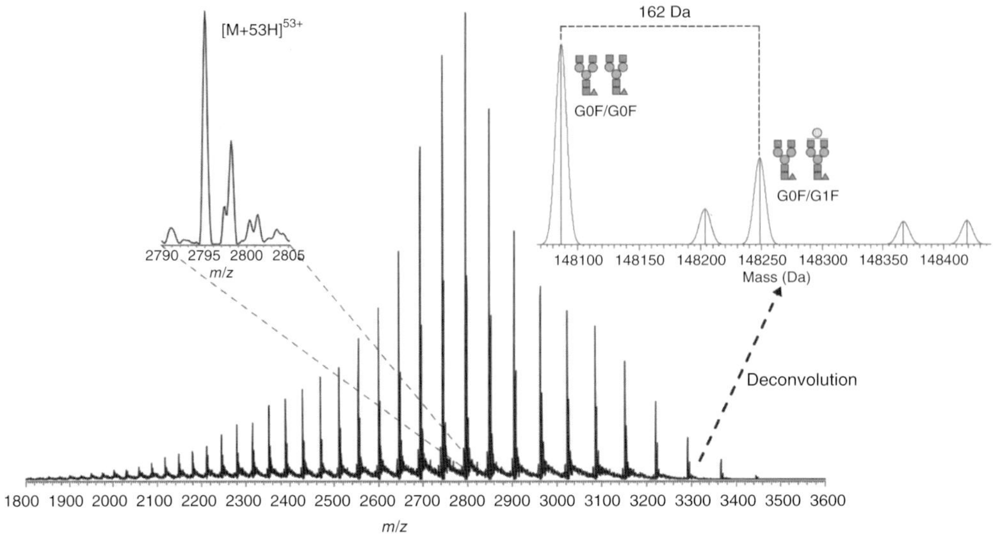

Figure 3.12 Intact protein (~150 kDa monoclonal antibody, IgG) mass analysis with electrospray ionization high-field Orbitrap FTMS. Left inset shows an expanded segment of a broadband mass spectrum demonstrating the presence of multiple proteoforms (glycoforms). Right inset shows the deconvoluted mass spectrum and confirms the identity of the glycoforms.

protein–ligand interactions. Native mass spectrometry is a rapidly growing MS application area particularly useful for protein–protein and protein–drug complexes analysis [35–38]. For example, native MS allows observation of a dimer of SOD in the mass spectrum shown in Figure 3.11.

As depicted in Figure 3.3, analysis of heavy, >50 kDa, intact proteins does not directly benefit from high resolving power. In these cases, FTMS is operated as a low resolution, but higher throughput, mass spectrometer (Figure 3.12) [39, 40].

Although modern FT-ICR MS and the high-field Orbitrap FTMS allow isotopic level resolution even of molecular antibodies [41, 42], the main driving force behind FTMS-based analysis of intact proteins is unambiguous characterization of their posttranslational modifications [43]. Furthermore, in proteomics of complex protein mixtures, it is essential to provide a level of resolution required to distinguish between coeluting isobaric proteins. Therefore, further improvement in high-resolution MS is needed to provide robust and routine accurate mass measurements of large proteins.

3.6
Advances in Tandem Mass Spectrometry

Figure 3.13 shows an overview of the main techniques of ion activation and dissociation as employed in MS/MS for structure analysis of biological molecules,

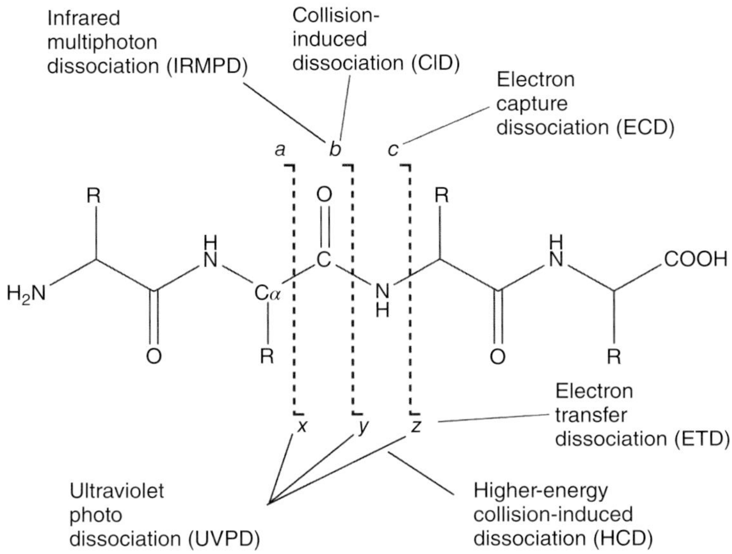

Figure 3.13 An overview of ion activation and dissociation methods for mass spectrometry-based biomolecular structure analysis.

for example, peptides and proteins [44]. The three bonds that form a peptide backbone in peptides and proteins can be cleaved by molecular ion interaction with electrons, photons, and neutral gas molecules. The fundamentals and characteristics of these interactions that form the basis for MS/MS can be obtained from numerous reviews [44–46]. The most widely employed MS/MS techniques for molecular structure analysis are those based on vibrational ion activation and dissociation – collision-induced dissociation (CID) achieved by low-energy multiple interaction of precursor ions with neutral gas molecules. The required gas, for example, nitrogen or helium, is introduced into the ion trap either continuously or by pulsed injection, Figure 3.14c [44].

The main dissociation channel in CID MS/MS is a cleavage of C_α-N peptide bond that yields sequence-specific *b*/*y* types of product ions. Similar peptide backbone cleavage is realized with HCD in Orbitrap FTMS (Figure 3.14c) [27]. Cleavage of the N-C_α peptide backbone bond leading to the formation of *c*/*z* type sequence-specific product ions can be achieved by precursor ion interaction with low-energy, ~1 eV, electrons (Figure 3.14b). These reactions are realized either by ion interaction directly with free electrons, as employed in electron capture dissociation (ECD) [47], or indirectly by electron transfer from radical anions, as takes place in electron transfer dissociation (ETD) [48, 49]. Finally, ion interaction with photons may also lead to ion activation and dissociation if functional groups in molecular structure allow for absorption of photons of a certain wavelength. Practically, photons of infrared (IR) [50] and ultraviolet (UV) [51] regions of electromagnetic spectrum are employed for MS/MS-based molecular structure analysis. Specifically, interaction of biomolecular ions with IR photons in the gas phase results in vibrational ion

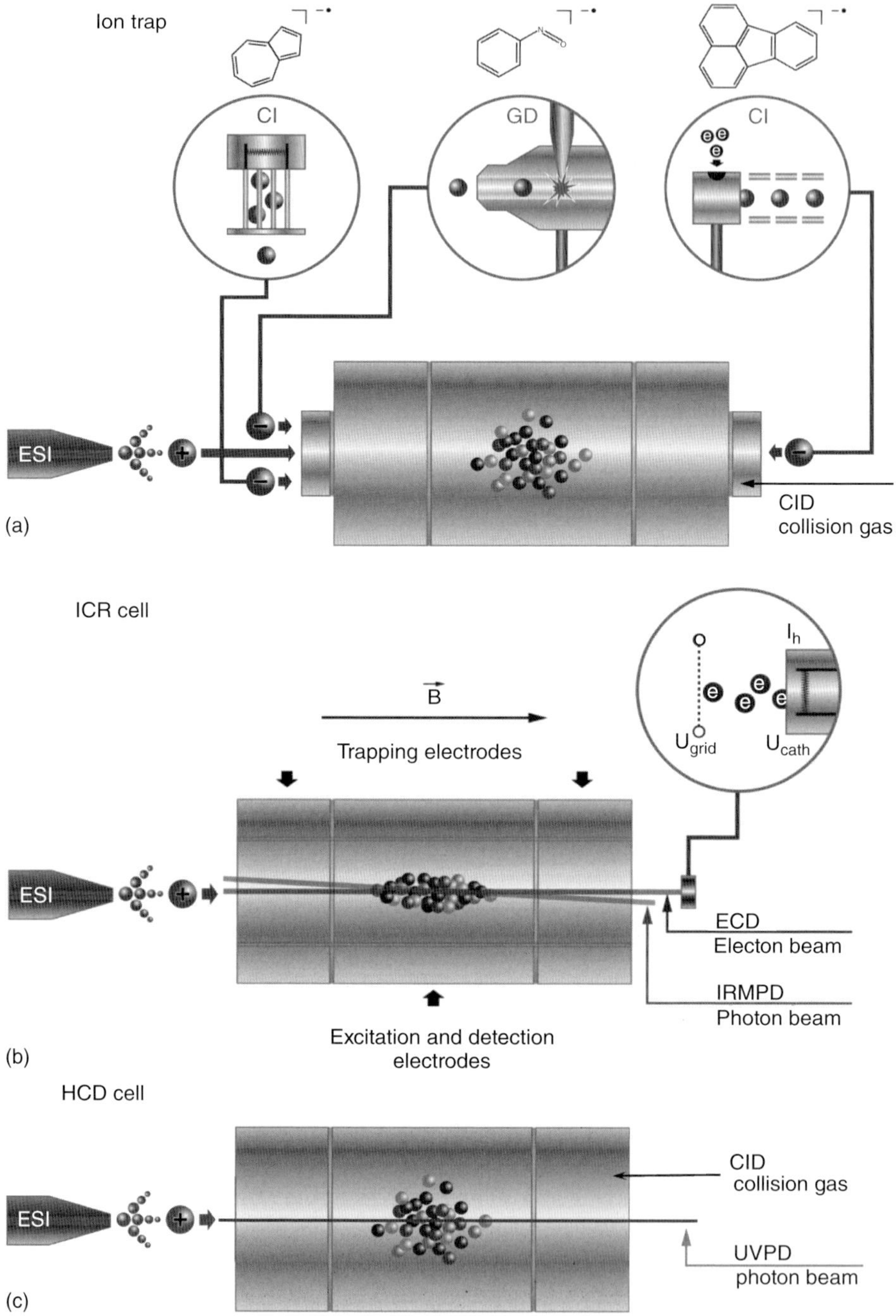

Figure 3.14 (a–c) Implementation of the main ion activation and dissociation techniques for MS/MS-based molecular structure analysis.

activation and dissociation, similarly to CID. This technique is known as infrared multiphoton dissociation (IRMPD) and can be realized in ion trap MS, for example, in the ICR cell of FT-ICR MS. Photons of the UV region are significantly more energetic than the IR photons and are absorbed by electronic subsystems of molecular functional groups. The number of chromophores capable of absorbing UV radiation increases with molecular size. Specifically, structure analysis of intact proteins with 193 nm UVPD pioneered by Brodbelt presents particular advantages for top-down proteomics [51].

3.7 Outlook: *Quo vadis* FTMS?

In FT-ICR MS, increasing magnetic field strength, to ~12–15 T [52], reduces the positive effect of dynamic harmonization provided by ParaCell (Figure 3.6) but on its own offers outstanding analytical capabilities with resolving powers routinely exceeding 1 000 000 at m/z 400. The ongoing initiative of developing 21 T FT-ICR MS should further push the boundaries of resolving power limitations. High-field and ultra-high-field Orbitrap FTMS strive to keep the company of high-resolution FT-ICR MS, but the routine resolving powers provided by these instruments fall short of 1 000 000 at m/z 400. Selected MR and FT TOFs already nowadays show performance at the level of standard Orbitrap FTMS. Resolving powers approaching 80 000 have been demonstrated with the optimized and fine-tuned standard reflectron-type TOF MS. Nevertheless, the routinely obtained resolving powers of reflectron TOF MS drop down to 20–40 000 when complex mixtures of ions are to be analyzed simultaneously. Therefore, the general trend of reaching the ultra-high levels of resolving powers is in increasing the frequency of ion oscillations in FTMS, achieved at higher electric and magnetic fields.

What application areas particularly benefit from high and ultra-high-resolution mass measurements? One of the obvious answers is the analysis of extremely complex mixtures of small molecules, for example, petroleomics, exemplified in Figure 3.10. Here, distinguishing between molecules that are as close in mass as an electron mass, ~0.5 mDa, is needed in a wide, up to 2000 m/z, mass range. The required level of performance is provided nowadays by high-field FT-ICR MS and FT-ICR MS equipped with ParaCell. In addition to FT-ICR MS platforms, high-field Orbitrap FTMS has recently entered this application field and demonstrates a high potential for becoming a routine petroleomics-grade platform. The achieved resolving powers allow elemental analysis of small molecules by providing not only extremely accurate mass measurements but also well-resolved isotopic fine structures of small molecules [53]. Increasing molecular size from small molecules to peptides puts higher demands on the resolving power (Figure 3.15) [54]. Nevertheless, it has already been shown that isotopic fine structure of peptides can be provided with a required level of resolution, mass, and abundance accuracy that may aid in peptide identification in MS-based bottom-up proteomics [22, 54].

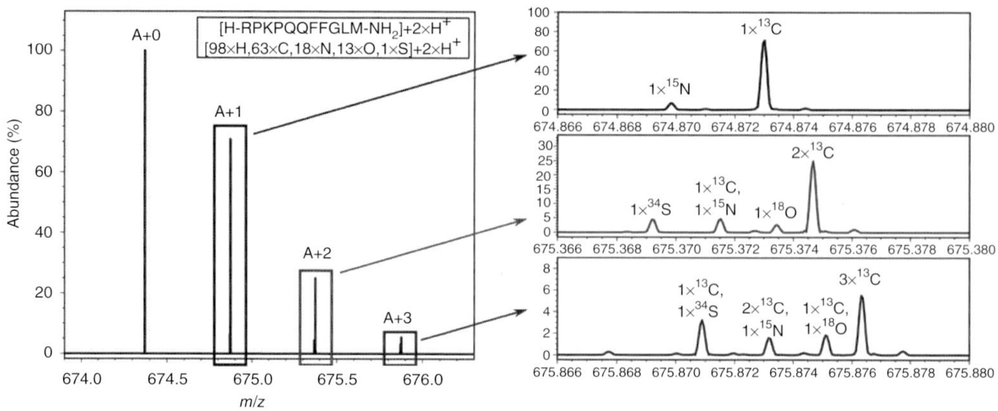

Figure 3.15 Isotopic fine structure of a doubly protonated peptide substance P (theoretical estimate).

On the other hand, reaching ultrahigh levels of resolving powers comes at an expense of the data acquisition speed, whereas increased electric and magnetic fields compensate this drawback only slightly (Figure 3.16). Why do we want to perform MS analysis faster? The answer is both the improved analysis of samples with high intra- and intermolecular complexity and the overall need for an increased throughput (number of samples analyzed per unit of time). Indeed, most modern MS-based experiments require online separation of complex mixtures of molecules before their introduction into a mass spectrometer. As a result, the mass measurement of each compound is time-constrained. For example, when metabolic systems (complex mixtures of small molecules) are analyzed, ultra-high-performance liquid chromatography (UHPLC) conditions are to be used. Each of the compounds under these conditions should be analyzed in ~2 s. To provide sufficiently accurate qualitative (MS/MS level) and quantitative (MS level) measurements, the spectral rate of minimum ~10–20 Hz (that translates into 10–20 mass spectra a second) is needed. Under conventional MS-based bottom-up proteomics conditions, each peptide elutes for ~10–30 s from the typically employed reversed-phase high-performance liquid chromatography (HPLC) system, but the time available for the analysis is reduced due to the coelution of other peptides. Furthermore, high-resolution MS and efficient MS/MS require substantial data acquisition times. Increasing molecular size to intact protein level, ~10–50 kDa, poses additional constraints on the spectral rate required for MS and MS/MS measurements. Therefore, there is an imperative need for accelerating high-resolution MS. The envisioned goal for the spectral rate of high, >100 000, resolution data is ~ 50–100 Hz. Importantly, this performance should not be achieved at an expense of the spectral dynamic range.

Accelerated high-resolution MS is of a particular importance for MS-based proteomics at all its levels: bottom-up, middle-down, and top-down (Figure 3.17). Nowadays, the absolute majority of MS-based applications for protein identification

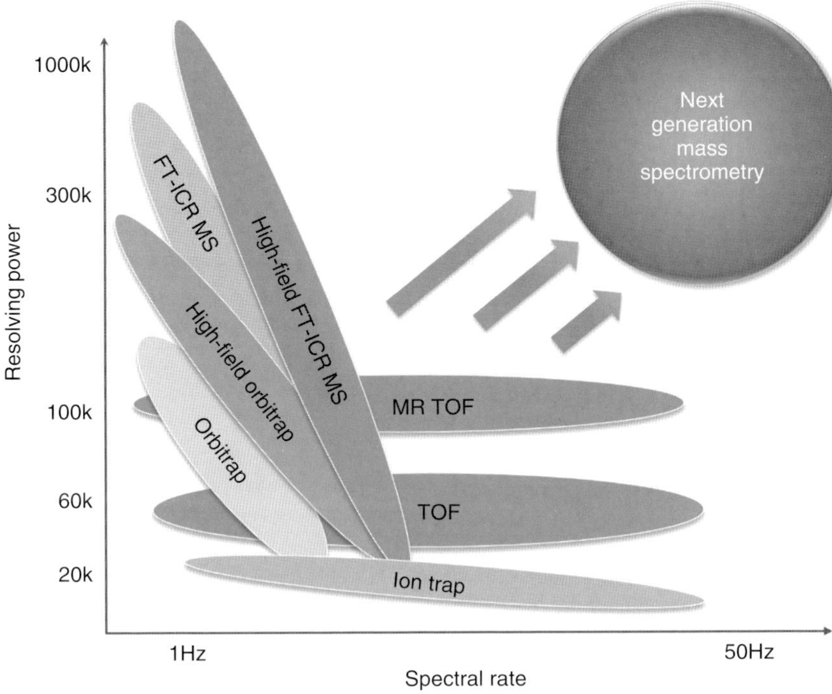

Figure 3.16 Compromise between data acquisition speed (spectral rate) and (approximate) resolving power (estimated at 400 *m/z* and with magnitude mode FT spectral representation) in modern mass spectrometry. Figure adapted based on the original idea of Arnd Ingendoh (Bruker Daltonics).

and quantification in proteomics are based on the bottom-up approach. That implies that proteins are in-solution, digested into small, ~0.6–3 kDa, peptides before their separation on (U)HPLC and further MS analysis. Theoretically, the method is very powerful and should provide almost complete information on the proteins constituting even the most complex biological systems. However, there are a number of drawbacks that limit this approach. One of them is that the information on intact proteins is missing. Therefore, the link between the identified peptide and its exact origin (an intact protein or a proteoform) [55] gets lost. The information provided is limited to the relation of a given peptide and a corresponding protein family (which may contain a number of proteoforms). Another drawback is the extremely high number of peptides that has to be analyzed to identify the maximum number of proteins present in a high dynamic range in the original sample. The size of the peptide pool exceeds the capabilities of any LC-MS system by an order of magnitude.

Reduction of a peptide pool complexity appears as a possible way of improvement. The corresponding approach is termed the middle-down proteomics [56–58]. Following this approach, proteins are digested into a pool of larger, ~3–15 kDa peptides, which significantly brings down the total number of peptides. The

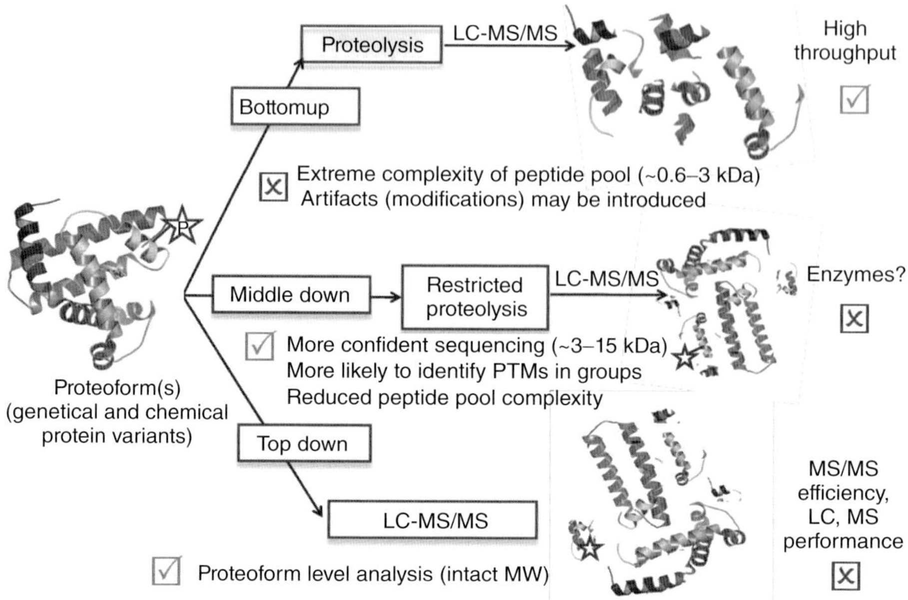

Figure 3.17 MS-based approaches for proteomics: bottom-up, middle-down, and top-down.

drawbacks here are the slightly reduced (down to 90–95%) potential coverage of a proteome and requirements on higher performance MS and MS/MS. The latter challenge can now be efficiently addressed by the modern LC and MS methods and techniques, as described in this chapter. Therefore, the rational application of high-resolution FTMS to proteomics can be envisioned in specifically targeting the middle-down proteomics approach.

Working at the level of intact proteins (proteoforms) provides fundamental benefits for protein structure analysis [43]. Most importantly, it allows for unambiguous characterization of protein modifications, including stoichiometry of such important posttranslational modifications as phosphorylation and glycosylation [59]. The corresponding MS-based approach is termed top-down mass spectrometry (applied to an isolated compound) or top-down proteomics (applied to a mixture of compounds). However, the technical challenges of top-down MS and proteomics limit its current application areas and capabilities.

Accelerating high-resolution FTMS is particularly important for advancing qualitative protein analysis by middle-down and top-down proteomics. In what concerns bottom-up proteomics, rapid high-resolution mass spectrometry would be beneficial specifically for peptide and protein quantification. For example, in some modern quantification methods, it is required to distinguish between isotopically labeled multiply charged peptides or isotopically labeled singly charged reporter small molecules that differ in mass by ~6–36 mDa.

Advanced signal processing of transients in FTMS appears as one of the possible routes for accelerating high-resolution MS. The intrinsic benefit of FTMS is the availability of a time-domain signal that may be approached with powerful signal processing methods being developed not only for MS but also for other areas of science and technology, for example, NMR spectroscopy and quantum physics. One of the obvious first steps is the representation of FT-derived mass spectra not in the conventional magnitude mode but in the absorption mode [60, 61]. The additional information required for this procedure is the phase function of a given mass spectrometer. When dealing with the electrostatic ion trap-based FTMS, including Orbitrap and TOF, the phase function can be received in a straightforward manner [8, 62, 63]. In these instruments, ions are injected as short packages into the ion trap and thus at a given moment in time all ions are in a phase coherence point (when phases of all ions are close to each other). Owing to the simplicity of the underlying mathematics in this case, the absorption mode-like mass spectra representation is routinely available for commercial Orbitrap FTMS. Owing to a different principle of ion excitation in FT-ICR MS, the phase function has a more complex nature. Nevertheless, the recent progress has made it possible to phase almost any FT-ICR mass spectrum in an automated way [60, 64]. Further improvement of method accuracy is expected in this area.

Non-FT time-domain signal processing is another option for accelerating FTMS. Indeed, owing to the fundamental principle of resolution limitation imposed by FT signal processing, it happens that the ultimate resolution level (determined by the physical limitations of ion cloud dynamics in a phase space) is not reached in FTMS even with absorption mode spectral representation. As it is known from 1D NMR spectroscopy, in addition to the FT signal processing (which is a spectral estimator), there are methods of the so-called super-resolution signal processing (for example, parameter estimator-based methods). Among them, filter diagonalization method (FDM) is a parameter estimator method that has been implemented in FTMS and demonstrates intriguing capabilities (Figure 3.18) [65].

The ability to provide high-resolution MS data from a significantly shorter transient, as indicated by Figure 3.18, addresses the challenge of developing rapid high-resolution MS. However, possible fundamental limitations of parameter estimator-based MS can significantly reduce its application area and thus require further investigation. One of the possible limitations of the method is peak regularity in a mass spectrum. The FDM signal processing provides most advantages over FT when the average peak density (number of peaks per a given frequency window) is significantly (by an order of magnitude) lower than the maximum peak density (minimum distance between any two peaks in the same frequency window). Therefore, the resolution advantage of FDM versus FT drops significantly when isotopic distribution of a protein regularly occupies the whole frequency window selected for signal processing (Figure 3.19).

Figure 3.11 shows a mass spectrum of overlapping isotopic distributions of a monomer and a dimer of a protein SOD obtained with Orbitrap FTMS when sufficiently long, ~768 ms, transient is processed with absorption mode-like FT (termed eFT algorithm by Thermo Scientific) [63]. When transient length is decreased to 530 ms and magnitude mode FT is employed for signal processing, the isotopic

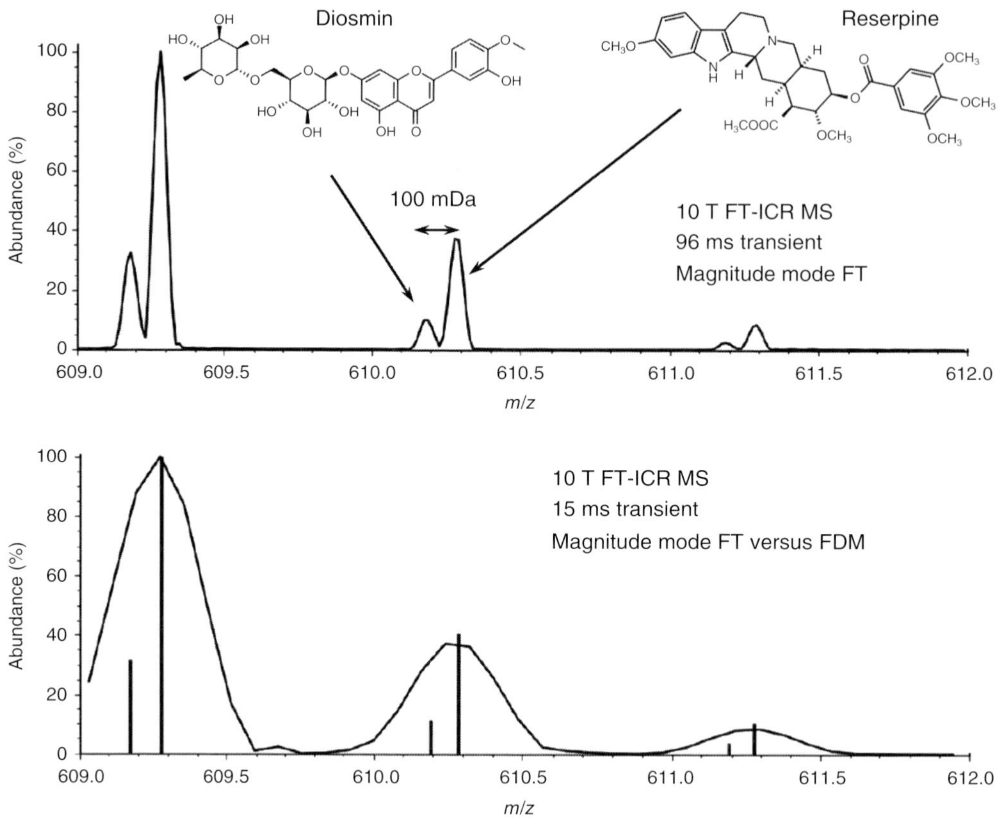

Figure 3.18 Comparison of FT and FDM-derived mass spectra of two isobaric compounds, diosmin and reserpine, obtained with 10 T FT-ICR MS from (a) 96 ms transient and (b) 15 ms transient. FDM solutions are shown as sticks.

distributions of a monomer and a dimer are not resolved (Figure 3.19). Application of FDM signal processing to the same 530 ms transient yields isotopically resolved mass spectra of both monomer and dimer, showing its superiority to the magnitude mode FT. However, the regular spacing of peaks in the *m/z* window in Figure 3.19 reduces the resolution level of FDM almost to the level of the absorption mode FT.

3.8 Summary and Future Issues

In this chapter, the following key points have been described:

1) FTMS in general and FT-ICR MS in particular continues to lead the mass spectrometry chart in terms of resolving power and mass accuracy for more then

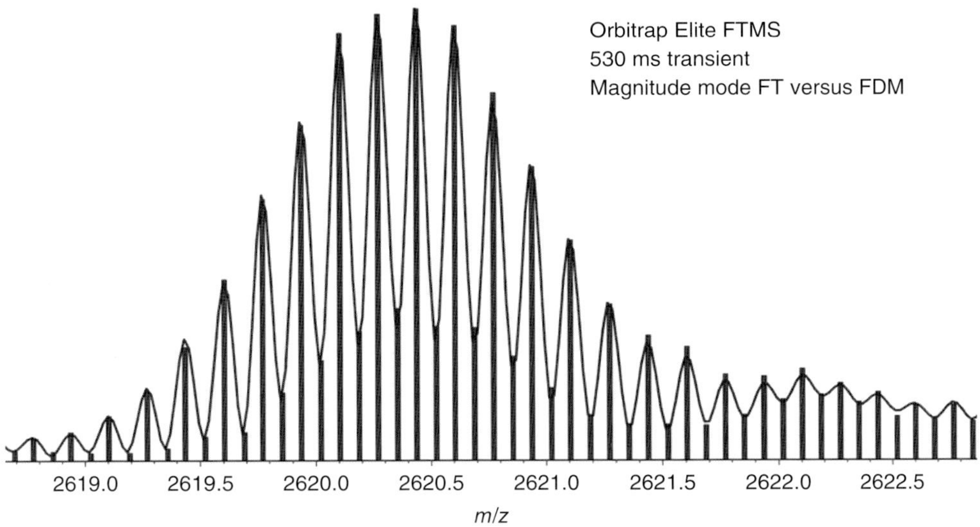

Figure 3.19 Comparison of FT (spectral representation) and FDM (parameter representation, stick plot) signal processing applied to Orbitrap FTMS transients of a protein superoxide dismutase. See Figure 3.11 for high-resolution Orbitrap FTMS of this protein.

40 years (since 1973). Orbitrap FTMS has joined the family of FTMS instruments in 2005 and substantially reinforced the positions of high-resolution mass spectrometry. Other electrostatic ion trap-based FTMS instruments, such as TOF FTMS, have already demonstrated their powerful analytical capabilities and may enter the market in the near future, but so far remain research-grade-only instruments.

2) Harmonized ICR cells have pushed the limits of resolving power in FT-ICR MS, specifically when low-to-medium strength magnetic fields are employed.
3) High and ultra-high electric field Orbitrap mass analyzers have been introduced to provide increased resolving powers. These advanced mass analyzers can now be employed to analyze extremely complex mixtures, for example, as in petroleomics and top-down proteomics.
4) The compromise between resolving power and spectral rate is the current FTMS challenge to be addressed. Both hardware (ICR cells with harmonics detection) and software (advanced signal processing) solutions have been proposed.
5) Time-domain, or transient, ion signals in FTMS provide possibilities for further improvements in FTMS performance via advanced signal processing. Absorption mode FT spectral representation has already replaced the magnitude mode FT-derived mass spectra on many FTMS platforms, including the commercial ones. It provides up to twofold increase in mass resolving power.
6) Further progress in FTMS development is expected from the non-FT signal processing methods, for example, FDM, which may be used to overcome the FT resolution limitation. These methods require substantial increase in

the computational power for data processing, which, albeit, should be readily accessible in the near future.

7) High-performance data acquisition systems with increased sampling rate, better synchronization, and faster data transfer enter the field of FTMS for fine-tuning of ion detection and improved data processing.

Acknowledgments

This work is the result of a joint effort of Biomolecular mass spectrometry laboratory at EPFL. Particularly, the input from Anton Kozhinov, Ünige Laskay, Luca Fornelli, and Konstantin Zhurov to the text and figures was crucial. Other group members, especially Konstantin Nagornov, Kristina Srzentić, and Daniel Ayoub, are gratefully acknowledged for the discussions. We thank Thermo Scientific and Bruker Daltonics for the technical support of our research program. The work was supported by the Swiss National Science Foundation (Projects 200021-125147/1 and 128357), the European Research Council (ERC Starting Grant 280271), and EPFL.

References

1. Zubarev, R.A. (2013) *Proteomics*, **13**, 723–726.
2. Michalski, A., Cox, J., and Mann, M. (2011) *J. Proteome Res.*, **10**, 1785–1793.
3. Sabido, E., Selevsek, N., and Aebersold, R. (2012) *Curr. Opin. Biotechnol.*, **23**, 591–597.
4. Marshall, A.G. and Hendrickson, C.L. (2008) *Annu. Rev. Anal. Chem.*, **1**, 579–599.
5. Xian, F., Hendrickson, C.L., and Marshall, A.G. (2012) *Anal. Chem.*, **84**, 708–719.
6. Marshall, A.G., Hendrickson, C.L., and Shi, S.D.H. (2002) *Anal. Chem.*, **74**, 252 A–259 A.
7. Marshall, A.G., Hendrickson, C.L., and Jackson, G.S. (1998) *Mass Spectrom. Rev.*, **17**, 1–35.
8. Scigelova, M., Hornshaw, M., Giannakopulos, A., and Makarov, A. (2011) *Mol. Cell. Proteomics*, 10. doi: 10.1074/mcp.M111.009431
9. Zubarev, R.A. and Makarov, A. (2013) *Anal. Chem.*, **85**, 5288–5296.
10. Makarov, A. (2000) *Anal. Chem.*, **72**, 1156–1162.
11. Zajfman, D., Rudich, Y., Sagi, I., Strasser, D., Savin, D.W., Goldberg, S., Rappaport, M., and Heber, O. (2003) *Int. J. Mass Spectrom.*, **229**, 55–60.
12. Gonin, M. (2003) Electrostatic ion trap mass spectrometers. US Patent 10/449328.
13. Hilger, R.T., Santini, R.E., and McLuckey, S.A. (2013) *Anal. Chem.*, **85**, 5226–5232.
14. Doussineau, T., Bao, C.Y., Clavier, C., Dagany, X., Kerleroux, M., Antoine, R., and Dugourd, P. (2011) *Rev. Sci. Instrum.*, **82**, 084104.
15. Contino, N.C. and Jarrold, M.F. (2013) *Int. J. Mass Spectrom.*, **345–347**, 153–159.
16. Verenchikov, A. (2010) Electrostatic trap mass spectrometer. Patent GB20100000649.
17. Satoh, T., Sato, T., and Tamura, J. (2007) *J. Am. Soc. Mass Spectrom.*, **18**, 1318–1323.
18. Marshall, A.G. (2000) *Int. J. Mass Spectrom.*, **200**, 331–356.
19. Tolmachev, A.V., Robinson, E.W., Wu, S., Kang, H., Lourette, N.M., Pasa-Tolic, L.,

and Smith, R.D. (2008) *J. Am. Soc. Mass Spectrom.*, **19**, 586–597.
20. Kaiser, N.K., Savory, J.J., McKenna, A.M., Quinn, J.P., Hendrickson, C.L., and Marshall, A.G. (2011) *Anal. Chem.*, **83**, 6907–6910.
21. Boldin, I.A. and Nikolaev, E.N. (2011) *Rapid Commun. Mass Spectrom.*, **25**, 122–126.
22. Nikolaev, E.N., Jertz, R., Grigoryev, A., and Baykut, G. (2012) *Anal. Chem.*, **84**, 2275–2283.
23. Kostyukevich, Y.I., Vladimirov, G.N., and Nikolaev, E.N. (2012) *J. Am. Soc. Mass. Spectrom.*, **23**, 2198–2207.
24. Denisov, E., Damoc, E., Lange, O., and Makarov, A. (2012) *Int. J. Mass Spectrom.*, **325–327**, 80–85.
25. Michalski, A., Damoc, E., Lange, O., Denisov, E., Nolting, D., Mueller, M., Viner, R., Schwartz, J., Remes, P., Belford, M., Dunyach, J.-J., Cox, J., Horning, S., Mann, M., and Makarov, A. (2012) *Mol. Cell. Proteomics*, **11**, 013698/1–013698/11.
26. Michalski, A., Damoc, E., Hauschild, J.P., Lange, O., Wieghaus, A., Makarov, A., Nagaraj, N., Cox, J., Mann, M., and Horning, S. (2011) *Mol. Cell. Proteomics*, 10. doi: 10.1074/mcp.M111.011015
27. Olsen, J.V., Macek, B., Lange, O., Makarov, A., Horning, S., and Mann, M. (2007) *Nat. Methods*, **4**, 709–712.
28. Marshall, A.G. and Rodgers, R.P. (2008) *Proc. Natl. Acad. Sci. U.S.A.*, **105**, 18090–18095.
29. Hsu, C.-S., Hendrickson, C.L., Rodgers, R.P., McKenna, A.M., and Marshall, A.G. (2011) *J. Mass Spectrom.*, **46**, 337–343.
30. Zhurov, K.O., Kozhinov, A.N., and Tsybin, Y.O. (2013) *Energy Fuel*, **27**, 2974–2983.
31. Rodgers, R.P. and McKenna, A.M. (2011) *Anal. Chem.*, **83**, 4665–4687.
32. Kekäläinen, T., Pakarinen, J.M.H., Wickström, K., Lobodin, V.V., McKenna, A.M., and Jänis, J. (2013) *Energy Fuel*, **27**, 2002–2009.
33. Kozhinov, A.N., Zhurov, K.O., and Tsybin, Y.O. (2013) *Anal. Chem.*, **85**, 6437–6445.
34. Smith, E.A., Brown, R.C., and Lee, Y.J. (2009) Abstracts, 44th Midwest Regional Meeting of the American Chemical Society, Iowa City, IA, October 21–24, 2009, p. MWRM-277.
35. Heck, A.J.R. (2008) *Nat. Methods*, **5**, 927–933.
36. Rosati, S., Thompson, N.J., and Heck, A.J.R. (2013) *TrAC, Trends Anal. Chem.*, **48**, 72–80.
37. Snijder, J., Rose, R.J., Veesler, D., Johnson, J.E., and Heck, A.J.R. (2013) *Angew. Chem. Int. Ed.*, **52**, 4020–4023.
38. Yin, S. and Loo, J.A. (2011) *Int. J. Mass Spectrom.*, **300**, 118–122.
39. Tsybin, Y.O., Fornelli, L., Stoermer, C., Luebeck, M., Parra, J., Nallet, S., Wurm, F.M., and Hartmer, R. (2011) *Anal. Chem.*, **83**, 8919–8927.
40. Fornelli, L., Damoc, E., Thomas, P.M., Kelleher, N.L., Aizikov, K., Denisov, E., Makarov, A., and Tsybin, Y.O. (2012) *Mol. Cell. Proteomics: MCP*, **11**, 1758–1767.
41. Valeja, S.G., Kaiser, N.K., Xian, F., Hendrickson, C.L., Rouse, J.C., and Marshall, A.G. (2011) *Anal. Chem.*, **83**, 8391–8395.
42. Shaw, J.B. and Brodbelt, J.S. (2013) *Anal. Chem.*, **85**, 8313–8318.
43. Tipton, J.D., Tran, J.C., Catherman, A.D., Ahlf, D.R., Durbin, K.R., and Kelleher, N.L. (2011) *J. Biol. Chem.*, **286**, 25451–25458.
44. Zhurov, K.O., Fornelli, L., Wodrich, M.D., Laskay, U.A., and Tsybin, Y.O. (2013) *Chem. Soc. Rev.*, **42**, 5014–5030.
45. McLuckey, S. and Mentinova, M. (2011) *J. Am. Soc. Mass. Spectrom.*, **22**, 3–12.
46. Cooper, H.J., Hakansson, K., and Marshall, A.G. (2005) *Mass Spectrom. Rev.*, **24**, 201–222.
47. Zubarev, R.A., Kelleher, N.L., and McLafferty, F.W. (1998) *J. Am. Chem. Soc.*, **120**, 3265–3266.
48. Good, D.M., Wirtala, M., McAlister, G.C., and Coon, J.J. (2007) *Mol. Cell. Proteomics*, **6**, 1942–1951.
49. Tsybin, Y.O., Quinn, J.P., Tsybin, O.Y., Hendrickson, C.L., and Marshall, A.G. (2008) *J. Am. Soc. Mass. Spectrom.*, **19**, 762–771.
50. Tsybin, Y.O., Witt, M., Baykut, G., Kjeldsen, F., and Hakansson, P. (2003) *Rapid Commun. Mass Spectrom.*, **17**, 1759–1768.

51. Shaw, J.B., Li, W., Holden, D.D., Zhang, Y., Griep-Raming, J., Fellers, R.T., Early, B.P., Thomas, P.M., Kelleher, N.L., and Brodbelt, J.S. (2013) *J. Am. Chem. Soc.*, **135**, 12646–12651.
52. Schaub, T.M., Hendrickson, C.L., Horning, S., Quinn, J.P., Senko, M.W., and Marshall, A.G. (2008) *Anal. Chem.*, **80**, 3985–3990.
53. Kim, S., Rodgers, R.P., and Marshall, A.G. (2006) *Int. J. Mass Spectrom.*, **251**, 260–265.
54. Miladinovic, S.M., Kozhinov, A.N., Gorshkov, M.V., and Tsybin, Y.O. (2012) *Anal. Chem.*, **84**, 4042–4051.
55. Smith, L.M. and Kelleher, N.L., Consortium Top Down Proteomics (2013) *Nat. Methods*, **10**, 186–187.
56. Cannon, J., Lohnes, K., Wynne, C., Wang, Y., Edwards, N., and Fenselau, C. (2010) *J. Proteome Res.*, **9**, 3886–3890.
57. Wu, C., Tran, J.C., Zamdborg, L., Durbin, K.R., Li, M.X., Ahlf, D.R., Early, B.P., Thomas, P.M., Sweedler, J.V., and Kelleher, N.L. (2012) *Nat. Methods*, **9**, 822–824.
58. Laskay, U.A., Srzentic, K., Fornelli, L., Upir, O., Kozhinov, A.N., Monod, M., and Tsybin, Y.O. (2013) *CHIMIA Int. J. Chem.*, **67**, 244–249.
59. Ge, Y., Rybakova, I.N., Xu, Q., and Moss, R.L. (2009) *Proc. Natl. Acad. Sci. U.S.A.*, **106**, 12658–12663.
60. Kilgour, D.P.A., Wills, R., Qi, Y.L., and O'Connor, P.B. (2013) *Anal. Chem.*, **85**, 3903–3911.
61. Qi, Y.L., Barrow, M.P., Li, H.L., Meier, J.E., Van Orden, S.L., Thompson, C.J., and O'Connor, P.B. (2012) *Anal. Chem.*, **84**, 2923–2929.
62. Hilger, R.T., Wyss, P.J., Santini, R.E., and McLuckey, S.A. (2013) *Anal. Chem.*, **85**, 8075–8079.
63. Lange, O. (2011) Methods and apparatus for producing a mass spectrum. US Patent 20110240841.
64. Xian, F., Hendrickson, C.L., Blakney, G.T., Beu, S.C., and Marshall, A.G. (2010) *Anal. Chem.*, **82**, 8807–8812.
65. Kozhinov, A.N. and Tsybin, Y.O. (2012) *Anal. Chem.*, **84**, 2850–2856.

4
Coherent Electronic Energy Transfer in Biological and Artificial Multichromophoric Systems

Elisabetta Collini

4.1
Introduction to Electronic Energy Transfer in Complex Systems

Electronic energy transfer (EET) or resonance energy transfer (RET) is a photophysical process whereby electronic excitation, resulting from photoexcitation of a molecule or of a distinct light-absorbing unit (chromophore), is transferred to nearby chromophores in a supramolecular structure [1].

A variety of systems have been studied in which it has been found that electronic excitation absorbed by one chromophore subsequently moves through space to another chromophore. At this point, fluorescence may be radiated or a photochemical reaction could be initiated. In multichromophoric systems excitation energy can be transferred over reasonably large distances – limited by the excited state lifetime – by a series of EET hops, which are collectively called energy migration. Examples of such multichromophoric systems include concentrated dye solutions [2], conjugated polymers [3], supramolecular assemblies [4], fluorescence-based sensors [5], as well as natural biological systems such as DNA [6] and components of the photosynthetic apparatus [7]. Figure 4.1 reports two examples (an artificial system and a biological complex) in which EET process plays a crucial role.

The mechanism of EET in such systems is, in general, well established and well reproduced by the traditional approach that makes use of Förster resonance energy transfer theory [2]. However, in recent years, many subtle aspects of the process, not accounted for in the traditional treatments, were found to be relevant – quantum coherence being one of them. Such subtleties have forced researchers to change the conventional way of thinking about EET and led to the introduction of substantial changes to this basic model [8].

Förster theory states that, when two molecules are separated by a distance large compared to their size, their electronic coupling can be represented approximately as a dipole–dipole interaction between electric dipole transition moments (this confers the typical $1/R^6$ distance dependence). Secondly, because the energy conservation must be ensured, transfer efficiency depends on the overlap of the emission spectrum of the donor and the absorption spectrum of the acceptor. Förster theory holds when the electronic coupling is weak (weak coupling limit),

Discovering the Future of Molecular Sciences, First Edition. Edited by Bruno Pignataro.

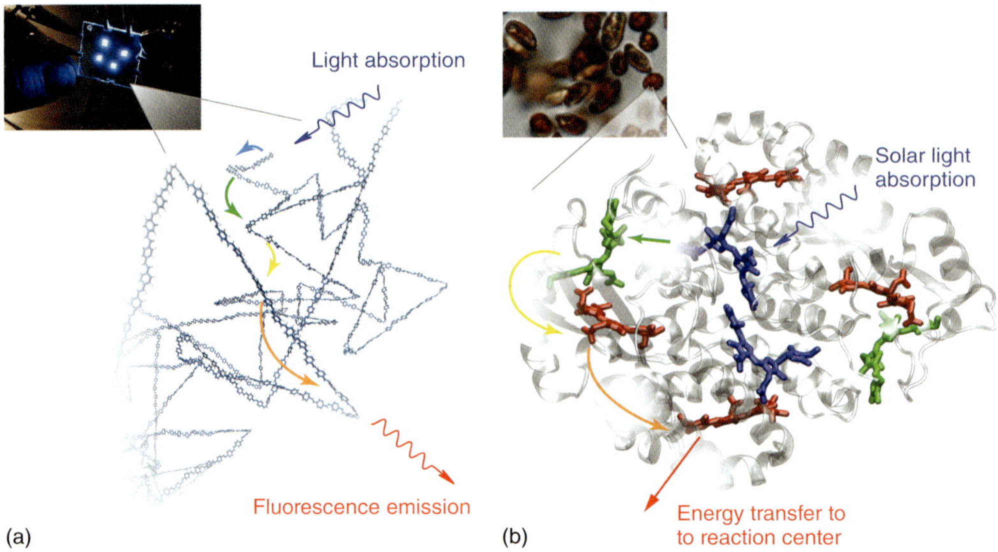

Figure 4.1 Examples of energy migration. (a) Conjugated polymers are characterized by disorder in the form of small-angle rotations that disrupt the π-conjugation, breaking the polymer into a series of electronically coupled chromophore units (conformational subunits). After photoexcitation, the energy migrates on a timescale of picoseconds to the longest and most planar conformational subunits, from where fluorescence is emitted. That process is indicated by the arrows. (b) In photosynthesis light is used to power reaction center proteins wherein electron transfer processes are used to harness the potential of light. Antenna proteins are employed to do most of the light absorption and funnel that excitation energy to proximate reactions centers.

meaning that it is small compared to the coupling to the environment. In this condition, the system explores the nuclear degrees of freedom more effectively than donor–acceptor coupling and, therefore, EET happens after complete vibrational equilibration at the photoexcited state of the donor. In this way, any quantum effect between the donor and acceptor excited states is completely quenched by the interaction with the vibrational degrees of freedom. The Förster model yields simple exponential dynamics and provides an excellent qualitative fit of the observed population transfer rates for most systems [9].

On the other hand, when the electronic coupling becomes stronger compared to the interaction with the environment, the so-called strong coupling limit is reached and EET acquires quantum coherent character. In this regime, the electronic coupling not only acts to transfer energy vectorially from donor to acceptor but also promotes a resonance of the excitation. The interference of the coherent forward and backward resonating excitation waves tends to build delocalized electronic eigenstates of the donor–acceptor pair. In that case, quantum coherence plays a part in the dynamical evolution of the electronic excitation after photoexcitation. This is the so-called wavelike transfer [10]. The differences in the two mechanisms are pictorially described in Figure 4.2.

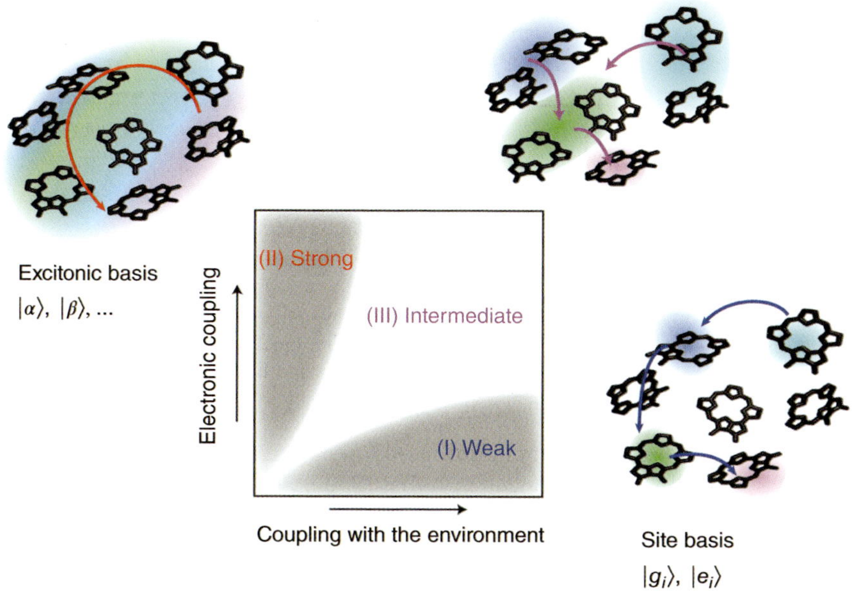

Figure 4.2 Schematic illustration of different EET regimes. The strength of electronic coupling increases along the vertical axis and that of interactions with the environment along the horizontal axis. Förster theory applies in region I, whereas the wavelike transport formalism can be applied to region II. In the intermediate region III, delocalized exciton formation and exciton–vibrational coupling have to be dealt with on an equal footing. The shaded colored areas symbolize the exciton extending over several monomers.

More puzzling is the intermediate coupling regime, when the electronic coupling and the interaction with the environment have the same magnitude and timescale. In this regime, the electronic coupling is large enough so that the excitation can be coherently shared among subsets of chromophores. These excitons provide a means of producing "new" chromophores from the existing molecular building blocks. These new chromophores take over the roles of donors and acceptors for energy transfer and can have a profound influence on energy transfer rates compared with a similar system where energy flows through localized states; rather than the excitation hopping incoherently from site to site, the system can coherently control the dynamics of energy migration so that the excitation travels in a wavelike manner, through the interference of multiple pathways.

This intermediate EET regime is of particular interest because it naturally interpolates between the classical (weak coupling, incoherent hopping mechanism) and the quantum (strong coupling, wavelike mechanism) limits. Moreover, the majority of the photosynthetic light-harvesting complexes (LHCs) as well as many artificial multichromophoric systems in conventional conditions belong to this regime. The intermediate coupling regime is the most challenging from the theoretical point of view as the approximations usually adopted to describe EET in the two limit cases do not hold in this regime. So far, several theoretical studies have tried to find a unifying formalism capable of interpolating between the weak and the strong coupling limits [8, 11–13].

4.2 The Meaning of Electronic Coherence in Energy Transfer

A convenient – yet not formal – way to introduce quantum coherence in EET is through the concept of interference. The concept of interference of pathways is peculiar to the quantum mechanical description of the energy migration. Suppose we want to determine the probability for excitation starting at D and ending up at A, D and A being our donor and acceptor, respectively. Let us indicate with $|d\rangle$ ($|a\rangle$) the state in which D (A) is in the excited state and A (D) in the ground state. Let us also introduce an operator $\widehat{k}$ corresponding to an observable. Suppose we have applied to the system a particular experiment sensitive to the observable $\widehat{k}$, which gives result k_A when excitation is on A and k_D when excitation is on D. We then have $\langle a|\widehat{k}|a\rangle = k_A$ and $\langle d|\widehat{k}|d\rangle = k_D$. According to classical mechanics, the average result of a large number of measurements on the system would be the weighted sum $|c_A|^2 k_A + |c_D|^2 k_D$, $|c_i|^2$ being the probability of finding the system in i. However, the individual probabilities do not add in this way for a system in a quantum mechanical superposition state. Instead, the expectation value is $\langle c_A a + c_D d|\widehat{k}|c_A a + c_D d\rangle$ where in addition to $|c_A|^2 k_A + |c_D|^2 k_D$ there is also another term $c_A^* c_D \langle a|\widehat{k}|d\rangle + c_D^* c_A \langle d|\widehat{k}|a\rangle$, which accounts for the interference between the so-called forward (D to A) and backward (A to D) propagation. The important result of the quantum mechanical nature of this transport process is that each pathway carries a phase that dictates how competing pathways interfere.

To understand this concept, an analogy can be formulated based on the double slit experiment (Figure 4.3). When a coherent light source such as a laser beam illuminates a thin plate pierced by two parallel slits, the wave nature of light causes the light waves passing through the two slits to interfere, producing bright and dark bands on a screen. This interference process is similar to that which forms the eigenstates representing shared excitation states between two molecules. However, if the frequency of the waves emerging from each slit fluctuates randomly and independently, analogously to the way that system–bath interactions cause the transition energies of the two molecules to fluctuate, the interference pattern will be less clear and it will completely disappear if the fluctuations are too severe. This is the so-called decoherence process: immediately after photoexcitation, the coherence starts dephasing because of small differences within the ensemble, and interaction of the system with its surroundings causes random fluctuations, which lead to the loss of memory of the initial electronic transition frequency distribution. The typical time associated with decoherence processes (dephasing time) is particularly critical to determine how relevant is coherence in the dynamics of the overall process: in order to have a significant influence, the quantum coherent state must persist on the same timescale as population terms [14].

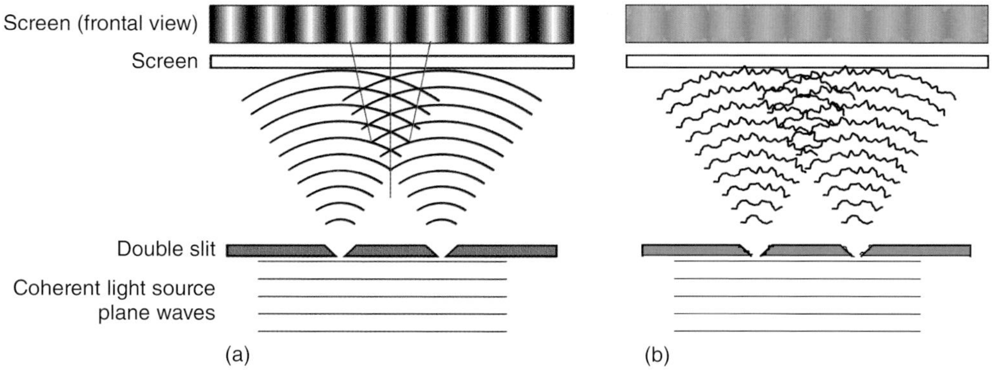

Figure 4.3 Effect of random fluctuations on the outcome of a double slit experiment. (a) In the absence of any perturbation, when a coherent light source such as a laser beam illuminates a thin plate pierced by two parallel slits, the light passing through the slits interferes, producing bright and dark bands on a screen. As a guide to the eye, red lines are traced in correspondence to the maxima (constructive interference). This interference process is similar to that which forms the eigenstates representing shared excitation states between two molecules. (b) In the presence of perturbations causing random and independent fluctuations of the frequency of the waves emerging from each slit, the interference pattern is less clear and it disappears if the fluctuations are too severe. This "decoherence" process is analogous to the fluctuations induced by the system-bath interactions in the transition energies of the two molecules.

The competition between electronic interactions, which "builds" superpositions, and decoherence, which destroys them, determines the EET dynamics. In summary, the degree of "quantumness" of the transport dynamics depends on a delicate balance between electronic properties of the system and its interactions with the environment.

For many years, scientists, although intrigued by the idea of quantum-driven processes, thought that the typical dephasing times associated with our "hot and wet" [15] world, were too fast to make electronic coherences survive enough to be relevant to the process, and their presence was treated mostly as some kind of academic curiosity. Until a few years ago, new technological advances in spectroscopy allowed recording direct experimental proof of long-lived quantum coherent dynamics in natural photosynthetic complexes [16–19]. These evidences rejuvenated the long-standing discussion regarding relationships between energy transfer efficiency and quantum features. In fact, several works have discussed the possibility that quantum coherent dynamics could boost the efficiency of biological light-harvesting systems and effectively funnel the energy transport. Within this framework, it was proposed that coherence could improve robustness against energetic disorder or dynamical perturbations [20, 21], be used as strategy to overcome energy traps [22] or as a mechanism that may speed up the search for the lowest energy level [16].

Even more bewildering is the possibility that the practical functionality of quantum coherent phenomena may be connected with the ability of natural systems to modulate their function under different environmental conditions. This point was recently raised by the observation that the experimentally detected coherences are not only long-lived in time, but are also long range in space as they seem to extend over several molecules across the whole antenna complex, including distant and weakly interacting pigments [23]. This would imply the possibility of an extremely sophisticated and fast communication between different parts of the system, capable of guaranteeing an effective regulation mechanism [24].

The effective role of electronic coherence in the biological light-harvesting processes is, however, still under debate [25]. Different opinions can be found and open questions are still the object of many discussions and investigations in the literature. To follow these discussions, it is important to stress the different meaning that the expression "electronic quantum coherence" may assume in the context of EET processes in multichromophoric biological and artificial systems [26].

At the simplest level, quantum coherence can refer to quantum superpositions of localized molecular excitations that occur naturally because electronic couplings between molecular excitations lead to delocalized eigenstates (excitons) among a subset of pigments, as illustrated in Figure 4.2 (type I). At a second more sophisticated level, coherence in light-harvesting and more generally in EET alludes to coherent dynamics of energy transfer, reflecting a superposition of excitonic states, created by the interaction with a laser pulse with suitable pulse width (type II). It is important to stress that the presence of type I coherence alone does not necessarily imply quantum transport, because, according to the generalized Förster theory [27], the excitation energy can migrate incoherently even between subsets of excitonically coupled chromophores.

4.3
Energy Migration in Terms of Occupation Probability: a Unified Approach

From the theoretical point of view, many attempts have been made to find an approach that can describe the EET in the intermediate coupling regime, extrapolating between the limiting cases of weak and strong electronic coupling, and elucidate the implications of quantum coherence. Most of them were recently reviewed in Ref. [8], and here only the highlights are outlined briefly.

A very useful starting point is the theory by Haken and Strobl, which studied EET dynamics in the intermediate coupling regime using a qualitative model easily solvable [28]. The main assumption of the model is that the effect of the environment on the electronic system can be described as white noise. The advantage of this approach is that it can be solved exactly to give helpful qualitative insights and it can be modified to explore, for example, local versus nonlocal bath fluctuations. Examples of modified Haken–Strobl models can be found in Refs [8, 29, 30].

For example, Bardeen and coworkers employed the standard Haken–Strobl model to address the question of whether quantum coherence may increase the efficiency of energy migration to a final trap and what the optimal conditions are. Interestingly, they found that the most efficient trapping does not require the longest coherence time and, furthermore, quantum coherence can increase or decrease the trapping efficiency depending on interference effects [30]. Recent theoretical work has further supported the conclusion that a combination of electronic coherence and dephasing optimizes migration and trapping of electronic excitation. This regime is called dephasing-assisted transport [15, 20, 21]. Dephasing-assisted transport is predicted to be especially significant for energetically disordered systems.

Plenio and coworkers recently made a step further in this direction introducing the concept of phonon antenna [31]. They proposed that coherent couplings could allow a pigment network to be spectrally "tuned" into configurations where the energy transport pathways may extract maximum noise strength from the environment and thus proceed faster. The strong mixing between electronic coherence and vibrational modes would give rise to mode-driven coherences, which will be prominent whenever vibrational modes have frequencies comparable to exciton energy differences of strongly coupled chromophores and have dephasing times on picosecond timescales [32].

In order to provide a unified approach that is capable of describing the salient features of EET in the different regimes within the same framework, the model described in Ref. [33] is outlined here. More sophisticated approaches may be found in the literature, as described above. However, the aim here is just to describe a simple model that is able to (i) quickly illustrate how the dynamics, the mechanism, and the role of quantum coherence depend on a delicate balance between electronic couplings (U) and interaction with the environment (γ) and (ii) allow a better understanding of the spectroscopic features connected with quantum effects.

Let us consider the simplest system of a dimer composed by a two-level donor (D) and acceptor (A). The initial state where both D and A are in the ground state is indicated by $|0\rangle$, whereas $|d\rangle$ ($|a\rangle$) represents the states in which D (A) is in the excited state and A (D) in the ground state (Figure 4.4).

The Hamiltonian for such a dimer can be written as:

$$\mathcal{H}(t) = H - E(t) \cdot V \tag{4.1a}$$

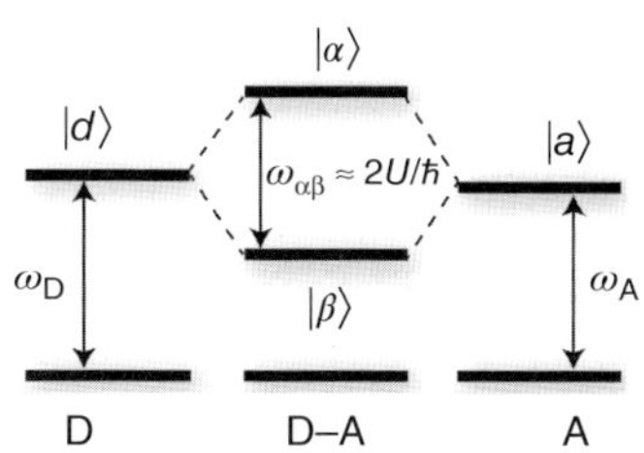

Figure 4.4 Schematic representation of the energy levels of a simple molecular dimer in the Heitler-London approximation. D = donor; A = acceptor. In the presence of nonnegligible electronic coupling U, the states $|d\rangle$ and $|a\rangle$ localized on D and A chromophores, respectively, mix to form delocalized excitonic states $|\alpha\rangle$ and $|\beta\rangle$. Applying the Heitler–London approximation, the ground state of the dimer is assumed to be the product of the ground state wave functions of the isolated molecules. This model may be easily generalized to the case of a heterodimer (Adapted from Refs [7, 10]).

$$H = H_0 + H_{ex} + H_{ex-ph} \tag{4.1b}$$

where V is the transition dipole operator, $E(t)$ is the electric field promoting transitions to excited states, and:

$$H_0 = \hbar\omega_D|d\rangle\langle d| + \hbar\omega_A|a\rangle\langle a| \tag{4.2a}$$

$$H_{ex} = U(|a\rangle\langle d| + |d\rangle\langle a|) \tag{4.2b}$$

$$H_{ex-ph} = h_0|0\rangle\langle 0| + h_d|d\rangle\langle d| + h_a|a\rangle\langle a| \tag{4.2c}$$

$\hbar\omega_D$ ($\hbar\omega_A$) is the energy of the state $|d\rangle$ ($|a\rangle$), U is the resonance coupling responsible for exciton formation, h_i ($i = 0, d, a$) represents the bath operator coupled to $|i\rangle$, describing the coupling of the system with vibrational degrees of freedom. The evolution of the system in time can be described by the density matrix $\rho(t)$ that at each time satisfies the Liouville equation. In general, the probability of finding the system in the state $|d\rangle$ or $|a\rangle$ at the time t (occupation probability) can be expressed as:

$$P_i(t) = Tr[\langle i|\rho(t)|i\rangle] \quad (i = a, d) \tag{4.3}$$

An expression for such probability can be derived using a quantum master equation. At this point, different approaches can be followed and different approximations can be applied. Following Ref. [11, 33], where a generalized master equation (GME) approach was applied, the probability $P_A(t)$ of finding the excitation on A at time t if at $t = 0$ it was on D, can be written as:

$$\frac{dP_A(t)}{dt} = \int_0^t dt_1 \; C_{ad}(t, t_1)[1 - P_A(t_1)] \tag{4.4}$$

where $C_{ad}(t,t_1)$ is the electronic energy-gap correlation function and defined as $C_{ad}(t, t_1) \equiv \hbar^2\langle\delta\omega_A(t)\delta\omega_D(t_1)\rangle$, where the energy gap fluctuation $\delta\omega_i(t)$ is a function describing the effects of the stochastic force exerted on the system transition energies by the bath.

When $i = j$, a "diagonal" correlation function is obtained, which correlates the energy gap function with itself. This quantity has a classical analog and it can be related to the spectral density by a Fourier transform [34]. The cross-correlation functions ($i \neq j$), which have no classical analog, are key quantities for understanding the effects of coherence in EET. When $C_{ad} = 0$, the fluctuations of the transition energies ω_D and ω_A are uncorrelated, and coherent EET cannot contribute to the overall transfer process unless the electronic coupling is strong [35]. This represents the usual assumption in theories for EET. The observation of coherent effects is connected with a nonvanishing value of C_{ad}.

When the bath correlation function is approximated as exponentially relaxing, and the high-temperature limit is assumed, then we can define γ as the dephasing strength, proportional to the inverse of the decoherence time τ_c. Then, using Equation 4.4, it is possible to define more clearly the two limiting regimes of EET shown in Figure 4.2. When $U \ll \gamma$ (weak coupling) and the time difference $t - t_1$

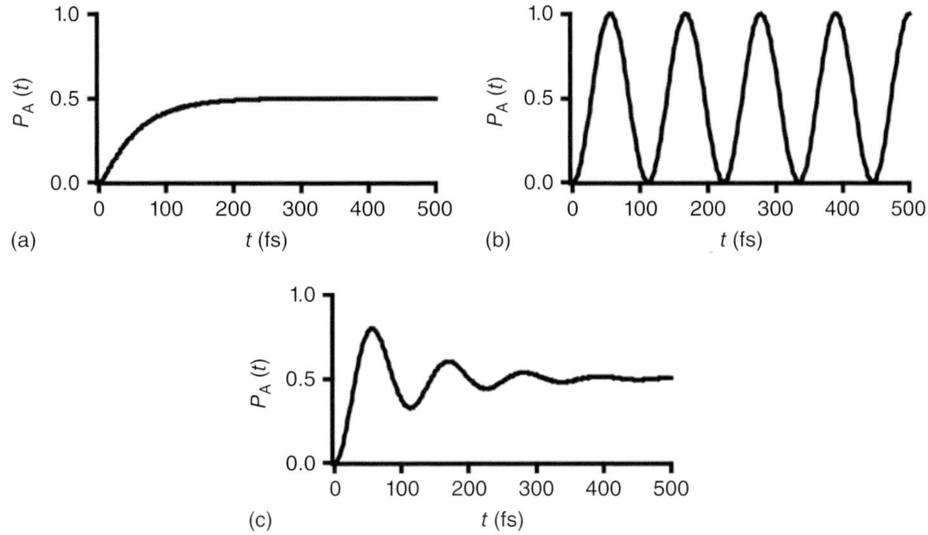

Figure 4.5 Comparison between different regimes of EET. The probability of finding the excitation on the acceptor state $P_A(t)$ is calculated using Equations 4.8a–4.8c with γ fixed at 150 cm^{-1} and $U = 0$, 1000, and 100 cm^{-1} for (a–c), respectively. In the weak coupling limit (a), the excitation energy is transferred incoherently and irreversibly between different sites. The probability $P_A(t)$ is an exponential function of t. In the strong coupling limit (b), the electronic state of the donor and the acceptor mix strongly to produce new delocalized states (excitons). Within these states the excitation energy is shared quantum mechanically between the donor and the acceptor excited states and the probability to find the excitation on the acceptor state $P_A(t)$ is an oscillating function with the period depending on the electronic coupling U. In the intermediate coupling limit (c) the excitation moves in space, yet part of the phase information is conserved (coherence effect). $P_A(t)$ shows a damped oscillatory behavior depending on the relative magnitudes of the electronic coupling U and the decoherence time τ_c.

is much bigger than the lifetime $\tau_c = \hbar/\gamma$ of the memory function, H_{ex} (Equation 4.2b) weakly perturbs the electronic properties of the system so that the localized functions $|d\rangle$ and $|a\rangle$ still represent a good basis set (site basis). The EET takes place in the weak limit regime and the GME converges to a Pauli master equation (Figure 4.5a), which leads to the often observed exponential evolution of the system toward an equilibrium between $P_A(t)$ and $P_B(t)$:

$$P_A(t) \approx \frac{1}{2}\left[1 - \exp\left(-\frac{4U^2}{\gamma\hbar}\right)\right] \tag{4.5}$$

On the other hand, if $U \ll \gamma$ and $t - t_1$ is much smaller than τ_c, H_{ex} promotes the formation of new delocalized states (excitons):

$$|\alpha\rangle = c_1|d\rangle + c_2|a\rangle \tag{4.6a}$$

$$|\beta\rangle = d_1|d\rangle - d_2|a\rangle \tag{4.6b}$$

where the coefficients simplify to $1/\sqrt{2}$ for a perfectly symmetric dimer. In this situation, the probability P_A oscillates indefinitely according to (Figure 4.5b):

$$P_A(t) \approx sin^2(Ut/\hbar) \tag{4.7}$$

with $P_D(t) = 1 - P_A(t)$. This shows that the excitation will coherently oscillate back and forth between chromophores D and A at the frequency $U/\hbar$ proportional to the energy difference $\hbar(\omega_D - \omega_A)$.

In the intermediate case, Kimura *et al.* derive [33]:

$$P_A(t) = \frac{1}{2}\left[1 - e^{-\gamma t/2\hbar}\left\{\cosh\left(\sqrt{\zeta t}\right) + \frac{\gamma}{2\hbar\sqrt{\zeta}}\sinh\left(\sqrt{\zeta t}\right)\right\}\right] \quad \text{for } U > \gamma/4 \tag{4.8a}$$

$$P_A(t) = \frac{1}{2}\left[1 - e^{-\gamma t/2\hbar}\left\{\cos\left(\sqrt{|\zeta|t}\right) + \frac{\gamma}{2\hbar\sqrt{|\zeta|}}\sin\left(\sqrt{|\zeta|t}\right)\right\}\right] \quad \text{for } U < \gamma/4 \tag{4.8b}$$

$$\zeta \equiv \frac{\gamma^2}{4\hbar^2} - \frac{4U^2}{\hbar^2} \tag{4.8c}$$

According to Equations 4.8a–4.8c, EET in the intermediate case can take place by two kinds of mechanisms. In the first case (Equation 4.8a), EET happens while retaining the oscillatory coherent character, until it is quenched owing to bath perturbations. In this condition, P_A shows a characteristic damped oscillating behavior (Figure 4.5c). In the second case (Equation 4.8b), no oscillations are recorded, but the EET takes place quickly, simultaneously with the vibrational relaxation. This is an example of nonequilibrium EET [36].

4.4 Experimental Detection of Quantum Coherence

The intuition that quantum principles, and quantum coherence in particular, may be relevant in the understanding and interpretation of life processes dates back to the beginning of the past century [37]. Only recently, however, technological advances in the field of spectroscopy allowed the experimental verification of such intuition and the demonstration that various quantum principles such as quantum tunneling, entanglement, and coherence in biological processes are not only an academic curiosity but possible design principles exploited by nature [31, 38, 39].

It is thus not a coincidence that the recent interest in quantum coherent EET mechanisms, demonstrated by the increasing number of papers devoted to the topic also in nonspecialist journals, was rejuvenated after the development of spectroscopic techniques capable of experimentally detecting such effects, 2D electronic spectroscopy (2DES) in particular. Nowadays, 2DES in all its different declinations is accepted as the more complete technique to experimentally confirm

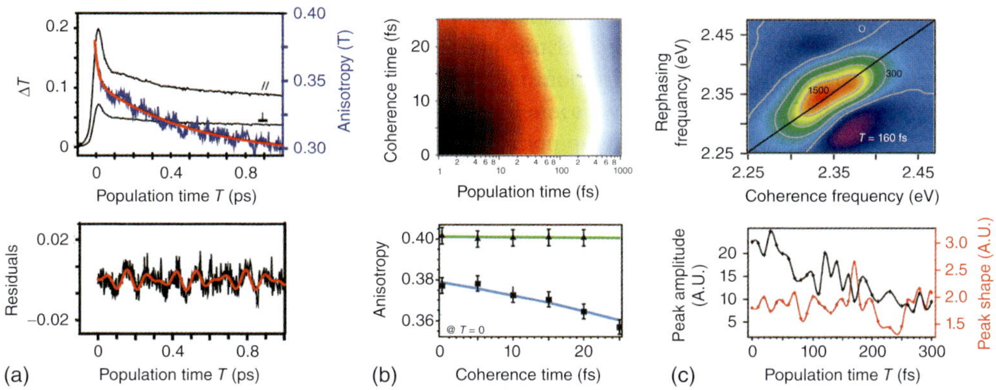

Figure 4.6 Overview of the main spectroscopic techniques employed for the detection of coherent mechanisms in EET applied to a sample solution of MEH-PPV in chloroform at room temperature [26, 40, 41]: (a) pump-probe anisotropy decay; (b) two-times anisotropy decay (TTAD); and (c) 2D photon echo (2DPE). In the upper panels, examples of the typical signal are reported whereas the lower panels highlight the signatures of electronic coherence in the corresponding signals. (a) Upper panel: In the pump-probe anisotropy decay technique, the signals measured with the pump and probe pulses having parallel (//) or perpendicular (⊥) polarization (black lines, left axis) are combined to calculate anisotropy (blue line, right axis) as $r(t) = [S_{//}(T) - S_{\perp}(T)]/[S_{//}(T) + 2S_{\perp}(T)]$. The red line is a bi-exponential fit of the anisotropy decay. Lower panel: Electronic coherences are manifested as oscillations of the anisotropy decay signal (here shown as residuals of the fit) as a function of the population time T. (b) Upper panel: In a typical TTAD measurement, anisotropy decay is plotted as a function of two time intervals, τ and T, defined as in Figure 4.7c. Lower panel: Electronic coherence is manifested through the presence of a decay of anisotropy as a function of τ. In the panel, the anisotropy decay as a function of τ is shown for the MEH-PPV sample (squares and blue line) and compared with a control sample solution of rhodamine 6G (triangles and green line). (c) Upper panel: Example of 2DPE spectrum (real part) recorded at $T = 160$ fs. Lower panel: Electronic coherences are manifested as anticorrelated oscillations in the amplitude (left axis, black line) and shape (right axis, red line) of the diagonal peak; the shape of the peak is defined as the ratio between the diagonal and antidiagonal widths at $1/e$ height

the presence and the role of quantum coherence in EET processes, both in biological as well as artificial systems. 2DES, however, is not the only technique capable of capturing coherence effects. Different experimental techniques whose response is able to manifest the signature of coherent EET were recently reviewed in Ref. [26] and summarized in Figure 4.6.

The possibility of experimentally detecting electronic coherence and quantum transport is intimately linked to the ability to excite the system with ultrashort (i.e., spectrally broad) laser pulses. Using the eigenstate description, a short pulse is indeed needed to prepare a superposition in which different eigenstates are excited in phase. When the pulse is not short enough, it excites many systems with different phases and the coherent oscillation disappears on averaging. In other words, in order to excite a particular superposition and be able to follow its evolution in time, the laser pulse must be sufficiently short compared with the characteristic period of the oscillation. To switch to the frequency domain, this means that to excite a coherent superposition of excitonically coupled states, these two states must lie within the bandwidth of the incoming pulse. Coherent oscillations may appear in femtosecond spectroscopy as quantum beats when two levels are excited coherently and emit to a common final level: the emission spectrum will oscillate with two-level frequency [34, 40].

The first indirect evidence for electronic coherence in EET was suggested by pump-probe anisotropy experiments in the late 1990s [41, 42]. The technique is, however, still employed [43, 44]. The presence of electronic coherence in such experiment is indicated by the appearance of oscillations in the anisotropy decay traces – but not in the isotropic decays! – with a frequency corresponding to the energy splitting between the two excitonic levels involved in the transfer (quantum beating). Such quantum beating can be correlated with the time evolution of a coherent superposition of excitonic states prepared by the laser pulse. The pump-probe anisotropy technique has the advantage to be relatively simple with respect to more sophisticated techniques such as 2DES (see below). The main drawbacks, however, are that (i) the technique is not completely general because of the geometrical constraints required for the detection of electronic beating in the anisotropy decay [41] and (ii) the presence of quantum beating can surely be associated with the presence of a type I coherence, but it does not necessary imply the presence of quantum transport of energy from a donor to an acceptor excited states (type II) [26].

Starting from the recognition of the great potentiality of the pump-probe anisotropy technique in capturing coherence effects, an advanced version of the experiment was then specifically designed to overcome its limitations and be sensitive to coherent energy transfer. The technique is called two-times anisotropy decay (TTAD) and can be roughly described as a "2D" version of a conventional anisotropy decay experiment (Figure 4.6b) [45, 43]. In a typical pump-probe anisotropy experiment, indeed, the first two interactions are simultaneous (τ is fixed to 0, see Figure 4.7a,b). The signal is then measured as a function of the delay time T between the pump ($E_1 + E_2$) and the probe (E_3). The principle of TTAD is to measure the anisotropy also as a function of τ in a configuration typical of the photon echo experiment (Figure 4.7c,d).

The "addition" of a second time axis τ allows following the evolution of the coherence formed after the interaction with the first field, when the system is in a superposition of ground and excited states. Anisotropy is then used to signal quantitatively that excitation coherence has been transferred. The TTAD experiment measures coherent EET directly and thereby probes the degree of coherence of the transfer. TTAD directly detects coherent EET because the transfer of electronic coherences from one chromophore to another is the only process that can promote a decay of the anisotropy along τ [3, 43]. Different from other techniques where the typical expected spectroscopic signatures of coherence are oscillations in the signal amplitude [26], here coherent EET is manifested as a decay of anisotropy as a function of τ. In this sense, TTAD provides much more direct

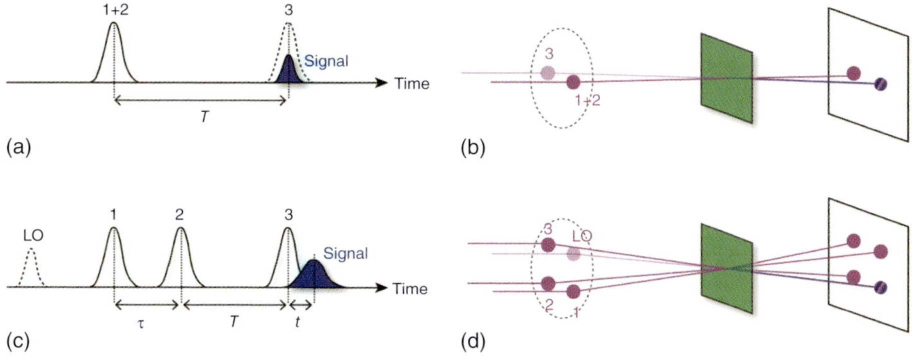

Figure 4.7 Definition of time intervals and excitation schemes in (a,b) pump-probe anisotropy decay, and (c,d) 2D electronic spectroscopy (2DES), and two-times anisotropy decay (TTAD). The time delays between pairs of pulses are conventionally named coherence time (τ), population or waiting time (T) and rephasing time (t), respectively. In the pump-probe experiment (a,b) the first two interactions (collectively indicated as "pump") happen simultaneously ($\tau = 0$) and the two exciting beams (E_1 and E_2) propagate along the same direction. The third interaction is with the "probe" beam (E_3), characterized by lower intensity. The signal (blue) is emitted in the same direction of the probe and it is measured as a function of the delay time T between pump and probe beams. In TTAD and 2DPE experiments (c,d), the beams impinge on the sample with the same geometry. The three fields interacting with the sample (E_1, E_2, and E_3) and a fourth beam used only for detection purposes (LO = local oscillator) are arranged at the vertices of a square. The signal is emitted in the same direction of the LO, and the measured quantity is indeed the interference between the signal and the LO (heterodyne detection). Although TTAD and 2DPE are characterized by the same beam geometry, in a TTAD experiment the signal is recorded as a function of T for different fixed values of τ and it is integrated for all t values, whereas in a 2DPE the signal is recorded as a function of τ and t for fixed values of T.

proof of quantum transport (type II coherence) than the conventional pump-probe anisotropy technique. Note, however, that the detection of anisotropy decay along τ is a sufficient but not a necessary condition for quantum energy transfer: while a decay of anisotropy as a function of τ can be directly related to the quantum transport, the absence of such a decay does not necessarily exclude the possible presence of a quantum transport phenomenon [26].

Despite the potentiality of anisotropy techniques, it must be recognized that the real turning point in the experimental detection of quantum coherence was reached with the development of 2DES methods, in particular the 2D photon echo technique (2DPE) (Figure 4.6c) [46]. Indeed, the capability of this technique to exploit the phase and coherence information in the time evolution of the optical polarization makes it sensitive to the presence of coherent mechanisms in the energy transfer process [46]. In 2007, 2DPE spectra recorded for the Fenna-Matthews-Olson (FMO) complex isolated from photosynthetic green sulfur bacteria showed a striking quantum beating pattern explained by Engel *et al.* to arise from long-lived coherent superpositions of electronic states [16]. Since then, 2DES techniques were extensively and successfully employed to characterize the light harvesting in many photosynthetic pigment-protein complexes [13]. The obtained results allowed finding evidences for the presence of long-lived electronic coherences both at cryogenic and physiologically relevant conditions in several biological complexes: the reaction center of purple bacteria [17], the light-harvesting complex II of superior plants (LCHII) [18], and phycobiliproteins of cryptophyte algae [19, 47]. Quantum beats, possibly a signature of electronic coherence, were also detected in 2D spectra of artificial systems such as bitubular J-aggregates of carbocyanine dyes [48] and a conjugated polymer [45], suggesting the possibility of engineering these effects also in artificial materials.

4.5 Electronic Coherence Measured by Two-Dimensional Photon Echo

2DPE, and more generally 2DES, is a four-wave mixing experiment in which three laser fields interact with the sample to induce a polarization [34, 49]. To understand how the technique works, a comparison with the more familiar pump-probe technique could be useful. In a typical pump-probe experiment, a short pump pulse from a femtosecond laser impulsively excites a fraction of molecules in an ensemble (e.g., a solution) to an electronic excited state. After some time delay T, a weak probe pulse records the changes in the absorption due to the action of the pump. The transmittance of the probe can be increased ("bleaching" or "stimulated emission") or decreased (excited state absorption) in different spectral regions. A typical pump-probe spectrum plots differential transmission as a function of probe frequency ω_t. An accessible way to introduce 2DPE spectroscopy is to think about how the pump-probe spectrum can be expanded into a 2D spectrum [50]. The new frequency axis ω_τ can be thought of as the distribution of frequencies excited by the pump pulse. This frequency–frequency correlation reveals information about mechanism as well as kinetics. The diagonal part of a 2DPE spectrum contains bands located at $\omega_t = \omega_\tau$ that show the positions of absorption features. The intensity of these bands changes as population decays to the ground electronic state, like the signal in pump-probe spectroscopy. The mechanism of the relaxation processes can be elucidated from the off-diagonal ($\omega_t \neq \omega_\tau$) peaks. For example, EET from one absorption band at energy ω_D (putting $\hbar = 1$) to another at lower energy ω_A produces a cross-peak at ($\omega_\tau = \omega_D; \omega_t = \omega_A$), revealing the kinetics by which the ω_D state relaxes to the ω_A state. If the energy transfer is downhill, then the cross-peak appears in the lower diagonal part of the 2D spectrum when it is plotted so that ω_τ is the abscissa and ω_t is the ordinate. The experiment is somewhat more sophisticated than the pump-probe analogy implies; the photon echo method correlates ω_t to ω_τ and, therefore, for instance, the mechanism of line broadening can be resolved [40, 51].

There are several excellent and instructive reviews that describe the experimental and theoretical details of 2DPE spectroscopy and provide a deeper understanding of the method [40, 49, 52]. Briefly, in a 2DPE experiment, three laser beams are incident on the sample and a fourth beam, the so-called local oscillator (LO), is used to perform heterodyne detection. Delay times between pairs of pulses are usually specified in the notation shown in Figure 4.7c. In the 2DPE experiment, for each fixed value of T, the τ delay is scanned from negative to positive time, which means that the time ordering of pulses changes from $E_2-E_1-E_3$ to $E_1-E_2-E_3$. At negative τ delays the nonrephasing diagrams generate the signal, whereas at positive τ delays (normal pulse time ordering) the rephasing diagrams generate the photon echo signal. The time-domain photon echo signal is measured using spectral interferometry [53]. By heterodyne-detecting the photon echo signal, amplitude and phase information can be recovered, enabling the data to be Fourier transformed with respect to the time intervals between the pulses to give the complex-valued 2DPE spectrum. Importantly, 2DPE combines the intuition of frequency-resolved pump-probe spectroscopy with the ability of the photon echo technique to correlate the electronic resonances before and after an evolution period, during which photophysical dynamics like energy transfer takes place.

One of the advantages of 2DPE in the quest of experimental proofs of quantum coherence is that it can directly reveal electronic couplings and energy transfer pathways by mapping coupled electronic states onto off-diagonal signals, far from the main diagonal congested region, where the main part of the system dynamics is concentrated. Therefore, 2DPE measurements can be exploited to obtain information about the electronic structure, that is, about energy, orientation, and spatial extent of electronic states, as well as about electronic couplings and energy transfer pathways of energy migration in multichromophoric systems [54, 55].

A second considerable advantage of 2DPE, which is in major part responsible for its rapid diffusion and development in the latest 5 years, is the sensitivity to coherent mechanisms in the energy transfer processes, manifested as oscillations of the signal amplitude at diagonal and off-diagonal positions with characteristic phase relationships and frequencies. Such oscillations are connected with the time evolution of coherences that evolves during the population time T [56].

Indeed, when the first two light-matter interactions produce a coherence $|i\rangle\langle j|$ or $|j\rangle\langle i|$, the signal evolution during the population time T is characterized by an amplitude of the radiated photon echo signal modulated, as a function of T, by the natural oscillation frequency of the coherence [26, 46]. For example, in the dimer system of Figure 4.4, the coherences $|\beta\rangle\langle\alpha|$ and $|\alpha\rangle\langle\beta|$ give rise to peaks above and below the diagonal, respectively, and their natural phase of oscillation are characterized by the opposite sign, $e^{-i\omega_{\alpha\beta}t}$ and $e^{+i\omega_{\alpha\beta}t}$.

These beats are strictly connected with the persistence of electronic coherence. The ability of the system to produce and maintain the coherence of these beats lies in its Hamiltonian. Hence the experiments elicit information that can help us understand how excitation energy is delocalized and funneled under any photoexcitation conditions.

It is worth thinking about the interpretation of excitonic beats during the period of an electronic coherence among the eigenstates $|\alpha\rangle$ and $|\beta\rangle$, recalling the concept of interference and the equations derived in Section 4.3. The presence of oscillations in the signal is indeed related to the oscillations in the occupation probability shown in Figure 4.5c.

In recent work, we have found evidence for quantum coherence dynamics in antenna proteins of marine algae called cryptophytes [19, 47, 57]. These light-harvesting antenna proteins belong to the widespread class of phycobiliproteins; water-soluble proteins containing tetrapyrrole chromophores covalently bind to the protein backbone. They represent the major peripheral light-harvesting proteins in cyanobacteria, red algae and cryptophytes [58]. In the latter, however, the phycobiliprotein antenna complexes are located in the thylakoid lumen [59]. Therefore, unlike purple bacteria and higher plants where all the light-harvesting proteins and photosystems are membrane bound, light-harvesting in cryptophytes occurs in the lumen [60], and energy is funneled into the membrane-bound photosystems, as illustrated in Ref. [61].

Figure 4.8 shows some results of 2DPE studies of one of these proteins, phycocyanin 645 (PC645) isolated from the cryptophyte organism *Chroomonas* CCMP270. The structural model of PC645, estimated from X-ray diffraction studies (Figure 4.8a), shows that the protein consists of four polypeptide chains arranged in a complex with an approximate twofold symmetry [59]. The absorption of light and funneling of the energy is carried out by eight chromophores in total, which are divided into three chemical types of bilins: two 15,16-dihydrobiliverdins (DBVs, blue) located in the center of the protein, two mesobiliverdins (MBVs, green), and four phycocyanobilins (PCBs, red) [61].

The chromophore variety that comprises cryptomonad biliproteins expands the spectral coverage of the antenna, as evident from the absorption spectrum (Figure 4.8b), that shows two quite broad absorption maxima at 585 nm (2.12 eV) and 645 nm (1.92 eV), as well a shoulder at 620 nm (2.00 eV). It was previously established that absorption at higher energies could be attributed mainly to the two DBVs. The DBVs are strongly coupled so their corresponding absorption bands are delocalized states labeled as DBV_- and DBV_+. The two MBVs and two of the four PCBs absorb in the central region of the spectrum, whereas the lower energy absorption peak is attributed to the remaining two PCBs, from which fluorescence emission is seen in isolated proteins. The eight chromophores are electronically coupled to each other, so despite our labeling system, the absorption bands do not really orrespond to isolated chromophore electronic transitions [47, 57, 61].

Figure 4.8c shows an example of a 2D spectrum recorded for isolated PC645 dispersed in aqueous buffer at room temperature. The 2D spectra contain signals arising from several different processes, and the separation of individual contributions is made difficult because of the broad line widths, which are particularly significant at room temperature. However, the position of the diagonal and off-diagonal features indicated in Figure 4.8c were estimated iteratively through the separated analysis of rephasing and nonrephasing parts of the signal and comparison with the results of quantum mechanical simulations [19, 57].

The more interesting features are the ones appearing in off-diagonal positions as they directly indicate coupling between exciton states. Three cross-peaks below the diagonal (lower diagonal cross-peaks) and three cross-peaks at the corresponding positions above the diagonal (upper diagonal cross-peaks) can be clearly distinguished. The amplitude of these peaks as a function of the population

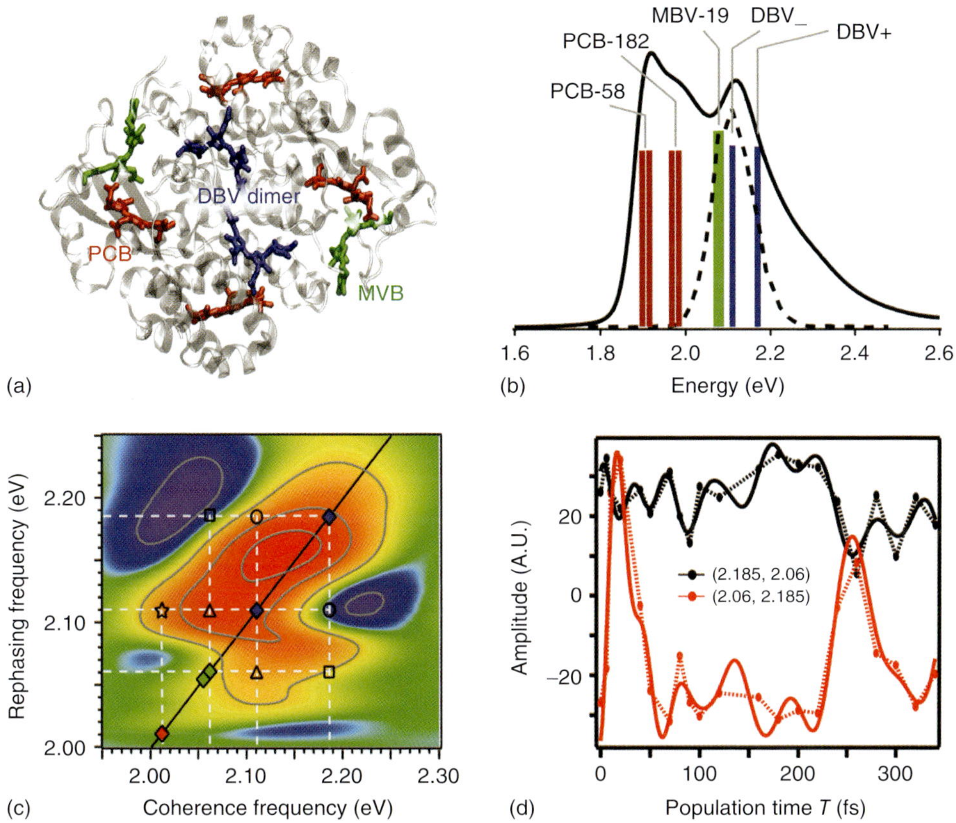

Figure 4.8 (a) Crystal structure of PC645 antenna protein determined to 1.4 Å resolution by X-ray crystallography showing the position of the eight light-harvesting bilins: DBV (blue), MBV (green), and PCB (red). (b) Electronic absorption spectrum of PC645 in aqueous buffer at room temperature. The colored lines illustrate the approximate absorption energies of the bilins, whereas the dashed black line represents the spectrum of the ultrafast laser pulse in the 2DPE experiment. (c) 2DPE spectrum (rephasing signal) for PC645 recorded at $T = 20$ fs showing the position of the diagonal and off-diagonal features. (d) Amplitude of the cross-peaks at positions indicated in the 2D map with open squares ($(\omega_\tau, \omega_t) = (2.185, 2.06)$ eV in black and $(\omega_\tau, \omega_t) = (2.06, 2.185)$ eV in red) as a function of time T. The dashed lines interpolate the data points (solid circles). The solid line is a fit to a sum of damped sine functions. For further info, see Refs [19, 57]

time T shows the presence of periodic oscillations with dominant frequencies corresponding to the energy differences between the exciton states. The relative phases of these electronic beats seem to support their assignment to electronic coherences (Figure 4.8d) [26, 46].

More recently, the 2DPE technique was also used to study the dynamics and the mechanism of energy transfer in other members of the phycobiliproteins antennas family, characterized by slightly different quaternary structure, arrangement and nature of chromophores, and light funnels [47, 62].

The main idea is to understand the possible functional significance of different structures in energy transfer, particularly with respect to the coherent driving of energy migration, and to investigate whether the emergence of quantum coherence in cryptophyte light-harvesting proteins is the result of some kind of evolutionary selection (S. J. Harrop, K. E. Wilk, R. Dinshaw, E. Collini, T. Mirkovic, C. Y. Teng, D. G. Oblinsksy, B. R. Green, K. Hoef-Emden, R. G. Hiller, G. D. Scholes and P. M. G. Curmi, Submitted, 2014.).

Coherent mechanisms in energy transfer are not restricted to biological systems possessing highly optimized energy funneling machinery, but similar dynamics were detected for a conjugated polymer at room temperature [3, 45, 58]. Conjugated polymers such as MEH-PPV (poly[2-methoxy,5-(2′-ethyl-hexoxy)-1,4-phenylenevinylene]) are fairly complex multichromophoric systems. The photophysics of MEH-PPV is derived from those of conformational subunits, also indicated as chromophores, in which the chain is broken by the influence of torsional disorder (Figure 4.9a). Each conformational subunit is electronically coupled to neighboring subunits, forming subtly delocalized collective states

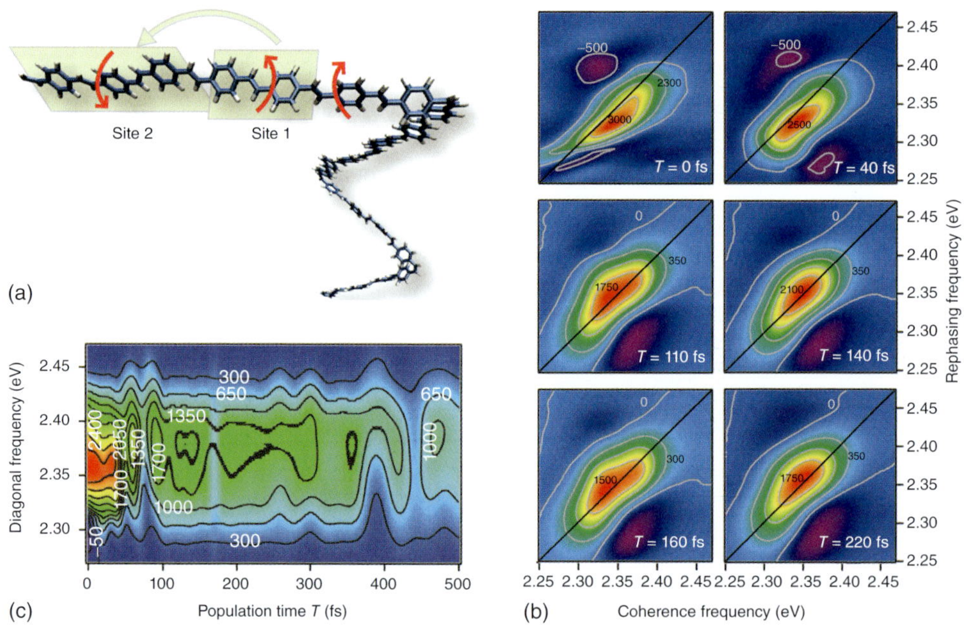

Figure 4.9 (a) Pictorial representation of *intra*chain energy transfer between two adjacent sites (conformational subunits) of a poly(*p*-phenylenevinylene) chain [65]. (b) Experimental 2DPE spectra (real part) of MEH-PPV in chloroform solution at room temperature for selected values of population time *T*. These plots are adapted from Ref. [40]. (c) Contour plot of the amplitude of the spectra along the diagonal line as a function of frequency and population time showing the beating behavior of the main diagonal peak [43].

[63]. After photoexcitation, energy is funneled to lower energy sites on the chain by EET before emission (Figure 4.1a). The mechanism of EET along the chain is strongly influenced by the polymer conformation because the disorder and electronic coupling between subunits depend critically on chain conformation [64]. In the specific case of MEH-PPV chloroform solutions considered in recent work, the EET mechanism is recognized to be mainly intrachain, that is, the energy migration proceeds along the backbone between adjacent segments, owing to the open, extended conformation assumed by the chain in a good solvent.

There are numerous studies of the dynamics of EET in conjugated polymers and their films, which are not reviewed here (see, for example, Ref. [66]). It has been found that energy migration occurs over multiple timescales from a few picoseconds to hundreds of picoseconds. Despite the dominance of fairly long EET timescales, quantum coherence can play a role in isolated polymer chains [26], and especially in films [67]. The 2DPE work summarized in Figure 4.9 is focused on the ultrafast regime (<300 fs) after photoexcitation. The main feature found in the 2DPE spectra (Figure 4.9b) is a positive, diagonally elongated peak, clearly signaling the presence of inhomogeneous broadening. No significant cross-peaks are observed in the temporal range examined, meaning that on these timescales excitonic coupling between different subunits and correlation among exciton states are absent or too small to produce a detectable cross-peak.

The absence of obvious cross-peaks in the 2DPE spectra measured for MEH-PPV (dilute solution in chloroform at room temperature) means that a direct measure of quantum coherence like the one described above for PC645 is not possible. Nonetheless, the interplay of coherent beats in different portions of the total diagonal peak gives rise to oscillations in the amplitude of the signal and in its shape, defined as the ratio of the diagonal and antidiagonal widths. For example, in Figure 4.9c, slices of the signal along the diagonal dimension are reported as a function of T.

In some work, the presence of beating patterns in the amplitude and shape of diagonal peaks characterized by anticorrelated phase was interpreted as evidence for electronic coherences [45, 47, 58]. More recent investigations, however, revealed that the attribution of beats to electronic coherence requires a more careful analysis [57, 68].

Looking at the oscillations in both samples (Figures 4.8 and 4.9), it must be noticed indeed that the beating behavior is complex. This is in general true for any other multichromophoric system, especially at room temperature where line broadening is significant, and overlapping bands partly obscure oscillating features. Moreover, it was recently highlighted that coherent nuclear motions could induce signals that may be mistaken for electronic coherences [68]. Nuclear coherences indeed have similar energies (and thus beating frequencies) to those of electronic coherences.

In the light of this, many authors are now reviewing 2D data trying to find a final attribution of the recorded beating frequencies in terms of vibrational or electronic coherences [26]. There is now a lot of debate about possible experimental protocols capable of unequivocally distinguishing between vibrational and electronic coherences. Some authors are also questioning the physical meaning of such a distinction, contesting the adiabatic framework, usually invoked in the development of EET models. The emerging picture is that the distinction between vibrational and electronic coherence is not so neat as it was initially thought [22]. Moreover, the possible presence of nuclear vibrations in close resonance with excitonic energy gaps contributing to the overall oscillating pattern is now arousing increasing interest. Such modes are indeed believed to be responsible for the activation of the energy transport and for the experimental detection of long-lived oscillations. A possible important implication of this close energy match is that vibrational modes can activate resonances that promote directed and effective energy transfer. For example, it was shown that coherent interactions between excitonic states and quasi-resonant fast vibrations could generate long-lived room temperature excitonic coherences and enhance population transfer [15, 31, 32].

4.6
Future Perspectives and Conclusive Remarks

Experimental proofs of persistent quantum coherence in energy transfer have been found in biological and artificial multichromophoric systems at ambient temperature through different spectroscopic techniques. To what extent are such coherences really relevant for the efficiency and the mechanism of biological and artificial EET processes? Would it be possible to implement quantum interference effects to control and optimize energy transfer pathways? These are the two main open questions that are challenging scientists working in the field.

To date, indeed, conclusive evidences are still lacking, and the presence, the role, and even the experimental methods to experimentally detect quantum effects are under scrutiny and debate. It is not clear at present what opportunities may be presented, but it will be fruitful anyway to pursue this idea, developing new experimental and theoretical tools capable of giving new answers.

The picture emerging from the data presented here suggests that one of the first requirements is a better understanding of the interactions between the electronic states and the vibrational modes of the chromophores and the surrounding medium in the so-called intermediate coupling regime. This represents now the frontier of the field, both theoretically and experimentally.

On the experimental side, it is essential to develop novel spectroscopic tools that can measure the influence of localized vibrations in channeling and/or enhancing EET more directly. Furthermore, conclusive signatures of the relevance of quantum behavior in conditions of biological relevance must be provided. Many groups around the world are already moving in this direction. For example, a recently proposed approach advocates the use of relatively narrow-band pulses, centered at suitable wavelengths capable of exciting specific coherence pathways and excluding population pathways from contributing to the signal [69, 70]. Another suggested solution is to implement the use of circularly polarized light in a 2DES scheme with the idea of exploiting the sensitivity of the circularly polarized based techniques, such as circular dichroism, to the nuclear distortions accompanying the excitation transport [71]. Moreover, the recently acquired technical capability to merge ultrafast spectroscopy and single-molecule detection made possible the development of single-molecule techniques, particularly promising to study the coherent dynamics in single complexes and to understand the physiological significance of quantum coherence in light harvesting [72].

In any case, it is important to highlight that even if proven nonrelevant in natural processes, the characterization of the role of quantum effects may be valuable if the underlying principle can be applied regardless of whether biology exploits it. The possibility of practically exploiting electronic coherence and quantum transport to enhance the performances of artificial devices represents itself as an achievement of considerable value, despite whether it mimics or not the natural process.

Acknowledgments

The European Research Council is gratefully acknowledged for support of this research providing financial support under the European Community's Seventh Framework Programme (FP7/2007–2013) with the ERC Starting Grant QUENTRHEL (grant agreement number 278560). EC acknowledges Prof. G.D. Scholes, Prof C. Ferrante, and Prof R. Bozio for their wise mentorship and useful discussions.

References

1. Scholes, G.D. (2003) *Annu. Rev. Phys. Chem.*, **54**, 57.
2. Förster, T. (1959) *Discuss. Faraday Soc.*, **27**, 7.
3. Hwang, I. and Scholes, G.D. (2010) *Chem. Mater.*, **23**, 610.
4. Balzani, V. and Scandola, F. (1991) *Supramolecular Photochemistry*, Elis-Horwood, Chichester.
5. Swager, T.M. (2007) *Chem. Rev.*, **107**, 1339.
6. Markovitsi, D., Ghustavsson, T., and Talbot, F. (2007) *Photochem. Photobiol. Sci.*, **6**, 717.
7. van Amerongen, H., van Grondelle, R., and Valkunas, L. (2000) *Photosynthetic Excitons*, World Scientific Publisher, Singapore.
8. Olaya-Castro, A. and Scholes, G.D. (2011) *Int. Rev. Phys. Chem.*, **30**, 49.
9. Parson, W.W. (2007) *Modern Optical Spectroscopy*, Springer-Verlag, Berlin, Heidelberg.
10. May, V. and Kühn, O. (2011) *Charge and Energy Transfer Dynamics in Molecular Systems*, 3rd edn, Wiley-VCH Verlag GmbH, Weinheim.
11. Kimura, A. and Kakitani, T. (2007) *J. Phys. Chem. A*, **111**, 12042.
12. Cheng, Y.C. and Silbey, R.J. (2006) *Phys. Rev. Lett.*, **96**, 028103.
13. Ishizaki, A., Calhoun, T.R., Schlau-Cohen, G.S., and Fleming, G.R. (2010) *Phys. Chem. Chem. Phys.*, **12**, 7319.
14. Scholes, G.D., Mirkovic, T., Turner, D.B., Fassioli, F., and Buchleitner, A. (2012) *Energy Environ. Sci.*, **5**, 9374.
15. Caruso, F., Chin, A.W., Datta, A., Huelga, S.F., and Plenio, M.B. (2009) *J. Chem. Phys.*, **131**, 105106.
16. Engel, G.S., Calhoun, T.R., Read, E.L., Ahn, T.-K., Mancal, T., Cheng, Y.-C., Blankenship, R.E., and Fleming, G.R. (2007) *Nature*, **446**, 782.

17. Lee, H., Cheng, Y.-C., and Fleming, G.R. (2007) *Science*, **316**, 1462.
18. Calhoun, T.R., Ginsberg, N.S., Schlau-Cohen, G.S., Cheng, Y.-C., Ballottari, M., Bassi, R., and Fleming, G.R. (2009) *J. Phys. Chem. B*, **113**, 16291.
19. Collini, E., Wong, C.Y., Wilk, K.E., Curmi, P.M.G., Brumer, P., and Scholes, G.D. (2010) *Nature*, **463**, 644.
20. Rebentrost, P., Mohseni, M., and Aspuru-Guzik, A. (2009) *J. Phys. Chem. B*, **113**, 9942.
21. Plenio, M.B. and Huelga, S.F. (2008) *New J. Phys.*, **10**, 113019.
22. Ishizaki, A. and Fleming, G.R. (2009) *Proc. Natl. Acad. Sci. U.S.A.*, **106**, 17255.
23. Scholes, G.D. (2011) *Nat. Phys.*, **7**, 448.
24. Fassioli, F. and Olaya-Castro, A. (2010) *New J. Phys.*, **12**, 085006.
25. Kassal, I., Yuen-Zhou, J., and Rahimi-Keshari, S. (2013) *J. Phys. Chem. Lett.*, **4**, 362.
26. Collini, E. (2013) *Chem. Soc. Rev.*, **42**, 4932.
27. Scholes, G.D., Jordanides, X.J., and Fleming, G.R. (2001) *J. Phys. Chem. B*, **105**, 1640.
28. Haken, H. and Strobl, G. (1973) *Z. Angew. Phys.*, **262**, 135.
29. Cao, J. and Silbey, R.J. (2009) *J. Phys. Chem. A*, **113**, 13825.
30. Gaab, K.M. and Bardeen, C.J. (2004) *J. Chem. Phys.*, **121**, 7813.
31. Chin, A.W., Huelga, S.F., and Plenio, M.B. (2012) *Philos. Trans. R. Soc. London, Ser. A*, **370**, 3638.
32. Chin, A.W., Prior, J., Rosenbach, R., Caycedo-Soler, F., Huelga, S.F., and Plenio, M.B. (2013) *Nat. Phys.*, **9**, 113.
33. Kimura, A., Kakitani, T., and Yamato, T. (2000) *J. Phys. Chem. B*, **104**, 9276.
34. Mukamel, S. (1995) *Principles of Nonlinear Optical Spectroscopy*, Oxford University Press, New York.
35. Kubo, R. (1969) *Adv. Chem. Phys.*, **15**, 101.
36. Jang, S., Jung, Y.J., and Silbey, R. (2002) *J. Chem. Phys.*, **275**, 319.
37. Schrödinger, E. (1944) *What is Life? The Physical Aspect of the Living Cell*, Cambridge University Press, Cambridge.
38. Scholes, G.D. (2010) *J. Phys. Chem. Lett.*, **1**, 2.
39. Lambert, N., Chen, Y.-N., Cheng, Y.-C., Li, C.-M., Chen, G.-Y., and Nori, F. (2013) *Nat. Phys.*, **9**, 10.
40. Jonas, D.M. (2003) *Annu. Rev. Phys. Chem.*, **54**, 425.
41. Savikhin, S., Buck, D.R., and Struve, W.S. (1997) *Chem. Phys.*, **223**, 303.
42. Arnett, D.C., Moser, C.C., Dutton, P.L., and Scherer, N.F. (1999) *J. Phys. Chem. B*, **103**, 2014.
43. Collini, E. and Scholes, G.D. (2009) *J. Phys. Chem. A*, **113**, 4223.
44. Smith, E.R. and Jonas, D.M. (2011) *J. Phys. Chem. A*, **115**, 4101.
45. Collini, E. and Scholes, G.D. (2009) *Science*, **323**, 369.
46. Pisliakov, A.V., Mancal, T., and Fleming, G.R. (2006) *J. Chem. Phys.*, **124**, 234505.
47. Wong, C.Y., Alvey, R.M., Turner, D.B., Wilk, K.E., Bryant, D.A., Curmi, P.M.G., Silbey, R.J., and Scholes, G.D. (2012) *Nat. Chem.*, **4**, 396.
48. Milota, F., Sperling, J., Nemeth, A., and Kauffmann, H.F. (2009) *Chem. Phys.*, **357**, 45.
49. Cho, M. (2008) *Chem. Rev.*, **108**, 1331.
50. Hamm, P., Lim, M.H., and Hochstrasser, R.M. (1998) *J. Phys. Chem. B*, **102**, 6123.
51. Beljonne, D., Pourtois, G., Silva, C., Hennebicq, E., Herz, L.M., Friend, R.H., Scholes, G.D., Setayesh, S., Müllen, K., and Brédas, J.L. (2002) *Proc. Natl. Acad. Sci. U.S.A.*, **99**, 10982.
52. Cho, M. (2009) *Two-Dimensional Optical Spectroscopy*, CRC Press, New York.
53. Dorrer, C., Belabas, N., Likforman, J.-P., and Joffre, M. (2000) *J. Opt. Soc. Am. B*, **17**, 1795.
54. Read, E.L., Schlau-Cohen, G.S., Engel, G.S., Wen, J.Z., Blankenship, R.E., and Fleming, G.R. (2008) *Biophys. J.*, **95**, 847.
55. Read, E.L., Schlau-Cohen, G.S., Engel, G.S., Georgiou, T., Papiz, M.Z., and Fleming, G.R. (2009) *J. Phys. Chem. B*, **113**, 6495.
56. Cheng, Y.C. and Fleming, G.R. (2008) *J. Phys. Chem. A*, **112**, 4254.
57. Turner, D.B., Dinshaw, R., Lee, K.-K., Belsley, M.S., Wilk, K.E., Curmi, P.M.G., and Scholes, G.D. (2012) *Phys. Chem. Chem. Phys.*, **14**, 4857.
58. Collini, E., Curutchet, C., Mirkovic, T., and Scholes, G.D. (2009) in *Energy*

Transfer Dynamics in Biomaterial Systems, vol. **93** (eds I. Burghardt, V. May, D.A. Micha, and E.R. Bittner), Springer-Verlag, Heidelberg, Berlin, p. 3.

59. Wilk, K.E., Harrop, S.J., Jankova, L., Edler, D., Keenan, G., Sharpes, F., Hiller, R.G., and Curmi, P.M.G. (1999) *Proc. Natl. Acad. Sci. U.S.A.*, **96**, 8901.

60. Gantt, E. (1979) *Biochemistry and Physiology of Protozoa*, Academic Press, New York, p. 121.

61. Mirkovic, T., Doust, A.B., Kim, J., Wilk, K.E., Curutchet, C., Mennucci, B., Cammi, R., Curmi, P.M.G., and Scholes, G.D. (2007) *Photochem. Photobiol. Sci.*, **6**, 964.

62. Collini, E. (2012) *Proc. SPIE, Quantum Opt. II*, **8440**, 84400A.

63. Scholes, G.D. and Rumbles, G. (2006) *Nat. Mater.*, **5**, 683.

64. Schwartz, B.J. (2003) *Annu. Rev. Phys. Chem.*, **54**, 141.

65. Brédas, J.-L. and Silbey, R. (2009) *Science*, **323**, 348.

66. Sariciftci, N.S. (ed) (1998) *Primary Photoexcitations in Conjugated Polymers: Molecular Exciton Versus Semiconductor Band Model*, World Scientific, Singapore.

67. Spano, F.C., Clark, J., Silva, C., and Friend, R.H. (2009) *J. Chem. Phys.*, **130**, 074904.

68. Tiwari, V., Peters, W.K., and Jonas, D.M. (2013) *Proc. Natl. Acad. Sci. U.S.A.*, **110**, 1203.

69. Richards, G.H., Wilk, K.E., Curmi, P.M.G., Quiney, H.M., and Davis, J.A. (2012) *J. Phys. B: At. Mol. Opt. Phys.*, **45**, 154015.

70. Richards, G.H., Wilk, K.E., Curmi, P.M.G., Quiney, H.M., and Davis, J.A. (2012) *J. Phys. Chem. Lett.*, **3**, 272.

71. Collini, E. (2013) *www.chimica.unipd.it/quentrhel* (accessed 21 November 2013).

72. Hildner, R., Brinks, D., Stefani, F.D., and van Hulst, N.F. (2011) *Phys. Chem. Chem. Phys.*, **13**, 1888.

5
Ultrafast Studies of Carrier Dynamics in Quantum Dots for Next Generation Photovoltaics

Danielle Buckley

5.1
Introduction

The majority of photovoltaics (PVs) on the market today are considered "first generation" and are single-junction silicon devices with market efficiencies around 15–20% [1]. Unfortunately, first generation devices suffer from losses in efficiency due to the (i) inability to absorb photons with energy less than the band gap; (ii) phonon cooling of electrons and holes with excess energy above the band gap; (iii) radiative recombination of electrons and holes; and (iv) trap-assisted recombination of electrons and holes.

Second generation devices are based on thin film designs and provide similar or even lower cell efficiencies than those of the first generation, the highest laboratory efficiencies being 20% using silicon and 29% with gallium arsenide [2]; but the lower material and manufacturing costs associated with thin films compared to that of first generation silicon wafers make them more appealing [1]. These devices also suffer from the efficiency losses listed above. Although some second generation devices utilize other semiconductor materials, the majority of first and second generation PVs currently on the market are designed using silicon, which has a band gap in the near-infrared, and do not efficiently use all of the energy from visible photons that occur in the peak of the solar spectrum. The solar spectrum is shown in Figure 5.1 (constructed from ASTM G173-03 data of Global Total Spectral Irradiance [3]) along with the band gaps of several key semiconductor materials.

Third generation PVs, also referred to as *next generation PVs*, aim to correct one or more of the efficiency losses found in the first and second generation devices, providing much higher cost efficiency. Next generation approaches to achieve this include utilizing multi-junction cells, intermediate band (IB) cells, hot carriers, multiple exciton generation (MEG), and spectrum conversion [4]. Multi-junction, or tandem, PVs have already been shown to increase power conversion efficiency, producing the current PV efficiency record of 44% that was reached in 2012 by the California company Solar Junction using a triple-junction concentrator cell design [2], but the current material and engineering challenges increase their cost and decrease their widespread appeal [5]. IB, MEG, and spectrum conversion cells aim

Discovering the Future of Molecular Sciences, First Edition. Edited by Bruno Pignataro.

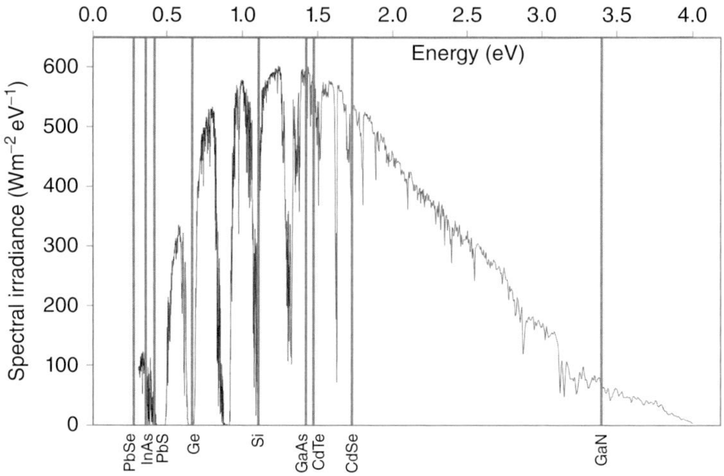

Figure 5.1 The AM1.5 Reference Spectrum reflects the standard global spectral irradiance and is shown here with band gaps of bulk semiconductors currently used in solar cells and those being studied for next generation devices.

to increase photocurrent while hot carrier cells would increase photovoltage. This chapter focuses on ultrafast studies of quantum dots that have the potential to contribute to the development of hot carrier and MEG cells.

5.2 Theoretical Limits

In 1961, Shockley and Queisser used a detailed balance limit to calculate a maximum thermodynamic efficiency of 33% for the conversion of unconcentrated solar irradiance into electrical free energy at room temperature (and an ultimate efficiency of 41% with a cell temperature of 0 K). This calculation assumes that in a defect-free bulk semiconductor, radiative recombination is the only route for the loss of charge carriers, only one electron–hole pair is produced per photon, and those excited electrons cool to the band edge via electron–phonon scattering [6].

A later calculation by Ross and Nozik [7] overcomes the Shockley–Queisser (S-Q) limit by assuming that excess photon energy is converted into electrical free energy rather than lost to cooling processes, producing a maximum efficiency of 66%. In order to achieve this efficiency, the absorber must maintain a high effective temperature compared to the lattice by not equilibrating with its environment. In this instance, solar photon conversion is increased via photovoltage if carriers are extracted before the excess kinetic energy is dissipated [8].

Another route to increased conversion efficiency is via photocurrent, producing two or more electron–hole pairs per single high energy photon [9, 10]. Hanna and Nozik use the detailed balance model but allow carrier multiplication (CM), which raises the maximum efficiency from S-Q's 33% to 44% [11].

Both ideal instances of optimizing solar photon conversion require that the relaxation of carriers be slower than other processes, such as carrier extraction. For increased photovoltage, carriers with excess energy must be extracted before cooling occurs. An increase in photocurrent requires that multiple electron–hole pairs be created and extracted before relaxation processes take place. These requirements make it necessary to understand electron and hole cooling dynamics and determine how they might be modified.

5.3 Bulk Semiconductors

Electronic energy levels in bulk semiconductors are spaced so closely together that they approximately form continuous bands. At thermal equilibrium and without photoexcitation, the highest occupied band, the valence band (VB), is normally full while the conduction band (CB) is normally empty. The VB contains the equilibrium population, forcing a carrier (electron or hole) to energetically overcome the band gap, E_g, in order to be promoted to the CB.

The absorption of a photon with energy greater than the band gap excites an electron from the VB to a CB, leaving behind a hole in the VB. The excess kinetic energy is then imparted to these carriers. If the carrier does not equilibrate with its environment, maintaining the excess energy in a nonequilibrium state with the lattice, it is considered a "hot" electron or hole. A hot carrier then either dissipates its excess energy via phonon cooling in order to relax back down to the band edge or undergoes impact ionization (II), transferring its excess energy to promote another carrier, depicted in Figure 5.2.

Ideally, impaction ionization occurs when a photon with as little as $2E_g$ is absorbed; but in reality, energy many times the band gap is required for II to occur in bulk semiconductors, making it inefficient until photon energies are in the ultraviolet region of the solar spectrum. The larger energy requirement is partly because the carriers, electron, and hole are uncoupled and act independently in II. Landsberg presents a quasi-classical approach to determine the minimum amount of energy in excess of the band gap necessary for a carrier to undergo II while conserving both energy and crystal momentum [12]. In the case of two parabolic bands, the excess energy, E_{excess}, is found to be

$$\frac{(2m_e + m_h)}{(m_e + m_h)} E_g$$

with m_e and m_h representing the effective masses of the electron and hole, respectively. For the case of bulk lead selenide, where the effective mass of the electron and hole are approximately equal, $E_{excess} = \frac{3}{2} E_g$ is needed in addition to the band gap energy for a carrier to undergo II; however, as $m_e \approx m_h$ in this example, the excess energy is equally distributed between the electron and hole, making an additional $\frac{3}{2} E_g$ necessary for impact ionization. The total $3E_g$ of excess energy required sets a threshold $h\nu = 4E_g$ for II in bulk PbSe, which is less than

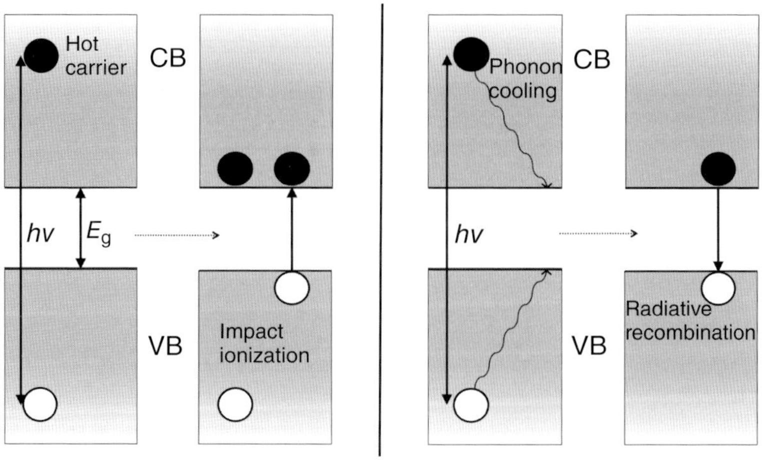

Figure 5.2 Excess energy absorbed in the bulk results in either the production of multiple electron–hole pairs through impact ionization or energy lost through phonon cooling before radiative recombination of the remaining electron and hole. The black and white circles represent negative electrons and positive holes, respectively.

the experimental value [13]. In addition to the large amount of energy required by conservation laws, phonon-carrier cooling occurs very quickly, in less than a picosecond, competing with the II process and further lessening its likelihood.

In order to increase solar cell efficiency by increasing photovoltage and/or photocurrent, phonon cooling must be circumvented so that hot carriers can be extracted or so that impact ionization is able to occur. Quantum dots offer the potential advantages of longer relaxation times and enhanced multiple electron–hole pair creation, making them a promising route to next generation PVs.

5.4 Semiconductor Quantum Dots

An electron–hole pair in a nanocrystal is spatially confined in three dimensions and can undergo very strong quantum confinement if the size of the crystal is smaller than its exciton Bohr radius. In bulk, the exciton Bohr radius is the closest distance at which an electron–hole pair can exist in a bound state, which is defined by

$$a_{\mathrm{B}} = \frac{4\pi\varepsilon\varepsilon_0\hbar^2}{\mu e^2}$$

$$\mu = \frac{m_{\mathrm{e}} m_{\mathrm{h}}}{(m_{\mathrm{e}} + m_{\mathrm{h}})}$$

where ε is the dielectric constant, μ defines the exciton reduced mass, and m_{e} and m_{h} are again the effective masses of the electron and hole.

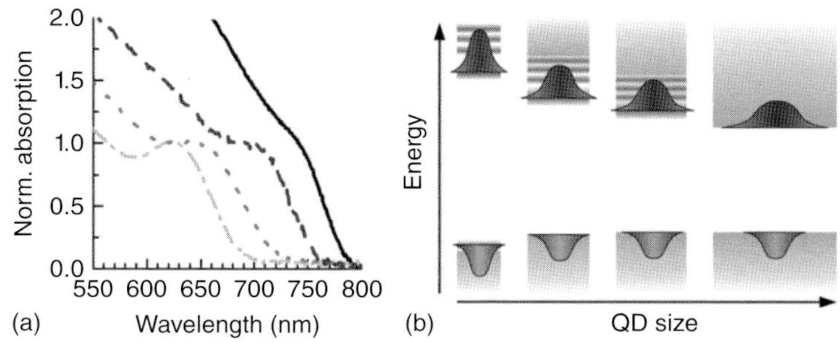

Figure 5.3 (a) Optical absorption spectrum showing the first absorption peak of colloidal CdTe quantum dots, increasing in diameter with increasing wavelength and (b) a representation of the quantization and increase in band gap with decreasing semiconductor size, including differences in wave function and confinement of the electron and hole due to their different effective masses. (Reprinted with permission from Rühle *et al.* Quantum-dot-sensitized solar cells. ChemPhysChem. Copyright 2010 Wiley-VCH [14].)

In quantum dots, these bound electron–hole pairs, or excitons, experience greatly enhanced Coulombic attraction due to their spatial proximity. The strong quantum confinement gives rise to the term quantum dot and creates discrete electronic energy levels within the VBs and CBs, rather than the continuum of each band found in bulk, creating a 3D particle-in-a-box atmosphere. The confinement also acts to "tune" the band gap to smaller or larger energies based on the semiconductor material and the size of the nanocrystal; as the size of the particle decreases, more energy is required for an electron to be excited from the VB to the CB. An example of this is shown in Figure 5.3 [14] for different sizes of cadmium telluride quantum dots. The first absorption peak represents the first optical transition, excitation from the highest level of the VB to the lowest level of the CB, and approximately corresponds to the band gap of the quantum dot.

During synthesis, most quantum dots are capped with long-chain organic ligands, such as oleic acid which attaches to the dot as oleate, for surface passivation. These ligands work well to prevent aggregation when dots are studied in colloidal form, but are too bulky and insulating when working with arrays as precursors to functional PVs. Studies have looked at ligand exchange, substituting shorter carbon chains and different functional groups (Figure 5.4) in place of the bulky organics. This effort aims to reduce dot-to-dot distance, increasing interdot coupling, and to better understand how different ligands may otherwise modify carrier dynamics. The influence of various ligands on CM and relaxation dynamics is discussed in a later section.

Another advantage of quantum dots for next generation PVs is the relatively simple and inexpensive synthesis procedure for making colloidal solutions. Although some materials are more challenging to work with than others, commonly studied dots, such as lead sulfide, lead selenide, and cadmium selenide, are fairly straightforward and inexpensive to synthesize. The reduced materials cost can help make

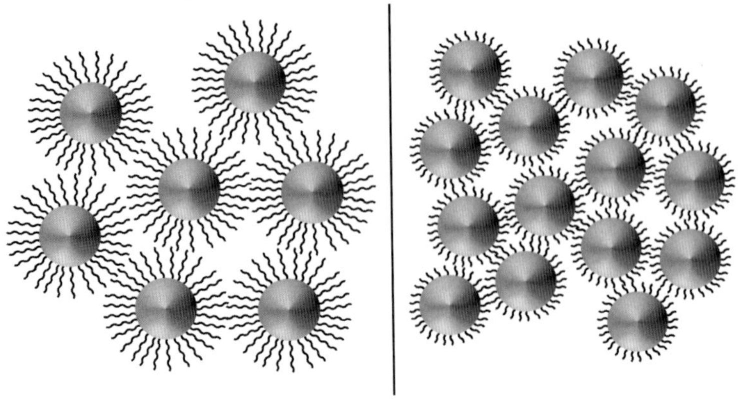

Figure 5.4 Quantum dots are passivated with long-chain organic ligands, such as oleate (a), but ligand exchanges replace these with shorter organic molecules, such as 1,2-ethanedithiol, which reduces interdot spacing (b).

quantum dot solar cells competitive in the market even if their single-junction cell efficiencies are less than those of more expensive materials.

5.4.1
Lead Chalcogenides

The IV–VI lead chalcogenide semiconductor quantum dots have perhaps received the most attention with regard to CM and applications in third generation PVs. Although lead selenide is more widely investigated, lead sulfide was recently used in the construction of a quantum dot solar cell with the highest certified power conversion yet, 7% [15]. For these reasons, information and results for these materials are summarized more completely in this chapter than for other semiconductor quantum dots.

Bulk lead chalcogenides include PbS, PbSe, and PbTe, all of which have rock-salt cubic lattice structure and similar conductance and VB curvature, generating electrons and holes with nearly equal effective masses [16]. This makes their electronic structures easier to work with compared to those of other semiconductors. In addition, PbS and PbSe have large Bohr exciton radii (a_B), 18 and 46 nm, respectively. The nanocrystal radius must be smaller than a_B for strong quantum confinement, but equally important, the individual electron and hole Bohr radii (a_e and a_h, respectively) must be confined by the size of the dot. a_e and a_h can be defined using a_B listed above, replacing μ with m_e or m_h.

In PbSe, $a_e = a_h = 23$ nm allowing dots that experience strong confinement effects to be easily synthesized, whereas other semiconductors may have a much smaller Bohr exciton radius or electron/hole Bohr radius. A small Bohr exciton radius, such as $a_B = 6$ nm in CdSe, allows only a limited range of sizes with strong confinement and a small electron or hole Bohr radius, such as InSb which has $a_B = 54$ nm but $a_h = 2$ nm, presents the same problem [17]. Figure 5.3b illustrates the effect of $a_e \neq a_h$ on the electronic energy levels and wave functions for cadmium telluride. The smaller effective mass of the electron is reflected in the larger spatial spread of the electron wave function as well as discrete energy levels associated with electron confinement at larger particle sizes than hole confinement [14].

5.5 Carrier Dynamics

A better understanding of fundamental quantum dot carrier dynamics can help researchers formulate ways to take advantage of the existing photophysics and design methods to modify them in order to achieve increased efficiency in PVs. In both MEG and hot carrier solar cells, relaxation mechanisms play a key role in increasing the number of carriers extracted and the time available to extract hot carriers.

5.5.1 Carrier Multiplication

As early as 1982, it was theorized that quantized structures would lead to greater efficiencies for PVs over those of their bulk counterparts [8], but it was not until 2004 [18] that researchers reported multiple electron–hole pair signatures that persisted at low laser intensity in ultrafast experiments on lead selenide quantum dots. This was attributed to the highly efficient CM from a single photon. It is believed that an increase in CM efficiency in quantum dots over bulk can be attributed to the relaxation of conservation of crystal momentum, which lowers the threshold energy necessary for CM.

Soon after the 2004 study by Schaller and Klimov, experiments on other semiconductor quantum dots, such as PbS [19], CdSe [20], and Si [21], reported similar persistent signals attributed to efficient production of multiple electron–hole pairs from one photon, referred to interchangeably in the literature as MEG or CM. But others have reported lower CM yields [22, 23] and offered arguments against more efficient CM in quantum dots compared to bulk [13], which has led to controversy over the topic.

5.5.2 Relaxation

In both bulk and quantum dot semiconductors, a single excited carrier will relax to the band edge before radiatively recombining over the course of nano- or even microseconds. Similarly, multiple excited carriers may cool to the band edge but then undergo Auger recombination (AR), one electron–hole pair will recombine and transfer its excess energy to a remaining carrier rather than emitting a photon, eventually leaving behind a single electron–hole pair. AR is the inverse of impact ionization and for the case of $n_e = n_h$ in defect-free, intrinsic bulk semiconductors, follows the rate equation [24]:

$$\text{Rate} = \frac{dn_e}{dt} = -Cn_e^3$$

where C is the Auger constant and n_e is the carrier density for both electrons and holes. C is based on the semiconductor material and includes an activation energy that is proportional to the band gap, which dominates in the case of direct-gap materials [25]. The above equation results in an instantaneous rate constant, $\tau(n_e)^{-1} = -Cn_e^2$, for AR. Large differences in the Auger rate constant, and therefore, the rate itself, are seen in bulk semiconductors with different band structures and especially between direct- and indirect-gap materials because the latter requires a phonon to take part in order to conserve crystal momentum [25]. This is illustrated in Figure 5.5 [26].

Electron–hole pair creation follows Poisson statistics in bulk semiconductors and it is assumed that initially, before recombination, quantum dots will also follow as:

$$P_k = \frac{\langle N_0 \rangle^k}{k!} e^{-\langle N_0 \rangle}$$

where P_k describes the fraction of dots in the excitation path with k excitons. $\langle N_0 \rangle$ is the average number of photons absorbed per quantum dot and is the product of the

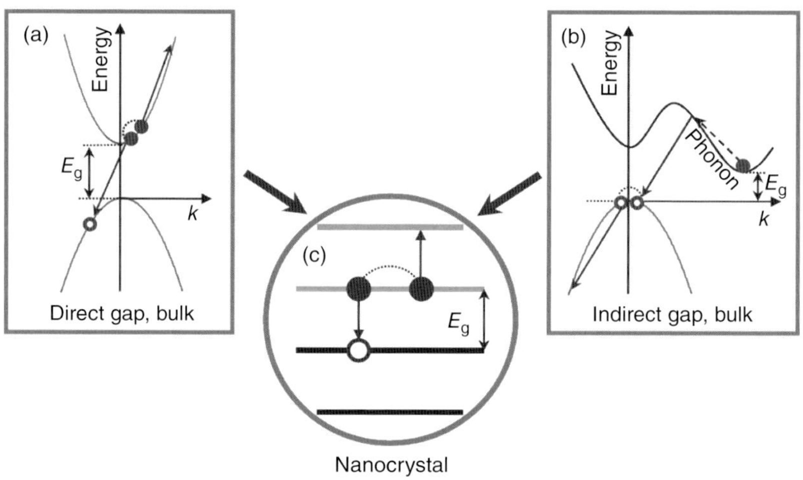

Figure 5.5 (a) Direct band gap AR occurs when an electron (solid) and hole (white) recombine and transfer excess kinetic energy to create a hot electron. (b) Indirect-gap semiconductors undergo phonon-assisted Auger recombination so that crystal momentum is conserved. (c) Quantum confinement relaxes crystal momentum conservation so that direct- and indirect-gap semiconductors are similar. (Reprinted with permission from Robel *et al.* Universal size-dependent trend in Auger recombination in direct-gap and indirect-gap semiconductor nanocrystals. Physical Review Letters 102(17), 2009, 177404. Copyright 2009 American Physical Society [26].)

absorption cross-section at the excitation wavelength and the density of incident photons per sample area. Allowing this assumption in nanocrystals suggests that AR occurs via independent carriers rather than excitons, but it should be noted that the bulk-like assumption of Poisson behavior is inconsistent with the expectation of strongly coupled excitons due to spatial confinement [39].

Applying the bulk approach to analyze Auger rates for nanocrystals requires adjusting the carrier density to a size-dependent *effective* density, which is the ratio of the average number of electron–hole pairs per quantum dot and the nanocrystal volume, ($\langle k\rangle/V_0$). This results in an initial relaxation time of the nanocrystal ensemble ($\tau_{\langle k\rangle}$), but discrete lifetimes (τ_k) are needed to model the multiexciton decay dynamics of an individual quantum dot with a definite number of excitons, k. In the limit of large $\langle k\rangle$, the bulk rate law for $n_e = \langle k\rangle/V_0$ arises from a Poisson distribution of k and individual Auger rates proportional to $k^2(k-1)$ [39].

Efforts have been made to find a model that accurately describes the multiexciton decay dynamics of individual dots using bulk-like [27, 28] and statistical [28, 29] approaches. For example, Klimov combines bulk and stochastic approaches to approximate the relationship between ensemble and discrete instantaneous initial rate constants in the limit of large $\langle k\rangle$ as $\tau_k \simeq \tau_{\langle k\rangle}\langle k\rangle^{-1}|_{\langle k\rangle=k}$. This results in an *effective* Auger constant from the approximate individual decay, $C_{\text{eff}} \simeq V_0^2(k^3\tau_k)^{-1}$ [28]. Robel *et al.* [26] apply this model and report a universal dependence of the *effective* biexciton Auger rate constant on nanocrystal volume, rather than band gap, across different semiconductor materials and band structures.

Barzykin and Tachiya present a stochastic model, evaluating both free carrier and excitonic treatment of AR, and assert that bulk assumptions underestimate the influence of electron–hole pair distribution in a quantum dot and can only be applied when the number of electron–hole pairs is much larger than what occurs in quantum dots [29]. For example, the 8-fold degeneracy of lead chalcogenides at band gap is still much too low for bulk-like assumptions of large k. Beard and Ellingson approximate this stochastic model in the case when biexciton lifetime is much less than single exciton lifetime ($\tau_1 \gg \tau_2$) and the average number of photons absorbed per quantum dot is less than 2 ($\langle k\rangle < 2$). Using these assumptions, they report a universal dependence on nanocrystal volume similarly as that of Robel *et al.* [26]. Figure 5.6 [26, 31] illustrates the size-dependent trends reported for quantum dots with very similar band structures, such as the lead chalcogenides in Figure 5.6a, as well as those with very different band structures in Figure 5.6b.

The fundamental physics behind the reported universal size-dependence is not completely understood but is presumed to be in part due to the relaxation of conservation laws that result from quantum confinement. Klimov postulates that this mitigating effect may negate the need for phonon-assisted AR in indirect-gap materials and remove the band gap dependent activation barrier for direct-gap semiconductors [32].

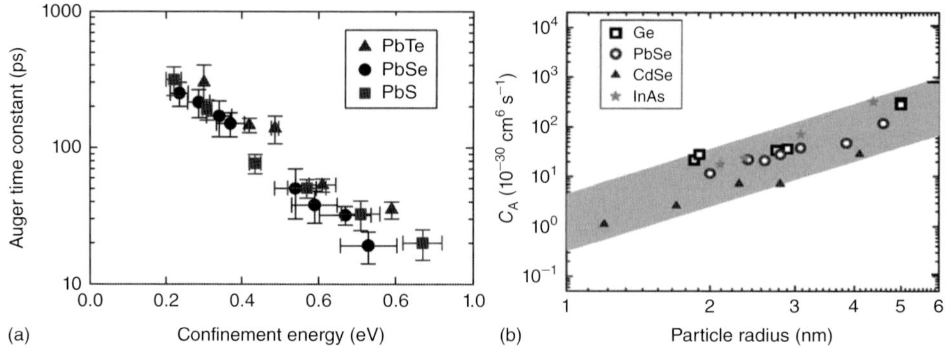

Figure 5.6 (a) E_c-dependent Auger relaxation trends for semiconductors with similar band structures, PbTe, PbSe, and PbS, where confinement energy is $E_c = E_g - E_{bulk}$ and the Auger time constant refers to τ_2 in the text. (Reprinted with permission from Padilha *et al.* Carrier multiplication in semiconductor nanocrystals: Influence of size, shape, and composition. Accounts of Chemical Research. Copyright 2013 American Chemical Society [31].) (b) Size-dependent trends are shown for semiconductors with very different electronic structures. Germanium is an indirect-gap semiconductor while PbSe, CdSe, and InAs have direct-gaps; however, CdSe is a wide-gap material while PbSe and InAs have very narrow band gaps. (Reprinted with permission from Robel *et al.* Universal Size-Dependent Trend in Auger Recombination in Direct-Gap and Indirect-Gap Semiconductor Nanocrystals. Physical Review Letters. Copyright 2009 American Physical Society [26].)

5.6 Ultrafast Techniques

Different spectroscopic techniques are employed to study the dynamics of quantum dots, such as transient absorption (TA), time-resolved terahertz spectroscopy (TRTS), and time-resolved photoluminescence (TRPL). In each case, ultrafast pulses are used to excite, or "pump," a sample with energy at or above the band gap and the subsequent probe or resulting emission provides information about carrier dynamics. Differences among these techniques include the ability to provide information about intraband versus interband dynamics, the sensitivity that can be achieved, and the information that can be gathered in addition to carrier dynamics.

5.6.1 Pump-Probe

A typical TA experimental setup is shown in Figure 5.7 and includes an ultrafast pump pulse produced from a Ti:Sapphire (Ti:Sapph) oscillator that can be used to excite a sample at 800 nm or can be amplified to pump an optical parametric amplifier (OPA). The OPA (in some cases a noncollinear optical parametric amplifier, NOPA, is used) can be tuned to different wavelengths in the visible and/or near-infrared, depending on the apparatus. The weaker probe pulse is generated in a similar way, but at a lesser power, and the wavelength chosen

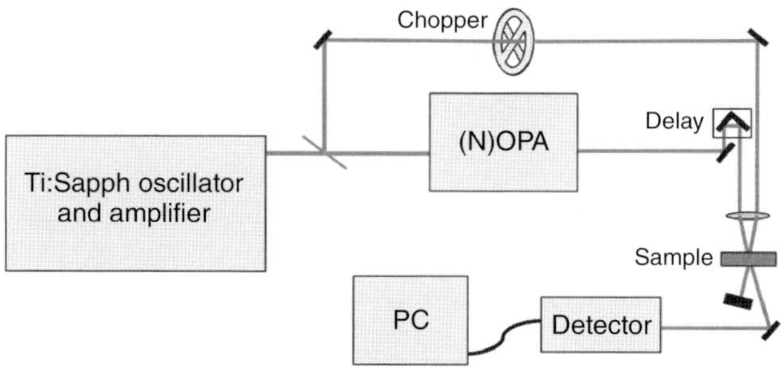

Figure 5.7 A basic representation of a TA experiment in which the Ti:Sapph output is split by a beamsplitter and one arm is routed to excite, or pump, the sample while the other pumps a OPA (or NOPA). The (N)OPA output is variably delayed and then used to probe pump-induced changes in the sample before entering the detector.

depends on the experiment; the probe is commonly tuned red of the excitation pulse but can also be degenerate with the energy of the pump. The probe is time delayed with respect to the pump and routed to a detector where the change in the probe with and without the pump is monitored at varying time delays as a measure of population decay.

In the case of TRTS, the change in the THz electric field is monitored and can provide information about photoconductivity when the pump is held constant and the probe is variably time delayed. To monitor carrier dynamics, the probe is held at a constant time delay while varying the pump delay. In both instances, the change in the THz electric field is detected by recombining the probe with a pulse train from the oscillator. The setup for this experiment is similar to that shown in Figure 5.7 but replaces the OPA with a nonlinear crystal to produce THz pulses via optical rectification, a type of difference frequency generation, and allows variable delay of either the pump or the probe. It also requires an additional pulse path to originate from the oscillator, be routed around the sample, and recombined with the probe for THz detection.

Pump-probe experiments on quantum dots commonly pump with photon energy above the band gap and probe with energy at the first exciton peak or in a region red of the band gap. The initial pump pulse results in the population of CB states and the subsequent relaxation of these carriers is monitored with the probe pulse. In these instances, the wavelength of the probe determines which population dynamics are being monitored. Different regions are probed to collect information about either interband or intraband relaxation dynamics; interband describes transitions between the CB and VB while intraband refers to transitions that occur within one of the bands: (i) tuning the probe to the first exciton peak, or anywhere blue of it but still red of the pump, provides interband population decay information; (ii) probing at energies red of the first exciton peak, in the near- or mid-infrared, monitors intraband dynamics of the carrier with the smaller effective mass and, therefore, larger interband spacing; and (iii) probing in the far-infrared with THz pulses provides information on intraband dynamics for carriers with small intraband spacing, often the hole when studying semiconductors such as CdTe with $m_e < m_h$.

5.6.2
Photoluminescence

In the case of TRPL, upconversion photoluminescence (uPL) is often used [22, 23]. A pulse, generated directly from a Ti:Sapph system (oscillator and amplifier), doubled with a nonlinear optical crystal or generated from an OPA, excites the sample and then the subsequent emission is collected and routed to a nonlinear crystal where it is mixed with another, variably delayed, Ti:Sapph or otherwise generated pulse. The resulting signal is spectrally filtered with a monochromator and routed to a detector. An alternative is to spectrally disperse the emission and then temporally resolve it using a streak camera, rather than mixing it with the delayed pulse before collection. The temporal resolution in these experiments can be worse than in TA experiments, a few picoseconds compared to tens or hundreds of femtoseconds, but provides a background-free measurement that is useful when working at the very low fluences required for CM experiments.

5.6.3
Relaxation Times

Pump-probe and TRPL studies of quantum dots, such as those described here, produce signals that can be described by a sum of exponentials, or in the case of biexciton decay

$$A_1 e^{-\left(\frac{t}{\tau_1}\right)} + A_2 e^{-\left(\frac{t}{\tau_2}\right)}$$

where τ_2 and τ_1 signify biexciton and single exciton decay time, respectively. The biexciton decay component is distinctly fast in comparison to the single exciton decay of the signal, which varies with the size and semiconductor being studied but is typically on the order of nano- to microseconds.

5.7
Quantum Efficiency

In order to quantify pump-probe results into efficiencies of multiple electron–hole pair production, signal amplitudes with excitation above and well below the energetic threshold for a particular material are compared [18]. Since nonradiative AR takes place on a much faster timescale than radiative decay, the signal at very long times is assumed to reflect the decay of a single electron–hole pair while early times are associated with the decay of multiple electron–hole pairs.

Multiple electron–hole pairs per quantum dot can be produced either via multiple photon absorption or via CM after single photon absorption. In order to distinguish between the two events using ultrafast spectroscopic techniques, photon energy

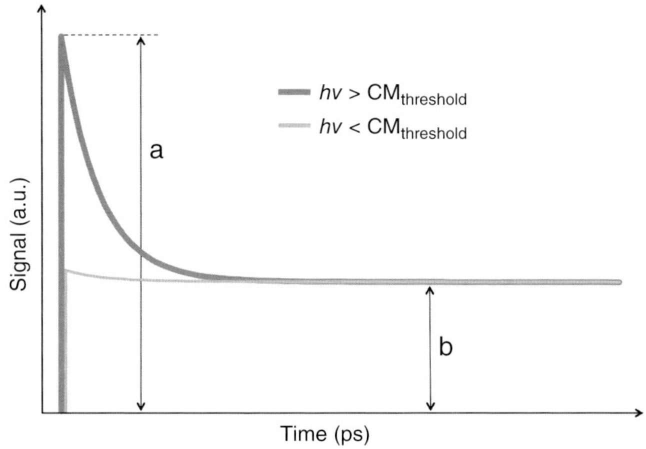

Figure 5.8 Representation of a typical trace found with TA when exciting above the CM threshold (dark gray) and below the CM threshold (light gray), both with low pump fluence. **a** and **b** are used to extrapolate quantum yield, as described in the text.

and laser fluence, or incident photons per sample area, are important parameters. Figure 5.8 is a representation of traces obtained through TA using low fluence but varying photon energy to be less than and greater than the energetic threshold for CM.

The signal generated at low fluence with energy less than the threshold (light gray) remains fairly constant and reflects long decay times corresponding to the radiative rate of single electron–hole recombination. For example, the energetic threshold for PbSe to undergo CM is approximately $3E_g$ [18], meaning that a single absorbed photon with less than this energy will not produce multiple electron–hole pairs; however, if the low energy is maintained but the fluence is raised to ensure that multiple photons are absorbed by a single quantum dot, then multiple electron–hole pairs can be created at this energy. In the latter instance, the shape of the signal when scaled to the same value at long times will resemble the dark gray trace as shown in Figure 5.8. The large amplitude at early times reflects the multiple electron–hole pairs compared to the lower amplitude at later times, which matches temporal dynamics of the single electron–hole pair decay of the light gray trace.

With this in mind, CM experiments excite quantum dot samples with energy above the threshold but at fluences that should allow, at most, absorption of only one photon per quantum dot. Then, the signal at short times is compared to that at long times and the number of electron–hole pairs is extrapolated. Referring back to the PbSe example, if the photon energy is at least $3E_g$, then CM should be measured with very low fluence, when the probability of absorbing a photon is very low so that at most one photon is absorbed per quantum dot, and should resemble the dark gray trace shown in Figure 5.8. The ratio of signals at early and late times (**a**/**b**) is assumed to represent the quantum efficiency (QE), or yield in TA experiments, which refers to the average number of electron–hole pairs produced per absorbed photon [18]. Obtaining quantum efficiencies from TRPL data is more complicated and usually includes extracting parameters for the particular system being studied by modeling data on the multiphoton excitation of multiple electron–hole pairs [22, 33]. QE can be plotted against photon energy per quantum dot band gap ($h\nu/E_g$) or ($h\nu$) to compare the efficiencies of different sizes of a particular dot or to compare the dot(s) with the bulk.

5.7.1 Quantum Yield Arguments

A question of whether the quantum yield should be plotted against ($h\nu/E_g$) or ($h\nu$) is part of the previously mentioned controversy surrounding CM efficiency in quantum dots versus bulk materials. Some [23, 34] stress the importance of comparing the quantum yield to the incoming photon energy per quantum dot band gap, or reduced photon energy, in order to assess the efficiency of nanocrystals of different sizes and semiconductor materials. The yield versus ($h\nu/E_g$) comparison, shown in Figure 5.9a, demonstrates the ability of quantum dots to undergo CM with a smaller percentage of their band gap energy than that required by bulk, $3E_g$ versus $6E_g$ for PbSe [23], reflecting an increase in efficiency.

Others [22, 35] view the importance to be the photon energy needed to achieve CM regardless of band gap, and, therefore, size, because this value determines if a device will undergo CM under solar illumination and potentially make an efficient PV (shown in Figure 5.9b). While Nair *et al.* agree that CM in quantum dots can potentially provide more energy for PVs simply due to the higher energy of their band gap, they dispute the increase in CM efficiency that has been claimed by others. In addition, they assert that differences in the underlying photophysics of quantum dots versus bulk materials are best evaluated through quantum yield versus $h\nu$.

Figure 5.9 [23] demonstrates the different trends observed by analyzing the QE versus (Figure 5.9a) $h\nu/E_g$ or simply (Figure 5.9b) $h\nu$ as well as by changing the y-axis to reflect different "figures of merit," (Figure 5.9c) and (Figure 5.9d). McGuire *et al.* explain that from a device design perspective, the product of the QE and band gap is necessary and appropriate to evaluate the power conversion efficiency of the resulting PV; further, a CM figure of merit, the ratio of real and ideal multiexciton yields for absorption of a photon with energy $h\nu$, is employed in order to compare experimental results to ideal theoretical performance.

Whichever analytical approach is used impacts the resulting QEs causing some [36] to assert that quantum dot CM efficiencies are equal to or less than those of bulk semiconductors when excited with the same photon energy while others [37] emphasize that quantum dot efficiencies exceed those of bulk with fewer multiples of the band gap necessary to achieve CM.

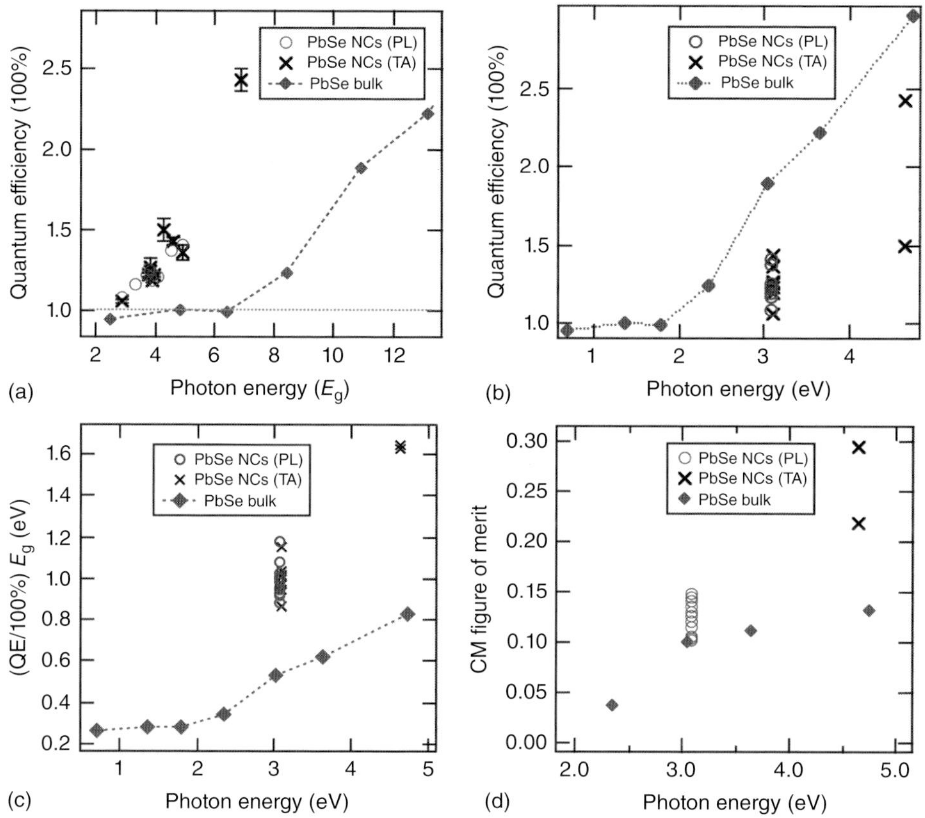

Figure 5.9 In each panel, the circles represent data obtained from TRPL, the crosses are from PL data, and the diamonds are bulk values that the authors extrapolated from reference [13]. (a) Denotes an increase in quantum efficiency of quantum dots over bulk when evaluated against the normalized photon energy while (b) shows bulk PbSe to be more efficient when absolute photon energy is considered. (c) and (d) further evaluate different perspectives by maintaining the absolute photon energy in the *x*-axis but using values that determine power conversion efficiency in a PV device and compare real and ideal multiexciton yields in the *y*-axis. (Reprinted with permission from McGuire *et al.* Apparent versus true carrier multiplication yields in semiconductor nanocrystals. Nano Letters. Copyright 2010 American Chemical Society [23].)

5.7.2
Experimental Considerations

Considerations other than different analytic approaches must also be taken into account, such as sample preparation, photodegradation, and surface effects, all of which contribute to the question of CM efficiency in quantum dots. Researchers have endeavored to address photocharging and photodegradation of quantum dot samples that can occur at even low excitation fluences, to find consistent approaches in the synthesis and handling of samples, and also to understand surface traps and the influence of ligands.

Early experiments used stationary samples [18, 19], which are susceptible to photocharging and degradation, but more recent data has been taken with flowing [38], stirring [22], and spinning [39] colloidal samples. McGuire *et al.* [23] studied the difference in dynamics between static and stirred colloidal solutions of PbSe and found additional short timescales in static samples as well as signal increase at early times and decrease at late times. The change in signal amplitude for static solutions can explain some of the previous reports of extremely high CM efficiencies [40]; nonetheless, the ratio of early to late time signals of the stirred samples exhibit some increase in the CM of the PbSe quantum dots over that of bulk in McGuire's analysis.

Quantum dot arrays, or well-ordered thin films, are direct precursors to functional PV devices but challenges exist in obtaining data on refreshed arrays, which cause most such experiments to be performed on static samples. Fortunately, novel techniques have recently been developed to deal with this dilemma better, such as a low-noise, high-speed, air-free, rotating sample cell. Initial TA experiments on lead selenide arrays using such a rotating cell found no difference in the decays of refreshed and static samples, but further studies are needed to confirm and expand on these results [41].

5.8 Ligand Exchange and Film Studies

Depending on the material of the quantum dot being studied, different bulky organic ligands are normally attached during synthesis and left on the dot for colloidal studies. For example, oleate is commonly used on lead chalcogenide colloids but exchanged for shorter ligands in arrays. Although it varies from ligand to ligand, a common procedure for the exchange includes coating a substrate with oleate-capped dots by spin-casting, dip-coating, or drop-casting and then dipping the substrate into a dilute solution containing the desired ligand for some period of time. The amount of time necessarily varies with the new ligand; while 1,2-ethanedithiol (EDT) requires only a few minutes [42], certain amines are treated over a 24-h period [43]. Many iterations of dipping the substrate in the oleate-capped solution followed by the shorter ligand solution may be necessary in order to achieve the necessary optical density for ultrafast studies.

Determining structural and optical changes that occur after treating arrays provides important information for spectroscopists when performing ultrafast studies. For example, Talapin and Murray observed the evolution of a redshift in the absorption spectrum of PbSe treated with hydrazine in acetonitrile [44]. Similarly, Law *et al.* replaced the oleate ligand on PbSe quantum dots with various amines and found the expected redshift of the first absorption peak, but treatments with pure hydrazine and pure pyridine caused considerable nanocrystal growth so

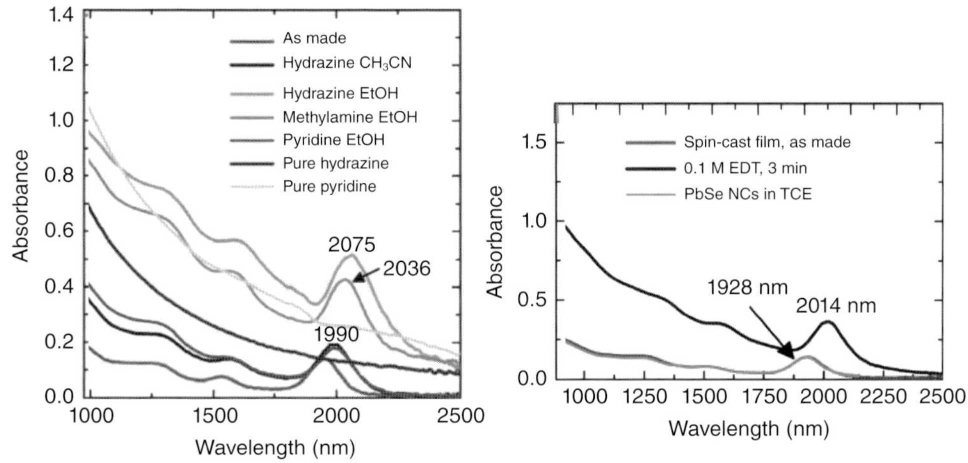

Figure 5.10 Absorption spectra of colloidal and spin-cast oleate-capped PbSe quantum dots with various ligand treatments, which redshift the first transition peak. (Adapted with permission from Law *et al.* Structural, Optical, and Electrical Properties of PbSe Nanocrystal Solids Treated Thermally or with Simple Amines. Journal of the American Chemical Society. Copyright 2008 American Chemical Society (a) [43] and Luther *et al.* Structural, optical, and electrical properties of self-assembled films of PbSe nanocrystals treated with 1,2-ethanedithiol. ACS Nano. Copyright 2008 American Chemical Society (b) [42].)

that the first absorption peak was not even observed on the absorption spectra out to 2500 nm (Figure 5.10a) [43].

The redshift of the first exciton peak occurs consistently for PbSe after ligand exchange treatment, as well as in other semiconductor quantum dots, and has been attributed to (i) an increased dipole–dipole interactions of neighboring dots [45]; (ii) polarization effects due to dielectric changes in the environment [46]; or (iii) a combination of increased interdot radiative coupling and increased interdot electronic nonradiative coupling [42].

Beard *et al.* [47] followed similar methods as those listed above to study ligand exchange but also looked at CM efficiencies and lifetimes that different treatments produced. This study built on their previous work with EDT [42, 48] (absorption spectra shown in Figure 5.10b) as well as hydrazine and methylamine [49, 50] by performing exchanges on arrays of two different sizes of PbSe quantum dots.

The data from this study are shown in Figures 5.11 and 5.12 [47] and illustrate differences in the carrier dynamics between colloidal solutions and various film treatments. In most cases, the authors find a decrease in CM efficiency with the ligand exchange but an increase in Auger decay times; however, ultrafast studies alone may not absolutely predict the success of a particular ligand, such as EDT, in producing greater CM QE. Although EDT was shown to extremely reduce efficient CM in both sizes of these PbSe dots, this ligand treatment has enabled successful Schottky- [48] and hetero-junction [51] solar cells.

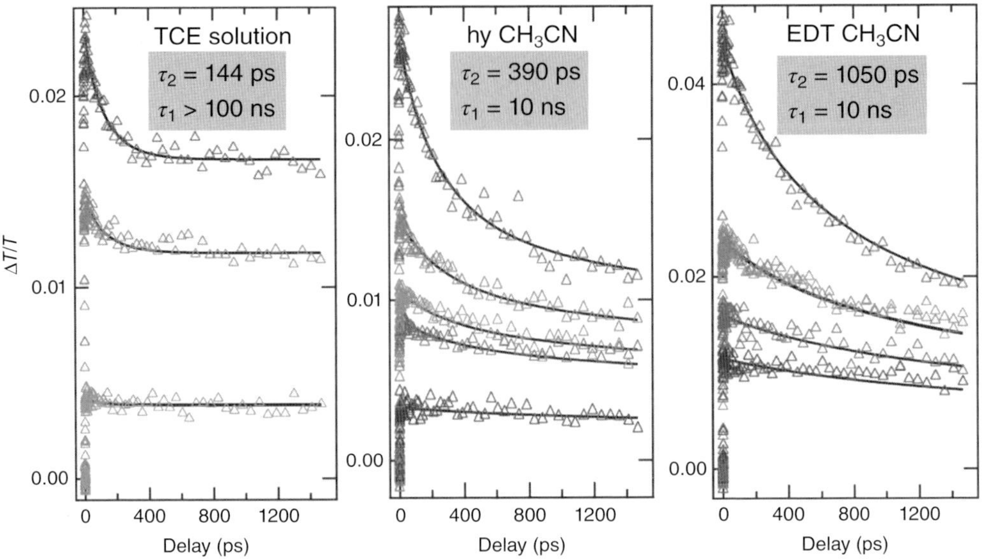

Figure 5.11 Transient absorption data of 7.4 nm diameter colloidal oleate-capped PbSe quantum dots in trichloroethylene (TCE), a hydrazine (hy) treated PbSe array, and a 1,2-ethanedithiol (EDT) treated PbSe array. All were taken at increasing pump fluence but below the CM threshold energy, at $1.6E_g$, and probed at the first exciton peak. (Reprinted with permission from Beard *et al.* Variations in the quantum efficiency of multiple exciton generation for a series of chemically treated PbSe nanocrystal films. Nano Letters. Copyright 2009 American Chemical Society [47].)

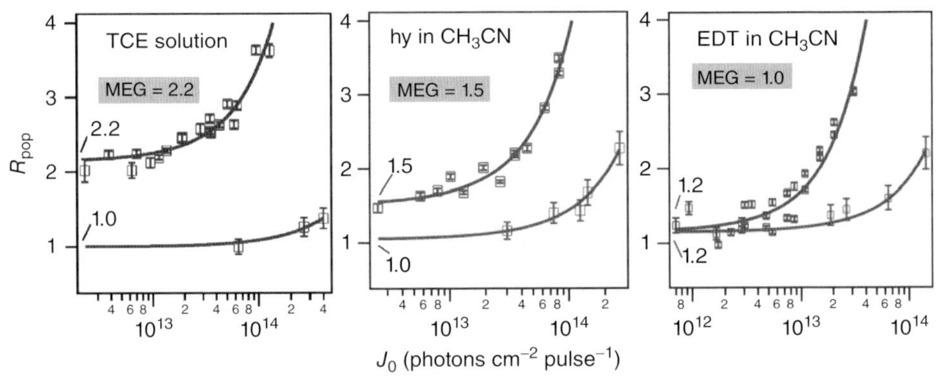

Figure 5.12 J_0 indicates the fluence and R_{pop} represents the ratio of the signal at short versus long times, or (**a**/**b**) from Figure 5.8, and is shown for excitation below and above the CM threshold. The hy treated array shows a reduction in MEG compared with isolated colloidal PbSe while the EDT treated array indicates no MEG at all. (Reprinted with permission from Beard *et al.* Variations in the quantum efficiency of multiple exciton generation for a series of chemically treated PbSe nanocrystal films. Nano Letters. Copyright 2009 American Chemical Society [47].)

More recently, researchers have looked to halides to passivate and protect quantum dot surfaces [52–54] and have found much higher electron mobilities than those associated with organic ligands. Carrier mobility describes the distance that an electron or hole will travel under the influence of an electric field and is an important parameter in designing next generation quantum dot PVs. A higher mobility increases the possibility of extracting a carrier within its lifetime and increasing the photon conversion efficiency.

Halide treatments also appear to produce fewer and shallower trap states compared to even the most ideal organic ligand, which may in turn contribute to high carrier mobility [52]. In addition, Bae *et al.* [54] reported that treatment with molecular chlorine increased the decay time in TRPL experiments and retained these timescales even after several weeks of air-exposure, although higher treatment concentrations resulted in the addition of a fast, early time component. However, like most other experiments on quantum dot arrays, the samples were static so that photocharging and photodegradation must be considered when interpreting the results.

5.9
Conclusions

This chapter highlighted the results of ultrafast spectroscopic studies of quantum dots but many other techniques have been used to characterize and otherwise analyze semiconductor nanocrystals, including photocurrent studies that report efficiencies over 100% in PbS [55] and PbSe [56] devices, providing evidence of MEG-enhancement and furthering the argument for improved CM efficiency in quantum dots over bulk. The "boom" in nanocrystal research over the past decade has resulted in substantial improvements to quantum dot solar cell efficiencies over a relatively short period of time. In only a few years, researchers have addressed the inconsistencies in data associated with sample preparation, data collection techniques, and data analysis. Although some argument on these issues remain, the resulting discussion and research has had a tremendous impact on the field of nanotechnology and contributed to the rapid advancement of quantum dot PV devices.

Acknowledgments

Special thanks to Prof. David Jonas for his valuable input. During the preparation of this chapter, the author was supported by the Center for Advanced Solar Photophysics, an Energy Frontier Research Center funded by the U.S. Department of Energy, Office of Science, Office of Basic Energy Sciences.

References

1. Green, M.A. (2006) *Third Generation Photovoltaics: Advanced Solar Energy Conversion*, Springer.
2. National Renewable Energy Laboratory, U.S.D.o.E (2013) Research Cell Efficiency Records, *www.nrel.gov/ncpv/images/efficiency˙chart.jpg* (accessed 26 November 2013).
3. National Renewable Energy Laboratory, U.S.D.o.E (2013) Reference Solar

Spectral Irradiance, *http://rredc.nrel.gov/solar/spectra/am1.5/*.

4. Semonin, O.E., Luther, J.M., and Beard, M.C. (2012) Quantum dots for next-generation photovoltaics. *Mater. Today*, **15** (11), 508–515.
5. Luque, A. (2011) Will we exceed 50% efficiency in photovoltaics? *J. Appl. Phys.*, **110** (3), 031301.
6. Shockley, W. and Queisser, H. (1961) Detailed balance limit of efficiency of p-n junction solar cells. *J. Appl. Phys.*, **32** (3), 510–519.
7. Ross, R. and Nozik, A. (1982) Efficiency of hot-carrier solar energy converters. *J. Appl. Phys.*, **53** (5), 3813–3818.
8. Boudreaux, D.S., Williams, F., and Nozik, A.J. (1980) Hot carrier injection at semiconductor-electrolyte junctions. *J. Appl. Phys.*, **51** (4), 2158–2163.
9. Kolodinski, S., Werner, J., Wittchen, T., and Queisser, H. (1993) Quantum efficiencies exceeding unity due to impact ionization in silicon solar cells. *Appl. Phys. Lett.*, **63** (17), 2405–2407.
10. Landsberg, P., Nussbaumer, H., and Willeke, G. (1993) Band-band impact ionization and solar cell efficiency. *J. Appl. Phys.*, **74** (2), 1451–1452.
11. Hanna, M. and Nozik, A. (2006) Solar conversion efficiency of photovoltaic and photoelectrolysis cells with carrier multiplication absorbers. *J. Appl. Phys.*, **100** (7), 074510.
12. Landsberg, P.T. (1991) *Recombination in Semiconductors*, Cambridge University Press.
13. Pijpers, J., Ulbricht, R., Tielrooij, K., Osherov, A., Golan, Y., Delerue, C., Allan, G., and Bonn, M. (2009) Assessment of carrier-multiplication efficiency in bulk PbSe and PbS. *Nat. Phys.*, **5** (11), 811–814.
14. Rühle, S., Shalom, M., and Zaban, A. (2010) Quantum-dot-sensitized solar cells. *ChemPhysChem*, **11** (11), 2290–2304.
15. Ip, A., Thon, S., Hoogland, S., Voznyy, O., Zhitomirsky, D., Debnath, R., Levina, L., Rollny, L., Carey, G., and Fischer, A. (2012) Hybrid passivated colloidal quantum dot solids. *Nat. Nanotechnol.*, **7** (9), 577–582.
16. Ravich, Y.I. and Khokhlov, D. (2003) in *Lead Chalcogenides: Physics and Applications* (ed. D. Khokhlov), Taylor and Francis, New York, p. 3.
17. Wise, F.W. (2000) Lead salt quantum dots: the limit of strong quantum confinement. *Acc. Chem. Res.*, **33** (11), 773–780.
18. Schaller, R. and Klimov, V. (2004) High efficiency carrier multiplication in PbSe nanocrystals: implications for solar energy conversion. *Phys. Rev. Lett.*, **92** (18), 186601.
19. Ellingson, R., Beard, M., Johnson, J., Yu, P., Micic, O., Nozik, A., Shabaev, A., and Efros, A. (2005) Highly efficient multiple exciton generation in colloidal PbSe and PbS quantum dots. *Nano Lett.*, **5** (5), 865–871.
20. Schaller, R., Petruska, M., and Klimov, V. (2005) Effect of electronic structure on carrier multiplication efficiency: comparative study of PbSe and CdSe nanocrystals. *Appl. Phys. Lett.*, **87** (253), 102.
21. Beard, M., Knutsen, K., Yu, P., Luther, J., Song, Q., Metzger, W., Ellingson, R., and Nozik, A. (2007) Multiple exciton generation in colloidal silicon nanocrystals. *Nano Lett.*, **7** (8), 2506–2512.
22. Nair, G., Geyer, S., Chang, L.Y., and Bawendi, M. (2008) Carrier multiplication yields in PbS and PbSe nanocrystals measured by transient photoluminescence. *Phys. Rev. B*, **78** (12), 125325.
23. McGuire, J., Sykora, M., Joo, J., Pietryga, J., and Klimov, V. (2010) Apparent versus true carrier multiplication yields in semiconductor nanocrystals. *Nano Lett.*, **10** (6), 2049–2057.
24. Klann, R., Hofer, T., Buhleier, R., Elsaesser, T., and Tomm, J.W. (1995) Fast recombination processes in lead chalcogenide semiconductors studied via transient optical nonlinearities. *J. Appl. Phys.*, **77** (1), 277–286.
25. Haug, A. (1988) Band-to-band auger recombination in semiconductors. *J. Phys. Chem. Solids*, **49** (6), 599–605.
26. Robel, I., Gresback, R., Kortshagen, U., Schaller, R., and Klimov, V. (2009) Universal size-dependent trend in auger recombination in direct-gap and

indirect-gap semiconductor nanocrystals. *Phys. Rev. Lett.*, **102** (17), 177404.
27. Klimov, V.I. (2000) Quantization of multiparticle auger rates in semiconductor quantum dots. *Science*, **287** (5455), 1011–1013.
28. Klimov, V., Mcguire, J., Schaller, R., and Rupasov, V. (2008) Scaling of multiexciton lifetimes in semiconductor nanocrystals. *Phys. Rev. B*, **77** (19), 195324.
29. Barzykin, A. and Tachiya, M. (2007) Stochastic models of charge carrier dynamics in semiconducting nanosystems. *J. Phys.: Condens. Matter*, **19** (6), 065105.
30. Beard, M. and Ellingson, R. (2008) Multiple exciton generation in semiconductor nanocrystals: toward efficient solar energy conversion. *Laser Photonics Rev.*, **2** (5), 377–399.
31. Padilha, L.A., Stewart, J.T., Sandberg, R.L., Bae, W.K., Koh, W.K., Pietryga, J.M., and Klimov, V.I. (2013) Carrier multiplication in semiconductor nanocrystals: influence of size, shape, and composition. *Acc. Chem. Res.*, **46** (6), 1261–1269.
32. Klimov, V.I. (2010) Multiexciton phenomena in semiconductor nanocrystals, in *Nanocrystal Quantum Dots*, vol. 2 (ed. V. Klimov), CRC Press.
33. McGuire, J., Joo, J., Pietryga, J., Schaller, R., and Klimov, V. (2008) New aspects of carrier multiplication in semiconductor nanocrystals. *Acc. Chem. Res.*, **41** (12), 1810–1819.
34. Beard, M. (2011) Multiple exciton generation in semiconductor quantum dots. *J. Phys. Chem. Lett.*, **2** (11), 1282–1288.
35. Nair, G., Chang, L.Y., Geyer, S., and Bawendi, M. (2011) Perspective on the prospects of a carrier multiplication nanocrystal solar cell. *Nano Lett.*, **11** (5), 2145–2151.
36. Pijpers, J., Ulbricht, R., Tielrooij, K., Osherov, A., Golan, Y., Delerue, C., Allan, G., and Bonn, M. (2009) Assessment of carrier-multiplication efficiency in bulk PbSe and PbS. *Nat. Phys.*, **5** (11), 811–814.
37. Beard, M., Midgett, A., Hanna, M., Luther, J., Hughes, B., and Nozik, A. (2010) Comparing multiple exciton generation in quantum dots to impact ionization in bulk semiconductors: implications for enhancement of solar energy conversion. *Nano Lett.*, **10** (8), 3019–3027.
38. Midgett, A.G., Hillhouse, H.W., Hughes, B.K., Nozik, A.J., and Beard, M.C. (2010) Flowing versus static conditions for measuring multiple exciton generation in PbSe quantum dots. *J. Phys. Chem. C*, **114** (41), 17486–17500.
39. Cho, B., Peters, W., Hill, R., Courtney, T., and Jonas, D. (2010) Bulk-like hot carrier dynamics in lead sulfide quantum dots. *Nano Lett.*, **10** (7), 2498–2505.
40. Schaller, R., Sykora, M., Pietryga, J., and Klimov, V. (2006) Seven excitons at a cost of one: redefining the limits for conversion efficiency of photons into charge carriers. *Nano Lett.*, **6** (3), 424–429.
41. Hill, R.J. (2013) Enabling two-dimensional fourier transform electronic spectroscopy on quantum dots. PhD thesis. University of Colorado, Boulder.
42. Luther, J.M., Law, M., Song, Q., Perkins, C.L., Beard, M.C., and Nozik, A.J. (2008) Structural, optical, and electrical properties of self-assembled films of PbSe nanocrystals treated with 1, 2-ethanedithiol. *ACS Nano*, **2** (2), 271–280.
43. Law, M., Luther, J., Song, Q., Hughes, B., Perkins, C., and Nozik, A. (2008) Structural, optical, and electrical properties of PbSe nanocrystal solids treated thermally or with simple amines. *J. Am. Chem. Soc.*, **130** (18), 5974–5985.
44. Talapin, D. and Murray, C. (2005) PbSe nanocrystal solids for n-and p-channel thin film field-effect transistors. *Science*, **310** (5745), 86–89.
45. Döllefeld, H., Weller, H., and Eychmüller, A. (2002) Semiconductor nanocrystals assemblies: experimental pitfalls and a simple model of particle-particle interaction. *J. Phys. Chem. B*, **106** (22), 5604–5608.
46. Leatherdale, C. and Bawendi, M. (2001) Observation of solvatochromism in CdSe colloidal quantum dots. *Phys. Rev. B*, **63** (16), 165315.
47. Beard, M., Midgett, A., Law, M., Semonin, O., Ellingson, R., and

Nozik, A. (2009) Variations in the quantum efficiency of multiple exciton generation for a series of chemically treated PbSe nanocrystal films. *Nano Lett.*, **9** (2), 836–845.

48. Luther, J., Law, M., Beard, M., Song, Q., Reese, M., Ellingson, R., and Nozik, A. (2008) Schottky solar cells based on colloidal nanocrystal films. *Nano Lett.*, **8** (10), 3488–3492.

49. Luther, J., Beard, M., Song, Q., Law, M., Ellingson, R., and Nozik, A. (2007) Multiple exciton generation in films of electronically coupled PbSe quantum dots. *Nano Lett.*, **7** (6), 1779–1784.

50. Murphy, J.E., Beard, M.C., and Nozik, A.J. (2006) Time-resolved photoconductivity of PbSe nanocrystal arrays. *J. Phys. Chem. B*, **110** (50), 25455–25461.

51. Choi, J., Lim, Y.F., Santiago-Berrios, M., Oh, M., Hyun, B.R., Sun, L., Bartnik, A., Goedhart, A., Malliaras, G., and Abruña, H. (2009) PbSe nanocrystal excitonic solar cells. *Nano Lett.*, **9** (11), 3749–3755.

52. Tang, J., Kemp, K., Hoogland, S., Jeong, K., Liu, H., Levina, L., Furukawa, M., Wang, X., Debnath, R., and Cha, D. (2011) Colloidal-quantum-dot photovoltaics using atomic-ligand passivation. *Nat. Mater.*, **10** (10), 765–771.

53. Ning, Z., Ren, Y., Hoogland, S., Voznyy, O., Levina, L., Stadler, P., Lan, X., Zhitomirsky, D., and Sargent, E. (2012) All-inorganic colloidal quantum dot photovoltaics employing solution-phase halide passivation. *Adv. Mater.*, **24** (47), 6295–6299.

54. Bae, W., Joo, J., Padilha, L., Won, J., Lee, D., Lin, Q., Koh, W., Luo, H., Klimov, V., and Pietryga, J. (2012) Highly effective surface passivation of PbSe quantum dots through reaction with molecular chlorine. *J. Am. Chem. Soc.*, **134** (49), 20160–20168.

55. Sambur, J., Novet, T., and Parkinson, B. (2010) Multiple exciton collection in a sensitized photovoltaic system. *Science*, **330** (6000), 63–66.

56. Semonin, O., Luther, J., Choi, S., Chen, H.Y., Gao, J., Nozik, A., and Beard, M. (2011) Peak external photocurrent quantum efficiency exceeding 100% via MEG in a quantum dot solar cell. *Science*, **334** (6062), 1530–1533.

6
Micro Flow Chemistry: New Possibilities for Synthetic Chemists

Timothy Noël

6.1
Introduction

Since the urea synthesis by Wöhler in the nineteenth century, organic synthetic chemistry has revolutionized the way organic molecules are constructed. Being a vibrant research branch within chemistry, new useful reactions and reactivity patterns are continuously described in the literature. These new synthetic methodologies have a clear impact on the pharmaceutical and other industries, in addition to the biologically active molecules, supramolecular frameworks, organic materials, and polymers. It is, therefore, surprising that the common laboratory techniques and equipment have not changed essentially for nearly two centuries.

In the past decade, continuous processing using microreactor technology has gained a growing interest from the chemical and processing engineering community [1, 2]. The ability to exert a high degree of control over reaction/residence time and other process parameters in continuous-flow microreactors has resulted in an enhanced reproducibility when compared to traditional batch techniques. Owing to small dimensions (typically 100–1000 μm), this technology provides reduced safety hazards and high surface-to-volume ratios that enable fast heat and mass transfer. Consequently, harsh and unusual process conditions (e.g., high temperatures and pressures) can be achieved. These conditions are called Novel Process Windows and constitute reaction conditions far from the common laboratory practices [3]. In addition, quick scalability without extensive optimization can be achieved in these devices by employing prolonged operation times or by a numbering-up strategy [4].

Similar to all new emerging technologies, the evolution from batch to continuous processing of chemical reactions faces some challenges and has met with skepticism. One of the major issues is that most chemists are more familiar with batch processes in round-bottom flasks. Changing to continuous-flow microreactors requires a different mindset from the practitioner. In addition, certain engineering skills with regard to fluid dynamics, process analytical technology, and process control are needed and will facilitate success in developing efficient flow protocols. However, mastering these skills requires a certain time investment and can be overwhelming. Another issue is the initial investment to implement

Discovering the Future of Molecular Sciences, First Edition. Edited by Bruno Pignataro.

flow chemistry in the research environment. This financial barrier can be partly overcome by the increasing popularity of flow chemistry that offers affordable microreactor solutions (such as capillary tubing in PTFE (polytetrafluoroethylene), stainless steel, Hastelloy, or copper). The competition between suppliers also lowers the cost of the more advanced commercial microfluidic platforms and accessories. The handling of solids in microflow constitutes one of the major hurdles for micro flow chemistry. The formation of solids usually leads to irreversible blockage of the microchannels. Dealing with such problems often requires a rethinking of the reaction conditions. Several solutions to deal with solid-forming reactions have been described recently [5].

In this chapter, I give an overview of the most important advantages of micro flow chemistry for the organic synthetic chemist: how it can impact a typical organic synthetic process; what the key factors are that can increase the productivity and selectivity of a chemical reaction, and, finally, how I see the field evolving in the near future. It is not my aim to provide an exhaustive review of all the different reaction types that have been transferred to continuous-flow processing. For this, I refer to the specialized reviews that highlight such aspects in detail [1].

6.2 Characteristics of Micro Flow – Basic Engineering Principles

6.2.1 Mass Transfer – the Importance of Efficient Mixing

Owing to small length scales, mass transfer (mixing) can be significantly accelerated in microreactors. Two dimensionless numbers are relevant to describe these mass transport phenomena in fluid flow: Reynolds number (*Re*, Equation 6.1) and Péclet number (*Pe*, Equation 6.2). The Reynolds number (ρ = density; v = mean fluid velocity; d = distance; and μ = dynamic viscosity) describes the ratio between the inertial forces and the viscous forces and is used to characterize the different flow regimes (laminar or turbulent flow). Because of the small dimension of microreactors, the fluid flow is constrained to the laminar regime ($Re < 1$), which means that all fluid elements are flowing in parallel layers. Therefore, it can be said that mixing in microreactors is achieved based on diffusion. The Péclet number (v = mean fluid velocity; d = hydrodynamic diameter; and D = molecular diffusivity) describes the ratio of mass transport owing to convection and diffusion. Again, it can be easily understood that mixing is dependent on diffusion.

$$Re = \frac{\rho v d}{\mu} \tag{6.1}$$

$$Pe = \frac{vd}{D} \tag{6.2}$$

For a binary mixture, two streams combined in a tee-mixer can stream in parallel, and mixing between the two can be achieved by diffusion. The diffusion time can be estimated via Equation 6.3 (t = diffusion time; d = hydrodynamic diameter; and

D = molecular diffusivity). This equation demonstrates that the diffusion time is directly proportionate to the square of the diffusion path length and inversely proportional with the molecular diffusivity. The shorter the diffusion path length, the faster a uniform concentration is obtained. For example, a 500 μm channel needs approximately 30 s to achieve homogeneous mixing. Shortening the diffusion path length can accelerate the homogeneous mixing significantly, for example, 50 μm microchannels yield a mixing time of 0.3 s. This insight has led to engineer highly specialized micromixers that shorten the diffusion path by splitting the two miscible streams in multilamellae and subsequently recombine them (Figure 6.1). Hence, complete mixing can be achieved in microseconds and enable fast reactions to be completed in a matter of seconds (so-called Flash Chemistry) [6].

$$t = \frac{d^2}{2D} \tag{6.3}$$

It is important to note that mixing does not take into account the reaction time [7]. The relative ratio between reaction rate and mass transfer via diffusion is given by the dimensionless Damköhler number (Equation 6.4, χ = a coefficient depending on kinetics and flow ratios; τ = residence time). When Da is larger than 1, concentration gradients exist of the reactor, which can lead to a significant by-product formation.

$$Da = \frac{\text{rate of reaction}}{\text{rate of diffusion}} = \frac{\chi\ dt^2}{4\tau D} \tag{6.4}$$

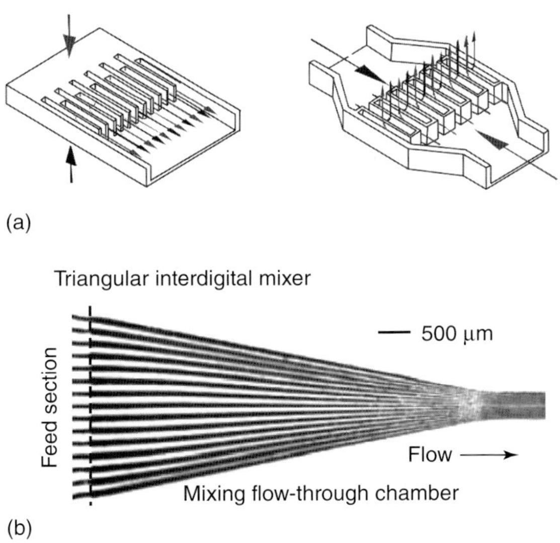

Figure 6.1 (a) Schematic representation of interdigital micromixers. (b) Flow pattern in an triangular interdigital micromixer with a flow distribution zone and a focusing zone. Mixing is done between two aqueous streams of which one stream contains a dye. Reprinted with permission from [3f]. Copyright (2004) John Wiley & Sons.

Micromixers are particularly useful to tune the selectivity of competitive consecutive reactions. The selectivity in this reaction class is determined by the kinetics of the two reactions as well as the mixing efficiency. Hereby, the timescale for mixing can be larger than the reaction timescale, which leads to the so-called disguised chemical selectivity. The Friedel–Crafts aminoalkylation of aromatic compounds with *N*-acyliminium ions constitutes an example of this principle (Scheme 6.1) [8]. Poor mixing leads to the formation of more dialkylated products than expected from the kinetically based selectivity. This can be explained by the formation of a greater local concentration of the monoalkylated product as a consequence of inefficient mixing. The excess of monoalkylated product can subsequently react with an excess of the *N*-acyliminium ions. A multilamination type micromixer can be used to establish rapidly a homogeneous mixture and lead to an impressive selectivity improvement for the monoalkylated product.

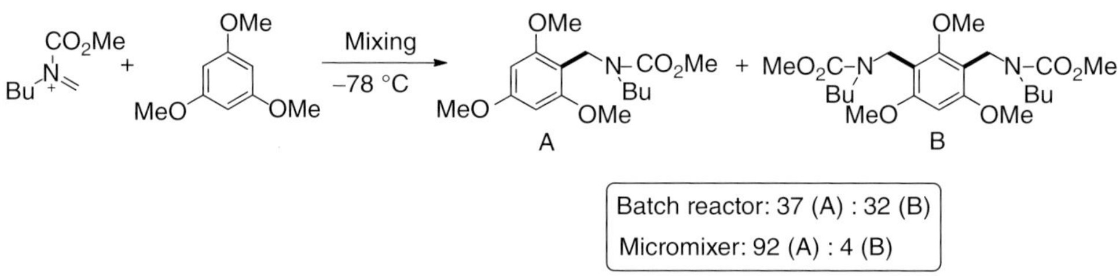

Scheme 6.1 Comparison between macromixing in a batch reactor and micromixing in a multilamination type micromixer for the Friedel–Crafts aminoalkylation of 1,3,5-trimethoxybenzene with *N*-acyliminium ions.

6.2.2
Heat Transfer – the Importance of Efficient Heat Management

Efficient heat transfer is a crucial property that needs to be considered by the chemical engineer when selecting an appropriate reactor type. Excellent heat removal is of great importance for highly exothermic reactions to avoid thermal runaways and significant by-product formation. To cope with these potential hazards, exothermic reactions are typically performed in large-scale reactor vessels under suboptimal reaction conditions, for example, at lower reagent concentrations or at lower temperatures. Owing to high surface-to-volume ratios encountered in microreactors, heat can be rapidly dissipated to the environment and avoid the formation of local hot spots. Moreover, an isothermal operation over the entire reactor length can be achieved in microfluidics.

The synthesis of ionic liquids is an example of the very exothermic reaction in which the presence of hot spots and uncontrolled overheating leads to a lower quality of the final product. The reaction between 1-methyl-imidazole and diethyl sulfate ($\Delta H = -130\ \mathrm{kJ\ mol^{-1}}$) was performed in a microreactor and cooled by means of heat pipes connected to he reactor [9]. Heat pipes are sealed tubes in which a phase transition (liquid to vapor) facilitates the heat removal from the reactor. Thus, the cooling rate is automatically adjusted to the heat released from the reaction and is therefore ideally suited to control runaway reactions.

Careful temperature control is also required in reactions where a small increase in temperature results in a degradation of the target product. An example of this is the single aryl halogen–lithium exchange of 1,2-dibromobenzene and subsequent quenching with an electrophile (Scheme 6.2) [10]. The hot spot formation in batch leads to the generation of a significant amount of benzyne and concomitant by-products. To make the transformation synthetically useful, reaction temperatures of −100 °C are required in batch. A microfluidic system consisting of micromixers and stainless steel capillaries resulted in the formation of the desired product in high yields (>70%) at higher reaction temperatures (−75 °C). Crucial for this result was the careful temperature control (Figure 6.2) and the rapid mixing enabled by the micromixers.

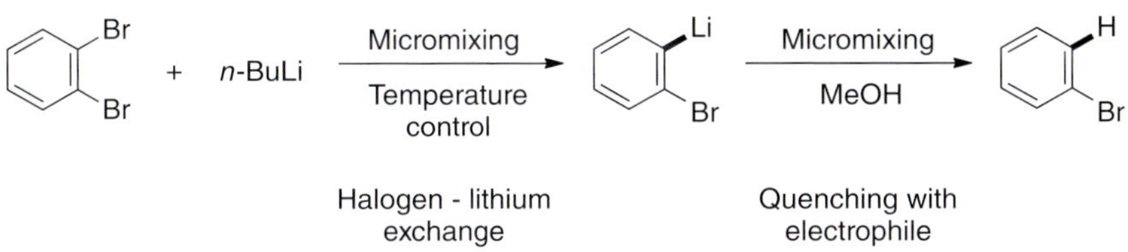

Scheme 6.2 Generation of *ortho*-bromophenyllithium and subsequent quenching with methanol to yield bromobenzene.

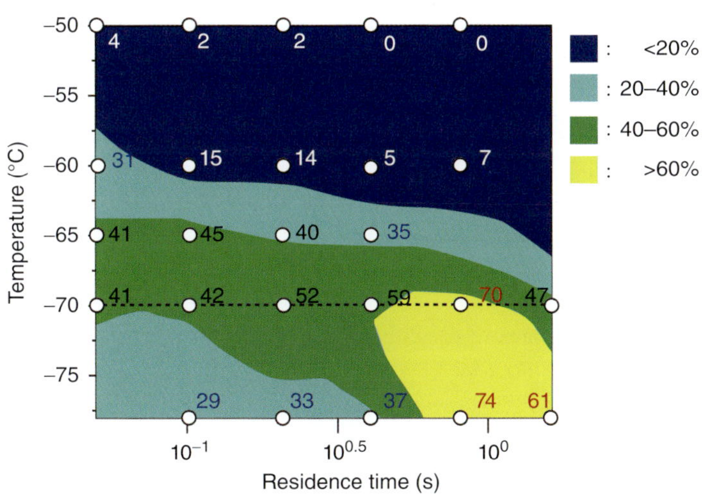

Figure 6.2 Contourplot that depicts the correlation between the reaction parameters (i.e., reaction temperature and reaction/residence time) and the yield of bromobenzene. Reprinted with permission from [10]. Copyright (2007) American Chemical Society.

6.2.3
Multiphase Flow

Multiphase flow constitutes the combination of two or more phases, such as gas, liquid, and solid phases, in a single immiscible flow. This involves gas–liquid flow, liquid–liquid, and liquid–solid flows, or a combination thereof. Liquid–solid flow is discussed in Sections 6.4 and 6.6. Gas–liquid reactions and biphasic liquid–liquid reactions are very common in the chemical industry. One of the main parameters that governs the reaction rate is the mass transfer rate of chemical reactants from one phase to the other. To facilitate this interfacial mass transfer, it is important to maximize the contact area between the two phases. In batch, this is typically done by using mechanically stirred vessels. However, one of the main drawbacks of stirred tank reactors is the poorly defined interfacial contact area (typically 100–1000 $m^2\ m^{-3}$ for liquid–liquid reactions), which leads to temperature and concentration gradients and can have its impact on the product selectivity. Microreactors have emerged as an enabling technology to achieve such multiphase reactions [11]. Owing to the small length scales, they provide large and well-defined interfacial areas (typically 10 000–50 000 m^2/m^3 for liquid-liquid reactions) and reduced axial dispersion allowing to operate the reactor as an ideal plug-flow reactor. For gas-liquid reactions, no headspace is present in continuous-flow microreactors. Depending on the flow rate, different flow regimes can be achieved in microreactors (Figure 6.3), and fine-tuning the interfacial area and thus reaction selectivity is also possible [12].

One of the most studied flow regimes in microreactors is the segmented flow, also denoted as Taylor or slug flow [13]. This flow regime is characterized by liquid slugs and interslugs. One of the main advantages of this flow regime is its excellent mixing capacity that originates from an internal circulation within these slugs (Figure 6.4) [14]. An intriguing example where slug flow was used to study the reaction kinetics was theso-called on water reactions [15]. Although the reactants are insoluble in water, a significant acceleration in reaction rate is observed when the organic reactants are suspended in water [16]. Generally, it is very difficult to achieve reproducible interfacial areas (which depends on aqueous to organic phase ratio, reactor geometry, and impeller speed) in batch reactors. However,

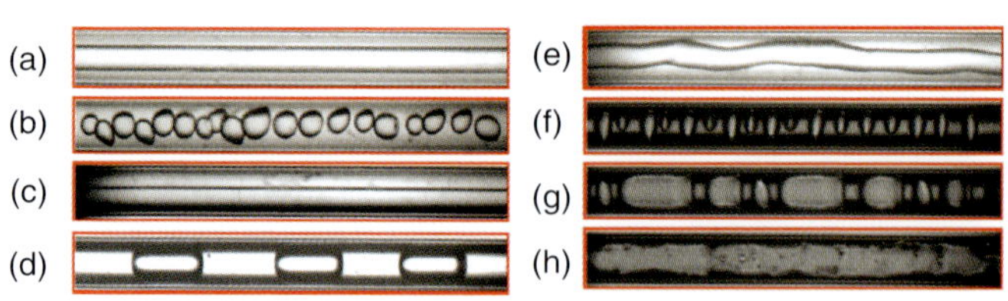

Figure 6.3 Different flow patterns in biphasic liquid–liquid flow encountered in microreactors: (a) annular flow, (b) bubbly flow, (c) parallel flow, (d) slug or segmented or Taylor flow, (e) wavy annular flow, (f) inverted bubbly flow, (g) inverted slug flow, and (h) inverted annular flow. Reprinted with permission from [12]. Copyright (2012) American Chemical Society.

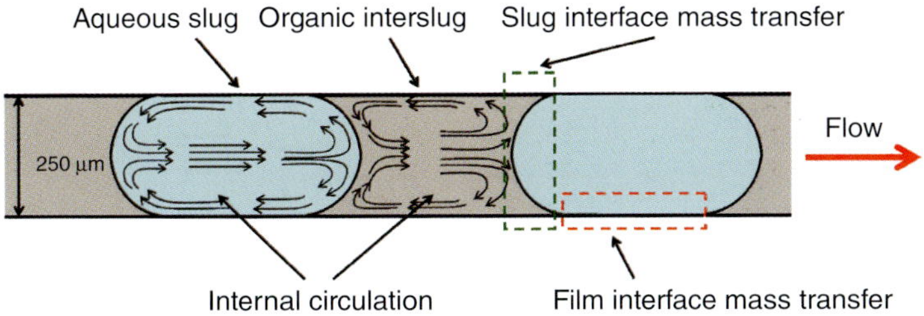

Figure 6.4 Schematic representation of liquid–liquid segmented flow. Internal circulation within the slugs provides excellent mixing. Reprinted with permission from [14]. Copyright (2010) American Chemical Society.

by changing the flow rates and the internal diameter of the microreactor, the surface-to-volume ratio can be changed in a reliable fashion. A linear correlation between conversion and surface area was found and suggested that the interfacial contact area was a crucial parameter for the observed rate enhancements.

Bubbly flow or dispersed flow is another interesting flow regime for liquid–liquid reactions. Of course, larger interfacial areas are possible (>150 000 m^2 m^{-3}), but the rapid coalescence of the bubbles can result in decreased surface-to-volume ratios. A very effective way to obtain stable dispersed flows is to use micro packed-bed reactors [17, 18]. Such a reactor design had been used to facilitate biphasic Suzuki–Miyaura reactions between heteroaryl halides and heteroarylboronic acids [19]. The reactor consisted of a stainless steel tube packed with stainless steel spheres (60–125 μm). Excellent yields were obtained within 3 min in the synthesis of a wide array of heterocyclic compounds.

Many different microreactor types have been developed to facilitate gas–liquid reactions and are often very similar to the ones used for liquid–liquid reactions [20]. One of the most recent designs involves the use of gas permeable membranes to facilitate gas–liquid mass transfer [20a]. Tube-in-tube reactors have been developed as an operationally simple membrane microreactor and consists of a pair of concentric capillaries in which the central capillary is made of a gas permeable membrane (Teflon AF-2400). The Mizoroki–Heck reaction with gaseous ethylene was performed in such a tube-in-tube reactor and furnished the desired products in good yield (Scheme 6.3) [21].

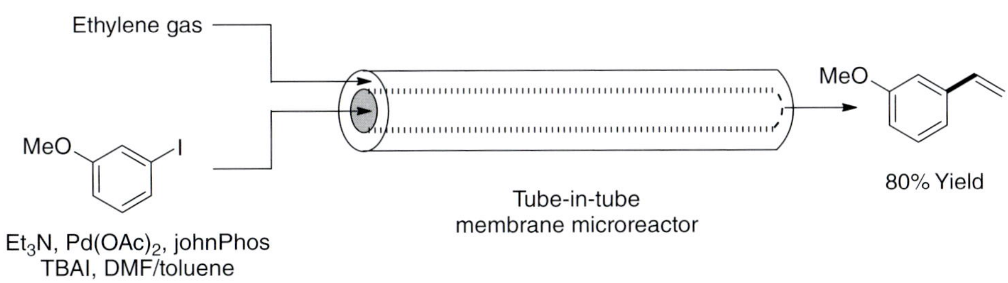

Scheme 6.3 Pd-catalyzed Mizoroki–Heck vinylation in a tube-in-tube microreactor system.

6.3
Unusual Reaction Conditions Enabled by Microreactor Technology

6.3.1
High-Temperature and High-Pressure Processing

The direct correlation between temperature and reaction rate as described by the Arrhenius equation (Equation 6.5) (k = rate constant, A = pre-exponential factor, E_a = activation energy, R = universal gas constant, and T = temperature) shows that an increase in temperature will lead to an increase in reaction rate. In batch, temperature increases are limited by the boiling point of the solvent at atmospheric pressure. However, by using pressure control valves, solvents can be heated above their boiling point in microreactors [22]. This phenomenon is also called superheating. As such, sluggish reactions can be significantly accelerated under superheated reaction conditions in a microreactor.

$$k = A\,e^{-E_a/RT} \tag{6.5}$$

In the absence of catalysts, sigmatropic rearrangements, such as the Claisen rearrangement, require elevated reaction temperatures (>200 °C) to initiate the reaction. Such harsh reaction conditions are feasible only in batch when employing specially designed reactors, such as autoclaves. The complexity of operations to perform reactions in such autoclaves prevents rapid screening of process parameters. In flow, several reaction parameters, for example, residence time, temperature, pressure, and concentration, can be rapidly adjusted in a safe and reliable manner. It was found that the best solvent for the Claisen rearrangement of allyl phenyl ether in flow was *n*-butanol. The target compound could be obtained within 4 min at 300 °C and 100 bar [23]. It should be noted that the reaction order in several protic solvents did not match with the observations in batch experiments. This could be explained by taking solvent expansion effects into account. The reaction could also be performed under solvent-free conditions (Figure 6.5) as both the starting material and the product were liquids.

The use of high pressure can lead to significant reaction rate accelerations in gas–liquid reactions as a consequence of the increased solubility of gaseous compounds in the liquid phase. The palladium-catalyzed aminocarbonylation allows for the synthesis of aryl amides under a carbon monoxide (CO) atmosphere and in the presence of a nitrogen nucleophile. The use of microreactor technology allows to vary the temperature and CO pressure rapidly while keeping the total amount of CO to an absolute minimum (Figure 6.6) [24]. The latter is especially interesting from a safety perspective. It was found that two compounds could be obtained, that is, amide and α-ketoamide, and their ratio was dependent on the temperature and CO pressure. High reaction temperatures lead to the formation of amide, whereas higher CO pressures give access to the α-ketoamide.

Another interesting application of high-temperature and high-pressure microreactors is to enlarge the processing conditions to the use of supercritical reaction conditions. Especially, the use of supercritical carbon dioxide ($scCO_2$) in combination

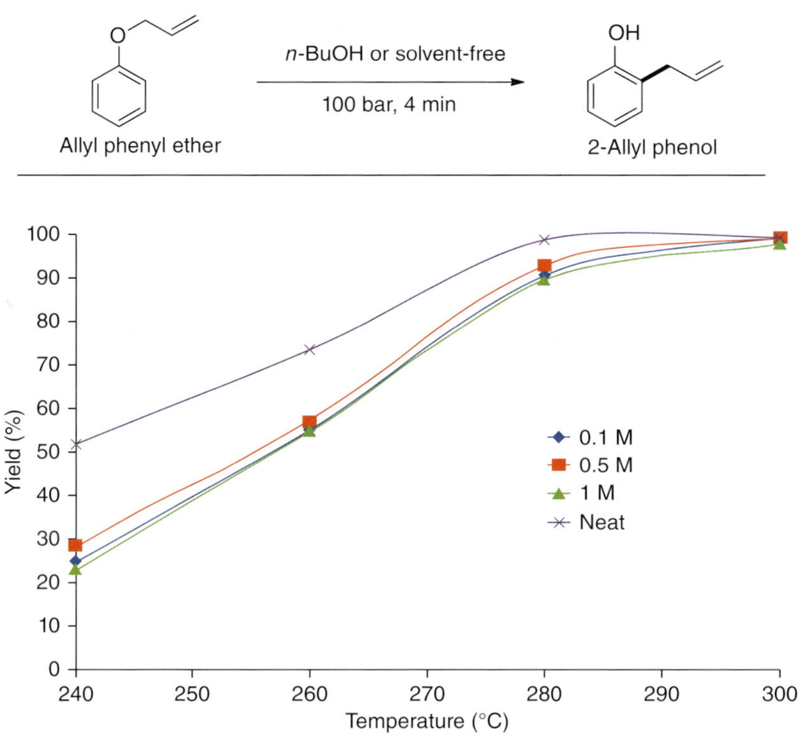

Figure 6.5 Rapid evaluation of the effects of concentration and reaction temperature on the conversion of allyl phenyl ether in a high-temperature, high-pressure microreactor. Reprinted with permission from [23]. Copyright (2013) Elsevier.

with microreactor technology has attracted a significant amount of attention [25]. $scCO_2$ provides excellent diffusivities comparable to those obtained in gases, making it attractive for performing gaseous reactions in one single supercritical phase, for example, oxidations, hydroformylations, and hydrogenations. A microfluidic system to generate singlet oxygen ($^1O_2$) by means of photoredox catalysis in $scCO_2$ has been developed [26]. Hereto, a Sapphire tubular microreactor, was exposed to irradiation from high-power visible LEDs. The presence of a photosensitizer allowed to produce $^1O_2$ efficiently, which was used subsequently for the oxidation of citronellol.

6.3.2
Use of Hazardous Intermediates – Avoiding Trouble

The use of microreactors to process toxic and other hazardous compounds in a safe manner has received particular attention. As the volumes of microreactors are typically very small (several hundreds of microliters), only limited amounts can be spilled into the environment in case of reactor failure. One strategy to minimize the risks is to generate the toxic reagent and utilize it immediately in a follow-up

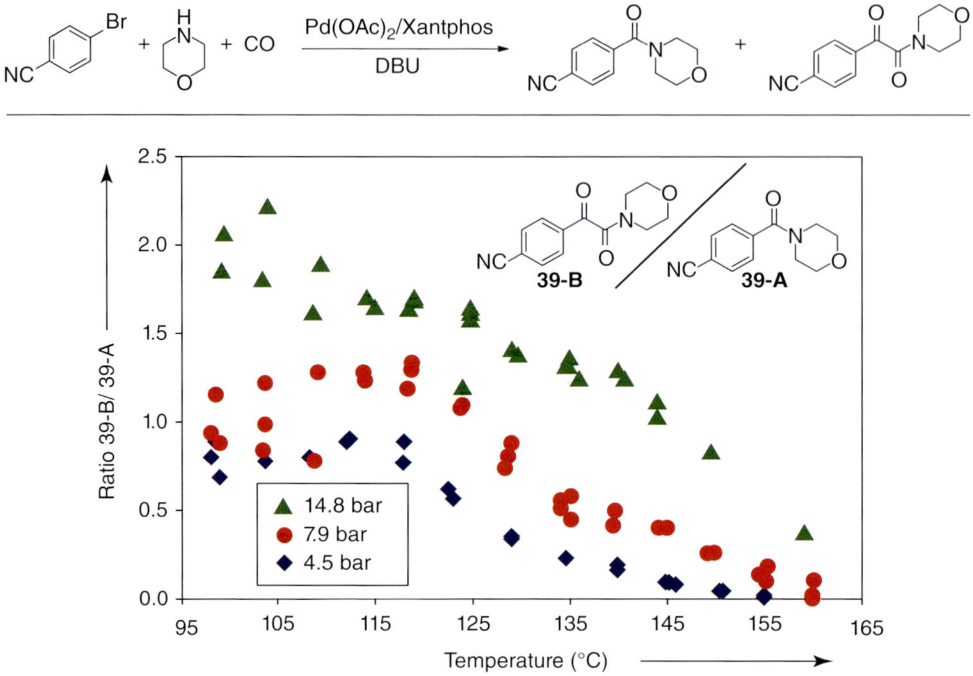

Figure 6.6 Rapid evaluation of the effects of reaction temperature and pressure on the product distribution of the Pd(0)-catalyzed aminocarbonylation in a high-temperature, high-pressure microreactor. Reprinted with permission from [24]. Copyright (2007) John Wiley & Sons.

reaction. As such, the required amount can be produced just-in-time on site, which prevents the need for storage of such hazardous compounds. A nice illustration of this is the generation of diazomethane, a common methylating agent in synthetic chemistry, which is carcinogenic and explosive. Diazomethane can be generated *in situ* in a membrane microreactor by treating *N*-methyl-*N*-nitroso-*p*-toluene sulfonamide (Diazald) with aqueous KOH (Figure 6.7) [27]. The diazomethane is

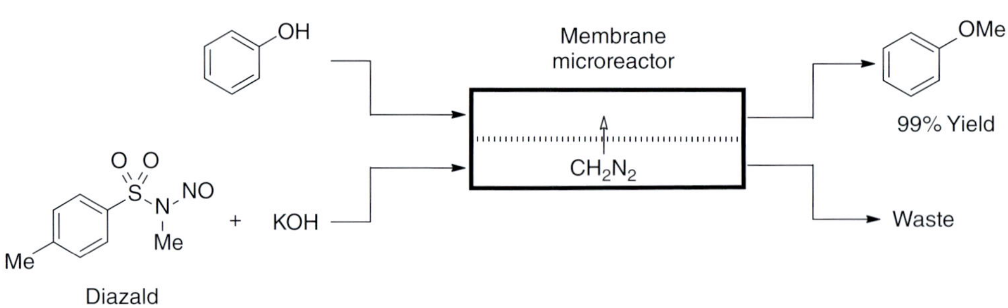

Figure 6.7 Continuous production of diazomethane and subsequent reaction in a membrane microreactor.

subsequently transported from the aqueous side through the polydimethylsiloxane (PDMS) membrane to the organic side where the methylation reaction takes place. In addition, other toxic reagents, such as hydrazoic acid [28], hydrogen cyanide [29], diazonium salts [30], and phosgene [31] have been continuously generated and consumed in a microreactor.

The direct synthesis of hydrogen peroxide (H_2O_2) starting from H_2 and O_2 over a palladium catalyst is an environmentally benign and atom efficient method. However, it involves the handling of an explosive gas mixture, making it a hazardous operation on a macroscale. However, the amounts present in the microreactor are below the explosion limits and do not represent enough material to destroy the reactor itself. A multichannel microchemical reactor filled with Pd/C as a catalyst allowed to selectively produce H_2O_2 at pressures of 2–3 MPa [32]. It was found that increases in yield could be obtained by placing several microreactors in a series.

6.3.3
Photochemistry

The use of light energy to initiate chemical reactions has a long tradition in synthetic chemistry [33]. It provides a clean and traceless energy source to activate molecules and is therefore embraced by the green chemistry principles. Nevertheless, the use of this versatile energy source has been hampered in the chemical industry because of its limited scale-up potential and the need to use high energy light sources (e.g., UV light) to activate efficiently organic molecules. Continuous-flow microreactor technology has impacted photochemistry because of the precise control over the residence/reaction time, the potential to irradiate the complete reaction and the possibility to scale-up photochemical reaction conditions [34].

A ey parameter in the design of novel photochemical microreactors is the choice of reactor material. The reactor material should be transparent for the targeted wavelength for the activation of the molecules and should prevent the light scattering of the irradiation. Glass microreactors are used quite often. Although Quartz is quite expensive, it can be used in both the UV and the visual spectral range (>170 nm). In addition, Pyrex (>275 nm), Corex (>260 nm), and Vycor (>220 nm) are interesting materials for the production of microfluidic channels. One of the main drawbacks of these devices is their incompatibility with strong bases and HF due to chemical etching of the material. Perfluoropolymers are interesting alternatives as they are cheap and transparent, and they provide a high chemical resistance. For example, perfluoroalkoxy polymers (PFA) allow transmission of the incident light in the visual range (91–96% transmission) and in the UV range (77–91% transmission from 250 to 400 nm).

In recent years, photoredox catalysis has become an important new incentive for the development of novel synthetic methodologies activated by light energy [35]. Some of the employed photoredox catalysts can be activated by visible light and transform the energy to electric potential energy that can be directly exploited in single electron transfer (SET) redox reactions. The combination of photoredox catalysis and microreactor devices allowed to accelerate the reaction rate with a

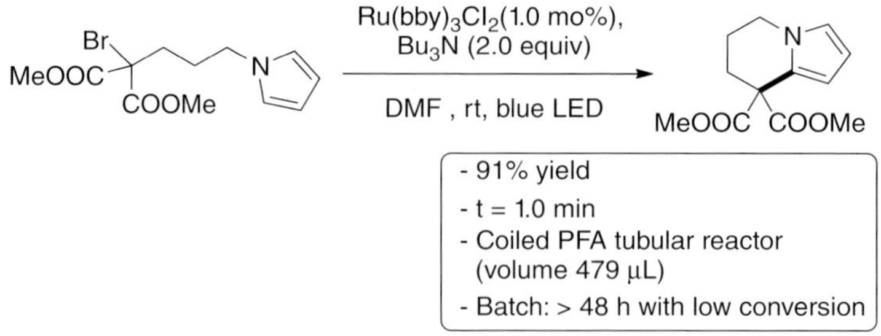

Scheme 6.4 Photoredox catalyzed internal cyclization of 1-pyrrolyl alkylbromide in a microreactor irradiated by a blue LED array.

few orders of magnitude [36]. Performing the intramolecular cyclization of 1-pyrrolyl alkylbromide in continuous-flow afforded the desired product in 1 min, while the corresponding batch reaction required more than 48 h (Scheme 6.4) [37]. Moreover, the scale up of the reaction conditions was difficult in batch and provided poor conversions. By continuously introducing the reagents into the microreactor assembly, the reaction could be scaled up without changing the reaction conditions.

The scale-up potential of photochemical reactions in microreactors is especially advantageous for the synthesis of complex molecules with interesting biological properties. Once the drug makes it to the market, it is important to have a reproducible and scalable synthesis to address the market needs. For photochemical reactions, this is hard to achieve in batch because of the limited penetration of the incident light in large reactor vessels. Artemisinin constitutes an effective treatment against malaria and possesses a synthetically challenging structure (Scheme 6.5). As it is one of the main treatments against malaria, it is important to have access to this compound in large quantities. Key in the total synthesis of this compound are the final steps, which involve a photochemical singlet oxygen induced ene reaction, followed by a Hock cleavage and the addition of triplet oxygen that triggers a cascade of reactions to yield the desired artemisinin. These three crucial steps

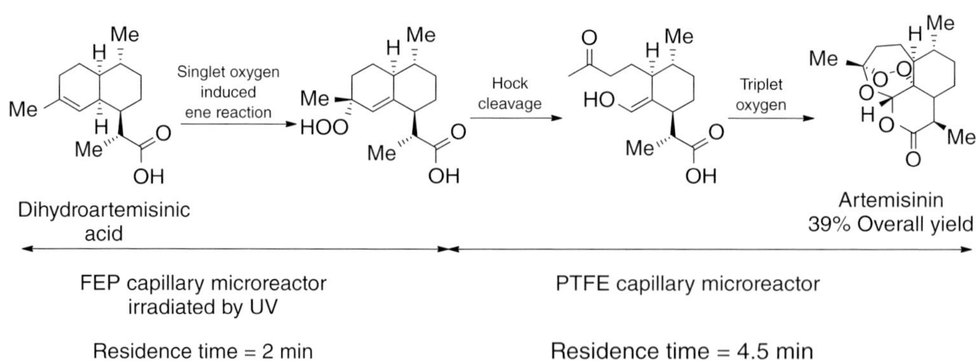

Scheme 6.5 Continuous-flow synthesis of artemisinin starting from dihydroartemisinic acid. FEP, fluorinated ethylene propylene copolymer and PTFE, polytetrafluoroethylene.

were integrated into one single continuous-flow procedure [38]. The first step could be performed in a FEP (fluorinated ethylene propylene) capillary microreactor in which the reagents were combined with oxygen gas and tetraphenylporphyrin as a photosensitizer. The microreactor was irradiated with UV light to generate the singlet oxygen species. Excellent yields were obtained in the microreactor setup (91% conversion and 75% yield). The addition of the two following steps in the continuous-flow setup allowed to produce artemisinin in 39% overall yield (Scheme 6.5). This system would allow to produce 200 g artemisinin per day. To meet the worldwide demand for artemisinin medication, an array of 1500 microreactors would be required.

6.4
The Use of Immobilized Reagents, Scavengers, and Catalysts

Immobilized reagents, scavengers, and catalysts are chemical reactive species that are bound to a solid support material. Such supported reagents can be loaded into a micro packed-bed reactor and, consequently, enables a simplified separation of the target compounds from the reagents [39]. An additional advantage is the possibility of using an excess of reagents to drive the reaction to completion. This principle was demonstrated in the palladium(0)-catalyzed fluorination of aryl triflates in flow [40]. It was found that an increasing amount of cesium fluoride (CsF), which is insoluble under the reaction conditions, resulted in a significant acceleration of the fluorination protocol. Cesium fluoride was loaded in a micro packed-bed reactor and the other reagents were directed over the CsF bed (Figure 6.8). All reactions could be completed within a residence time of 20 min. The advantages of performing this reaction in continuous-flow are (i) handling of large amounts of CsF without having to use specially designed mixers in batch to enable stirring the

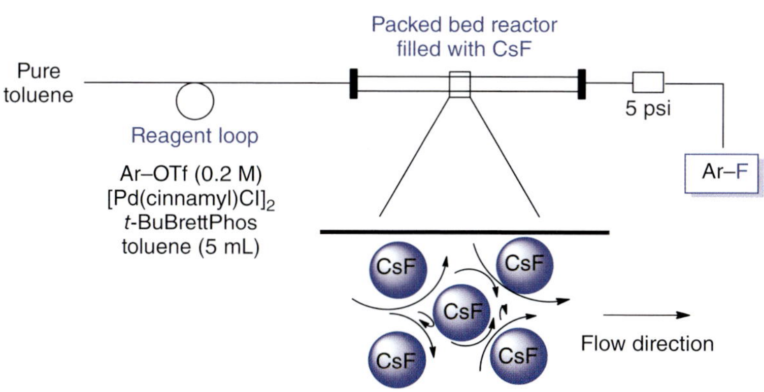

Figure 6.8 Pd-catalyzed fluorination of aryl triflates in a micro packed-bed reactor filled with cesium fluoride. Reprinted with permission from [40]. Copyright (2011) John Wiley & Sons.

slurry; (ii) potential to transfer this chemistry out of the glove box. CsF is highly hygroscopic and needs to be handled in a glove box. The packing of the reactor still requires a glove box, but once capped properly the reactor could be stored and on the benchtop; and (iii) excellent mixing and rapid heat transfer that leads to an additional acceleration compared to batch experiments.

The use of immobilized enzymes are interesting for applications in industrial biotechnology as it facilitates the reuse and recycling of the enzyme. Moreover, it has been shown that in certain cases enzyme immobilization leads to an enhanced stability [41]. The combination with microreactor technology adds several additional advantages [42]. Novozym 435, which constitutes an immobilized version of Candida antartica lipase B (CAL B) on an acrylic resin, was utilized for the transesterification of ethyl butyrate with 1-butanol in a micro packed-bed reactor [43]. It was found that both external and internal transport limitations were present under conventional batch conditions and resulted in a lower reaction rate. These transport limitations could be largely overcome in a micro packed-bed reactor because of a short diffusion path length (Figure 6.9), which makes that the supply of reactants to the catalytic site of the enzyme is not hampered. In addition, high catalyst loadings were employed in continuous flow. The use of such high loadings were difficult to employ in batch and resulted in stirring problems and particle damage because of mechanical shear forces. Continuous operation of micro packed-bed reactor did not result in the deactivation of the immobilized enzyme for more than 12 h, making it therefore suitable for long-term operation [44].

Metal catalyzed reactions are common in a wide variety of synthetic transformations. Hereto, a myriad of homogeneous catalysts have been developed, which display high activities and selectivities by a proper selection of the ligand–metal

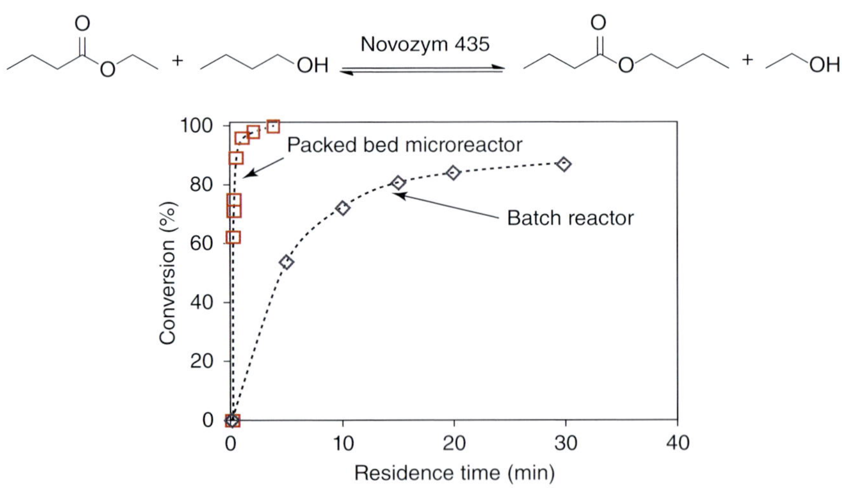

Figure 6.9 Transesterification of ethyl butyrate with 1-butanol in a micro packed-bed reactor: conversion in function of time at 70 °C – a comparison between microreactor and batch reactor systems.

combination. However, on a large scale, the use of expensive homogeneous catalysts requires efficient catalyst recycling and recuperation solutions to become cost effective. The use of heterogeneous catalysts can provide a cost-efficient alternative and allow for facile catalyst recovery. A lot of research has been devoted toward the immobilization of homogeneous catalysts on mesoporous solid supports and their combination with continuous-flow micro reactor technology [1a, 45]. In certain cases, even the microreactor can be made out of catalytic material, for example, copper tubing. Such a copper microreactor has been used for the copper-catalyzed azide-alkyne cycloaddition [46]. The formation of the organic azides was achieved *in situ* by SN_2 reaction of alkyl halides with sodium azide. Subsequent cycloaddition with a suitable alkyne coupling partner was catalyzed by the catalytic wall microreactor and allowed for the rapid synthesis of a library of 30 triazoles. Notably, no additional ligand or copper metal was required for this reaction.

The combination of heterogeneous catalysis and continuous-flow technology constitutes a powerful merger, which allows to reach high turnover numbers and to recuperate the precious metal species. However, it should be thoroughly investigated whether the mechanism is truly heterogeneous or homogeneous. In the latter case, the immobilized catalyst bed or wall actually functions as a reservoir of the catalytically active species that is leached into solution. In batch, homogeneous catalysis is often confused with heterogeneous catalysis as the leached metal redeposits on the carrier at the end of the reaction, so-called boomerang mechanism. The true nature of the catalytic active species can be revealed by several tests, such as the three phase test [47]. In continuous flow, the catalytically active species is leached in a continuous liquid phase, which transports the metal downstream in the reactor and, after reaction, the metal is deposited on the reactor walls. This leaching-deposition cycle can be compared with a chromatography effect known from separation technology (Figure 6.10). As such, the metal is eventually leached out of the microreactor, which will result eventually in a complete deactivation of the catalyst bed.

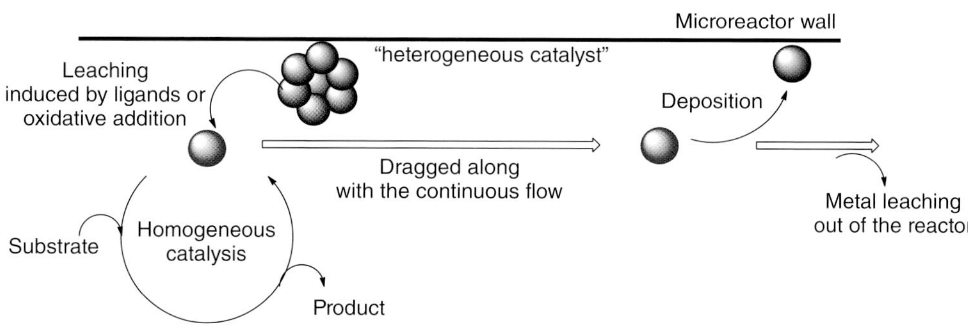

Figure 6.10 Chromatography effect encountered in continuous-flow reactors in which immobilized heterogeneous catalysts leach the catalytically active species into solution. After several leaching-deposition cycles, the catalyst will be removed from the reactor.

Notwithstanding the homogeneous nature of a given catalytic reaction, the use of immobilized catalysts in micro packed-bed reactor can still be interesting because of its operational flexibility. Moreover, the degree of leaching can be fine-tuned by engineering an optimal solid support, which binds the metal more strongly. Hence, contamination of the target product with trace amounts of metal can be minimized, simplifying the need for postprocessing of the reaction stream. An example of this involves the development of silica-supported palladium catalyst, which demonstrated a low leaching behavior. This supported catalyst was loaded in a micro packed-bed reactor and was used for the Suzuki–Miyaura cross-coupling reaction in flow [48]. The catalyst bed could be used for more than 8 h without noticeable decrease in catalytic activity. The leaching of palladium was minimized; in the organic phase, 10 ppb of Pd was detected while the aqueous phase contained 20 ppb of Pd. As such, the calculated turnover number of the catalyst was higher than 100. In addition, 30 consecutive Suzuki–Miyaura cross-coupling reactions could be performed without a need to change the catalyst bed, making this continuous-flow methodology interesting for the screening of reaction conditions in medicinal chemistry applications.

6.5
Multistep Synthesis in Flow

The total synthesis of complex organic molecules (e.g., active pharmaceutical ingredients) is a laborious and time-consuming undertaking, which involves multiple reaction steps with intermediate purification and isolation steps. The use of continuous-flow reactors allows to integrate these different stages of a typical synthetic process in one continuous, and often automated, process (i.e., process integration) [49]. Consequently, the time investment of the operator can be minimized by simplifying the synthetic procedure in continuous flow. Several strategies have been developed in order to combine a whole reaction sequence in an uninterrupted continuous-flow process.

One strategy involves the use of telescoping reactions, in which several reactions steps are combined in single flow without intermediate purification. This is a very straightforward and easy implemented strategy, which resembles the one-pot strategies in conventional batch procedures [50]. However, one should note that reagents used and impurities formed in the upstream reactions should be compatible with the downstream reaction conditions. This compatibility issue requires often a rethinking of the reaction conditions for the whole sequence. Such a strategy was used for the synthesis of 1-substituted benzotriazoles (Scheme 6.6) [30b]. Four reactions were connected in a sequence of three microreactors. The first reaction involves a nucleophilic aromatic substitution (S_NAr) in stainless steel capillary microreactor at elevated temperatures (160–180 °C). *N*,*N*-dimethylacetamide (DMA) and *n*-hexanol were utilized as a solvent mixture for the S_NAr reaction and were amenable toward downstream transformations. Next, hydrogen was added to obtain a gas–liquid segmented flow, which was introduced in a micro packed-bed

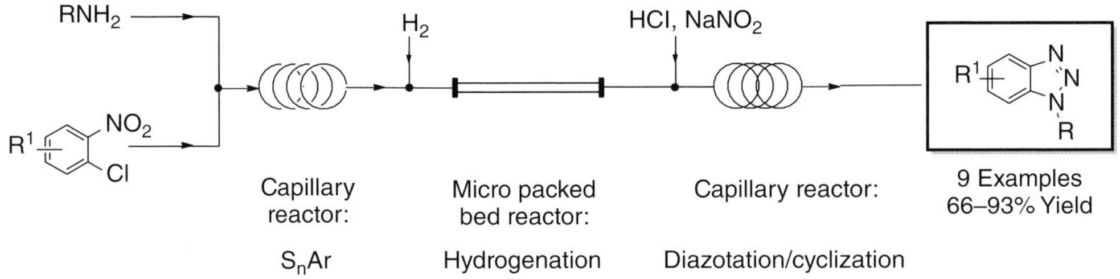

Scheme 6.6 Continuous-flow setup for the synthesis of 1-substituted benzotriazoles.

reactor. The reactor was filled with Pd/C and allowed for the hydrogenation of the nitro group. Next, diazotation and subsequent cyclization was achieved in PFA capillary microreactor by adding aqueous sodium nitrite and hydrochloric acid.

A second strategy deals with the use of immobilized reagents, catalysts, and scavengers in flow as discussed in the previous chapter. Hereby, the substrates are directed over cartridges filled with functionalized beads and by placing several cartridges in a series, subsequent transformations can be performed in a single-flow operation. This method is used to synthesize complex biologically active molecules in an automated manner without the need for further purification (Scheme 6.7) [51]. "Catch and release" strategies are employed to execute intermediate purifications and solvent switches. The need to periodically replace the reagent cartridges makes it most suitable for the preparation of small amounts of compound.

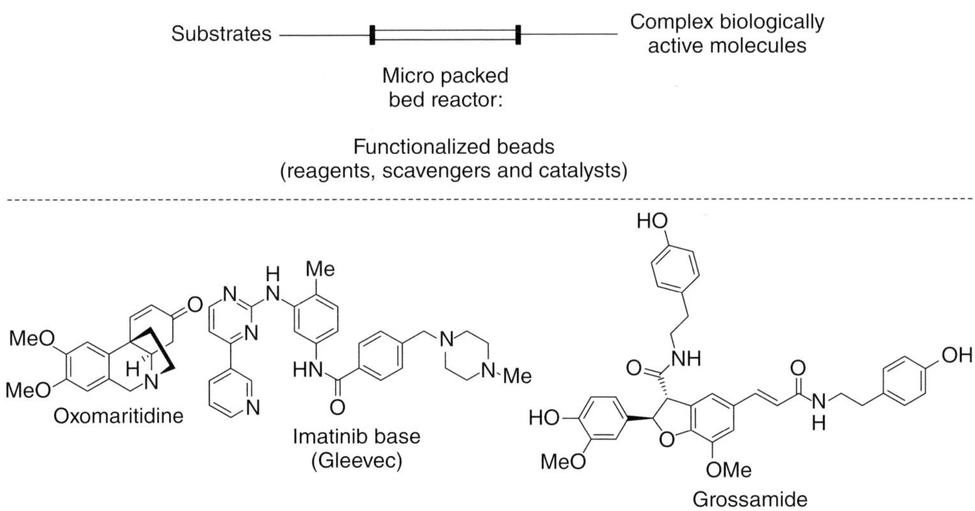

Scheme 6.7 Synthesis of complex biologically active compounds in flow via a functionalized beads strategy.

A third strategy constitutes the utilization of miniaturized unit operations to execute continuous purifications in flow. This is a very powerful approach as it allows to purify the reaction mixture without interruptions. However, the

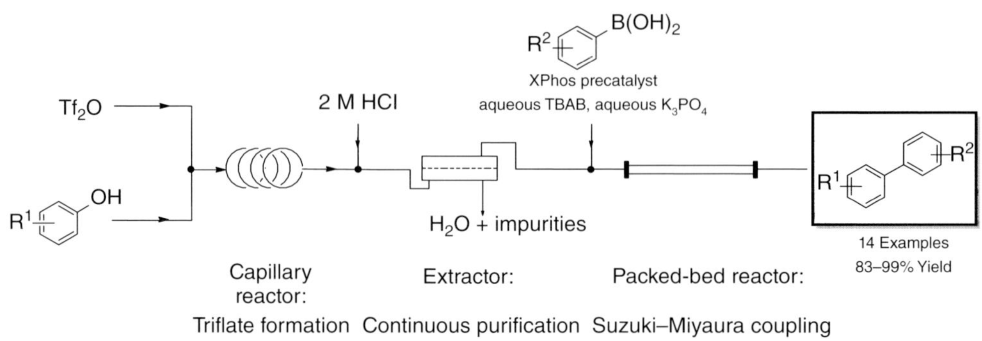

Scheme 6.8 Microfluidic assembly for the continuous-flow synthesis of biaryls enabled by a microfluidic extraction.

development of such miniaturized unit operations has been challenging because of the fact that interfacial forces dominate over gravitational forces. Microfluidic extractions [52], distillations [53], and simulated moving beds [54] have been combined with micro flow reactors to facilitate multistep syntheses or a continuous purification of the reaction stream. A microfluidic system, consisting of two microreactors and a microfluidic extraction unit, for the synthesis of biaryls starting from substituted phenols via a triflation/Suzuki–Miyaura sequence is shown in Scheme 6.8 [52a]. In the first PFA capillary reactor, substituted phenols were reacted with triflic anhydride to yield aryl triflates in quantitative yields. The combination of the subsequent Suzuki–Miyaura cross-coupling reaction without intermediate purification was not feasible. Small amounts of impurities generated in the first reaction step poisoned the palladium catalyst required for the Suzuki–Miyaura cross coupling. This problem could be avoided by using an extractive work-up. Hereto, the reaction stream exiting the first reactor was quenched with 2 M HCl, which facilitated convective mass transfer between the two phases and allowed for the efficient removal of the impurities. The segmented organic/aqueous flow was subsequently delivered to a phase separation device, which contained a Zefluor membrane and allowed to separate the two layers by capillary forces. The organic phase was combined with other reagents to enable the Suzuki–Miyaura cross-coupling reaction and introduced a micro packed-bed reactor to facilitate good mixing between the aqueous and organic phase.

6.6 Avoiding Microreactor Clogging

The combination of continuous-flow microreactors and solid materials (i.e., as a reagent, catalyst, or product) has proved to be a formidable challenge and represents a major barrier for the widespread use of microreactor technology in the chemical industry [5]. Owing to their small dimensions, these devices are rapidly blocked in the presence of any solid material and, often, this leads to irreparable damage

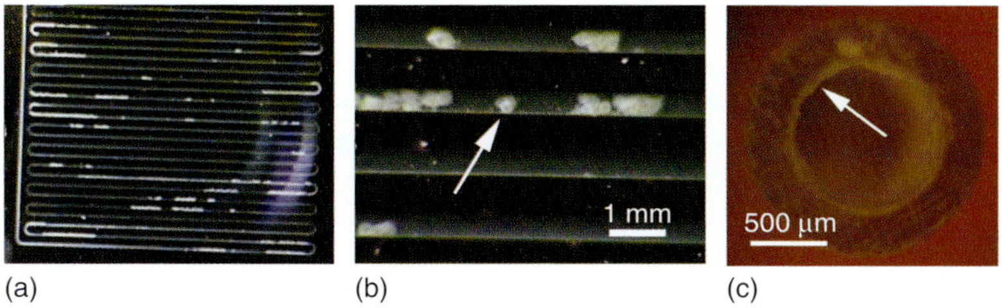

Figure 6.11 Microreactor clogging in Pd-catalyzed C–N cross-coupling reactions because of the generation of NaCl formation. (a) Deposition of the salt happens typically at sharp bends and toward the end of the microreator. (b) Agglomeration of NaCl precipitation leads to bridging. (c) Deposition of NaCl on the microreactor walls leads to constriction of the microreactor channel. Reprinted with permission from [56b]. Copyright (2010) American Chemical Society.

of the microreactor (Figure 6.11). A survey performed by researchers at Lonza demonstrated that about 50% of the reactions performed in the fine chemical and pharmaceutical industry could benefit from continuous-flow processing [55]. However, this figure had to be reduced as 63% of these reactions involved the use of solids and are therefore difficult to combine with microreactor technology. Recently, several strategies have been developed to enable such solids handling reactions in microreactors without the blockage of the microchannels.

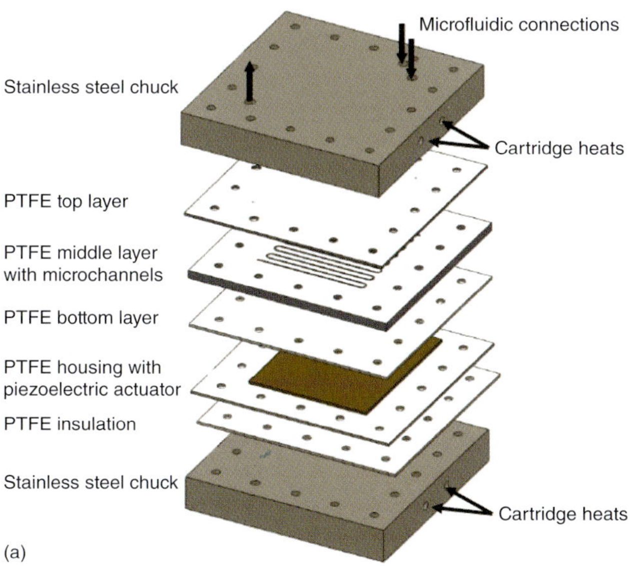

Frequency	Particle size (μm) d_{10}	d_{50}	d_{90}
No sonication	6.68	19.19	87.29
40 kHz	2.27	20.05	60.63
50 kHz	2.82	7.90	20.71
60 kHz	4.81	12.43	36.93

(b)

Figure 6.12 (a) Teflon stack microreactor for solids handling reactions. (b) Particle size distribution in the function of the applied ultrasonication frequency. d_{10}: 10% of the particles are smaller than this value, d_{50}: median particle diameter, and d_{90}: 90% of the particles are smaller than this value. Reprinted with permission from [60]. Copyright (2011) Royal Society of Chemistry.

Surface imperfections promote the attachment of solids at the surface of the microreactor wall and enable crystal growth (heterogeneous nucleation). Contact with the microreactor walls can be avoided when the solids forming reaction is performed in a droplet [57]. Also, nanoparticles [56] and polymer-based microstructures [58] have been synthesized employing this strategy.

Acoustic irradiation has been used for the breaking up of particle conglomerates in microchannels [59]. The use of acoustic irradiation creates small cavitations at the surface of the particle. On implosion of this cavitation bubble, the particle breaks apart following the high shear forces created by this implosion. As such, the particle size could be minimized as shown in Figure 6.12. Such small particles can be transported efficiently with the fluid through the microreactor without blocking the microchannels. It was found that the optimal frequency for particle breakup was 50 kHz. A Teflon stack microreactor with integrated piezoelectric

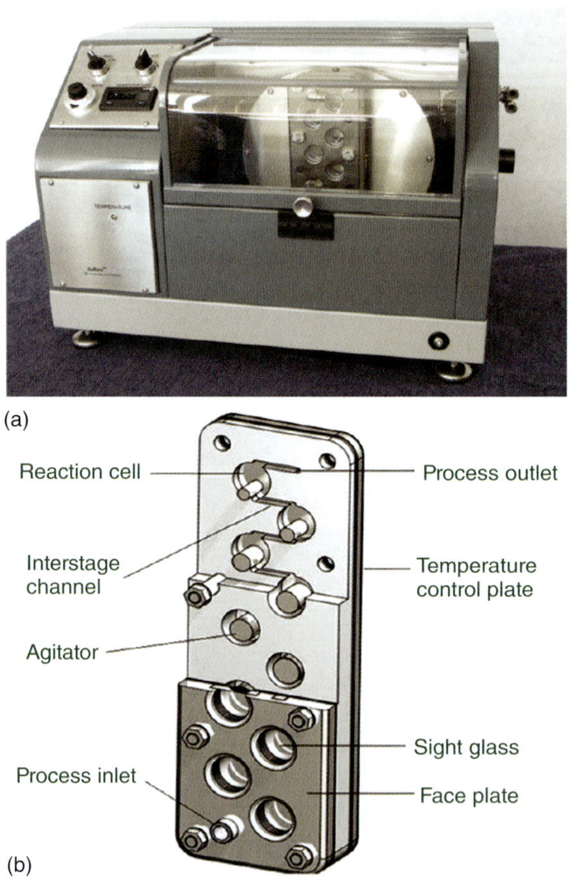

Figure 6.13 (a) A commercially available agitating cell reactor (Coflore ACR). (b) Sketch of the agitated cell reactor block. Reprinted with permission from [61]. Copyright (2011) American Chemical Society.

actuator was developed for solids forming reactions [60]. In this device, Pd-catalyzed C–N cross-coupling reactions could be performed without microreactor clogging.

Finally, mechanical agitation can also be employed for the processing of solid-forming reactions [61]. A commercially available agitating cell reactor (Coflore ACR) is depicted in Figure 6.13. This constitutes an assembly of continuously stirred tank reactors placed in the series. Slurries can be transported through the reactor by agitators that are mechanically shaken to keep the solids in suspension.

6.7 Reaction Screening and Optimization Protocols in Microreactors

The integration of spectroscopic detection within a microreactor results in the so-called integrated microreactor environments and facilitates online reaction monitoring [62]. This is of great importance to the pharmaceutical industry that makes use of such process analytical technology to guarantee the final drug quality by monitoring the entire production process [63]. When such spectroscopic detection systems are combined with logic and feedback control, numerous reactions can be performed in an automated manner, which allows for rapid reaction screening and reaction optimization. Such a system for "self-optimization" was used to optimize the Mizoroki–Heck reaction of 4-chlorobenzotrifluoride with 2,3-dihydrofuran (Scheme 6.9) [64]. The reaction was performed in a microreactor and the yield and selectivity was analyzed by an inline HPLC. The obtained data were analyzed by DoE software and feedback control was provided with regard to concentration and residence time. With only 19 experiments and a limited amount of reagents, optimal reaction conditions were obtained. These conditions were subsequently transferred to a mesoscale flow reactor system, which allowed a 50-fold scale up without further reoptimization.

Rapid screening of reaction conditions in combination with inline analysis in an automated manner allows to extract kinetic parameters from the studied reaction. The limited amount of chemicals required and the high accuracy because of the well-defined mass- and heat-transfer characteristics makes the use of microreactors the ultimate method for studying reaction kinetics [65]. *In situ* spectroscopic detection can also be combined with continuous-flow microreactors to characterize immobilized catalysts, which allows to elucidate mechanistic data under relevant reaction conditions [62a].

6.8 Scale-Up Issues – from Laboratory Scale to Production Scale

Leading a chemical route from laboratory to production scale is a crucial step in any production process. Scale up in batch reactor vessels is time-consuming and requires a careful investigation of mass- and heat-transfer issues on every scale,

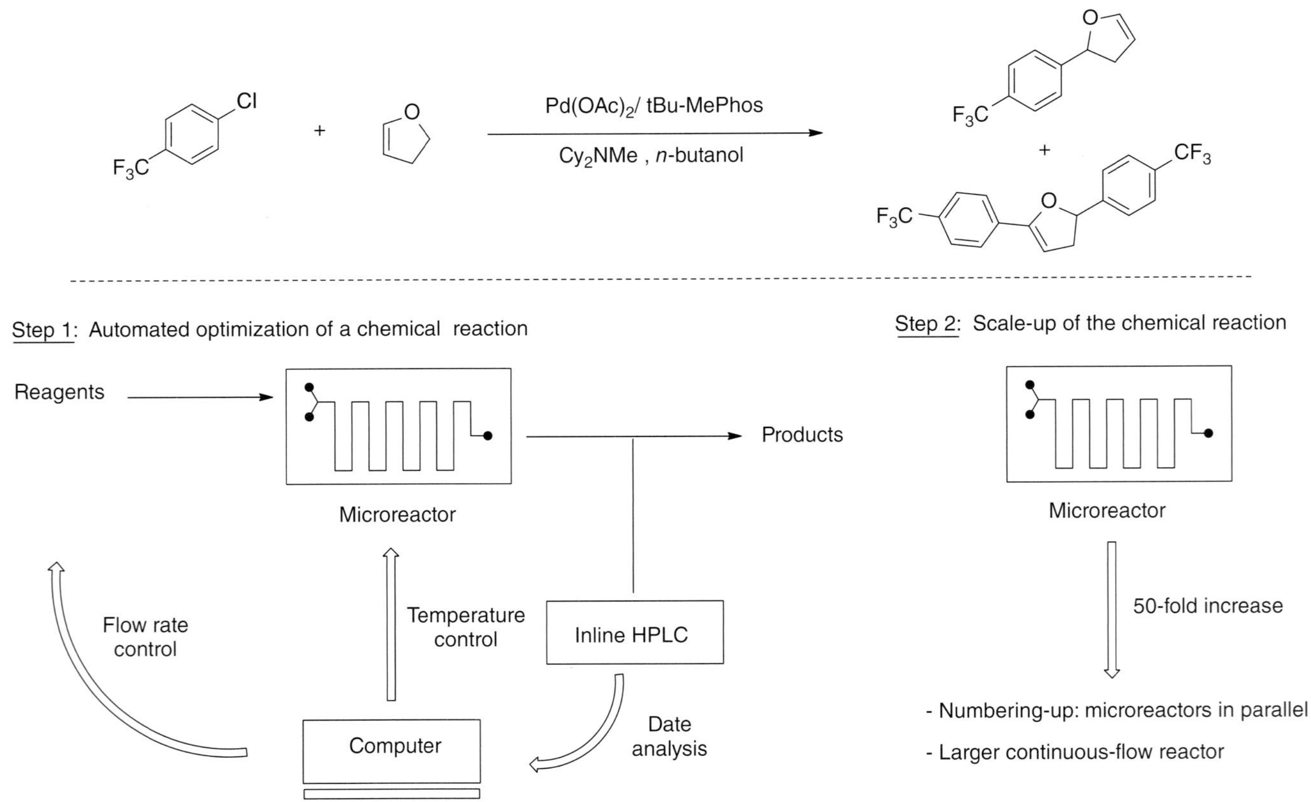

Scheme 6.9 Schematic representation of a microfluidic platform for the automated optimization of the Mizoroki–Heck reaction of 4-chlorobenzotrifluoride with 2,3-dihydrofuran and its subsequent scale up.

that is, laboratory scale, pilot scale, and production scale [66]. This is especially true for the upscaling of exothermic, toxic, and explosive reactions, which are often performed under suboptimal conditions at production scale to avoid any hazardous situations. The use of microreactor technology allows to scale up such hazardous reactions in a consistent manner to a production scale without the risk of a technology change [67]. This can be done in three ways: (i) longer operation times of the device, (ii) numbering-up by placing several reactors in parallel, and (iii) increasing the throughput by applying higher flow rates through the device.

Longer operation modes of the device is a straightforward way of scaling up in the laboratory environment [19]. Once the optimal continuous-flow conditions are established, the same device can be run by the introduction of the reagents for multiple hours until the desired amount of product has been produced. No reengineering of the device is required as the same reaction conditions are achieved at steady state for each run. This is especially interesting for scaling a reaction from several milligrams to the maximum of a few hundreds of grams.

Numbering-up of microchannels is the preferred method to scale out the performance of microreactors as required on an industrial scale. There are two ways of numbering-up, that is, internally and externally. External numbering-up is achieved by placing devices and their individual control units in parallel. This ensures that exactly same processing conditions are achieved in each reactor and, hence, provide a reliable way of scaling-up. Such parallel devices also allow to clean a single device whereas the other ones are still in operation. However, this represents a huge equipment cost and a larger building footprint. Internal numbering-up is achieved by using a single microstructured device with several microchannels in parallel. Flow distributors are used to direct the fluid over different channels. However, the realization of an equal performance via internal numbering-up is very difficult as large microstructured devices are far from isothermal [68]. These temperature differences result in different pressure drops over the microchannel and, therefore, also in a nonuniform flow distribution. This leads to an overall decrease of the reactor efficiency. A lot of research has been devoted to optimize the flow distribution [68]. Especially for gas-phase reactions and low viscosity liquids, good results have been obtained [69]. An internal numbering-up strategy was used for the scale out of the Fischer–Tropsch synthesis [70]. The kinetic data of the reaction was obtained in a single channel device. This information was used to build a pilot-scale microstructured device of 450 parallel microchannels and allowed to produce up to 2 l/day. Next, a cross-flow microstructured device was developed in which the reagents were directed over an immobilized heterogeneous catalyst, whereas boiling water was directed in cross-flow through the device to dissipate the generated heat and to maintain an isothermal operation of the reactor. This device ($0.6 \times 0.6 \times 0.15$ m) contains more than 10 000 microchannels and can generate up to 450 tons of product per year (Figure 6.14).

To avoid issues of flow maldistribution, the use of single channel devices is advantageous [71]. An increased throughput can be obtained by using higher flow rates. However, when keeping the channel dimensions constant, the increase of flow rates is limited by the increased pressure drop over the device. Therefore, the

Figure 6.14 Velocys manufacturing scale-up device for Fischer–Tropsch synthesis. This microstructured device contains more than 10 000 microchannels for reaction and cooling purposes. Reprinted with permission from [70]. Copyright (2010) John Wiley and Sons.

(a) (b) (c)

Figure 6.15 The Lonza FlowPlate TM concept for a step-wise scale up of continuous-flow reactions: (a) a single reactor plate with tangential mixing elements. (b) Laboratory scale reactor with a transparent view glass for optical inspection. (c) Lonza's plate stack reactors for step-wise scale up. Reprinted with permission from [71c]. Copyright (2012) John Wiley & Sons.

cross section of the microchannel should be enlarged while the mass- and heat-transfer characteristics are kept constant. Plate stack reactors, which allow for a modular and consistent scale up of the reaction conditions, have been developed by several commercial companies (Figure 6.15) [72, 73]. As such, a step-wise scale up of the continuous-flow reaction conditions can be achieved without encountering too many technical problems.

6.9 Outlook

Whether or not continuous-flow microreactors are going to replace entirely the traditional batch protocols and round bottomed flasks remains to be seen. However, one can hardly ignore the fact that the microreactor field has not evolved in the past couple of years. Owing to significant advances, the implementation of continuous-flow microreactors in both academia and industry has grown exponentially. In academia, microreactor technology provides elegant solutions for the so-called difficult reactions, for example, gas–liquid reactions, reactions under supercritical conditions, the handling of hazardous compounds, photochemical transformations, and the execution of harsh reaction conditions. The combination of multiple reaction steps in a single continuous-flow scheme provides opportunities to minimize the burden of the researcher. In addition, the technology also provides a tool to rapidly elucidate the reaction kinetics of a given chemical reaction. For the chemical industry, the use of microstructured devices shortens the time-to-market because of the enhanced safety procedures and reliable scale-up potential.

We anticipate substantial advancements and refinements in the coming years in all current research areas of micro process technology as delineated in this chapter. This will bring the technology further to maturity and will convince the last skeptics of continuous-flow technology. Especially, the development of microfluidic unit operations needs another push to facilitate continuous-flow purifications to meet the stringent requirements of the pharmaceutical industry. The combination of such purification systems with microreactors will lead to the synthesis of increasingly complex drug structures in a truly continuous manner. The implementation of automation and inline spectroscopic detection in micro process technology provides a unique potential for future reaction development. It would be interesting to see when the first drug structure, which has been entirely optimized from hit to lead in an automated microfluidic manner, will be launched on the market. Although this seems like a "science fiction" vision, recent advances have shown that significant advances have been made in this direction. Further improvements are, however, mandatory and require an interdisciplinary approach among synthetic chemists, process engineers, software developers, and others.

References

1. For selected reviews about microreactor technology: (a) Noël, T. and Buchwald, S.L. (2011) *Chem. Soc. Rev.*, **40**, 5010–5029; (b) Hartman, R.L., McMullen, J.P., and Jensen, K.F. (2011) *Angew. Chem. Int. Ed.*, **50**, 7502–7519; (c) Geyer, K., Gustafsson, T., and Seeberger, P.H. (2009) *Synlett*, 2382–2391; (d) Yoshida, J.-i., Nagaki, A., and Yamada, T. (2008) *Chem. Eur. J.*, **14**, 7450–7459; (e) Wiles, C. and Watts, P. (2008) *Eur. J. Org. Chem.*, **2008**, 1655–1671; (f) Mason, B.P., Price, K.E., Steinbacher, J.L., Bogdan, A.R., and McQuade, D.T. (2007) *Chem. Rev.*, **107**, 2300–2318.
2. For application of continuous-flow chemistry in the pharmaceutical industry: (a) Malet-Sanz, L. and Susanne, F. (2012) *J. Med. Chem.*, **55**, 4062–4098; (b) Proctor, L., Dunn, P.J., Hawkins, J.M., Wells, A.S., and Williams, M.T. (2010) in *Green Chemistry in the Pharmaceutical Industry* (eds P.J. Dunn, A.S. Wells, and M.T. Williams), Wiley-VCH Verlag GmbH, Weinheim, pp. 221–242.
3. For selected references about Novel Process Windows: (a) Hessel, V., Kralisch, D., Kockmann, N., Noël, T., and Wang, Q. (2013) *ChemSusChem*, **6**, 746–789; (b) Stouten, S.C., Noël, T., Wang, Q., and Hessel, V. (2013) *Aust. J. Chem.*, **66**, 121–130. doi: 10.1071/CH12465 (c)

Hessel, V., Vural-Gursel, I., Wang, Q., Noël, T., and Lang, J. (2012) *Chem. Eng. Technol.*, **35**, 1184–1204; (d) Hessel, V., Vural-Gursel, I., Wang, Q., Noël, T., and Lang, J. (2012) *Chem. Ing. Tech.*, **84**, 660–684; (e) Hessel, V. (2009) *Chem. Eng. Technol.*, **32**, 1655–1681; (f) Lob, P., Drese, K.S., Hessel, V., Hardt, S., Hofmann, C., Lowe, H., Schenk, R., Schonfeld, F., and Werner, B. (2004) *Chem. Eng. Technol.*, **27**, 340–345.

4. Anderson, N.G. (2012) *Org. Process Res. Dev.*, **16**, 852–869.
5. (a) Flowers, B.S. and Hartman, R.L. (2012) *Challenges*, **3**, 194–211; (b) Hartman, R.L. (2012) *Org. Process Res. Dev.*, **16**, 870–887.
6. (a) Yoshida, J.-I. (2008) *Flash Chemistry: Fast Organic Synthesis in Microsystems*, Wiley-Blackwell, Hoboken, NJ; (b) Yoshida, J.-I. (2010) *Chem. Rec.*, **10**, 332–341.
7. Nagy, K.D., Shen, B., Jamison, T.F., and Jensen, K.F. (2012) *Org. Process Res. Dev.*, **16**, 976–981.
8. Nagaki, A., Togai, M., Suga, S., Aoki, N., Mae, K., and Yoshida, J.-i. (2005) *J. Am. Chem. Soc.*, **127**, 11666–11675.
9. (a) Lowe, H., Axinte, R.D., Breuch, D., Hang, T., and Hofmann, C. (2010) *Chem. Eng. Technol.*, **33**, 1153–1158; (b) Ehm, N. and Lowe, H. (2011) *Org. Process Res. Dev.*, **15**, 926–930.
10. Usutani, H., Tomida, Y., Nagaki, A., Okamoto, H., Nokami, T., and Yoshida, J.-i. (2007) *J. Am. Chem. Soc.*, **129**, 3046–3047.
11. Gunther, A. and Jensen, K.F. (2006) *Lab Chip*, **6**, 1487–1503.
12. Jovanovic, J., Rebrov, E.V., Nijhuis, T.A., Kreutzer, M.T., Hessel, V., and Schouten, J.C. (2012) *Ind. Eng. Chem. Res.*, **51**, 1015–1026.
13. (a) Kuhn, S. and Jensen, K.F. (2012) *Ind. Eng. Chem. Res.*, **51**, 8999–9006; (b) Yue, J., Luo, L., Gonthier, Y., Chen, G., and Yuan, Q. (2009) *Chem. Eng. Sci.*, **64**, 3697–3708; (c) Yue, J., Chen, G., Yuan, Q., Luo, L., and Gonthier, Y. (2007) *Chem. Eng. Sci.*, **62**, 2096–2108; (d) Kreutzer, M.T., Kapteijn, F., Moulijn, J.A., and Heiszwolf, J.J. (2005) *Chem. Eng. Sci.*, **60**, 5895–5916; (e) Bercic, G. and Pintar, A. (1997) *Chem. Eng. Sci.*, **52**, 3709–3719.
14. Jovanovic, J., Rebrov, E.V., Nijhuis, T.A., Kreutzer, M.T., Hessel, V., and Schouten, J.C. (2010) *Ind. Eng. Chem. Res.*, **49**, 2681–2687.
15. Mellouli, S., Bousekkine, L., Theberge, A.B., and Huck, W.T.S. (2012) *Angew. Chem. Int. Ed.*, **51**, 7981–7984.
16. Narayan, S., Muldoon, J., Finn, M.G., Fokin, V.V., Kolb, H.C., and Sharpless, K.B. (2005) *Angew. Chem. Int. Ed.*, **44**, 3275–3279.
17. Naber, J.R. and Buchwald, S.L. (2010) *Angew. Chem. Int. Ed.*, **49**, 9469–9474.
18. Shang, M., Noël, T., Wang, Q., and Hessel, V. (2013) *Chem. Eng. Technol.*, **36**, 1001–1009.
19. Noël, T. and Musacchio, A.J. (2011) *Org. Lett.*, **13**, 5180–5183.
20. (a) Noël, T. and Hessel, V. (2013) *ChemSusChem*, **6**, 405–407; (b) Chen, G., Yue, J., and Yuan, Q. (2008) *Chin. J. Chem. Eng.*, **16**, 663–669.
21. Bourne, S.L., Koos, P., O'Brien, M., Martin, B., Schenkel, B., Baxendale, I.R., and Ley, S.V. (2011) *Synlett*, 2643–2647.
22. Razzaq, T. and Kappe, C.O. (2010) *Chem. Asian J.*, **5**, 1274–1289.
23. Kobayashi, H., Driessen, B., van Osch, D.J.G.P., Talla, A., Ookawara, S., Noël, T., and Hessel, V. (2013) *Tetrahedron*, **69**, 2885–2890.
24. Murphy, E.R., Martinelli, J.R., Zaborenko, N., Buchwald, S.L., and Jensen, K.F. (2007) *Angew. Chem. Int. Ed.*, **46**, 1734–1737.
25. For a concise review about flow reactions in $scCO_2$:Han, X. and Poliakoff, M. (2012) *Chem. Soc. Rev.*, **41**, 1428–1436.
26. Bourne, R.A., Han, X., Poliakoff, M., and George, M.W. (2009) *Angew. Chem. Int. Ed.*, **48**, 5322–5325.
27. Maurya, R.A., Park, C.P., Lee, J.H., and Kim, D.-P. (2011) *Angew. Chem. Int. Ed.*, **50**, 5952–5955.
28. Gutmann, G., Roduit, J.-P., Roberge, D., and Kappe, C.O. (2010) *Angew. Chem. Int. Ed.*, **49**, 7101–7105.
29. Acke, D.R.J. and Stevens, C.V. (2007) *Green Chem.*, **9**, 386–390.
30. (a) Chernyak, N. and Buchwald, S.L. (2012) *J. Am. Chem. Soc.*, **134**, 12466–12469; (b) Chen, M. and

Buchwald, S.L. (2013) *Angew. Chem. Int. Ed.*, **52**, 4247–4250.

31. Fuse, S., Tanabe, N., and Takahashi, T. (2011) *Chem. Commun.*, **47**, 12661–12663.
32. Inoue, T., Schmidt, M.A., and Jensen, K.F. (2007) *Ind. Eng. Chem. Res.*, **46**, 1153–1160.
33. For some recent reviews about photochemical reactions: (a) Bach, T. and Hehn, J.P. (2011) *Angew. Chem. Int. Ed.*, **50**, 1000–1045; (b) Hoffmann, N. (2008) *Chem. Rev.*, **108**, 1052–1103.
34. (a) Knowles, J.P., Eliott, L.D., and Booker-Milburn, K.I. (2012) *Beilstein J. Org. Chem.*, **8**, 2025–2052; (b) Oelgemoeller, M. (2012) *Chem. Eng. Technol.*, **35**, 1144–1152; (c) Erickson, D., Sinton, D., and Psaltis, D. (2011) *Nat. Photonics*, **5**, 583–590.
35. (a) Tucker, J.W. and Stephenson, C.R.J. (2012) *J. Org. Chem.*, **77**, 1617–1622; (b) Prier, C.K., Rankic, D.A., and MacMillan, D.W.C. (2013) *Chem. Rev.* doi: 10.1021/cr300503r
36. For a review about the merger between photoredox catalysis and microreactor technology: Noël, T., Wang, X., and Hessel, V. (2013) *Monogr. suppl. Ser. Chim. Oggi - Chem. Today: Organomet. Chem., Biocatal. Catal.*, **31**, 10–14.
37. Tucker, J.W., Zhang, Y., Jamison, T.F., and Stephenson, C.R.J. (2012) *Angew. Chem. Int. Ed.*, **51**, 4144–4147.
38. Levesque, F. and Seeberger, P.H. (2012) *Angew. Chem. Int. Ed.*, **51**, 1706–1709.
39. For reviews pertaining the combination of immobilized reagents and flow chemistry: (a) Ley, S.V. (2012) *Chem. Rec.*, **12**, 378–390; (b) Baxendale, I.R. and Ley, S.V. (2007) in *New Avenues to Efficient Chemical Synthesis-Emerging Technologies* (eds P.H. Seeberger and T. Blume), Springer-Verlag, Berlin, pp. 151–185; (c) Ley, S.V. and Baxendale, I.R. (2002) *Nat. Rev. Drug Discovery*, **1**, 573–586.
40. Noël, T., Maimone, T.J., and Buchwald, S.L. (2011) *Angew. Chem. Int. Ed.*, **50**, 8900–8903.
41. Sheldon, R.A. (2007) *Adv. Synth. Catal.*, **349**, 1289–1307.
42. For some recent reviews about the combination of immobilized enzymes and microreactors: (a) Bolivar, J.M., Wiesbauer, J., and Nidetzky, B. (2011) *Trends Biotechnol.*, **29**, 333–342; (b) Asanomi, Y., Yamaguchi, H., Miyazaki, M., and Maeda, H. (2011) *Molecules*, **16**, 6041–6059; (c) Marques, M.P.C. and Fernandes, P. (2011) *Molecules*, **16**, 8368–8401.
43. Dencic, I., de Vaan, S., Noël, T., Meuldijk, J., de Croon, M., and Hessel, V. (2013) *Ind. Eng. Chem. Res.*, **52**, 10951–10960.
44. Fu, H., Dencic, I., Tibhe, J., Sanchez Pedraza, C.A., Wang, Q., Noël, T., Meuldijk, J., de Croon, M., Hessel, V., Weizenmann, N., Oeser, T., Kinkeade, T., Hyatt, D., Van Roy, S., Dejonghe, W., and Diels, L. (2012) *Chem. Eng. J.*, **207-208**, 564–576.
45. (a) Frost, C.G. and Mutton, L. (2010) *Green Chem.*, **12**, 1687–1703; (b) Zhao, D. and Ding, K. (2013) *ACS Catal.* doi: 10.1021/cs300830x
46. Bogdan, A.R. and Sach, N.W. (2009) *Adv. Synth. Catal.*, **351**, 849–854.
47. (a) de Vries, J.G. (2006) *Dalton Trans.*, 421–429; (b) Balanta, A., Godard, C., and Claver, C. (2011) *Chem. Soc. Rev.*, **40**, 4973–4985; (c) Broadwater, S.J. and McQuade, D.T. (2006) *J. Org. Chem.*, **71**, 2131–2134; (d) Davies, I.W., Matty, L., Hughes, D.L., and Reider, P.J. (2001) *J. Am. Chem. Soc.*, **123**, 10139–10140.
48. Munoz, J., Alcazar, J., de la Hoz, A., and Diaz-Ortiz, A. (2012) *Adv. Synth. Catal.*, **354**, 3456–3460.
49. Webb, D. and Jamison, T.F. (2010) *Chem. Sci.*, **1**, 675–680.
50. For reviews about one-pot reactions in batch: (a) Ramon, D.J. and Yus, M. (2005) *Angew. Chem. Int. Ed.*, **44**, 1602–1634; (b) Mayer, S.F., Kroutil, W., and Faber, K. (2001) *Chem. Soc. Rev.*, **30**, 332–339; (c) Tietze, L.F. (1996) *Chem. Rev.*, **96**, 115–136.
51. For imatinib base: (a) Hopkin, M.D., Baxendale, I.R., and Ley, S.V. (2010) *Chem. Commun.*, **46**, 2450–2452; (b) Hopkin, M.D., Baxendale, I.R., and Ley, S.V. (2013) *Org. Biomol. Chem.*, **11**, 1822–1839; cBaxendale, I.R., Deeley, J., Griffiths-Jones, C.M., Ley, S.V., Saaby, S., and Tranmer, G.K. (2006) *Chem. Commun.*, 2566–2568; dBaxendale, I.R.,

Griffiths-Jones, C.M., Ley, S.V., and Tranmer, G.K. (2006) *Synlett*, 427–430.

52. (a) Varas, A.C., Noël, T., Wang, Q., and Hessel, V. (2012) *ChemSusChem*, **5**, 1703–1707; (b) Noël, T., Kuhn, S., Musacchio, A.J., Jensen, K.F., and Buchwald, S.L. (2011) *Angew. Chem. Int. Ed.*, **50**, 5943–5946; (c) Sahoo, H.R., Kralj, J.G., and Jensen, K.F. (2007) *Angew. Chem. Int. Ed.*, **46**, 5704–5708.
53. Hartman, R.L., Naber, J.R., Buchwald, S.L., and Jensen, K.F. (2010) *Angew. Chem. Int. Ed.*, **49**, 899–903.
54. O'Brien, A.G., Horvath, Z., Levesque, F., Lee, J.W., Seidel-Morgenstern, A., and Seeberger, P.H. (2012) *Angew. Chem. Int. Ed.*, **51**, 7028–7030.
55. Roberge, D.M., Ducry, L., Bieler, N., Cretton, P., and Zimmermann, B. (2005) *Chem. Eng. Technol.*, **28**, 318–323.
56. (a) Zhao, C.-X., He, L., Qiao, S.Z., and Middelberg, A.P.J. (2011) *Chem. Eng. Sci.*, **66**, 1463–1479; (b) Marre, S. and Jensen, K.F. (2010) *Chem. Soc. Rev.*, **39**, 1183–1202; (c) Abou-Hassan, A., Sandre, O., and Cabuil, V. (2010) *Angew. Chem. Int. Ed.*, **49**, 6268–6286.
57. Poe, S.L., Cummings, M.A., Haaf, M.P., and McQuade, D.T. (2006) *Angew. Chem. Int. Ed.*, **45**, 1544–1548.
58. (a) Bremond, N. and Bibette, J. (2012) *Soft Matter*, **8**, 10549–10559; (b) Shah, R.K., Kim, J.-W., Agresti, J.J., Weitz, D.A., and Chu, L.-Y. (2008) *Soft Matter*, **4**, 2303–2309.
59. (a) Noël, T., Naber, J.R., Hartman, R.L., McMullen, J.P., Jensen, K.F., and Buchwald, S.L. (2011) *Chem. Sci.*, **2**, 287–290; (b) Hartman, R.L., Naber, J.R., Zaborenko, N., Buchwald, S.L., and Jensen, K.F. (2010) *Org. Process Res. Dev.*, **14**, 1347–1357; (c) Horie, T., Sumino, M., Tanaka, T., Matsushita, Y., Ichimura, T., and Yoshida, J.-i. (2010) *Org. Process Res. Dev.*, **14**, 405–410.
60. Kuhn, S., Noël, T., Gu, L., Heider, P.L., and Jensen, K.F. (2011) *Lab Chip*, **11**, 2488–2492.
61. Browne, D.L., Deadman, B.J., Ashe, R., Baxendale, I.R., and Ley, S.V. (2011) *Org. Process Res. Dev.*, **15**, 693–697.
62. For reviews concerning the combination of spectroscopic detection and microreactors: (a) Yue, J., Schouten, J.C., and Nijhuis, T.A. (2012) *Ind. Eng. Chem. Res.*, **51**, 14583–14609; (b) McMullen, J.P. and Jensen, K.F. (2010) *Annu. Rev. Anal. Chem.*, **3**, 19–42.
63. Chew, W. and Sharratt, P. (2010) *Anal. Chem.*, **2**, 1412–1438.
64. McMullen, J.P., Stone, M.T., Buchwald, S.L., and Jensen, K.F. (2010) *Angew. Chem. Int. Ed.*, **49**, 7076–7080.
65. (a) McMullen, J.P. and Jensen, K.F. (2011) *Org. Process Res. Dev.*, **15**, 398–407; (b) Reizman, B.J. and Jensen, K.F. (2012) *Org. Process Res. Dev.*, **16**, 1770–1782.
66. Caygill, G., Zanfir, M., and Gavriilidis, A. (2006) *Org. Process Res. Dev.*, **10**, 539–552.
67. Kockmann, N., Gottsponer, M., Zimmermann, B., and Roberge, D.M. (2008) *Chem. Eur. J.*, **14**, 7470–7477.
68. Rebrov, E.V., Schouten, J.C., and de Croon, M.H.J.M. (2011) *Chem. Eng. Sci.*, **66**, 1374–1393.
69. Tonkovich, A., Kuhlmann, D., Rogers, A., McDaniel, J., Fitzgerald, S., Arora, R., and Yuschak, T. (2005) *Chem. Eng. Res. Des.*, **83**, 634–639.
70. Tonkovich, A.L. and Lerou, J.J. (2010) in *Novel Concepts in Catalysis and Chemical Reactors: Improving the Efficiency for the Future* (eds A. Cybulski, J.A. Moulijn, and A. Stankiewicz), Wiley-VCH Verlag GmbH, Weinheim, pp. 239–260.
71. (a) Kockmann, N., Gottsponer, M., and Roberge, D.M. (2010) *Chem. Eng. J.*, **167**, 718–726; (b) Kockmann, N. and Roberge, D.M. (2011) *Chem. Eng. Process.*, **50**, 1017–1026; (c) Kockmann, N. (2012) *Chem. Ing. Tech.*, **84**, 646–659.
72. Lavric, E.D. and Woehl, P. (2009) *Chem. Today*, **27**, 45–48.
73. Roberge, D.M., Gottsponer, M., Eyholzer, M., and Kockmann, N. (2009) *Chem. Today*, **27**, 8–11.

7
Understanding Trends in Reaction Barriers

Israel Fernández López

7.1
Introduction

Controlling the reactivity of molecules has been (and still is) a challenge for chemists. Apart from concepts such as atom-economy, an efficient chemical reaction ideally leads to the desired target molecule in quantitative or very high reaction yields without forming side products. The design of such efficient transformations necessarily implies an *a priori* detailed understanding of the physical factors that control the relative heights of the barriers associated with the different reaction pathways involved in the chemical reaction. However, in most cases, synthetic (and also theoretical) chemists optimize the reaction conditions *a posteriori* based on trial/error procedures to finally achieve the most efficient (i.e., optimized) transformation. This is mainly due to the fact that those physical factors governing the intrinsic reactivity of molecules in fundamental processes are incompletely understood prior to conducting the reaction of interest.

Despite that, theory has provided chemists with a good number of tools to understand and also predict the reactivity of molecules. For instance, Woodward–Hoffmann rules [1] and Fukui's frontier molecular orbital (FMO) theory [2] have become a powerful conceptual framework to interpret the reactivity and regioselectivity patterns in different pericyclic processes such as cycloaddition reactions. In the FMO theory, the interactions of the HOMO and LUMO of reactants are emphasized and the strongest interactions are suggested to occur between orbitals that are closest in energy and have the largest overlap [2, 3]. This means that smaller HOMO–LUMO gaps should lead to lower activation energies. However, only the reactants at their equilibrium geometries are considered within the widely used FMO theory. This is a rather crude assumption, because the interactions in the corresponding transition states or at any other point along the reaction coordinate are completely ignored. Other approaches such as conceptual density functional (DFT) and hard and soft acid and base (HSAB) theories [4], valence bond (VB) analyses [5], or Marcus theory [6] have also contributed to the current understanding of fundamental processes in chemistry theories.

Discovering the Future of Molecular Sciences, First Edition. Edited by Bruno Pignataro.

Nevertheless, the relatively recent introduction by Bickelhaupt and coworkers of the so-called *activation strain model* (ASM), also known as *distortion/interaction model* as proposed by Houk *et al.*, has allowed us to gain quantitative insight into the physical factors governing how the activation barriers arise in different chemical reactions. This model, which is based on accurate quantum chemical calculations, provides, in combination with the energy decomposition analysis (EDA) method, a robust methodology to explore the trends in reactivity within organic and organometallic chemistry.

In this chapter, we demonstrate the performance of the combined ASM/EDA method to explore and understand trends in reactivity in various fundamental types of reactions in organic chemistry. We mainly focus on recent contributions from our laboratories and briefly summarize illustrative highlights from the Houk and Bickelhaupt research groups.

7.2 Activation Strain Model and Energy Decomposition Analysis

As both methods have been described in detail in recent reviews [7, 8], herein we summarize only briefly their most relevant aspects.

7.2.1 Activation Strain Model

The ASM is a fragment-based approach to understanding chemical reactions, in which the height of the associated reaction barriers is described and understood in terms of the original reactants [7, 9–11]. This model is a systematic further development of the fragment approach that is being transferred from equilibrium structures to transition states as well as *nonstationary* points, for example, points along a reaction coordinate. Thus, the potential energy surface $\Delta E(\zeta)$ is decomposed, along the reaction coordinate ζ, into the strain $\Delta E_{strain}(\zeta)$ associated with deforming the individual reactants plus the actual interaction $\Delta E_{int}(\zeta)$ between the deformed reactants:

$$\Delta E(\zeta) = \Delta E_{strain}(\zeta) + \Delta E_{int}(\zeta)$$

The strain $\Delta E_{strain}(\zeta)$ is determined by the rigidity of the reactants and on the extent to which groups must reorganize in a particular reaction mechanism. Therefore, this geometrical deformation is characteristic for the reaction pathway under consideration. On the other hand, the interaction $\Delta E_{int}(\zeta)$ between the reactants depends on their electronic structure and on how they are mutually oriented as they approach each other. Thus, the latter term is related to the bonding capabilities and mutual interaction between the increasingly deformed reactants along the same pathway. It is the interplay between $\Delta E_{strain}(\zeta)$ and $\Delta E_{int}(\zeta)$ that determines if and at which point along ζ a barrier arises (Figure 7.1).

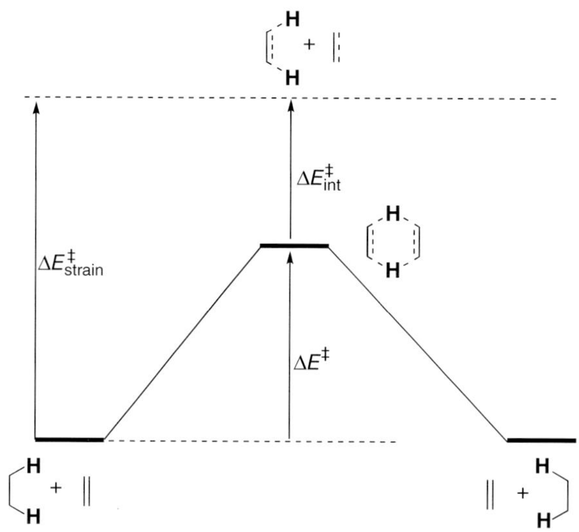

Figure 7.1 Illustration of the activation strain model using the double hydrogen atom transfer reaction from ethane to ethene.

The activation energy of a reaction $\Delta E^{\ddagger} = \Delta E(\zeta^{\mathrm{TS}})$ consists of the activation strain $\Delta E^{\ddagger}{}_{\mathrm{strain}} = \Delta E_{\mathrm{strain}}(\zeta^{\mathrm{TS}})$ plus the transition-state interaction $\Delta E_{\mathrm{int}}{}^{\ddagger} = \Delta E_{\mathrm{int}}(\zeta^{\mathrm{TS}})$:

$$\Delta E^{\ddagger} = \Delta E_{\mathrm{strain}}{}^{\ddagger} + \Delta E_{\mathrm{int}}{}^{\ddagger}$$

According to the above definitions, $\Delta E_{\mathrm{strain}}{}^{\ddagger}$ is the energy associated with deforming the reactants from their equilibrium geometries to the geometry they adopt in the transition state (TS). This term can, of course, be further divided into the individual contributions stemming from each reactant. Similarly, $\Delta E_{\mathrm{int}}{}^{\ddagger}$ stands for the actual interaction energy between the deformed reactants in the transition state.

This decomposition of the energy $\Delta E(\zeta)$ is carried out along the intrinsic reaction coordinate (IRC) (provided by the IRC method), that is, from the separate reactants (or from a weakly bonded reactant complex when it exists) to the reaction products via the corresponding transition state.

7.2.2
Energy Decomposition Analysis

The interaction energy between the reactants, $\Delta E_{\mathrm{int}}(\zeta)$, can be further decomposed in meaningful energy contributors with the help of the EDA method. The EDA [8, 12, 13], which was developed by Ziegler and Rauk [14] following a similar procedure suggested by Morokuma [15], has been proven to give important information about the nature of the bonding in main-group compounds [16], transition-metal complexes [17], as well as biological and supramolecular aggregates [18].

The EDA method employs a systematic procedure to evaluate bonding energies. The strategy is to divide the system of interest, AB, into fragments, for example, A and B, which are then recombined in three separate steps in order to obtain the energies of individual interactions. In EDA step (1), the fragments, A and B, with their geometries frozen as in AB are computed individually in appropriately selected spin states (e.g., valence states, which may not be the ground states) and then are superimposed with unrelaxed electron densities at the geometry of AB to give A′B′. This gives the quasi-classical electrostatic interaction, ΔE_{elstat}, as the energy difference between the original A + B and A′B′. This superposition usually lowers the energy because the total nuclear–electron attraction in most cases is larger than the sum of the nuclear-nuclear and electron–electron repulsion [16]. However, the resulting product wave function for this modified A′B′ species, that is, the Hartree wave function, violates the Pauli principle because electrons with same spin from two different fragments may occupy the same spatial region. In step (2), this situation is rectified by antisymmetrization and renormalization of the A′B′ wave function, thereby removing electron density, particularly from the AB bonding region where the overlap of the frozen densities is large. This step gives the Pauli repulsion term, ΔE_{Pauli}. It comprises the repulsive orbital interactions between closed-shells and is responsible for any steric repulsion between molecular fragments. In step (3), the molecular orbitals are relaxed. This allows the occupied and vacant orbitals to mix. The resulting electron delocalization gives the stabilizing orbital interaction term, ΔE_{orb}. The total interaction energy, ΔE_{int}, is the sum of the three terms:

$$\Delta E_{\text{int}} = \Delta E_{\text{elstat}} + \Delta E_{\text{Pauli}} + \Delta E_{\text{orb}}$$

Note that ΔE_{int} is not the same as a bond (or group) dissociation energy, as the additional energy related to the geometric relaxation of the A and B fragments (i.e., the ΔE_{strain} or ΔE_{prep} term) is not included. The orbital contribution, ΔE_{orb}, can be further partitioned into contributions by orbitals belonging to different irreducible representations of the point group of the interacting system (when applicable). This has been, for instance, crucial for the application of this method to the direct estimate of conjugation, hyperconjugation, and aromaticity in organic and organometallic compounds [19, 20].

7.3 Pericyclic Reactions

7.3.1 Double Group Transfer Reactions

Double group transfer (DGT) reactions are a general class of pericyclic reactions that involve the simultaneous migration of two atoms/groups from one compound to another in a concerted reaction pathway [21]. This definition includes textbook reactions such as the diimide reduction of double or triple bonds [22],

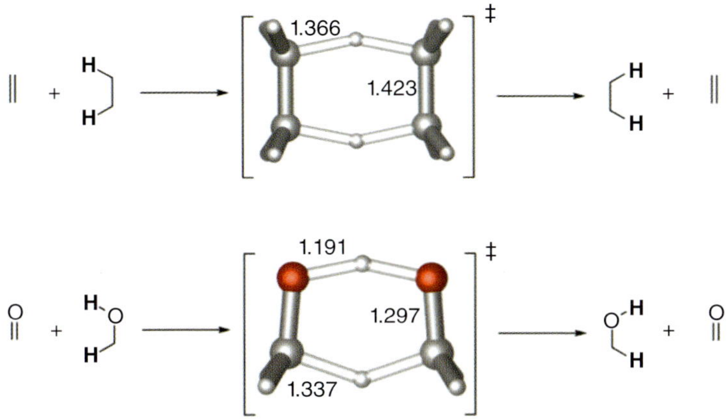

Figure 7.2 Representative DGT reactions: (a) ethane + ethene and (b) formaldehyde + methanol. Bond distances in the corresponding transition states are given in angstroms.

the Meerwein–Ponndorf–Verley reduction (MPV) of carbonyl groups [23], and some type II-dyotropic rearrangements that are characterized by the intramolecular migration of the two groups (generally hydrogen atoms) [24]. The archetypical DGT process is the thermally allowed concerted and synchronous transfer of two hydrogen atoms from ethane to ethylene that proceeds suprafacially on both reaction sites. According to the Woodward–Hoffmann rules [1], these $[{}_{\sigma}2_s + {}_{\sigma}2_s + {}_{\pi}2_s]$ transformations may be considered as thermally allowed pericyclic reactions that occur via a highly symmetric planar six-membered ring transition state as shown in Figure 7.2 [25–27].

The occurrence of a planar highly symmetric six-membered transition state, where the C–C and C–X bonds have a partial double-bond character and are equalized, is an indication of electronic delocalization within the plane containing the six electrons involved in the process. Therefore, these saddle points satisfy the so-called geometric criterion for aromaticity [28]. Indeed, these species exhibit strongly negative nucleus independent chemical shifts (NICS) [29] values (in the range of −10 to −25 ppm) thus fulfilling the magnetic criterion for aromaticity as well. It is proposed that the six electrons involved in the double hydrogen atom transfers lie approximately in the molecular plane and give rise to an appreciable ring current. In turn, this ring current promotes a strong diamagnetic shielding at the ring critical point leading to the observed high negative NICS values [25]. This special type of aromaticity, which is present in the transition states of DGT reactions, is known as *in-plane* aromaticity [30]. As expected for this kind of aromatic transition states, the variation of NICS values along the z-axis perpendicular to the molecular plane describes the usual "bell-shape curve" with a maximum NICS value at $z = 0$ Å, that is, at the (3,+1) ring critical point (Figure 7.3a). Further confirmation of the aromatic nature of the DGT reactions transition states is given by the plot of the induced current density. When the anisotropy of the induced

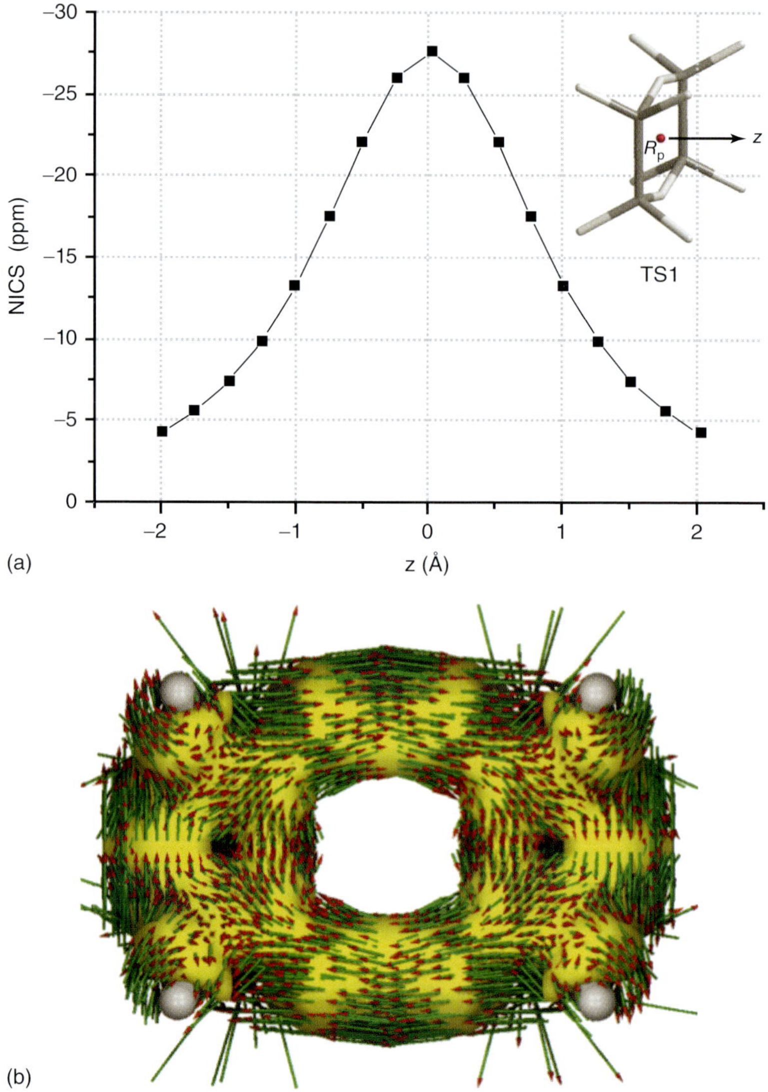

Figure 7.3 (a) AICD plot for the transition state of the DGT reaction between ethane and ethene. (b) Variation of the NICS values along the *z*-axis perpendicular to the molecular plane of this transition state.

current density (AICD) method, developed by Herges and coworkers [31], is applied to the saddle points of the above considered transformations, a strong and diatropic (i.e., aromatic) induced current is observed (Figure 7.3b).

Despite the aromatic character of these transition states, DGT reactions are associated with relatively high barriers (typically $\Delta E^{\ddagger} > 40\,\text{kcal mol}^{-1}$) [25, 26]. Indeed, it was found that ΔE continuously increases along the reaction coordinate with the concomitant gain in aromaticity, which seems contradictory if we consider that a gain in aromaticity is usually translated into a gain in stability [26]. This finding suggests that DGT reactions are controlled by a different factor that cannot be compensated by aromatic delocalization.

At this point, the combined ASM/EDA method becomes an extremely helpful tool to understand the origins of the high activation barriers computed for these pericyclic transformations [26]. Figure 7.4a shows the plot of the computed potential energy surface along the IRC trajectories, projected onto the distance r(H···C) for

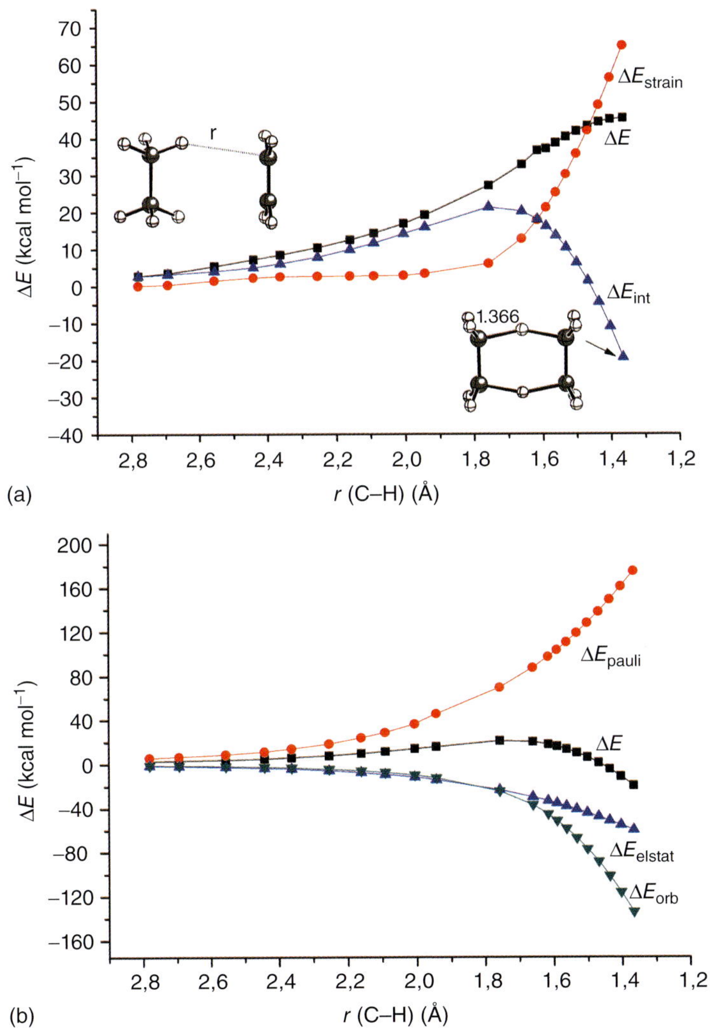

Figure 7.4 (a) Activation strain analysis of the parent DGT reaction between ethane and ethene along the reaction coordinate projected onto the forming C···H bond distance. (b) Corresponding decomposition of the interaction energy along the reaction coordinate.

the parent reaction between ethane and ethene together with the change of the energy contributions to $\Delta E(\zeta)$, namely, the strain $\Delta E_{strain}(\zeta)$ and the instantaneous interaction $\Delta E_{int}(\zeta)$ between the deformed reactants. We can see that at the early stages of the process the reaction profile ΔE monotonically becomes more and more destabilized as the reactants approach each other. Then, a sharp increase of ΔE occurs in the proximity of the transition-state region (i.e., at H–C distances in the range from 2.0 to 1.6 Å) leading to the observed high reaction barriers.

Interestingly, the interaction energy between the deformed reactants (ΔE_{int}) becomes destabilizing at long H–C distances and causes the net energy ΔE to increase as well (Figure 7.4a). This initial increase of ΔE_{int} can be traced to steric (Pauli) repulsion between the reactants in the early stages of the reaction. Thus, before anything else happens, the reactants approach, and overlap occurs between closed-shells, notably between C–H bonds of the H-donor and the π-system of the H-acceptor, which causes Pauli repulsion. In addition, the initial increase in ΔE is also caused by the ethane reactant, as it has to adopt the required eclipsed conformation to interact with the π-system of ethene. If we now further proceed along the DGT reaction coordinate, the trend in ΔE_{int} inverts at a certain point, after which this term becomes more and more stabilizing. This stabilization in the ΔE_{int} curve occurs shortly after the onset of the strain curve. Despite that, the reason that the overall energy ΔE still goes up until the transition state is of course also the increase in the destabilizing strain energy, which clearly compensates the stabilization provided by ΔE_{int}. This destabilization is ascribed to the breaking of the two C–H bonds in ethane, which turns into the dominant contribution to the strain term as the transition state is approached.

The EDA method provides further insight into the different contributors to the interaction energy between the deformed reactants. As seen in Figure 7.4b, it is clear that the dominant term causing the inversion in ΔE_{int} is the orbital interaction energy ΔE_{orb}, which is closely related to the in-plane aromatic delocalization at the proximities of the transition state. The reason for the onset of sizable orbital interactions at shorter H–C distances is that the original C–H bonds are beginning to elongate. This causes the associated σ^*_{C-H} orbitals to drop in energy and become more localized on the transferring hydrogen atoms. As a result, the HOMO–LUMO interactions between the π-electron system of the H-acceptor reactant and the σ^*_{C-H} orbitals of the H-donor become stronger, more stabilizing. Interestingly, although the electrostatic attraction ΔE_{elstat} is not the dominant bonding term, it is certainly far from being negligible as it contributes about 30% of the total attraction between the deformed reactants.

Trends in DGT reactivity are thus controlled by the energy needed to deform the reactants from their equilibrium geometries to the geometry they adopt in the transition state. This factor dominates the trend in the stabilizing interaction energy between the reactants and the gain in stability by aromaticity in the corresponding cyclic transition states. This suggests that it would be possible to design low barrier DGT reactions if we could enhance the interaction energy between the reactants and/or reduce the destabilizing contribution of the strain term. Indeed, both possibilities can be found in the literature: (i) in MPV reduction of carbonyl groups, there occurs the formation of an intramolecular hydrogen-bond that approximates both reactants making the interaction energy between them stronger; as a consequence, the computed barrier for this process drops to about 25 kcal mol^{-1} [26]. (ii) In intramolecular type II-dyotropic reactions in sesquinorbornanes [32], the initial geometry of the reactant resembles that of the corresponding transition state that significantly reduces the strain energy associated with the required deformation, leading to low barrier processes.

The above mentioned cases illustrate clearly the utility of the combined ASM/EDA method to understand the physical factors controlling the barrier heights of DGT reactions. But they also show that these factors can be tuned to design more favorable transformations prior to conducting the experiment.

7.3.2
Alder-ene Reactions

Alder-ene reactions [33] are transformations closely related to DGT reactions. In fact, these processes also involve the migration of a hydrogen atom from the ene reactant to enophile with concomitant C–X bond formation (Scheme 7.1) [34]. In addition, these reactions usually require highly activated substrates and/or high temperatures. Similar to DGT reactions, most Alder-ene reactions proceed in a concerted and synchronous fashion via six-membered aromatic transition states [35]. Only in those cases where the reaction is promoted by Lewis acids, the reaction mechanism can change from concerted to stepwise involving cationic intermediates [36].

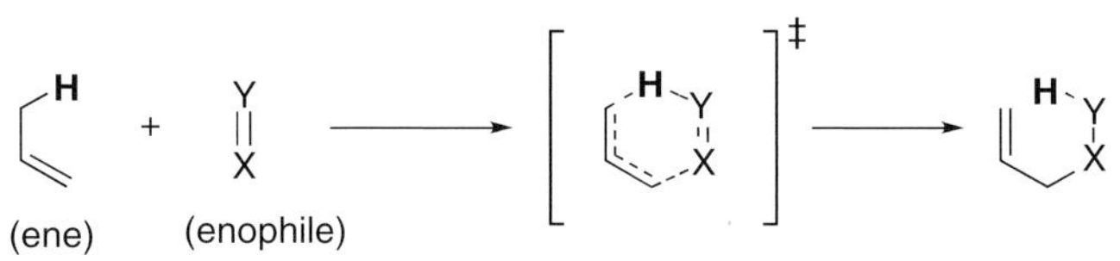

X=Y: C=C, C≡C,C=O, C=S, C=NH, C=PH

Scheme 7.1 Alder-ene reaction.

We also applied the ASM/EDA method to explore the trend in the reactivity of the Alder-ene reactions between propene and a series of different enophiles (Scheme 7.1) [37]. Despite the aromatic character of the corresponding transition states (NICS values ranging from −22 to −25 ppm), these processes are associated with relatively high barriers ($\Delta E^{\ddagger}$ up to 33 kcal/mol for the parent reaction between propene and ethene). It was found that the activation strain, $\Delta E^{\ddagger}_{\text{strain}}$, controls the height and the trend in these transformations. Moreover, the strain energy associated with the deformation of propene, $\Delta E^{\ddagger}_{\text{strain}}$(ene), is not only the main contributor to the total activation strain but also it varies more pronouncedly than the smaller $\Delta E^{\ddagger}_{\text{strain}}$(enophile) term. The reason is that in the course of the reaction, the ene reactant must break one of its C–H bonds, which is quite costly in terms of energy because C–H bonds are quite strong. However, no bond is broken in the enophile during the course of the geometrical rearrangement associated with the Alder-ene reaction.

Interestingly, our calculations indicate that barriers drop in particular, if third-period atoms become involved in the double bond of the enophile (computed $\Delta E^{\ddagger}$ of 21 and 15 kcal mol^{-1} for the Alder-ene reactions between propene and $H_2C{=}PH$ and $H_2C{=}S$, respectively). This is related to the enophile π^*-LUMO, which achieves a larger amplitude on the carbon atom (X = C in Scheme 7.1) and a smaller one on the heteroatom (Y) which also possesses more diffuse 3p lobes. As a consequence, this π^* molecular orbital becomes less suitably shaped for overlapping with the C–H bond of the hydrogen atom that is transferred from ene to enophile and a more asynchronous process occurs, that is, C–C bond formation runs somewhat more ahead, whereas C–H bond breaking lags a bit more behind. The direct consequence of this asynchronicity is the reduction of the activation strain (from the ene moiety) that translates into a lower activation barrier therefore.

7.3.3
1,3-Dipolar Cycloaddition Reactions

The Houk group has investigated the synthetically useful 1,3-dipolar cycloaddition reactions [38] by means of the ASM (or distortion/interaction model). Thus, [3+2]-cycloadditions between different 1,3-dipoles of the type $X{=}Y^+ - Z^-$ (X,Y,Z = first-row elements) and ethylene or acetylene as dipolarophiles have been studied (Scheme 7.2) [39, 40].

X = RC, R_2C, RN, R_2N, O
Y = N, NR, O
Z = RC, R_2C, RN, R_2N, O

Scheme 7.2 1,3-Dipolar cycloaddition reactions studied by Ess and Houk (see references [39] and [40]).

It was found that the barrier heights for the cycloadditions of a given 1,3-dipole with ethylene and acetylene have the same activation despite very different reaction thermodynamics and FMO energy gaps. It appears to be the energy to distort the initial reactants to the transition-state geometry that is the major factor controlling the reactivity for cycloadditions of 1,3-dipoles with alkenes or alkynes. In the majority of cases, the activation strain (about 80%) arises from deformation of the 1,3-dipole due to angle change associated with achieving the required product-like structure to narrow the FMO gaps and increase intermolecular orbital overlap. Only in those cases in which the activation strain is nearly constant for different reactions (for instance, the reactions between diazonium dipoles and a set of related substituted alkenes), the FMO interaction energies become large enough to modulate the trend in reactivity.

Moreover, the Houk group has also explored the reaction dynamics of this transformation [41]. Thus, trajectories were propagated in order to ascertain the contributions to the activation barriers from reactant vibration, rotation, and relative translation. In good agreement with the above commented significance of the deformation of the dipole, it was found that the dipole bending modes are extremely important. In fact, the reaction requires a large amount of vibrational excitation in the dipole bending modes in order to occur and these modes contribute greatly to the transition-state energy.

The importance of the strain energy in 1,3-dipolar cycloadditions was also stressed by comparing the [3+2]-reactions of phenyl azide with acytelene and cyclooctyne [42]. Whereas the strain-free acetylene cycloaddition proceeds with a computed activation barrier of 16.2 kcal mol^{-1}, the cyclooctyne "strain-promoted" cycloaddition is kinetically easier (activation barrier of only 8.0 kcal mol^{-1}). This has been ascribed to decreased distortion energy in cyclooctyne ($\Delta\Delta E^{\ddagger}_{strain} = 4.6$ kcal mol^{-1}) and phenyl azide ($\Delta\Delta E^{\ddagger}_{strain} = 4.5$ kcal mol^{-1}) to achieve the geometry of the corresponding transition state.

A similar finding, that is, the bending of the 1,3-dipole as the main factor controlling the activation barriers, was found by us in the [3+2]-cycloaddition reaction between heteroallenes and triple bonds [43].

Very recently, Gornitzka, Escudié, and coworkers reported that heteroallene **1** is readily converted into the tricyclic compound **3** when reacted with methyl acetylenedicarboxylate (Scheme 7.3) [44]. This process is suggested to proceed

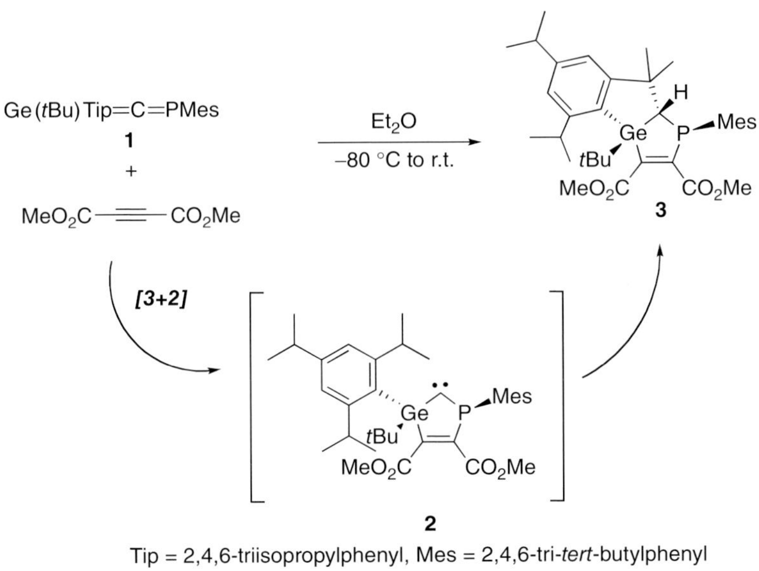

Scheme 7.3 Reaction of heteroallene 1 and acetylenedicarboxylate leading to 3 (see reference [44]).

via the carbene intermediate **2** through a concerted [3+2]-cycloaddition reaction. Carbene **2** is then transformed into compound **3** by the insertion of the carbenic carbon atom into the C–H bond of an ortho isopropyl group of the Tip (2,4,6-$^{i}Pr_3C_6H_3$) group on the germanium atom.

This interesting transformation prompted us to carry out a comparative study on the effect of group 14 elements in heteroallenes $H_2E{=}C{=}PH$ (E = C to Pb) in their [3+2]-cycloaddition reaction toward acetylene [43]. It was found that all processes

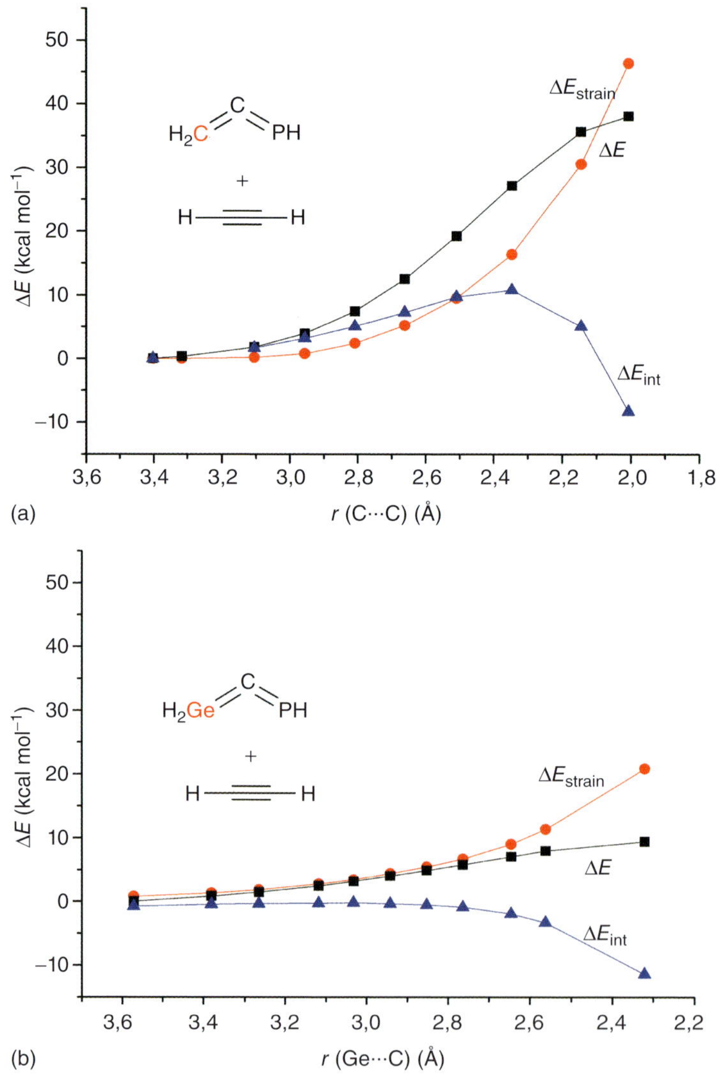

Figure 7.5 Activation strain analysis for the [3+2]-cycloaddition reaction between acetylene and (a) $H_2C{=}C{=}PH$ and (b) $H_2Ge{=}C{=}PH$.

occur concertedly through C_s-symmetric and in-plane aromatic transition states (NICS values in the range of −10 to −21 ppm). Despite that, the corresponding reaction barriers drop significantly from E = C (nearly 50 kcal mol^{-1}) to E = Si-Pb (about 20 kcal mol^{-1}).

The activation strain analysis plot for the reaction involving $H_2C{=}C{=}PH$ (Figure 7.5a) resembles that for double group transfer reactions (see above). Thus, the reaction profile ΔE raises monotonically as the reactants approach each other and a sharp increase of ΔE occurs in the proximity of the transition-structure region, leading to the observed high reaction barrier. The interaction energy, which is destabilizing at the beginning of the process, becomes stabilizing at the proximities of the transition state. However, it cannot compensate the strong destabilizing effect of the strain energy (Figure 7.5a). In contrast, in the reaction involving $H_2Ge{=}C{=}PH$ as phosphallene, the interaction energy between the deformed reactants remains practically unaltered at long Ge···C distances and then smoothly becomes stabilizing in the vicinity of the transition states (Figure 7.5b).

As seen clearly in Figure 7.5, the interaction energy is not very different in both reactions (about −10 kcal mol^{-1}), which indicates that the strain energy is the major factor controlling the barriers of the [3+2]-cycloadditions. Indeed, an excellent linear relationship between the computed activation barriers, ΔE_a, and the total activation strain energies ($\Delta E_{strain}{}^{\ddagger}$) was found (correlation coefficient of 0.9998 and standard deviation of 0.3 kcal mol^{-1}, Figure 7.6). Despite that, it becomes obvious that this deformation energy is clearly lower when E = Si-Pb. The reason is that heteroallene

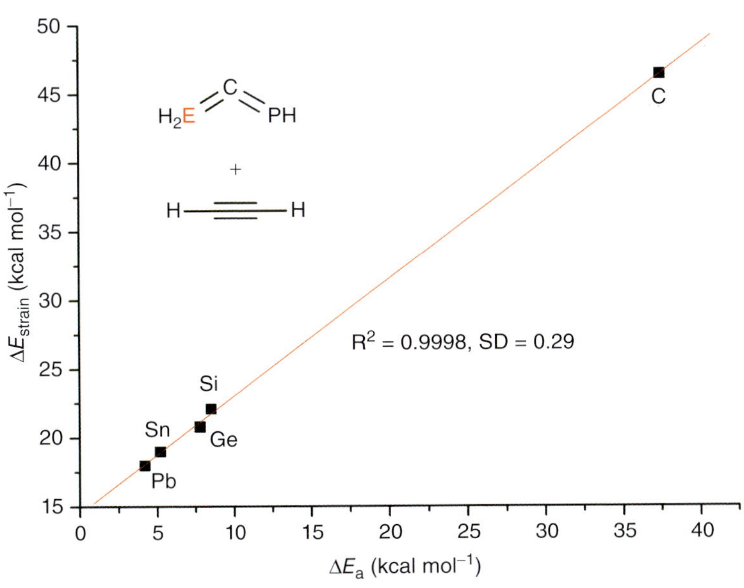

Figure 7.6 Linear relationship between $\Delta E^{\ddagger}_{strain}$ and ΔE_a for the 1,3-dipolar reaction between phosphallenes $H_2E{=}C{=}PH$ (E = group 14 element) and acetylene.

$H_2C{=}C{=}PH$, which possesses a practically linear equilibrium geometry (C=C=P angle of 174.8°), must be bent significantly in the transition state (C=C=P angle of 120.8°). At variance with this, the heteroallenes with a heavier group 14 element *E* do already possess a bent equilibrium geometry that fits into the transition-state structure better and therefore requires less deformation. As a consequence, the latter compounds undergo a much more facile [3+2]-cycloaddition toward acetylene.

7.3.4
Diels-Alder Reactions

An example of the utility of the ASM to understand and predict trends in reaction barriers has been reported by Paton, Houk, and coworkers in a recent combined experimental-computational study on the Diels–Alder reactivities of cycloalkenones and cyclic dienes [45].

Scheme 7.4 shows the experimental results of the Diels–Alder reaction between cyclopentadiene and cyclohexenone, cyclopentenone, and cyclobutenone [45, 46]. Whereas the reactions involving cyclohexenone and cyclopentenone require high temperatures (150 °C) and prolonged reaction times (one day) to produce the corresponding [4+2]-cycloadduct in moderate yields, cyclobutenone leads to the reaction product in good reaction yield in only 1 h and at room temperature.

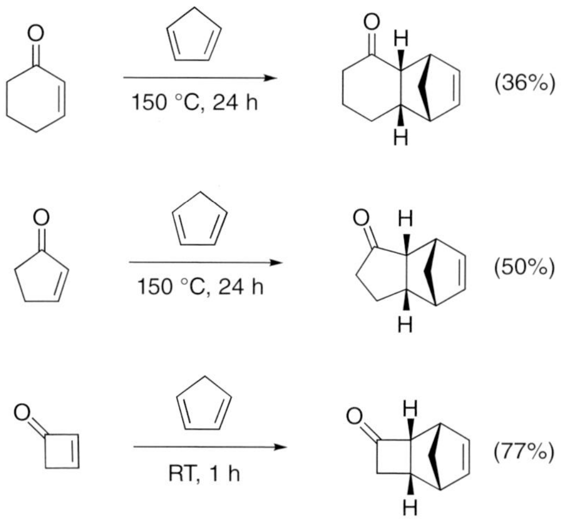

Scheme 7.4 Diels-Alder reactions between cycloalkenones and cyclopentadiene.

Inspection of the corresponding transition states suggests that these transformations are asynchronous concerted processes. The computed activation barriers for the reactions involving cyclohexenone and cyclopentenone are similar to that computed for the acyclic analog (i.e., pent-3-ene-2-one) but clearly higher than the activation barrier involving cyclobutanone. Interestingly, cyclopropenone is predicted to be even more reactive in view of the computed much lower reaction barrier.

Similarly to [3+2]-cycloaddition reactions (see above), a clear linear relationship between activation barriers and distortion energies was found (correlation coefficient of 0.93). In these Diels–Alder reactions, where the interaction energy between the deformed reactants is nearly constant, the distortion in the diene is related to the energy of bringing the diene termini into a geometry that maximizes overlap with the dienophile termini. On the other hand, the distortion of the dienophile is associated with the bending of C–H bonds out of the plane of the C=C bonds to which they are attached. The force constants for the bending of the alkene C–H bonds out of plane are reduced significantly by angle strain in cyclobutenone and cyclopropenone, which translates into reduced activation barriers. This trend in reactivity finds its origin in the larger *s* character in the C–H bond and the fact that the smaller internal angle in the small rings is more appropriate for the pyramidal transition structure.

7.4 Nucleophilic Substitutions and Additions

7.4.1 S_N2 Reactions

The Bickelhaupt group has profusely studied the bimolecular nucleophilic substitution (S_N2) reaction, a fundamental process in organic, inorganic, and biological systems. In a series of papers [47–50], different aspects of this transformation have been considered: nucleophilicity of X^-, leaving-group ability of Y, role of the electrophilic center A, effect of substituents R as well as solvent effects (Scheme 7.5).

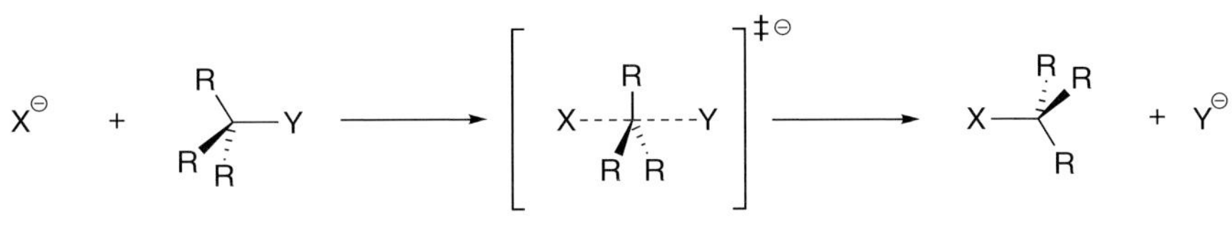

Scheme 7.5 Bimolecular nucleophilic substitution reaction.

By means of the ASM, it was revealed that the activation barrier of nucleophilic substitutions at carbon atoms is steric in nature. This is mainly due to the pentacoordinate structure of the corresponding transition state in which five substituents try to approach the relatively small central carbon atom. The height of the S_N2 activation barrier strongly depends on electronic effects, such as the mutual capabilities of the reactants and their internal bonding or rigidity [47–50]. Thus, it was found that stronger C–Y(leaving-group) bonds translate into more destabilizing strain energies leading to higher activation barriers. The nucleophilicity is also clearly related to the electron donating capability of the nucleophile: higher-energy *np* atomic orbital on X^- is reflected in more stabilizing interaction energies with the substrate, which leads to lower barriers. ASM analyses furthermore show that backside S_N2 is in general favored over frontside S_N2 because of (i) the sterically less favorable proximity, in the latter, of the larger and more electronegative nucleophile and leaving group; and (ii) the fact that the nucleophile lone-pair HOMO overlaps and interacts more favorably with the large backside lobe of the substrate's σ^*_{C-Y} LUMO than with this orbital's frontside region that features the nodal surface stemming from the antibonding combination between C and Y [47c].

The situation changes dramatically when the substitution reaction occurs at the larger silicon atom, which allows for more space between the five substituents in the pentacoordinate saddle point. The steric congestion occurring at the carbon atom decreases when silicon is involved. Consequently, the central barrier disappears, turning the transition state into a stable intermediate [47]. Despite that, the usual "carbon behavior," that is, reaction through a central barrier, reappears as the steric hindrance around the silicon atom increases, which further supports the steric nature of the barriers associated with S_N2 reactions.

Additional studies by other research groups on S_N2 reactions using the ASM and EDA methods can also be found in the literature [51–54].

7.4.2
Nucleophilic Additions to Arynes

The regioselectivity of nucleophilic additions on arynes has been explored in a recent joint experimental-computational study [55]. It was reported that 4,5-indolynes, generated from silyl triflates, preferentially produce 5-substituted adducts (Scheme 7.6) [55, 56]. This regioselectivity has been ascribed to the lower distortion energy exhibited by the favored transition state (i.e., associated with the nucleophilic addition on C5). Indeed, the initial 4,5-indolyne already possesses a distorted geometry, particularly at the C-3a position, which is in part relieved with the C5-attack but increased by the C4-addition. As a result, the activation barrier for the C4-attack is higher than that computed for the C5-addition, leading to the observed regioselectivity.

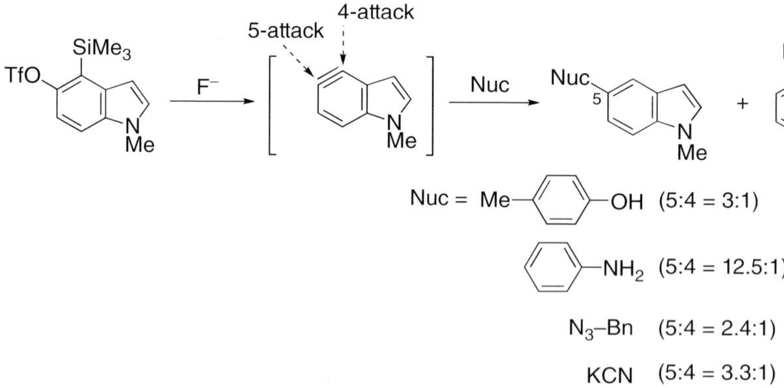

Scheme 7.6 Nucleophilic additions to arynes (see reference [55]).

This study has been extended to different arynes. In general, it was concluded that the unsymmetrical distortion present in ring-fused benzynes biases nucleophilic attack to the flatter, more electropositive end of the aryne.

7.5 Unimolecular Processes

One of the main limitations of the ASM has been that it was originally conceived for understanding bimolecular processes that correspond to a two-fragment picture (similarly to the transformations described above). Despite that, we have successfully expanded the scope of ASM to a new formulation for unimolecular reactions very recently.

Thus, we have focused on type I 1,2-dyotropic reactions, a particular class of DGT reactions (see above) in which two atoms or groups migrate simultaneously interchanging their relative molecular positions [24e, 57]. Nowadays, this kind of process has become quite a useful synthetic tool in the construction of organic and organometallic compounds, including complex natural products [24e]. In our study, we have considered the 1,2-shift of vicinal atoms and groups X along $H_2C{=}CH_2$, which acts as static scaffold (Scheme 7.7a) [58]. This transformation, which proceeds concertedly through a four-membered ring transition state [58, 59], has been used as a model of the experimentally observed mutarotation of vicinal dibromides in steroideal systems reported by Grob and Winstein (Scheme 7.7b) [60].

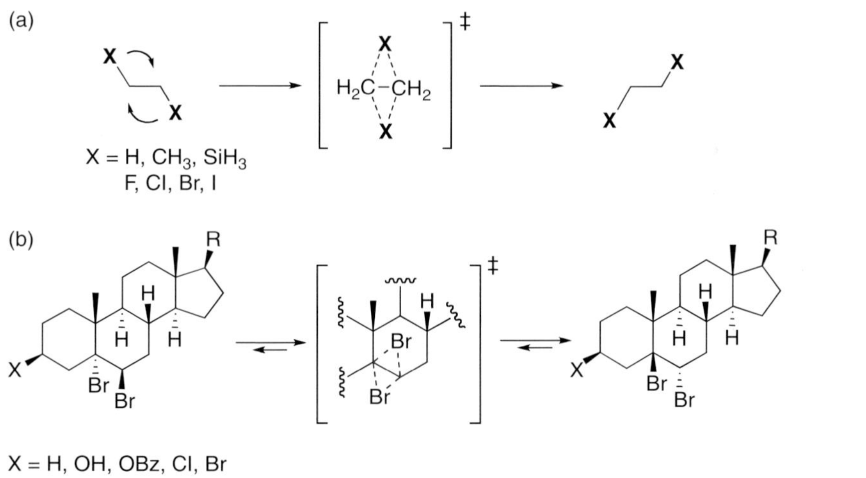

X = H, OH, OBz, Cl, Br
R = $CH(CH_3)CHYCHYCHZCH(CH_3)_2$ (Y=H, Br; Z=H, Et)

Scheme 7.7 Type I 1,2-dyotropic reactions.

Our type I 1,2-dyotropic reactions can be conceived as the interconversion between two (very strongly bound!) reactant complexes of $X_2 + H_2C{=}CH_2$. In fact,

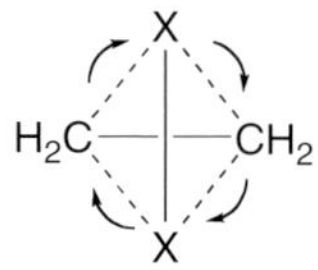

Figure 7.7 Schematic representation of the rotation of the [X···X] fragment (or "reactant") relative to the $H_2C{=}CH_2$ fragment in type I-dyotropic reactions.

the progress of the reaction indeed strongly resembles a rotation of the [X···X] fragment (or "reactant") relative to the $H_2C{=}CH_2$ fragment (or "reactant"), as shown schematically in Figure 7.7. This approach turns out to provide a detailed insight into the trends in activation energies by separating them into trends in X_2 and $H_2C{=}CH_2$ rigidity and C–X bonding. The latter is directly determined by the electronic structure and bonding capability of the migrating groups X. Therefore, in this picture, the barrier of the 1,2-dyotropic reaction arises from the *change* in the strain of and in interaction between X_2 and $H_2C{=}CH_2$ as one goes from $X{-}CH_2{-}CH_2{-}X$ to the corresponding transition state. Therefore, in this particular case, $\Delta E^{\ddagger} = \Delta\Delta E_{strain}{}^{\ddagger} + \Delta\Delta E_{int}{}^{\ddagger}$.

Our calculations indicate that the migratory aptitude of the considered groups/atoms in the selected 1,2-shifts increases in the order $H < CH_3 < SiH_3 \ll F < Cl < Br < I$. Indeed, the process can be considered as not feasible when a hydrogen atom, a methyl, or a SiH_3 group is involved in the dyotropic movement (activation barrier > 100 kcal mol^{-1}). However, the computed activation barriers associated with the migration of halogen atoms are much lower, observing the lowest activation barriers for X = Br and I (32 and 25 kcal mol^{-1}, respectively) [58].

The ASM analyses show that the change in the interaction energies, $\Delta\Delta E_{int}$, between the CH_2CH_2 and [X···X] fragments from the reactants to their corresponding transition states is clearly destabilizing. This is mainly due to the partial dissociation of the C–X bond in the transition state. On the other hand, the change in the strain energies, $\Delta\Delta E_{strain}$, is comparatively small and stabilizing in the considered type I-dyotropic reactions. This is ascribed to the fact that in the respective saddle points, the ethane fragment can adopt an almost planar geometry that closely resembles its intrinsically preferred ethylene structure. In view of these findings, it can be concluded that the weakening in the interaction energy ΔE_{int}, which derives from partial C–X bond breaking in the transition state, constitutes the major factor controlling the barrier of the 1,2-dyotropic migration. Indeed, a very good linear relationship (correlation coefficient of 0.99 and standard deviation of 6.7 kcal mol^{-1}) is found when plotting the computed activation barriers $\Delta E^{\ddagger}$ *versus* the change in the transition-state interaction energy ($\Delta\Delta E_{int}{}^{\ddagger}$, see Figure 7.8). Therefore, the trends in reactivity on variation of X can be rationalized in terms of how sensitive the C–X interaction is toward adopting the transition-state geometry.

The low barrier in the case of migrating halogen atoms can be ascribed to an additional stabilizing donor-acceptor orbital interaction between the halogen lone pairs and the π^* orbital of ethylene fragment, which completes the pericyclic circuit and which is absent for H, CH_3, and SiH_3. Finally, the decrease of the activation barrier when going down form F to I is in part due to a better and

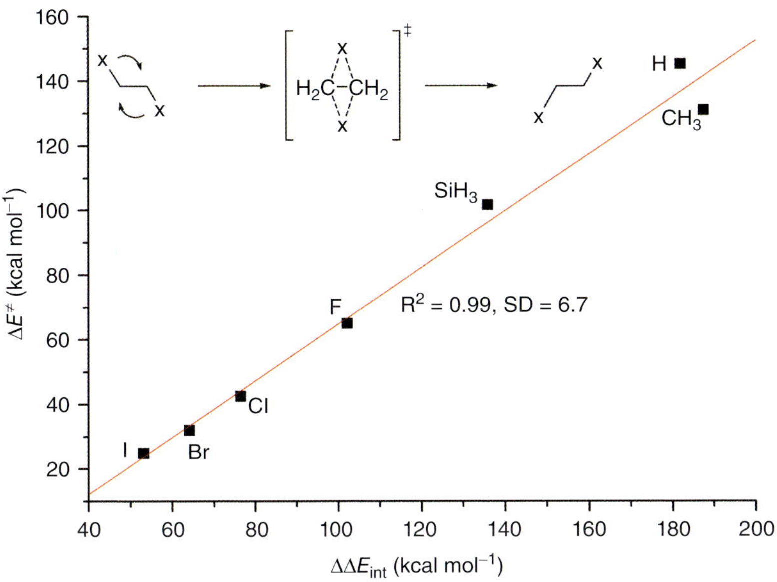

Figure 7.8 Linear relationship between $\Delta\Delta E^{\ddagger}_{int}$ and $\Delta E^{\ddagger}$ for the studied type I-dyotropic rearrangements.

better <HOMO(CH_2CH_2)—LUMO(X···X)> overlap as the halogen atomic orbitals become spatially more extended (i.e., as we descend in the halogen group).

7.6 Concluding Remarks

In this chapter, we have demonstrated how the combined ASM/EDA method can provide qualitative insight into the physical factors that control the barrier heights and reactivity trends of chemical transformations. This method, which is based on accurate quantum chemical calculations, is anchored in the interplay between the strain (or distortion) energy, $\Delta E_{strain}(\zeta)$, and the mutual interaction energy, $\Delta E_{int}(\zeta)$, between the deformed reactants, which determines if and at which point along the reaction coordinate ζ, a barrier arises.

Using illustrative applications, we have shown that the methodology is quite robust and fully applicable to any chemical reaction. Thus, fundamental processes in organic chemistry such as addition and substitution reactions or pericyclic processes can be understood in terms of the properties of the initial reactants. Despite that, this method is not restricted to organic reactions. It can also be applied to transformations involving transition metals in the fields of catalysis and organometallic chemistry, as recently demonstrated [61]. Although the ASM was originally developed for bimolecular processes corresponding to a two-fragment picture, it has been successfully extended to cover unimolecular processes as well.

This recent and valuable "up-date" of the method opens doors to explore any known (or unknown) chemical reaction. Therefore, there are no limitations for the applicability of this methodology.

It becomes obvious that the insight provided by the ASM/EDA method can be used *a posteriori* to interpret the outcome of a chemical transformation and, more importantly, also *a priori* to rationally design more efficient processes before the experiment.

Acknowledgments

Financial support from the Spanish MINECO (CTQ2010-20714-C02-01/BQU and Consolider-Ingenio 2010, CSD2007-00006) and CAM (S2009/PPQ-1634) is gratefully acknowledged.

References

1. Woodward, R.B. and Hoffmann, R. (1970) *The Conservation of Orbital Symmetry*, Verlag Chemie GmbH, Weinheim, p. 141.
2. (a) Fukui, K. and Fujimoto, H. (1967) *Bull. Chem. Soc. Jpn.*, **40**, 2018; (b) Fukui, K. and Fujimoto, H. (1969) *Bull. Chem. Soc. Jpn.*, **42**, 3399; (c) Fukui, K. (1970) *Fortschr. Chem. Forsch.*, **15**, 1; (d) Fukui, K. (1971) *Acc. Chem. Res.*, **4**, 57; (e) Fukui, K. (1982) *Angew. Chem., Int. Ed. Engl.*, **21**, 801.
3. (a) Fleming, I. (1976) *Frontier Orbitals and Organic Chemical Reactions*, Wiley-Interscience, New York; (b) Houk, K.N. (1975) *Acc. Chem. Res.*, **8**, 361; (c) Houk, K.N. (1977) in *Pericyclic Reactions*, vol. 2 (eds A.P. Marchand and R.E. Lehr), Academic Press, New York, p. 181.
4. (a) Geerlings, P., De Proft, F., and Langenaeker, W. (2003) *Chem. Rev.*, **103**, 1793; (b) Ess, D.H., Jones, G.O., and Houk, K.N. (2006) *Adv. Synth. Catal.*, **348**, 2337; (c) Geerlings, P., Ayers, P.W., Toro-Labbé, A., Chattaraj, P.K., and de Proft, F. (2012) *Acc. Chem. Res.*, **45**, 683.
5. (a) Pross, A. and Shaik, S.S. (1983) *Acc. Chem. Res.*, **16**, 363; (b) Hoffmann, R., Shaik, S.S., and Hiberty, P.C. (2003) *Acc. Chem. Res.*, **36**, 750; (c) Shaik, S.S. and Hiberty, P.C. (2007) *A Chemist's Guide to Valence Bond Theory*, Wiley-Interscience, Hoboken, NJ.
6. Marcus, R.A. (1964) *Annu. Rev. Phys. Chem.*, **15**, 155.
7. Activation Strain Model: (a) van Zeist, W.J. and Bickelhaupt, F.M. (2010) *Org. Biomol. Chem.*, **8**, 3118; (b) Bickelhaupt, F.M. (1999) *J. Comput. Chem.*, **20**, 114.
8. (a) von Hopffgarten, M. and Frenking, G. (2012) *WIREs Comput. Mol. Sci.*, **2**, 43; (b) Lein, M. and Frenking, G. (2005) in *Theory and Applications of Computational Chemistry* (eds E.D. Clifford, G. Frenking, S.K. Kwang, and E.S. Gustavo), Elsevier, Amsterdam, p. 291.
9. de Jong, G.T. and Bickelhaupt, F.M. (2007) *ChemPhysChem*, **8**, 1170–1181.
10. Diefenbach, A., de Jong, G.T., and Bickelhaupt, F.M. (2005) *Mol. Phys.*, **103**, 995.
11. van Stralen, J.N.P. and Bickelhaupt, F.M. (2006) *Organometallics*, **25**, 4260.
12. (a) Bickelhaupt, F.M. and Baerends, E.J. (2000) *Rev. Comput. Chem.*, **15**, 1; (b) te Velde, G., Bickelhaupt, F.M., Baerends, E.J., van Gisbergen, S.J.A., Fonseca Guerra, C., Snijders, J.G., and Ziegler, T. (2001) *J. Comput. Chem.*, **22**, 931; (c) Mitoraj, M.P., Michalak, M., and Ziegler, T. (2009) *J. Chem. Theory Comput.*, **5**, 962.
13. Some aspects of the EDA method have been questioned; see the rebuttal by Weinhold, F. (2003) *Angew. Chem. Int. Ed.*, **42**, 4188 to the paper by

Bickelhaupt, F.M. and Baerends, E.J. (2003) *Angew. Chem. Int. Ed.*, **42**, 4183.

14. Ziegler, T. and Rauk, A. (1977) *Theor. Chim. Acta*, **46**, 1.
15. Morokuma, K. (1971) *J. Chem. Phys.*, **55**, 1236.
16. (a) Esterhuysen, C. and Frenking, G. (2004) *Theor. Chem. Acc.*, **111**, 381; Erratum: 2005, **113**, 294. (b) Kovács, A., Esterhuysen, C., and Frenking, G. (2005) *Chem. Eur. J.*, **11**, 1813. (c) Krapp, A., Bickelhaupt, F.M., and Frenking, G. (2006) *Chem. Eur. J.*, **12**, 9196.
17. Frenking, G., Wichmann, K., Fröhlich, N., Loschen, C., Lein, M., Frunzke, J., and Rayón, V.M. (2003) *Coord. Chem. Rev.*, **238–239**, 55–82.
18. See, for example: (a) Fonseca Guerra, C., Bickelhaupt, F.M., Snijders, J.G., and Baerends, E.J. (1999) *Chem. Eur. J.*, **5**, 3581–3594; (b) Fonseca Guerra, C., van der Wijst, T., and Bickelhaupt, F.M. (2006) *Chem. Eur. J.*, **12**, 3032–3042; (c) Fonseca Guerra, C., Szekeres, Z., and Bickelhaupt, F.M. (2011) *Chem. Eur. J.*, **17**, 8816; (d) Fonseca Guerra, C., Zijlstra, H., Paragi, G., and Bickelhaupt, F.M. (2011) *Chem. Eur. J.*, **17**, 12612.
19. Conjugation and Hyperconjugation: (a) Cappel, D., Tüllmann, S., Krapp, A., and Frenking, G. (2005) *Angew. Chem. Int. Ed.*, **44**, 3617; (b) Fernández, I. and Frenking, G. (2006) *Chem. Eur. J.*, **12**, 3617; (c) Fernández, I. and Frenking, G. (2006) *J. Org. Chem.*, **71**, 2251; (d) Fernández, I. and Frenking, G. (2006) *Chem. Commun.*, 5030; (e) Fernández, I. and Frenking, G. (2007) *J. Phys. Chem. A*, **111**, 8028; (f) Fernández, I. and Frenking, G. (2007) *J. Org. Chem.*, **72**, 7367.
20. Aromaticity: (a) Fernández, I. and Frenking, G. (2007) *Faraday Discuss.*, **135**, 403; (b) Fernández, I. and Frenking, G. (2007) *Chem. Eur. J.*, **13**, 5873; (c) Pierrefixe, S.C.A.H. and Bickelhaupt, F.M. (2007) *Chem. Eur. J.*, **13**, 6321; (d) Pierrefixe, S.C.A.H. and Bickelhaupt, F.M. (2008) *J. Phys. Chem. A*, **112**, 12816; (e) Pierrefixe, S.C.A.H. and Bickelhaupt, F.M. (2008) *Aust. J. Chem.*, **61**, 209; (f) Wang, Y., Fernández, I., Duvall, M., Wu, J.I.-C., Li, Q., Frenking, G., and Schleyer, P.v.R. (2010) *J. Org. Chem.*, **75**, 8252; (g) Fernández, I., Duvall, M., Wu, J.I.-C., Schleyer, P.v.R., and Frenking, G. (2011) *Chem. Eur. J.*, **17**, 2215.
21. Sankararaman, S. (2005) *Pericyclic Reactions – A Textbook: Reactions, Applications and Theory*, Wiley-VCH Verlag GmbH, Weinheim, pp. 326–329, and references therein.
22. (a) Hünig, S., Müller, H.R., and Thier, W. (1965) *Angew. Chem., Int. Ed. Engl.*, **4**, 271; (b) Franck-Neumann, M. and Dietrich-Buchecker, C. (1980) *Tetrahedron Lett.*, **21**, 671; (c) Garbisch, E.W. Jr.,, Schildcrout, S.M., Patterson, D.B., and Sprecher, C.M. (1965) *J. Am. Chem. Soc.*, **87**, 2932.
23. This process was proposed to evolve through a cyclic six-membered transition state at high temperatures. See: Sominsky, L., Rozental, E., Gottlieb, H., Gedanken, A., and Hoz, S. (2004) *J. Org. Chem.*, **69**, 1492.
24. (a) Reetz, M.T. (1972) *Angew. Chem., Int. Ed. Engl.*, **11**, 130; (b) Reetz, M.T. (1973) *Tetrahedron*, **29**, 2189; (c) Reetz, M.T. (1977) *Adv. Organomet. Chem.*, **16**, 33; (d) Houk, K.N., Li, Y., McAllister, M.A., O'Doherty, G.A., Paquette, L.A., Siebrand, W., and Smedarchina, Z.K. (1994) *J. Am. Chem. Soc.*, **116**, 10895; For a recent review on dyotropic reactions, see:(e) Fernández, I., Cossío, F.P., and Sierra, M.A. (2009) *Chem. Rev.*, **109**, 6687.
25. (a) Fernández, I., Sierra, M.A., and Cossío, F.P. (2007) *J. Org. Chem.*, **72**, 1488; (b) Frenking, G., Cossío, F.P., Sierra, M.A., and Fernández, I. (2007) *Eur. J. Org. Chem.*, **2007**, 5410.
26. Fernández, I., Bickelhaupt, F.M., and Cossío, F.P. (2009) *Chem. Eur. J.*, **15**, 13022.
27. Fernández, I. and Cossío, F.P. (2010) *Curr. Org. Chem.*, **14**, 1578.
28. Schleyer, P.v.R. and Jiao, H. (1996) *Pure Appl. Chem.*, **68**, 209, and references therein.
29. Chen, Z., Wannere, C.S., Corminboeuf, C., Puchta, R., and Schleyer, P.v.R. (2005) *Chem. Rev.*, **105**, 3842.
30. Cossío, F.P., Morao, I., Jiao, H., and Schleyer, P.v.R. (1999) *J. Am. Chem. Soc.*, **121**, 6737.

31. (a) Herges, R. and Geuenich, D. (2001) *J. Phys. Chem. A*, **105**, 3214; (b) Geuenich, D., Hess, K., Köhler, F., and Herges, R. (2005) *Chem. Rev.*, **105**, 3758.
32. (a) Mackenzie, K., Howard, J.A.K., Mason, S., Gravett, E.C., Astin, K.B., Liu, S.X., Batsanov, A.S., Vlaovic, D., and Maher, J.P. (1993) *J. Chem. Soc., Perkin Trans. 2*, **6**, 1211; (b) Paquette, L.A., O'Doherty, G.A., and Rogers, R.D. (1991) *J. Am. Chem. Soc.*, **113**, 7761. See also reference [24d].
33. (a) Alder, K., Pascher, F., and Schmitz, A. (1943) *Chem. Ber.*, **76B**, 27; (b) Alder, K. and Noble, T. (1943) *Chem. Ber.*, **76B**, 54; (c) Alder, K. and Schmidt, A.C.-H. (1943) *Chem. Ber.*, **76B**, 183.
34. Selected applications of the Alder-ene reaction to natural product synthesis: (a) Xia, Q. and Ganem, B. (2001) *Org. Lett.*, **3**, 485; (b) Zhang, J.-H., Wang, M.-X., and Huang, Z.-T. (1998) *Tetrahedron Lett.*, **39**, 9237; (c) Barriault, L. and Deon, D.H. (2001) *Org. Lett.*, **3**, 1925.
35. (a) Loncharich, R.J. and Houk, K.N. (1987) *J. Am. Chem. Soc.*, **109**, 6947; (b) Thomas, B.E. IV,, Loncharich, R.J., and Houk, K.N. (1992) *J. Org. Chem.*, **57**, 1354; (c) Paderes, G.D. and Jorgensen, W.L. (1992) *J. Org. Chem.*, **57**, 1904; (d) Achmatowicz, O. and Bialecka-Florjancyzk, E. (1996) *Tetrahedron*, **52**, 8827; (e) Houk, K.N., Beno, B.R., Nendel, M., Black, K., Yoo, H.Y., Wilsey, S., and Lee, J. (1997) *J. Mol. Struct. (THEOCHEM)*, **398–399**, 169.
36. For instance, see: Snider, B.B. and Ron, E. (1985) *J. Am. Chem. Soc.*, **107**, 8160.
37. Fernández, I. and Bickelhaupt, F.M. (2012) *J. Comput. Chem.*, **33**, 509.
38. Padwa, A. and Pearson, W.H. (eds) (2002) *Synthetic Applications of 1,3-Dipolar Cycloaddition Chemistry Toward Heterocycles and Natural Products*, John Wiley & Sons, Inc., New York.
39. Ess, D.H. and Houk, K.N. (2007) *J. Am. Chem. Soc.*, **129**, 10646.
40. Ess, D.H. and Houk, K.N. (2008) *J. Am. Chem. Soc.*, **130**, 10187.
41. Xu, L., Doubleday, C.E., and Houk, K.N. (2009) *Angew. Chem. Int. Ed.*, **48**, 2746.
42. Ess, D.H., Jones, G.O., and Houk, K.N. (2008) *Org. Lett.*, **10**, 1633.
43. Fernández, I., Bickelhaupt, F.M., and Cossío, F.P. (2011) *J. Org. Chem.*, **76**, 2310.
44. Ghereg, D., André, E., Sotiropoulos, J.-M., Miqueu, K., Gornitzka, H., and Escudié, J. (2010) *Angew. Chem. Int. Ed.*, **49**, 8704.
45. Paton, R.S., Kim, S., Ross, A.G., Danishefsky, S.J., and Houk, K.N. (2011) *Angew. Chem. Int. Ed.*, **50**, 10366.
46. For previous Diels–Alder studies with cyclopentenone through cyclooctenones, see: (a) Karthikeyan, M., Kamakshi, R., Sridar, V., and Reddy, B.S.R. (2003) *Synth. Commun.*, **33**, 4199; (b) Fringuelli, F., Pizzo, F., Taticchi, A., Halls, T.D.J., and Wenkert, E. (1982) *J. Org. Chem.*, **47**, 5056.
47. (a) van Bochove, M.A., Swart, M., and Bickelhaupt, F.M. (2006) *J. Am. Chem. Soc.*, **128**, 10738; (b) Bento, A.P. and Bickelhaupt, F.M. (2007) *J. Org. Chem.*, **72**, 2201; (c) Bento, A.P. and Bickelhaupt, F.M. (2008) *Chem. Asian J.*, **3**, 1783; (d) van Bochove, M.A. and Bickelhaupt, F.M. (2008) *Eur. J. Org. Chem.*, **2008**, 649.
48. Alternative perspectives on the steric nature of the S_N2 barriers: (a) Pierrefixe, S.C.A.H., van Stralen, S.J.M., van Stralen, J.N.P., Fonseca Guerra, C., and Bickelhaupt, F.M. (2009) *Angew. Chem. Int. Ed.*, **48**, 6469; (b) Pierrefixe, S.C.A.H., Poater, J., Im, C., and Bickelhaupt, F.M. (2008) *Chem. Eur. J.*, **14**, 6901; (c) Pierrefixe, S.C.A.H., Fonseca Guerra, C., and Bickelhaupt, F.M. (2008) *Chem. Eur. J.*, **14**, 819.
49. Poater, J., Visser, R., Solà, M., and Bickelhaupt, F.M. (2007) *J. Org. Chem.*, **72**, 1134.
50. Bento, A.P. and Bickelhaupt, F.M. (2008) *J. Org. Chem.*, **73**, 7290.
51. Galabov, B., Nikolova, V., Wilke, J.J., Schaefer, H.F., and Allen, W.D. (2008) *J. Am. Chem. Soc.*, **130**, 9887.
52. Fábián, A., Ruff, F., and Farkas, Ö. (2008) *J. Phys. Org. Chem.*, **21**, 988.
53. Wu, X.-P., Sun, X.-M., Wei, X.-G., Ren, Y., Wong, N.-B., and Li, W.-K. (2009) *J. Chem. Theory Comput.*, **5**, 1597.
54. (a) Fernández, I., Frenking, G., and Uggerud, E. (2009) *Chem. Eur. J.*, **15**, 2166; (b) Fernández, I., Frenking, G.,

and Uggerud, E. (2010) *J. Org. Chem.*, **75**, 2971.

55. Cheong, P.H.-Y., Paton, R.S., Bronner, S.M., Im, G.-Y.J., Garg, N.K., and Houk, K.N. (2010) *J. Am. Chem. Soc.*, **132**, 1267.

56. Bronner, S.M., Bahnck, K.B., and Garg, N.K. (2009) *Org. Lett.*, **11**, 1007.

57. Reetz, M.T. (1972) *Angew. Chem., Int. Ed. Engl.*, **11**, 129. See also references [24b,c].

58. Fernández, I., Bickelhaupt, F.M., and Cossío, F.P. (2012) *Chem. Eur. J.*, **18**, 12395.

59. Fernández, I., Sierra, M.A., and Cossío, F.P. (2006) *Chem. Eur. J.*, **12**, 6323.

60. (a) Grob, C.A. and Winstein, S. (1952) *Helv. Chim. Acta*, **35**, 782; For related examples, see: (b) Mulzer, S. and Brüntrup, A.G. (1979) *Angew. Chem., Int. Ed. Engl.*, **18**, 793; (c) Black, T.H., Hall, J., and Sheu, R.G. (1988) *J. Org. Chem.*, **53**, 2371; (d) Black, T.H., Eisenbeis, S.H., McDermott, T.S., and Maluleka, S.L. (1990) *Tetrahedron*, **46**, 2307; (e) Purohit, V.C., Matla, A.S., and Romo, D. (2008) *J. Am. Chem. Soc.*, **130**, 10478.

61. Selected examples: (a) Diefenbach, A. and Bickelhaupt, F.M. (2001) *J. Chem. Phys.*, **115**, 4030; (b) Diefenbach, A., de Jong, G.T., and Bickelhaupt, F.M. (2005) *J. Chem. Theory Comput.*, **1**, 286; (c) van Zeist, W.J., Visser, R., and Bickelhaupt, F.M. (2009) *Chem. Eur. J.*, **15**, 6112; (d) Ess, D.H. (2009) *J. Org. Chem.*, **74**, 1498.

Part II
Materials, Nanoscience, and Nanotechnologies

Discovering the Future of Molecular Sciences, First Edition. Edited by Bruno Pignataro.

8
Molecular Metal Oxides: Toward a Directed and Functional Future

Haralampos N. Miras

8.1
Introduction

Inorganic chemistry is advancing rapidly at its frontiers, especially where interdisciplinary research efforts perfuse through the boundaries of well-defined disciplines such as life sciences, condensed-matter physics, materials science, and environmental chemistry. This has been demonstrated quite elegantly over the years by different research groups investigating the properties and novel synthetic methodologies of numerous metal oxide based materials with a plethora of archetypes ranging from 1D and 2D to the 3D frameworks of zeolitic compounds and the diverse structural morphologies of the molecular metal oxides.

More specifically, the science of molecular metal oxides or polyoxometalates (POMs) [1] has attracted the attention of research groups over the years, with their plethora of unique archetypes with applications ranging from catalysis [2] and medicine [3] to molecular electronics [4], magnetism [5], and energy [6]. It is worth noting that researchers' interest to POM chemistry was initially triggered in the early 1990s when Pope and Müller [7] summarized in their detailed review article published in 1991 the attractive features and the potential of this unique class of compounds. The development of POM chemistry during the following years was rapid accompanied by an explosion in the number and complexity of structurally characterized POM compounds, see Figure 8.1. This explosion in characterized POM-based materials is directly related to the developments in instrumentation and employed synthetic approaches [8]. In terms of technique development, fast and high-resolution single crystal data collection has allowed the area to blossom in an unimaginable way. Moreover, advances in spectroscopic techniques such as electrospray-ionization mass spectrometry (ESI-MS) and hetero-nuclear NMR allowed the researchers to bridge the gap between solution and solid state of complex self-assembled chemical systems and reveal mechanistic aspects of the underlying chemistry [9]. However, despite the increase in the number of reported species and the investigation of their physical properties, it is frustratingly difficult to reveal the crucial information regarding the selection rules that trigger their

Discovering the Future of Molecular Sciences, First Edition. Edited by Bruno Pignataro.

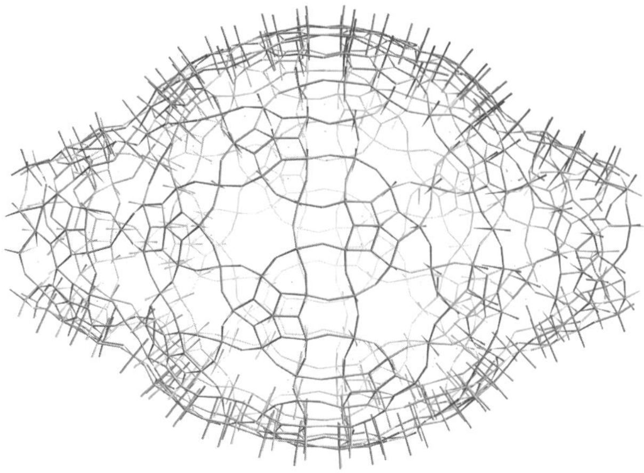

Figure 8.1 Wired-frame representation of the largest POM structure (Mo_{368}) reported so far, synthesized under one-pot reaction conditions (O: small gray spheres). The size of this cluster is 5.4 nm and is comparable to the size of small proteins such as the carbonic anhydrase II.

self-organization in a controlled fashion and even more difficult to design specific routes leading to predefined or emergence of new properties and phenomena.

Therefore, given the enormous challenge in understanding and controlling the self-assembly process for a range of self-assembled cluster-based architectures, new synthetic approaches have been employed over the past decade in an effort to shed light on the underlying processes of high nuclearity POMs as well as the generation and identification of novel building block libraries in the reaction mixtures [10]. This chapter discusses the most recent developments and how the combination of new technologies and synthetic approaches can offer a novel route in overcoming the present difficulties, revealing important mechanistic information, which ultimately leads to the engineering of functionality and emergence of unique occurrences and phenomena. Furthermore, this will allow us to take real control over the self-assembly processes of complex chemical systems and open the door for further discoveries toward a well-established and directed functional future.

8.2
New Technologies and Analytical Techniques

The development of instrumentation in the past two decades such as high-resolution detectors, high-intensity X-ray sources, and computer processing power promoted fast and routine single crystal data collection and consequently allowed the area to accelerate to the point that the bottle neck has moved to the crystallization of new compounds rather than the time taken for data collection and initial structure solution.

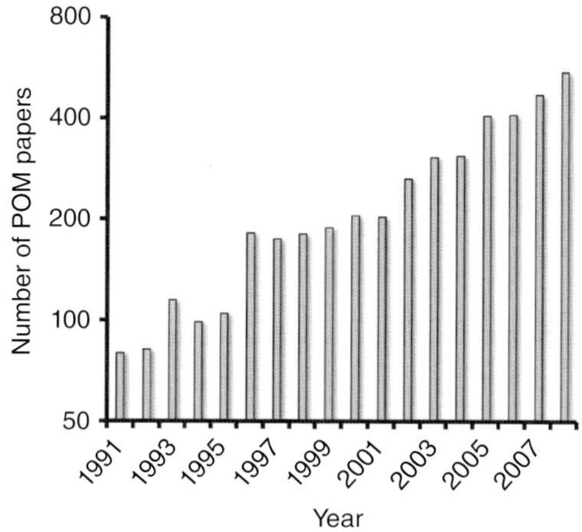

Figure 8.2 The number of publications based on the investigation of polyoxometalates plotted against the year.

This development is reflected not only in the increased number of papers (not including patents) published each year, which discuss at least some aspect of POM chemistry (Figure 8.2), but also in the complexity and size of the characterized architectures (Figure 8.3) [11], with sizes ranging from 1 nm to the impressive size of biological molecules as in the case of the $\{Mo_{368}\}$ species (Figure 8.1) with

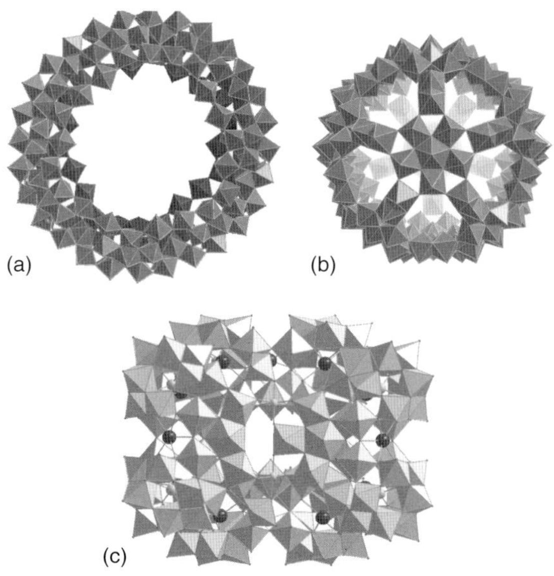

Figure 8.3 Representative examples of complex architectures reported over the past two decades; (a) $\{Mo_{154}\}$ molybdenum blue (MB) ring structure, (b) $\{Mo_{132}\}$ Keplerate archetype, and (c) $\{W_{119}Se_8Fe_2\}$ heteropolytungstate cluster. Color code: Mo, dark gray polyhedral; W, light gray polyhedral; Se, black spheres; and Fe, light gray spheres.

the size of 5.4 nm comparable to the molecule of carbonic anhydrase II with 260 residues and molecular weight of 29.6 kDa (5.6 nm) [12].

The precise structural identification of nanosized species led to the realization of a few key building blocks that exist in solution under specific experimental conditions and could be used as fundamental units for the construction of large architectures and potentially the emergence of new properties. Even though this starting point was crucial for the development that followed, it would not be feasible without bridging the obvious gap between solution and solid state. During this effort, the conventional spectroscopic techniques present significant drawbacks. For example, NMR is of limited use when the symmetry of the assembled architecture is high [13], when the structures are labile or paramagnetic, and for nuclei that have poor receptivity. In addition, the reaction mixtures are far too complicated to extract any useful information regarding the nature and availability of the building block libraries, as well as the preferable mechanistic routes followed during the formation of the polynuclear clusters. Therefore, given the enormous challenge in understanding and controlling the self-assembly process for a range of self-assembled POM-based architectures, high-resolution time-of-flight (TOF) mass spectrometry has been employed over the past decade [14] in an effort to unveil crucial information associated with the assembly-disassembly process of high nuclearity POMs as well as identification of novel reactive and intermediate species in reaction mixtures.

ESI-MS has proven to be a multipurpose and powerful tool [15] for the investigation of complex systems in terms of structural architecture as well as the nature of the reaction mixture. In a few cases it helped the research groups to probe the protonation state and the number of heteroatoms trapped in molecular capsules such as the Dawson clusters. For example, the $[H_mW_{18}O_{60}X_n]^{y-}$ (where X = As, Sb, Bi) [16] cluster have been known for three decades, with an approximate formulation of $n = 1$, but their precise composition could not be confirmed unambiguously due to crystallographic disorder of the heteroatoms over two positions in a single cluster. However, utilization of ESI-MS was crucial for the in-depth investigation of the Sb-based heteropolytungstates $[H_mSb_nW_{18}O_{60}]^{y-}$ [17]. During the course of these studies it was revealed that the correct formulation involves one hetero-ion disordered over two positions. This situation can be compared with the discovery of the $[H_mP_nW_{18}O_{62}]^{y-}$ family, which was also reported with $n = 1$.

In other cases, ESI-MS studies proved to be very useful for the discovery and identification of new species in solution such as in the case of isopolytungstates and isopolyniobates where it was possible to identify and characterize the species in solution and in solid state, expanding these families of compounds that consist of a limited number of species. In 2008, two new members of tungsten-based iso-POMs were discovered ($Na_{12}[H_4W_{22}O_{74}]\cdot 31H_2O$ and $Na_{18}[H_{10}W_{34}O_{116}]\cdot 47H_2O$) [18] while later, in 2010, the largest isopolyoxoniobate, $\{Nb_{27}\}$, was identified and characterized [19]. Moreover, the technique demonstrated its efficacy also in more complicated cases where either we have mixed metal systems [14a, 20] or nanosized clusters [21]. For example, in the case of mixed metal systems, it was possible to identify in the reaction mixture not only the intact cluster that exhibit a unique

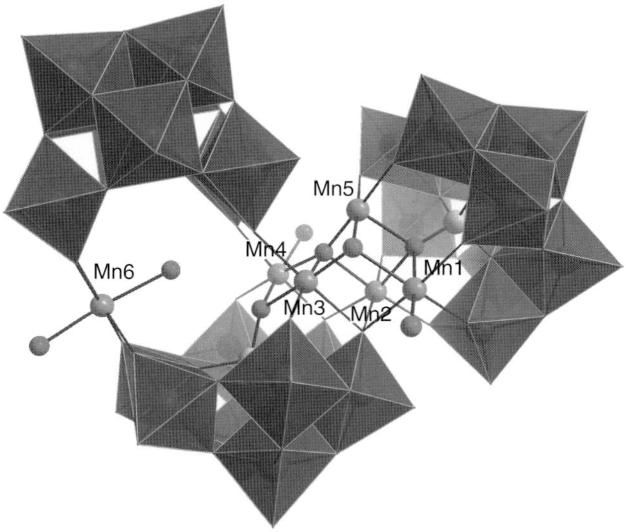

Figure 8.4 Representation of the $[Mn^{III}_2Mn^{II}_4(\mu_3\text{-}O)_2(H_2O)_4(B\text{-}\beta\text{-}SiW_8O_{31})(B\text{-}\beta\text{-}SiW_9O_{34})(\gamma\text{-}SiW_{10}O_{36})]^{18-}$ cluster showing three inequivalent silicotungstate Keggin fragments, $\{SiW_8\}$, $\{SiW_9\}$, and $\{SiW_{10}\}$, and the appended $\{Mn_4O_4\}$ cubane core. Color scheme: WO_6 (dark gray polyhedra); Mn (gray spheres); and O (dark gray spheres).

archetype of Mn-cubane core trapped by inequivalent lacunary Keggin building units [22], but also the available building blocks and isomers that exist in solution, Figure 8.4.

Furthermore, we demonstrated recently that the application of the ESI-MS technique in the case of Palladium-based systems allowed the observation in the gas phase of an unprecedented nanosized $\{Pd_{84}\}$ wheel [23], Figure 8.5, as well as in smaller species, giving additional information on the self-assembly process of such large and complex architectures. Moreover, it highlighted the symmetry-building block correlation in this category of cluster systems and the similarities with the molybdenum blue chemistry. Finally, it demonstrated that it is possible to predict the assembly of nanoscale architectures, based on an established minimal building block set and symmetry number, utilizing a set of "magic numbers" for these molecular nanoparticles.

Most importantly, the ESI-MS technique has been proven to be a powerful tool for the mechanistic investigation of self-assembled systems. A recent example described the real-time monitoring of the reaction mixture of $[\alpha\text{-}Mo_8O_{26}]^{4-}$, coordination of Mn^{III}, and subsequent coordination of two tris(hydroxymethyl)aminomethane molecules (TRIS), to form the symmetrical Mn-Anderson cluster $TBA_3[MnMo_6O_{18}((OCH_2)_3CNH_2)_2]$ tetrabutyl ammonium (TBA), which gave us important information on the disassembly/reassembly processes that take place during the reaction as well as the available building block library that exist under the experimental conditions. This piece of work [24] is very important as the

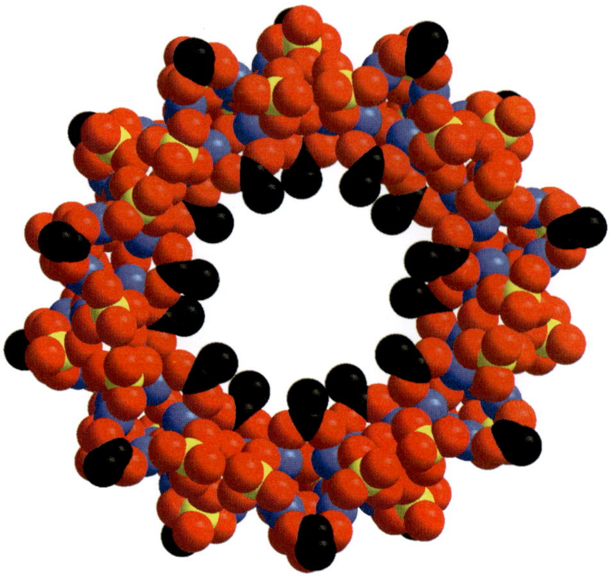

Figure 8.5 Space filling representation of the $\{Pd_{84}\}$ cluster. Color scheme: Pd – blue, O – red, P – yellow, and C – black.

organic-inorganic POM-hybrids can further be used as secondary building blocks for the construction of nanoscaled hybrid-POM architectures, making finally the field of nanoscale functional materials more accessible for further exploration.

8.3 New Synthetic Approaches

As we discussed briefly above, the development of modern techniques such as ESI-MS was crucial in bridging the gap between solution and solid state while the use of advanced diffractometers equipped with fast and high-resolution detectors allowed the accurate identification of complex structural features of the synthesized nanosized molecules. Nevertheless, that was not sufficient for the accurate control of the final architecture and functionality of the isolated materials. This is due to the fact that the metal oxide chemical systems are governed by self-assembly processes. The self-assembly is an exciting occurrence that governs how simple building blocks [25] can be organized spontaneously into complex architectures [26] depending on the experimental conditions [27] often to such a degree that total control is never easily achieved [11b]. This can be daunting because extremely small changes in reaction conditions can drive the self-assembly process toward a totally different direction [28]. This is another reason why the development of the molecular metal oxides family was based mainly on serendipitous results obtained from earlier studies that were usually lacking the element of design. Consequently, the development of POM-based materials and exploitation of their potential was dependant, at least up to the beginning of the previous decade, on the constructive use of the results obtained by serendipity, in order to maximize the desirable outcome and our understanding of the selection rules that govern these chemical systems.

8.3.1 The Building Block Approach

Generally, the approaches used in the synthesis of POM-based clusters are simple requiring, in most of the cases, just one step (one-pot syntheses), during which acidification of an aqueous solution of molybdates, tungstates, or vanadates initiates a condensation process leading from the initial organization of low nuclearity metallic species (primary building block, BBs) toward the formation of larger archetypes. The aggregation process though is controlled by a long list of experimental variables, which should be taken into account for the synthesis of a given POM archetype, such as (i) concentration/type of metal oxide anion, (ii) pH, (iii) ionic strength, (iv) heteroatom type/concentration, (v) presence of additional ligands, (vi) disintegrating environment, (vii) temperature and pressure of reaction (e.g., microwave, hydrothermal, refluxing), (viii) counter-ion and metal-ion effect, and (ix) processing methodology (one-pot, nonequilibrium conditions). Moreover, the development of POM chemistry has led to the realization that the isolated species could act as a set of transferable building blocks that can be reliably utilized in the formation of larger architectures and potentially give rise to new properties and, finally, new functional materials. The first key point at this stage is the use of reliable synthetic routes for the regeneration of the previously observed building blocks in the reaction mixture and their further assembly to new architectures. This goal has been achieved to some extent and is reflected by the observed diversity of complex structures reported, which are based on a specific building unit, for example, the lacunary Keggin or Dawson type building block. More specifically, utilization of the tungsten based lacunary species as primary building blocks gave rise to a range of interesting structures with nuclearities ranging from $\{W_{18}\}$ to $\{W_{224}\}$ [28, 29] in the presence of transition metals (e.g., manganese) giving compounds with interesting magnetic properties (Figure 8.6).

In a similar manner, other groups have utilized effectively preformed building blocks for the construction of new species that exhibit new properties. For example, Mizuno *et al.* have reported the incorporation of the catalytic active M–OH–M (M = W, Zn, or V) units in POM-based lacunary building blocks, such as $\{\gamma\text{-SiW}_{10}\}$ [30] for the construction of catalytically active POMs opening a new research avenue for POM species in catalysis [31]. Moreover, Proust *et al.* have reported the syntheses of a few high valence metal–nitrido POMs, such as $[(Ru^{VI}RN)_2(SiW_{10}O_{38})]^{6-}$, by either photo oxidation of the metal–azido precursor or ligand exchanging the metal–nitrido complex giving solid evidence of the influence of POM-based ligands on the reactivity of high-valence metal nitride units [32]. Also, recently, various groups managed to trap magnetic cores that exhibit interesting magnetic

Figure 8.6 Polyhedral representation of $[Mn^{III}{}_{40}P_{32}W^{VI}{}_{224}O_{888}]^{144-}$ shows that all the fragments are joined together through Mn^{III}–O=W bridges. The entire cluster adopts an idealized S_4 symmetry, with the principal axis coinciding with the fourfold axis of the central $\{P_8W_{48}\}$ wheel. The WO_6 units are shown in gray polyhedra and the W/Mn disordered positions are dark gray polyhedra. Counter ions and solvent molecules have been omitted for clarity. Color code: Mn^{III}, black spheres; and O, dark gray spheres.

behavior. For example, two novel $\{Mn_{19}\}$ and $\{Co_{16}\}$ magnetic clusters that exhibit single molecule magnet (SMM) behavior [5b, 33], $[Mn_{19}(OH)_{12}(SiW_{10}O_{37})_6]^{34-}$ and $[\{Co_4(OH)_3PO_4\}_4(PW_9O_{34})_4]^{28-}$, have been isolated by Kortz *et al.* utilizing the Keggin tungstosilicate $\{SiW_{10}\}$ and tungstophosphate $\{PW_9\}$ lacunary fragments, respectively, as robust building blocks for the stabilization of the magnetic cores. In a similar fashion, Kögerler *et al.* reported a magnetically responsive material where its SMM properties can be switched "on" and "off," triggered by the modification of the ligands coordinated to the magnetic core [34].

8.3.2
Generation of Novel Building Block Libraries

It is obvious that in every case, the research groups used effectively well-established procedures for the regeneration of specific type of building blocks that were used as transferable synthons for the construction of larger architectures and isolation of materials with interesting properties, for example, in catalysis, magnetism, and so on. Even though this synthetic approach proved to be very useful, there is still an obvious limitation in terms of the diversity of the available building blocks that can be generated based on prior observations. In order to go one step further, the research groups developed new synthetic approaches that gave rise to new libraries of building blocks that deviate substantially from the common observed archetypes such as the lacunary versions of Keggin, Dawson, and Lindqvist structures. The main synthetic approaches that have been developed significantly over the past 10 years were based on the structure directing properties of ligands present in solution ("shrink-wrapping" effect), hydrothermal/ionothermal synthesis, and use of new template anions.

8.3.2.1 **Shrink-Wrapping Effect**

The POM compounds are polyanions and, consequently, their negative charge is balanced by the available cations in the reaction mixture. The role of the charge balancing can be played by a series of charged organic ligands such as protonated amines that define the cationic environment that the anion is assembled. Taking into consideration the above observation, it becomes apparent that the cations are able to direct and control the formation of a specific moiety [35]. For example, Cronin *et al.* demonstrated that the use of triethanolamine (TEA) in a solution of tungstates can generate new libraries of building blocks and at the same time direct their assembly in a controlled manner, giving rise to new clusters such as the $\{W_{36}\}$ [36], which is the largest isopolytungstate reported so far, Figure 8.7. Interestingly, in the absence of TEA, the assembly process favors the formation of the common Keggin archetype. In a similar manner, the use of either TEA or hexamethylenetetramine (HTMA) as structure directing agents in a molybdate solution under reducing conditions can direct the assembly process toward the

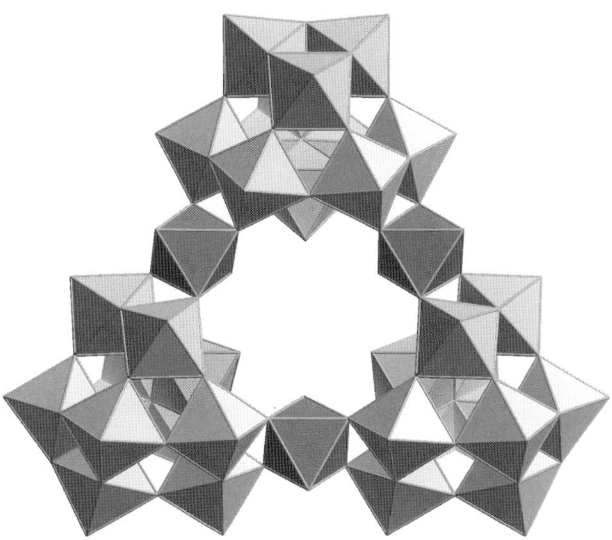

Figure 8.7 Polyhedral representation of $\{(H_2O)_4K\subset[H_{12}W_{36}O_{120}]\}^{11-}$. The $\{W_{11}\}$ building units are highlighted as light gray polyhedral and three $\{W_1\}$ linkers as dark gray polyhedra. This view reveals the central cavity formed that is occupied by a potassium cation (not shown here). Counterions and solvent molecules are omitted for clarity.

formation of sulfite templated $\{Mo_{18}(SO_3)_2\}$ Dawson architecture or a novel isopolymolybdate $\{Mo_{16}\}$ species [27].

8.3.2.2 Hydrothermal and Ionic Thermal Synthesis

Besides the conventional solution synthesis, solvothermal and ionothermal synthesis are two other methods that are used quite often for the generation of novel building block libraries and direct the assembly process toward the formation of novel POM-based compounds. The use of aqueous or organic media (e.g., acetonitrile, methanol, and pyridine) limits the reaction temperature during the course of conventional synthesis. Owing to the above limitations, the use of Teflon auto-claves during the solvothermal process gives the opportunity to reach higher temperatures and higher pressures for the same reaction mixture. Under these conditions, metastable or intermediate phases can be synthesized, which normally lead to kinetically controlled products, such as the "basket-shaped" cluster $[P_6Mo_{18}O_{73}]^{11-}$ [37]. However, solvothermal methods are based on the generation of significant autogenous pressure when the reaction mixture is heated in a sealed container, which introduce an intrinsic weakness: the general reproducibility of the reactions requires perfect control of the reaction parameters while the reaction temperature is still limited due to safety concerns. On the other hand, the ionothermal synthesis, which was recently employed by Wang *et al.*, Pakhomova *et al.*, and others, has been started to be used widely by POM chemists [38]. During this approach, an ionic liquid acts as a solvent, potential template, and structure directing agent in a similar fashion to the shrink-wrapping strategy discussed above, while at the same time we can apply to the system much higher temperatures. Use of a wide range of ionic liquids can offer a powerful tool for the generation of new BB libraries and induce the assembly of new molecular metal oxide species.

8.3.2.3 Novel Templates: XO_3 and XO_6-Templated POMs

Traditionally, the existence of a template in the reaction mixture seems to be necessary for the generation of the primary building blocks that can be used further for the synthesis of POM structures. The numerous examples reported in the literature usually make use one of the phosphates, silicates, sulfates, germanates, and so on, which usually occupy the cavity of lacunary species and adopt a tetrahedral geometry [39]. In the past decade, a new approach was developed, which involves the use of templates that exhibit pyramidal geometry and can be used for the generation of fundamentally new libraries of building blocks, which gave rise to a plethora of novel structural motifs and interesting properties. More specifically, the utilization of the pyramidal sulfite, selenite, and tellurite anions (XO_3^{2-}) as inorganic ligands, which are mild reducing (in the case of sulfite anion) and structure directing agents (because of the lone pair of electrons), introduced the necessary diversity to the POM systems and allowed access to new libraries of building units as well as the control of the associated assembly process and engineering of functionality [14b, 20].

Moreover, the origins of the observed constructive interaction of the XO_3^{2-} anions with metal oxide systems were based on the plethora of coordination modes

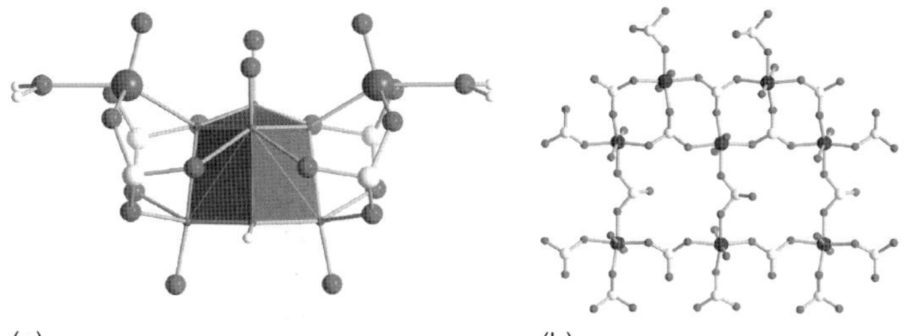

Figure 8.8 Ball-and-stick representations of (a) $[(V^{IV}O)_6(\mu_4\text{-}O)_2(\mu_3\text{-}OH)_2(\mu_3\text{-}SO_3)_4(H_2O)_2]^{2-}$ and (b) $[(V^{IV}O)(SO_3)_{1.5}(H_2O)]^-$ sulfite POMs. Color code: V, gray big spheres and polyhedron; O, small dark gray spheres; and S, white spheres.

adopted by the template anions. In 2003, Kabanos *et al.* reported a series of new archetypes following the generation of new libraries of BBs that led to the assembly of new molybdenum and vanadium based structures, Figure 8.8 [40]. In both cases, the isolated polyanions are fully reduced to the V and IV oxidation state for molybdenum and vanadium compounds, respectively, with archetypes ranging from spherical molecules to open frameworks and 1D chains with interesting modular magnetic properties. A few years later, Cronin *et al.* expanded this synthetic approach and reported a series of new clusters incorporating sulfites, selenites, and tellurites as templates and structure directing agents giving rise to unique complex archetypes [41, 42] (Figure 8.9) and materials with interesting thermochromic properties [43].

In a similar manner, the use of the XO_6 type of template, which is rarely used for the construction of nanosized POMs, has given rise to a series of new species. More specifically, Cronin *et al.* demonstrated the first example of nonclassic Dawson POM $[H_4W_{19}O_{62}]^{6-}$ [44]. The cluster consists of the typical $\{W_{18}O_{54}\}$ cage framework as in the classical Dawson structure ($[W_{18}O_{54}(XO_4)_2]^{y-}$, X = P, S) but the two tetrahedral $XO_4^{\,n-}$ heteroanions are replaced with either an octahedral or a trigonal prismatic $WO_6^{\,6-}$ anion and two protonated μ_3-oxido ligands. The stabilization of the $\{WO_6\}$ moiety in a trigonal prismatic coordination environment is unprecedented in polyoxotungstate chemistry. Following the same synthetic approach, the authors reported the isolobal molecular cages of tungstatoperiodate and tungstatotellurate compounds, $[H_nW_{18}O_{56}(XO_6)]^{m-}$ (X = I^{VI} or Te^{VI}), which incorporate either high-valent $I^{VII}O_6$ or the $Te^{VI}O_6$ [45].

8.3.3 POM-Based Networks

The building block approach has been used very efficiently in the preparation of extended modular frameworks that incorporate inorganic building blocks. Again, in this case, the assembly of the primary building units have been directed by the fine control of the experimental variables as well as the presence of structure directing agents such as cations and charged organic moieties. For

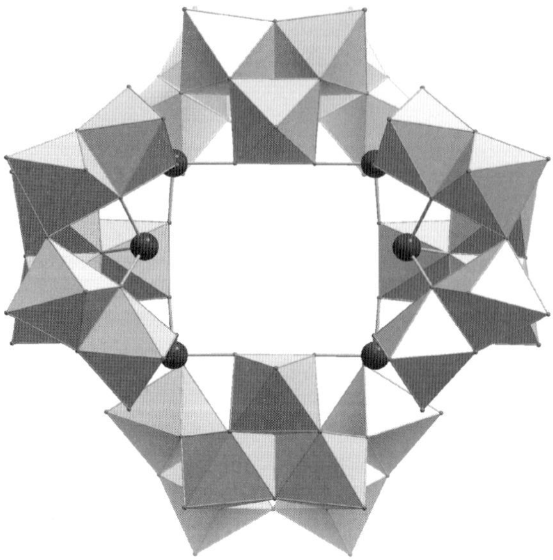

Figure 8.9 Polyhedral representation of compound $\{W_{28}Te_8O_{112}\}$ where four $\{W_7O_{25}(TeO_3)\}$ fragments combine with four $\{TeO_4\}$ linkers forming the square saddle-shaped architecture. All cations and solvent water molecules are omitted for clarity. Polyhedra: WO_6; gray spheres: oxygen; small dark gray spheres: Te; large black spheres.

example, Cronin *et al.* reported in 2008 the first modular 3D POM framework $[(C_4H_{10}NO)_{40}(W_{72}Mn_{12}O_{268}X_7)n]$ (X = Si or Ge) that is constructed by substituted Keggin-type building blocks forming a material that can undergo a reversible redox process in solid state under spatially ordered redox change of the framework [46]. Extension of this work showed that it is possible to construct stable extended inorganic frameworks that can undergo reversible single crystal to single crystal (SC–SC) transformations retaining their structural integrity following modular kinetics while it has been realized for the first time the molecular alloy concept [47]. Another interesting example reported by the same group is the face-directed assembly of a ring-shaped building unit, $[P_8W_{48}O_{184}]^{40-}$, with manganese linkers yielding a three-dimensional extended framework architecture based on a truncated cuboctahedron [48]. The 1 nm diameter pores of the $\{P_8W_{48}\}$ structural building unit lead to approximately spherical 7.24 nm^3 cavities with cation exchange properties.

Moreover, a few novel 3D polyoxometalate based metal organic frameworks (POMOF) materials have been reported recently, which have made use of the primary POM-based building units in the reaction mixture and the structure directing properties of organic ligands [49]. For example, Dolbecq *et al.* have developed a new family of POMOFs using ε-Keggin POMs as building blocks. This POM has the general formula $\{\varepsilon\text{-}PMo^{V}{}_8Mo^{VI}{}_4O_{40-x}(OH)_xM_4\}$ ($M = Zn^{II}$, La^{III}) and contains an ε-Keggin core capped by four metallic ions. These Keggin moieties are versatile building blocks that can be used either as anions (M = Zn, $x = 0$) or cations (M = Zn, La, $x = 3$–5). The ε-Keggin ion has a remarkable ability to react

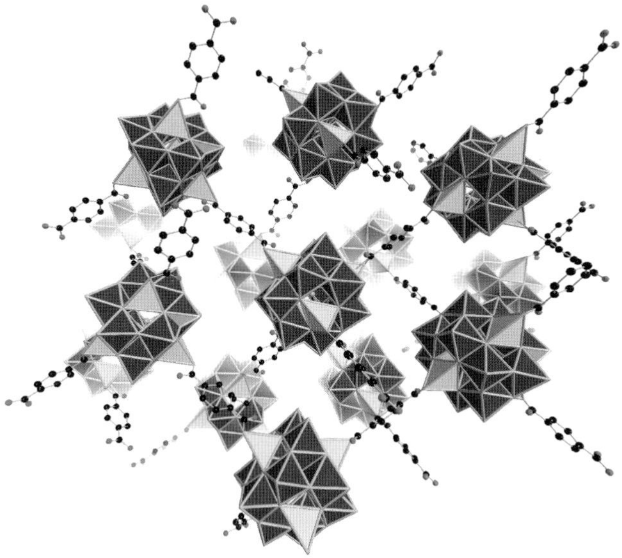

Figure 8.10 Representation of the $(TBA)_3[PMo^V{}_8Mo^{VI}{}_4O_{36}(OH)_4Zn_4][C_6H_3(COO)_3]_{4/3}$ POMOF composite. The POM is shown in polyhedral representation and the organic ligands are represented in ball-and-stick mode. TBA cations and hydrogen atoms are omitted for clarity. Color code: Mo, black polyhedra; O, small dark gray spheres; Zn, light gray polyhedra; and C, black spheres.

successfully with a variety of organic linkers such as bipyridine, benzenedicarboxylic acid (BDC), or imidazole (Figure 8.10) [50]. Specifically, the Zn-ε-Keggin possesses a tetrahedral shape in which four Zn^{II} cations are exposed in a regular tetrahedral arrangement in a fashion similarly to oxygen atoms in $SiO_4{}^{2-}$ anion, leading to the formation of a new family of electro-active POMOF catalysts. The grafting of the triangular 1,3,5-benzene tricarboxylate linkers (denoted trim) on this ε-Keggin POMs capped by Zn^{II} ions, formed in situ under hydrothermal conditions, has generated three novel POMOFs.

8.4 Continuous Flow Systems and Networked Reactions

As discussed above, self-assembly processes are highly dependent on the experimental conditions often to such a degree that total control is never easily achieved. The previously discussed methodologies have given an improved degree of control over the chemistry of self-assembled chemical systems, which gradually allowed the design of materials with specific functionality. Nevertheless, there is still a lot of missing information regarding the underlying chemistry that is very often masked by the self-assembly processes and the formation mechanisms of POM-based complex systems. For the above reasons, we introduced recently a flow reactor system approach to explore both the mechanism and the synthesis of complex POM clusters. This approach allows the generation of a gradient of the experimental variable of interest (pH, reducing environment, concentration, etc.) inside the reactor, which promotes the chemical evolution of the system in a controlled manner. During this process, it is possible to unveil crucial information regarding the mechanism of the system under investigation by trapping intermediate species as in the case of the molybdenum blue family where we managed to study the steps underlying the assembly of previously characterized molybdenum oxide wheel 2.6 nm in diameter. The crystallization of the intermediate structure revealed a $\{Mo_{36}\}$ cluster that appears to template the assembly of the surrounding MB wheel. The transient nature of the template was demonstrated by its ejection after the wheel was reduced to its final electronic state [51]. Extending the flow chemistry approach, we demonstrated the applicability of this technique, using an automated flow process with multiple batch crystallization for the screening and scaling-up of the syntheses of manganese-based single molecule magnets as well as those of a selection of exemplary POMs. Screening of the synthetic conditions was achieved by programing a multiple pump reactor system to sequentially vary reaction parameters, thus exploring a large area of parameter space and identifying the successful flow conditions for product isolation. The continuous application of these flow conditions provided a direct route to "scale-up," ultimately resulting in large quantities of phase-pure material in a much shorter time frame than conventionally possible [52]. In a similar manner, the linear flow system has been used in the case of thiometalate chemistry where the accelerated synthesis and crystallization of the first thiometalate-based Möbius strip molecule within 24 h was reported [53].

The linear flow reactor system has been proven an exceptionally effective approach for the scale-up and discovery of new clusters. The development of the flow system that followed shortly after opened the door for further exploration, discoveries, and ultimately better control of complex self-assembled systems. The new implementation of the flow reactor system allowed the networking of fundamentally different "one-pot" reactions. The advantage of this alternative system allows the interaction of libraries of building blocks that are not possible to coexist under the same experimental conditions in a specific chemical environment during the course of one-pot reactions. The interaction of two or three fundamentally different building block libraries allowed the discovery of unprecedented architectures and more effective use of different building blocks as a set of transferable synthons for the construction of larger clusters in a controlled fashion. Specifically, the application of the networked reactor system (NRS) to the synthesis of an unknown family of metal-containing isopolyoxotungstates (iso-POTs) in the presence of templating transition metals such as Co^{2+}, by screening networks of one-pot reactions led to the discovery of new clusters in a reproducible way, allowing one-pot reactions to be probed or expanded over a number of reaction vessels, rather than relying on one single vessel. The use of the NRS led to the discovery of a 4 nm species, $Na_{16}(DMAH)_{72}[H_{16}Co_8W_{200}O_{660}(H_2O)_{40}]\cdot ca600H_2O \equiv \{W_{200}Co_8\}$ (DMAH = protonated dimethylamine) [54], and it represents the largest discrete polyoxotungstate cluster characterized so far, Figure 8.11.

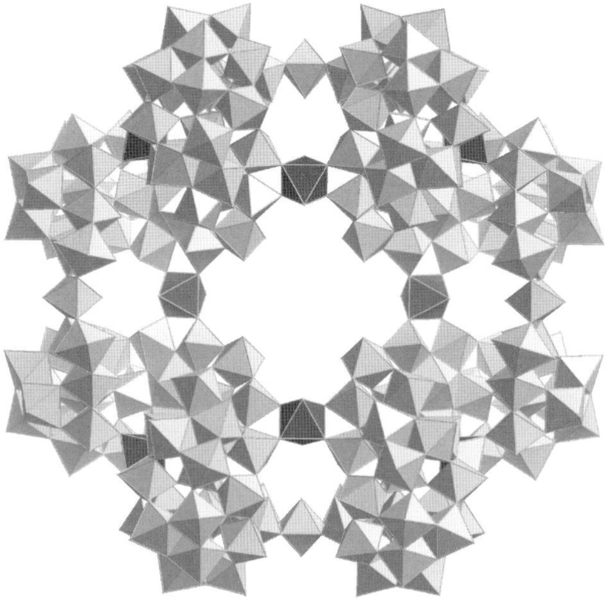

Figure 8.11 Representation of the $Na_{16}(DMAH)_{72}[H_{16}Co_8W_{200}O_{660}(H_2O)_{40}]\cdot ca600H_2O \equiv \{W_{200}Co_8\}$ cluster. Cations and hydrogen atoms are omitted for clarity. Color code: W, light gray polyhedral; and Co, black polyhedra.

The potential of the NRS methodology is transformative due to its ability to explore one-pot reactions as configurable modules by controlling the chemistry at the nanoscale using macroscale parameters. Moreover, it provides the synthetic chemists with an efficient tool to scan the parameter space of a complex chemical system in a programed way, leading to an overall far better control of the underlying chemical processes.

8.5 3D Printing Technology

The use of three-dimensional (3D) printing technologies promises to bypass sophisticated manufacturing centers and simplify the way that materials are turned into functional devices. Even though 3D printing technology is employed for the production of highly specialized electronic and microfluidic devices, the potential for using 3D-printed easily configurable reactors for chemical discovery has not been explored. In an effort to demonstrate the efficacy of new technologies in chemical discovery and design of functional materials, Cronin *et al.* showed recently the potential for further control of complex chemical systems by implementing the 3D technology in a chemical discovery platform. More specifically, the group showed the general applicability of 3D printed architectures for the formation and crystallization of two new inorganic nanoclusters of formula

$(C_2H_8N)_mNa_n[W_{19}M_2O_{61}Cl(SeO_3)_2(H_2O)_2]Cl_2{\cdot}xH_2O$ (where M = Co(II) or Mn(II)), the synthesis of the organic heterocyclic compound $C_{21}H_{17}BrN_2O$ and the real-time in situ spectroelectrochemistry during the reduction of phosphomolybdic acid (PMA). In addition, they demonstrated high level of control over the outcome of the reaction of 4-methoxyaniline with 5-(2-bromoethyl)phenanthridinium bromide (from 80% $C_{22}H_{20}N_2O$ to >90% $C_{22}H_{19}BrN_2O$) by altering only the reactor's architecture, following the same experimental procedure [55], and consequently showing that coupling digital design technology with 3D printing constitutes an experimental parameter that can be optimized systematically in order to promote chemical reactions toward the desirable direction.

An extension of this approach is the application of the 3D printing methodology for the fabrication of configurable millifluidic devices for the designed synthesis of POMs under precise control, and manipulation of the reaction environment demonstrating the suitability of the technique for implementation in lab-on-a-chip devices. In this case, a UV-Vis spectrometer was used for in-line monitoring of the initial formation of the $[Mo_{36}O_{112}(H_2O)_{16}]$, $\{Mo_{36}\}$, cluster on acidification of molybdate solution and its consequent transformation to the molybdenum blue $[Mo_{154}O_{462}H_{14}(H_2O)_{70}]_{142}$, $\{Mo_{154}\}$, wheel on introduction of the reducing agent into the microfluidic reactor [56].

In all cases, the ease and modularity to the reactor design and synthesis of a specific material showed clear advantages over traditional techniques (e.g., glass-blowing), and the range of materials that can be printed has great potential for creating reactionware for material manufacturing processes. Using this approach, it should be feasible to fabricate reactors at macro- and microscale, which exhibit predefined control over reagent mixing sequences, flow rates, and in-line analyses into the reactor design. Combining the disciplines of synthetic chemistry, molecular modeling, and chemical engineering in a low-cost platform allows not only the easier exploration and monitoring of complex chemical systems but also the control of the nanoscale properties (architectures and functionality of the material) via parameterization of the macroscale parameters (flow rates and design of the reactor).

8.6 Emergent Properties and Novel Phenomena

The previously discussed novel synthetic approaches and techniques allowed the synthetic chemists to move from serendipitous and observation based methodologies to designed approaches. Moreover, a deep understanding of the underlying chemistry that is masked by the self-assembly of POM-based chemical systems has allowed the generation and detailed identification of new building block libraries and better control over self-assembly processes as well as the way different libraries can interact constructively. This is reflected by a plethora of unprecedented architectures and the emergence of intriguing properties and new phenomena reported over the past years, setting the scene for the engineering of materials with innovative functionalities.

8.6.1
Porous Keplerate Nanocapsules –Chemical Adaptability

In the past decade, Müller *et al.* have reported the functionality of nanosized Keplerate-based porous molecular capsules under specific conditions and their efforts to design artificial systems that mimic some properties of species of the biological world such as the intracellular response to extracellular molecular signals in the case of enzymatic reactions within the cell. The group studied in a lot of cases similar responsive properties of the $(NH_4)_{42}$ $[Mo_{72}{}^{VI}Mo_{60}{}^{V}O_{372}(CH_3COO)_{30}(H_2O)_{72}]$, $\{Mo_{132}\}$, keplerate capsules [57]. The discovery of such a cluster, particularly as it is porous, water soluble, and has such a large cavity, represents one of the most significant findings in recent years [58], especially because this discovery has gone on to reveal a whole family of related Keplerates.

The configurable properties of these capsules, such as (i) the overall charge, (ii) the related pore sizes, and (iii) the internal surface functionalities which can be either hydrophilic or hydrophobic, have been studied extensively. Also, it has been demonstrated that the control of the above variables alters the overall behavior and reactivity of the capsule and shows a design approach for the construction of materials with predefined functionality. For example, their charges can be tuned on a wide range by exchanging the 30 internal ligands for differently charged ones (e.g., $AcOO^-$ or $SO_4{}^{2-}$, etc.), which consequently tunes their affinity for cation uptake [59]. The nature of the ligand can modify further the hydrophilicity ($SO_4{}^{2-}$) or the hydrophobicity (CH_3COO^-, $CH_3CH_2COO^-$) of the capsule's interior space. Finally, the sizing of the Keplerate spheres has been demonstrated by replacing the $\{Mo^V{}_2\}$ linkers with mononuclear ones such as Mo^V, V^{IV}, Fe^{III}, or Cr^{III}. This modification led the shrinking of not only the whole cluster but also the pore sizes formed (Figure 8.12). Interestingly, the porous surface of this molecular nanosponge reveals a dual function; it not only allows the exchange/separation of small charged

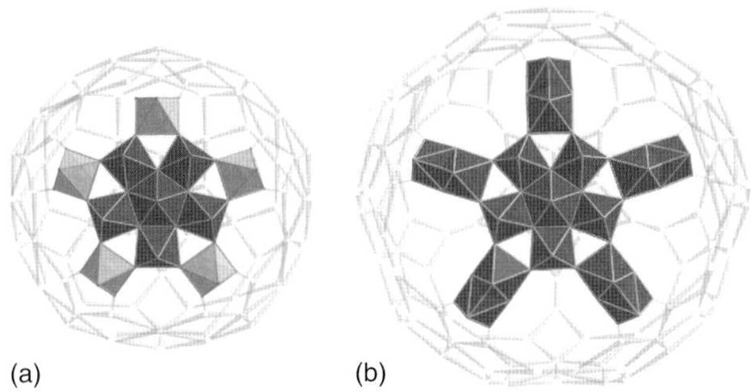

Figure 8.12 "Sizing" of the Keplerate-type nanospheres: comparison of the clusters (a) $\{Mo_{72}M_{30}\}$ and (b) $\{Mo_{132}\}$ showing the size "shrinking" when the dinuclear $\{Mo_2{}^V\}$ linkers are replaced by the mononuclear $\{M\}$ ones, which also leads to smaller pores. Top: polyhedral representation of the clusters; Color code: Mo, black polyhedra; M (Fe, Cr, or V), gray polyhedra.

molecules and metal ions (H_3O^+, Li^+, Na^+, Cs^+, Ce^{3+}, Pr^{3+}) by passing from the solution through the pores to the interior space but also has receptor properties comparable with those of the classical crown ethers, which, consequently, can be reversibly "blocked" by metal cations of appropriate ionic radius or via hydrogen bond formation between protonated molecules and oxygen bridging atoms at the rim of the pore. This finding not only demonstrates the controlled ion separation, entrapment, and release under well-defined conditions but also opens the door for a special type of chemistry, that is, coordination chemistry under confined conditions. Moreover, in the presence of different concentrations of metal ions such as Pr^{3+}, it is shown that the cations can enter the internal space of the nanosponge capsule and occupy specific sites forming two solids inside the cavity, a dodecahedron and an icosidodecahedron, while the metal centers are interestingly found with different coordination environments that are formed by sulfate and encapsulated water ligands [60]. As the concentration of the Pr^{3+} increases, the overall negative charge of the cluster as well as the electrochemical gradient across the pores decreases, forcing the additional cations to occupy a position above the entrance of the pores and consequently gate them. Thus the uptake is controlled by a negative feedback mechanism, which is a phenomenon comparable to the one observed in biological systems.

Furthermore, the Keplerate-based nanocapsules have also given the opportunity to study the structures formed by water molecules. Even though it is the simplest chemically known compound, it gives rise to a plethora of structural motifs that are difficult to study [61]. The use of nanocapsules as crystallization flasks has given the advantage of confining different kinds of water structures in well-defined spaces. The study of different density water structures controlled by the chemical environment under confined conditions is extremely important as very similar processes have been observed in biological systems, for example, similar water formations are found in cells, especially above protein surfaces [62].

Taking into consideration the above discussed findings, Müller *et al.* introduced the term *chemical adaptability* for this type of chemical response. A similar behavior has been reported recently by the same group for the family of molybdenum blue species [63], in which the term refers to the variability of connectivity for molybdenum oxide building blocks in solution, leading to a wide range of chemical pathways and structures. This is reflected by the special properties of the dynamic libraries containing Mo-based species that may form and break reversibly, allowing a continuous change in the observed architectures by the reorganization of building blocks, which is a unique occurrence for inorganic chemical systems.

8.6.2
Transformation of POM Structures at Interfaces – Molecular Tubes and Inorganic Cells

The intrinsic anionic nature of molecular metal oxides along with their diverse architectures, modular redox activity, as well as their multiple functionalities mean that they can reveal new and unexpected properties. For example, Cronin *et al.*

discovered recently that it is possible to fabricate POM-based tubular structures *via* an osmotic process resembling a type of material morphogenesis process [64]. The self-assembly of these cluster anions into a molecular tube via a "chemical garden" [65] type mechanism is the first step toward constructing a network. The fabrication process is initiated by an ion exchange and aggregation that occurs after slow dissolution of single crystals of POM immersed into a solution containing large organo-cations (dihydroimidazophenanthridinium, DIP). Moreover, they showed that microtubes can be spontaneously grown from crystals of POM-based materials with variable growth rates (1–100 μm/s) and with high aspect ratios (>10 000). Interestingly, the same group also demonstrated the possibility of controlling the growth of the micro tubular formations utilizing "optical tweezers." The flow patterns are generated via a laser heating process, which can easily be reconfigured on the fly either by direct user input or autonomous computer control, to produce a variety of architectures. The controllable micronetworks naturally form hollow tubes, allowing material to flow through them during and after fabrication. This discovery opens up exciting prospects for macroscopic control of surface patterning as well as the development of multicomponent composite materials with predefined device functionality (Figure 8.13).

Furthermore, the same group demonstrated the fabrication of POM-based membrane structures with well-defined boundaries. In this case, the formation of membrane material can be achieved *via* an "extrusion/exchange" mechanism of a solution containing large metal oxide anions along with small cations into a solution containing the organic cations accompanied by small anions [66]. The use of POM building blocks is very interesting because these can induce functionality

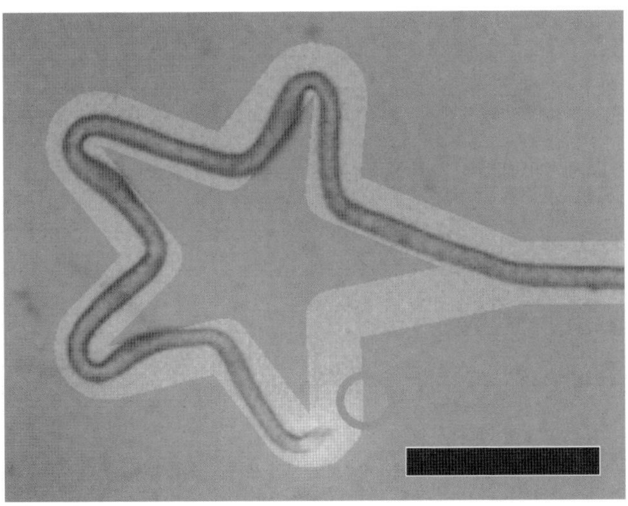

Figure 8.13 Automated control of the laser spot is possible; by using image analysis to track the end of the tube and keep the laser spot a constant distance ahead of the growth (circle). This enables us to assemble structures along predefined tracks (pale overlay). Scale bar: 500 mm.

to the formed inorganic chemical cells (iCHELLs), including redox, catalytic, photochemical, and magnetic properties [10, 11]. The above methodology can produce robust, spontaneously repairing membranous iCHELLs with diameters that range from 50 mm to cell-like compartments of several millimeters. The membranous structures exhibit physical properties "inherited" by their molecular building blocks, such as redox activity or chiral structure, while also being able to partition chemical components within a system by injecting an aqueous solution of one component through a nozzle into an aqueous solution of the other, forming a closed compartment. This is an exciting finding that paves way for the fabrication of diverse and easily configurable functional structures that allow designing and building membrane-based devices with well-defined compartmentalized spaces at the microscale.

8.6.3
Controlled POM-Based Oscillations

As discussed earlier, POM chemistry has been developed to the point where substantial control has been achieved and an engineering philosophy can be employed. Taking advantage of the accumulated knowledge of the metal oxide systems and use of informative techniques such as ESI-MS, we reported recently the first POM-based chemical oscillation process, which is a new occurrence in the POM chemistry. Even though chemical oscillations have been observed in the past – that is, Belousov–Zhabotinsky and Briggs–Rauscher reaction – they were based solely on redox processes. In this case, we discovered the first chemical guest exchange oscillation associated with major structural rearrangement [67]. More specifically, we reported a redox-driven oscillatory template exchange that caused the exchange of the two XO_4^{3-} heteroatom guests (denoted as "P" and "V" for $X = P^V$ and V^V, respectively) contained within the $\{M_{18}O_{54}(XO_4)_2\}^{6-}$ capsule for two complete oscillation cycles ($P2 \rightarrow V2 \rightarrow P2 \rightarrow V2 \rightarrow P2$) before being "chemically" damped. This was due to the thermodynamic stability of Keggin species that form spontaneously in the reaction mixture and drive the equilibria toward an undesirable direction and consequently stop further template exchange taking place. Interestingly, we demonstrated that it is possible to reset the system by reforming the active lacunary species, $\{M_9X\}$, in solution allowing up to four further complete cycles to be observed, as shown by in situ UV-vis spectroscopy experiments. In this work, we postulated that the template exchange proceeds via the opening and closing of the cluster capsule and showed that this process is driven by a competition between reduction and oxidation of the molecular capsule. The discovery was made by the combined use of electron paramagnetic resonance (EPR) spectroscopy and ESI-MS. In addition, the MS studies have proven to be a powerful tool for the investigation of this system allowing us to follow the whole process in real time by stopping and sampling the mixture at given time intervals, revealing crucial information regarding the mechanism and the nature of intermediate molecular fragments that take part in the process, Figure 8.14.

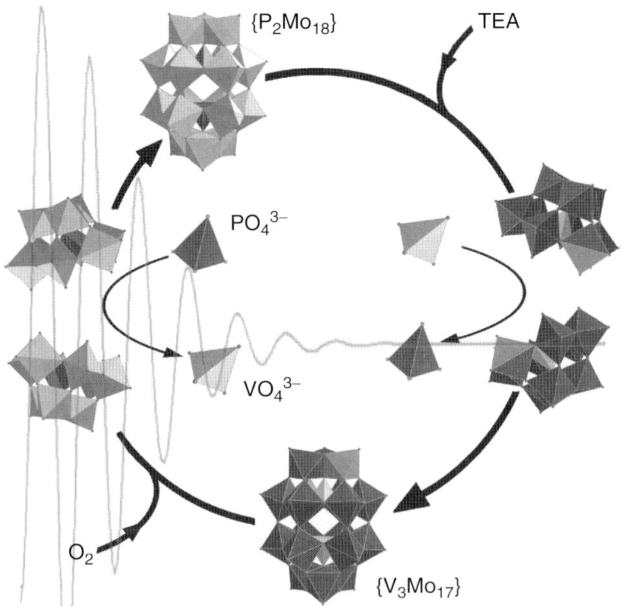

Figure 8.14 The reaction cycle described here, showing the redox-driven guest exchange reaction. Color scheme: PO_4^{3-} templated cluster, gray; VO_4^{3-} templated cluster, black; PO_4^{3-} black tetrahedron; VO_4^{3-} gray tetrahedron; reduced V, dark gray polyhedron. The data point to the dissociation of the $\{M_{18}X_2\}$ cluster into two $\{M_9X\}$ halves.

Furthermore, the solution studies showed that the original capsule dissociates into smaller fragments even though the main species in solution were identified as the two halves of the original molecular capsule. Moreover, the existence of a reduced vanadium center on the cap of the VO_4^{3-} templated cluster has been proven extremely helpful for monitoring the process with EPR spectroscopy, which revealed a similar oscillatory pattern, in good agreement with the ESI-MS data.

The unique occurrence discussed here shows the potential of POM-based capsules for the design of "smart" molecules and responsive materials. This observation opens up interesting prospects for the exploration of complex chemical processes involving coupled structural rearrangements driven away from equilibrium using a type of "redox-metabolism." Moreover, it is quite an interesting fact that this system showed a high level of adaptability – first observed in molybdenum blue/brown chemistry by Müller *et al.* – which seems to be a novel inherent property probably for POM-based species in general. This is directly associated with the unique properties of the dynamic libraries containing the related molybdenum-based synthons, $\{XMo_9\}$, that may form and break reversibly, allowing a continuous change in composition by the reorganization of building blocks driven by the changes in their chemical environment.

8.7
Conclusions and Perspectives

There is no doubt that POMs are extremely diverse, and their dynamic nature associated with their unique chemistry often leads to new discoveries and novel phenomena. The field of molecular metal oxides is now entering into a new period whereby it is possible to design and control both the structure as well as the function of the systems. The obvious endless structural diversity coupled with the adaptive nature of the molecular metal oxides paves the way for further development of synthetic approaches that deviate substantially from the traditional methodologies and take into consideration the adaptability of the systems. These new approaches will be used to access new building block libraries that will lead to the formation of novel nanostructured materials and functions not accessible from traditional synthetic and processing techniques. Moreover, the control and the correlation of the structure with the function is the crucial point that will promote the engineering of functionality utilizing a design approach. Finally, the ultimate challenge will be the fundamental understanding and control of these new properties that will lead toward the development of a new generation of functional molecular metal oxide systems.

References

1. Pope, M.T. and Yamase, T. (eds) (2002) *Polyoxometalate Chemistry for Nanocomposite Design*, Kluwer Academic Publishers, Dordrecht.
2. Mizuno, N., Yamaguchi, K., and Kamata, K. (2005) *Coord. Chem. Rev.*, **249**, 1944.
3. Rhule, J.T., Hill, C.L., and Judd, D.A. (1998) *Chem. Rev.*, **98**, 327.
4. Compain, J.-D., Mialane, P., Dolbecq, A., Mbomekalle, I.M., Marrot, J., Sécheresse, F., Rivière, E., Rogez, G., and Wernsdorfer, W. (2009) *Angew. Chem. Int. Ed.*, **48**, 3077.
5. (a) Ritchie, C., Ferguson, A., Nojiri, H., Miras, H.N., Song, Y.F., Long, D.-L., Burkholder, E., Murrie, M., Kögerler, P., Brechin, E.K., and Cronin, L. (2008) *Angew. Chem. Int. Ed.*, **47**, 5609; (b) Ibrahim, M., Lan, Y., Bassil, B.S., Xiang, Y., Suchopar, A., Powell, A.K., and Kortz, U. (2011) *Angew. Chem. Int. Ed.*, **50**, 4708.
6. (a) Yin, Q.S., Tan, J.M., Besson, C., Geletii, Y.V., Musaev, D.G., Kuznetsov, A.E., Luo, Z., Hardcastle, K.I., and Hill, C.L. (2010) *Science*, **328**, 342; (b) Sartorel, A., Carraro, M., Scorrano, G., De Zorzi, R., Geremia, S., McDaniel, N.D., Bernhard, S., and Bonchio, M. (2008) *J. Am. Chem. Soc.*, **130**, 5006.
7. Pope, M.T. and Müller, A. (1991) *Angew. Chem., Int. Ed. Engl.*, **30**, 34.
8. Hill, C.L. (ed.) (1998) *Chem. Rev.* (Special Issue on Polyoxometalates), **98**, 1.
9. Rosnes, H., Yvon, C., Long, D.-L., and Cronin, L. (2012) *Dalton Trans.*, **41**, 10071.
10. Miras, H.N., Yan, J., Long, D.-L., and Cronin, L. (2012) *Chem. Soc. Rev.*, **41**, 7403.
11. (a) Long, D.-L., Burkholder, E., and Cronin, L. (2007) *Chem. Soc. Rev.*, **36**, 105; (b) Long, D.-L., Tsunashima, R., and Cronin, L. (2010) *Angew. Chem. Int. Ed.*, **49**, 1736.
12. Müller, A., Beckmann, E., Bögge, H., Schmidtmann, M., and Dress, A. (2002) *Angew. Chem. Int. Ed.*, **41**, 1162.
13. (a) Seeber, G., Kögerler, P., Kariuki, B.M., and Cronin, L. (2004) *Chem. Commun.*, 1580; (b) Seeber, G., Kögerler, P.,

Kariuki, B.M., and Cronin, L. (2002) *Chem. Commun.*, 2912.

14. (a) Miras, H.N., Long, D.-L., Kögerler, P., and Cronin, L. (2008) *Dalton Trans.*, 214; (b) Miras, H.N., Corella Ochoa, M.N., Long, D.-L., and Cronin, L. (2010) *Chem. Commun.*, 8148; (c) Walanda, D.K., Burns, R.C., Lawrance, G.A., and Von Nagy-Felsobuki, E.I. (1999) *Inorg. Chem. Commun.*, **10**, 487; (d) Miras, H.N., Stone, D.J., McInnes, E.J.L., Raptis, R.G., Baran, P., Chilas, G.I., Sigalas, M.P., Kabanos, T.A., and Cronin, L. (2008) *Chem. Commun.*, 4703.
15. (a) Ohlin, C.A., Villa, E.M., Fettinger, J.C., and Casey, W.H. (2008) *Angew. Chem. Int. Ed.*, **47**, 5634; (b) Walanda, D.K., Burns, R.C., Lawrance, G.A., and Von Nagy-Felsobuki, E.I. (1999) *J. Chem. Soc., Dalton Trans.*, 311; (c) Alyea, E.C., Craig, D., Dance, I., Fisher, K., Willett, G., and Scudder, M. (2005) *CrystEngComm*, **7**, 491; (d) Miras, H.N., Wilson, E.F., and Cronin, L. (2009) *Chem. Commun.*, 1297.
16. (a) Jeannin, Y. and Martin-Frére, J. (1979) *Inorg. Chem.*, **18**, 3010; (b) Ozawa, Y. and Sasaki, Y. (1987) *Chem. Lett.*, **185**, 923; (c) Rodewald, D. and Jeannin, Y. (1999) *C.R. Acad. Sci., Ser. IIc: Chim.*, **2**, 161.
17. Long, D.-L., Streb, C., Song, Y.-F., Mitchell, S.G., and Cronin, L. (2008) *J. Am. Chem. Soc.*, **130**, 1830.
18. Miras, H.N., Yan, J., Long, D.-L., and Cronin, L. (2008) *Angew. Chem. Int. Ed.*, **47**, 8420.
19. Tsunashima, R., Long, D.-L., Miras, H.N., Gabb, D., Pradeep, C.P., and Cronin, L. (2010) *Angew. Chem. Int. Ed.*, **49**, 113.
20. Corella-Ochoa, M.N., Miras, H.N., Long, D.-L., and Cronin, L. (2012) *Chem. Eur. J.*, **18**, 13743.
21. Miras, H.N., Zang, H.Y., Long, D.-L., and Cronin, L. (2011) *Eur. J. Inorg. Chem.*, **2011**, 5105.
22. Mitchell, S.G., Molina, P.I., Khanra, S., Miras, H.N., Prescimone, A., Cooper, G.J.T., Winter, R.S., Brechin, E.K., Long, D.-L., Cogdell, R.J., and Cronin, L. (2011) *Angew. Chem. Int. Ed.*, **50**, 9154.
23. Xu, F., Miras, H.N., Scullion, R.A., Long, D.-L., Thiel, J., and Cronin, L. (2012) *Proc. Natl. Acad. Sci. U.S.A.*, **109**, 11609.
24. Wilson, E.F., Miras, H.N., Rosnes, M.H., and Cronin, L. (2011) *Angew. Chem. Int. Ed.*, **50**, 3720.
25. Steed, J.W. and Atwood, J.L. (2000) *Supramolecular Chemistry*, John Wiley & Sons, Ltd, Chichester.
26. Sanchez, G., de Soler-Illia, G.J., Ribot, F., Lalot, T., Mayer, C.R., and Cabuil, V. (2001) *Chem. Mater.*, **13**, 3061.
27. Long, D.-L., Kögerler, P., Farrugia, L.J., and Cronin, L. (2003) *Angew. Chem. Int. Ed.*, **42**, 4180.
28. Mal, S.S. and Kortz, U. (2005) *Angew. Chem. Int. Ed.*, **44**, 3777.
29. (a) Contant, R. and Teze, A. (1985) *Inorg. Chem.*, **24**, 4610; (b) Bassil, B.S., Dickman, M.H., Römer, I., von der Kammer, B., and Kortz, U. (2007) *Angew. Chem. Int. Ed.*, **46**, 6192; (c) Kortz, U., Hussain, F., and Reicke, M. (2005) *Angew. Chem. Int. Ed.*, **44**, 3773; (d) Keita, B., de Oliveira, P., Nadjo, L., and Kortz, U. (2007) *Chem. Eur. J.*, **13**, 5480; (e) Reinoso, S., Giménez-Marqués, M., Galán-Mascarós, J.R., Vitoria, P., and Gutiérrez-Zorrilla, J.M. (2010) *Angew. Chem. Int. Ed.*, **49**, 8384; (f) Fang, X., Kögerler, P., Furukawa, Y., Speldrich, M., and Luban, M. (2011) *Angew. Chem. Int. Ed.*, **50**, 5212.
30. Suzuki, K., Kikukawa, Y., Uchida, S., Tokoro, H., Imoto, K., Ohkoshi, S.-I., and Mizuno, N. (2012) *Angew. Chem. Int. Ed.*, **51**, 1597.
31. Mizuno, N., Uchida, S., Kamata, K., Ishimoto, R., Nojima, S., Yonehara, K., and Sumida, Y. (2010) *Angew. Chem. Int. Ed.*, **49**, 9972.
32. (a) Besson, C., Mirebeau, J.-H., Renaudineau, S., Roland, S., Blanchard, S., Vezin, H., Courillon, C., and Proust, A. (2011) *Inorg. Chem.*, **50**, 2501; (b) Besson, C., Musaev, D.G., Lahootun, V., Cao, R., Chamoreau, L.-M., Villanneau, R., Villain, F., Thouvenot, R., Geletii, Y.V., Hill, C.L., and Proust, A. (2009) *Chem. Eur. J.*, **15**, 10233; (c) Izzet, G., Ishow, E., Delaire, J., Afonso, C., Tabet, J.C., and Proust, A. (2009) *Inorg. Chem.*, **48**, 11865.
33. Bassil, B.S., Ibrahim, M., Al-Oweini, R., Asano, M., Wang, Z., van Tol, J.,

Dalal, N.S., Choi, K.-Y., Ngo Biboum, R., Keita, B., Nadjo, L., and Kortz, U. (2011) *Angew. Chem. Int. Ed.*, **50**, 5961.
34. Fang, X., Kögerler, P., Speldrich, M., Schilder, H., and Luban, M. (2012) *Chem. Commun.*, **48**, 1218.
35. Pradeep, C.P., Long, D.-L., and Cronin, L. (2010) *Dalton Trans.*, **39**, 9443.
36. Long, D.-L., Abbas, H., Kögerler, P., and Cronin, L. (2004) *J. Am. Chem. Soc.*, **126**, 13880.
37. Zhang, X.M., Wu, H.S., Zhang, F.Q., Prikhod'ko, A., Kuwata, S., and Comba, P. (2004) *Chem. Commun.*, 2046.
38. (a) Cooper, E.R., Andrews, C.D., Wheatley, P.S., Webb, P.B., Wormald, P., and Morris, R.E. (2004) *Nature*, **430**, 1012; (b) Pakhomova, A.S. and Krivovichev, S.V. (2010) *Inorg. Chem. Commun.*, **13**, 1463; (c) Lin, S., Liu, W., Li, Y., Wu, Q., Wang, E., and Zhang, Z. (2010) *Dalton Trans.*, **39**, 1740.
39. (a) Keggin, J.F. (1934) *Proc. R. Soc. London, Ser. A*, **144**, 75; (b) López, X., Bo, C., and Poblet, J.M. (2002) *J. Am. Chem. Soc.*, **124**, 12574; (c) Streb, C., Ritchie, C., Long, D.-L., Koegerler, P., and Cronin, L. (2007) *Angew. Chem. Int. Ed.*, **46**, 7579; (d) Mitchell, S.G., Khanra, S., Miras, H.N., Boyd, T., Long, D.-L., and Cronin, L. (2009) *Chem. Commun.*, 2712; (e) Leclerc-Laronze, N., Marrot, J., and Herve, G. (2005) *Inorg. Chem.*, **44**, 1275; (f) Boyd, T., Mitchell, S.G., Miras, H.N., Long, D.-L., and Cronin, L. (2010) *Dalton Trans.*, **39**, 6460.
40. (a) Manos, M.J., Miras, H.N., Tangoulis, V., Woolins, J.D., Slawin, A.M.Z., and Kabanos, T.A. (2003) *Angew. Chem. Int. Ed.*, **42**, 425; (b) Miras, H.N., Raptis, R.G., Baran, P., Lalioti, N., Harrison, A., and Kabanos, T.A. (2005) *C.R. Chim.*, **8**, 957.
41. Gao, J., Yan, J., Beeg, S., Long, D.-L., and Cronin, L. (2012) *Angew. Chem. Int. Ed.*, **51**, 3373.
42. Gao, J., Yan, J., Beeg, S., Long, D.-L., and Cronin, L. (2013) *J. Am. Chem. Soc.*, **135**, 1796.
43. Long, D.L., Kögerler, P., and Cronin, L. (2004) *Angew. Chem. Int. Ed.*, **43**, 1817.
44. Long, D.-L., Kögerler, P., Parenty, A.D.C., Fielden, J., and Cronin, L. (2006) *Angew. Chem. Int. Ed.*, **45**, 4798.
45. (a) Long, D.L., Song, Y.F., Wilson, E.F., Kögerler, P., Guo, S.X., Bond, A.M., Hargreaves, J.S.J., and Cronin, L. (2008) *Angew. Chem. Int. Ed.*, **47**, 4384. (b) Yan, J., Long, D.-L., Wilson, E.F., and Cronin, L. (2009) *Angew. Chem. Int. Ed.*, **48**, 4376.
46. (a) Ritchie, C., Streb, C., Thiel, J., Mitchell, S.G., Miras, H.N., Long, D.-L., Boyd, T., Peacock, R.D., McGlone, T., and Cronin, L. (2008) *Angew. Chem. Int. Ed.*, **47**, 6881; (b) Thiel, J., Ritchie, C., Streb, C., Long, D.-L., and Cronin, L. (2009) *J. Am. Chem. Soc.*, **131**, 4180.
47. Thiel, J., Ritchie, C., Miras, H.N., Streb, C., Mitchell, S.G., Boyd, T., Ochoa, M.N.C., Rosnes, M.H., McIver, J., Long, D.-L., and Cronin, L. (2010) *Angew. Chem. Int. Ed.*, **49**, 6984.
48. Mitchell, S.G., Streb, C., Miras, H.N., Boyd, T., Long, D.-L., and Cronin, L. (2010) *Nat. Chem.*, **2**, 308.
49. (a) Mialane, P., Dolbecq, A., Lisnard, L., Mallard, A., Marrot, J., and Secheresse, F. (2002) *Angew. Chem. Int. Ed.*, **41**, 2398; (b) Dolbecq, A., Dumas, E., Mayer, C.R., and Mialane, P. (2010) *Chem. Rev.*, **110**, 6009.
50. Marleny Rodriguez-Albelo, L., Rabdel Ruiz-Salvador, A., Sampieri, A., Lewis, D.W., Gomez, A., Nohra, B., Mialane, P., Marrot, J., Secheresse, F., Mellot-Draznieks, C., Biboum, R.N., Keita, B., Nadjo, L., and Dolbecq, A. (2009) *J. Am. Chem. Soc.*, **131**, 16078.
51. (a) Miras, H.N., Cooper, G.J.T., Long, D.-L., Bögge, H., Müller, A., Streb, C., and Cronin, L. (2010) *Science*, **327**, 72; (b) Miras, H.N., Richmond, C.J., Long, D.-L., and Cronin, L. (2012) *J. Am. Chem. Soc.*, **134**, 3816.
52. Richmond, C.J., Miras, H.N., de la Oliva, A.R., Zang, H., Sans, V., Paramonov, L., Makatsoris, C., Inglis, R., Brechin, E.K., Long, D.-L., and Cronin, L. (2012) *Nat. Chem.*, **4**, 1037.
53. Zang, H., Miras, H.N., Yan, J., Long, D.-L., and Cronin, L. (2012) *J. Am. Chem. Soc.*, **134**, 11376.
54. de la Oliva, A.R., Sans, V., Miras, H.N., Yan, J., Zang, H., Richmond, C.J., Long, D.-L., and Cronin, L. (2012) *Angew. Chem. Int. Ed.*, **51**, 12759.

55. Symes, M.D., Kitson, P.J., Yan, J., Richmond, C.J., Cooper, G.J.T., Bowman, R.W., Vilbrandt, T., and Cronin, L. (2012) *Nat. Chem.*, **4**, 349.
56. Kitson, P., Rosnes, M., Sans, V., Dragone, V., and Cronin, L. (2012) *Lab Chip*, **12**, 3267.
57. Müller, A., Kögerler, P., and Bögge, H. (2000) *Struct. Bond.*, **96**, 203.
58. Müller, A., Krickemeyer, E., Bögge, H., Schmidtmann, M., and Peters, F. (1998) *Angew. Chem., Int. Ed. Engl.*, **37**, 3359.
59. (a) Müller, A., Fedin, V.P., Kuhlmann, C., Bögge, H., and Schmidtmann, M. (1998) *Chem. Commun.*, 927; (b) Mitra, T., Miró, P., Tomsa, A.-R., Merca, A., Bögge, H., Avalos, J.B., Poblet, J.M., Bo, C., and Müller, A. (2009) *Chem. Eur. J.*, **15**, 1844; (c) Schäffer, C., Todea, A.M., Bögge, H., Petina, O.A., Rehder, D., Haupt, E.T.K., and Müller, A. (2011) *Chem. Eur. J.*, **17**, 9634.
60. Müller, A., Zhou, Y., Bögge, H., Schmidtmann, M., Mitra, T., Haupt, E.T.K., and Berkle, A. (2006) *Angew. Chem. Int. Ed.*, **45**, 460.
61. Müller, A., Krickemeyer, E., Bögge, H., Schmidtmann, M., Botar, B., and Talismanova, M.O. (2003) *Angew. Chem. Int. Ed.*, **42**, 2085.
62. (a) Cho, C.H., Singh, S., and Robinson, G.W. (1996) *Faraday Discuss.*, **103**, 19; (b)Gregory, R.B. (ed.) (2005) *Protein-Solvent Interactions*, Marcel Dekker, New York.
63. Müller, A., Merca, A., Al-Karawi, A.J.M., Garai, S., Bögge, H., Hou, G., Wu, L., Haupt, E.T.K., Rehder, D., Haso, F., and Liu, T. (2012) *Chem. Eur. J.*, **18**, 16310.
64. Ritchie, C., Cooper, G.J.T., Song, Y.-F., Streb, C., Yin, H., Parenty, A.D.C., MacLaren, D.A., and Cronin, L. (2009) *Nat. Chem.*, **1**, 47.
65. (a) Collins, C., Zhou, W., Mackay, A.K., and Klinowski, J. (1998) *Chem. Phys. Lett.*, **286**, 88; (b) Cartwright, J.H.E., Garci-Ruiz, J.M., Novella, M.L., and Otálora, F. (2002) *J. Colloid Interface Sci.*, **256**, 351.
66. Cooper, G.J.T., Kitson, P.J., Winter, R.S., Zagnoni, M., Long, D.-L., and Cronin, L. (2011) *Angew. Chem. Int. Ed.*, **50**, 10373.
67. Miras, H.N., Sorus, M., Hawkett, J., Sells, D.O., McInnes, E.J.L., and Cronin, L. (2012) *J. Am. Chem. Soc.*, **134**, 6980.

9
Molecular Metal Oxides for Energy Conversion and Energy Storage

Andrey Seliverstov, Johannes Forster, Johannes Tucher, Katharina Kastner, and Carsten Streb

9.1
Introduction to Molecular Metal Oxide Chemistry

Metal oxides are inorganic compounds with immense potential for the development of new energy conversion and storage systems. They feature a variety of catalytic and photocatalytic properties with applications in fundamental *and* applied research [1]. Metal oxides are subdivided into classical solid-state oxides and molecular metal oxide clusters, so-called polyoxometalates (POMs). POMs can be considered the molecular analogs of solid-state metal oxides and offer a fascinating range of structures and properties [2–4].

9.1.1
Polyoxometalates – Molecular Metal Oxide Clusters

POMs are molecular metal oxide cluster compounds that are typically formed in self-assembly reactions by oligo-condensation of small oxometalate precursors, for example, VO_3^-, MoO_4^{2-}, or WO_4^{2-}. If the cluster consists only of the metal centers and oxo ligands, the species is referred to as an *isopolyoxometalate*. Often, the clusters contain an internal templating anion such as SO_4^{2-}, PO_4^{3-}, or SiO_4^{4-}, or a heteroelement is incorporated into the cluster shell, see Figure 9.1. These cluster species are referred to as *heteropolyoxometalates*. During the cluster assembly, precise control of secondary reaction parameters such as solvent, solution pH, temperature, and pressure, redox-agents, or counterions allows the adjustment of the final cluster architecture [2–4]. However, it should be noted that the bottom-up assembly of predetermined cluster architectures is still a major challenge. Therefore, new POM cluster syntheses are often driven by empirically derived reaction control parameters.

The most well-studied subclasses of POMs are based on group 5 and 6 metal oxoanions such as vanadates, molybdates, and tungstates [5]. Molybdate and tungstate chemistry is dominated by the formation of heteropolyoxometalates $[X_xM_yO_z]^{n-}$ (M = Mo, W), where heteroelements X are incorporated into the

Discovering the Future of Molecular Sciences, First Edition. Edited by Bruno Pignataro.

$$12\ WO_4^{2-} + PO_4^{3-} + 24\ H^+ \longrightarrow [PW_{12}O_{40}]^{3-} + 12\ H_2O$$

(a)

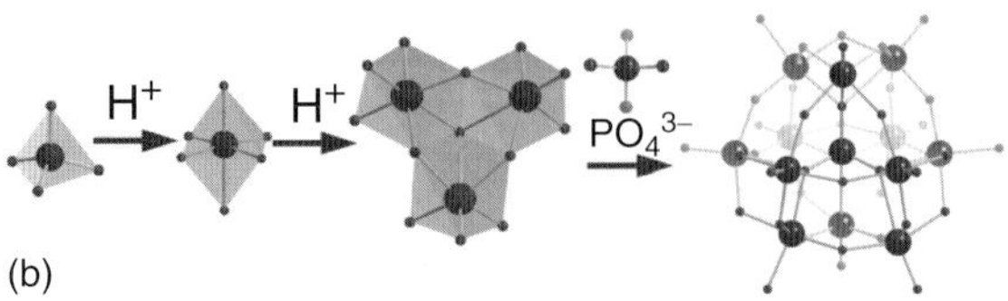

(b)

Figure 9.1 (a) Formal reaction equation showing the condensation reaction between 12 tungstate units and a phosphate anion under acidic conditions, giving the phosphate-templated prototype Keggin anion $[PW_{12}O_{40}]^{3-}$. (b) Schematic illustration of the building blocks involved in the Keggin assembly. (i) A tungstate anion undergoes a coordination shell expansion (tetrahedral to octahedral) on acidification (protonation of the oxo ligands). (ii) Formal aggregation of three octahedral tungstate units into a tri-nuclear so-called triad unit $[W_3O_{13}]$ (note, this is a formally assigned unit, and has not been isolated to date). (iii) Phosphate-templated aggregation of four triads, forming the final Keggin anion $[PW_{12}O_{40}]^{3-}$.

cluster shell, leading to the formation of highly stable and reactive molecular units. Prime example is the so-called Keggin anion, $[XM_{12}O_{40}]^{n-}$ (X = B, Si, P, etc.; M = Mo, W), which is easily accessible from aqueous acidic solutions containing an oxometalate precursor and the corresponding templating anion, see [5]. Further, molybdate and tungstate solution chemistry is characterized by the aggregation of very large metal oxo clusters with up to 368 metal centers combined in one molecular unit [6]. In contrast, vanadate chemistry is more focused on small to medium-sized clusters, typically featuring between 4 and 40 vanadium centers [7]. Vanadates feature a large structural variety as the V centers can adopt tetrahedral $[VO_4]$, square pyramidal $[VO_5]$, and octahedral $[VO_6]$ coordination geometries [7]. In contrast, the metal centers in Mo- and W-based clusters are typically found in an octahedral $[MO_6]$ coordination environment, thereby somewhat restricting the accessible cluster shells.

Arguably the most important aspect in POM chemistry is the ability to incorporate a wide range of heterometals into the cluster shell, thus giving access to a large number of cluster derivatives with tuneable physicochemical properties. This concept is briefly exemplified in Figure 9.2: Starting from the intact Keggin anion

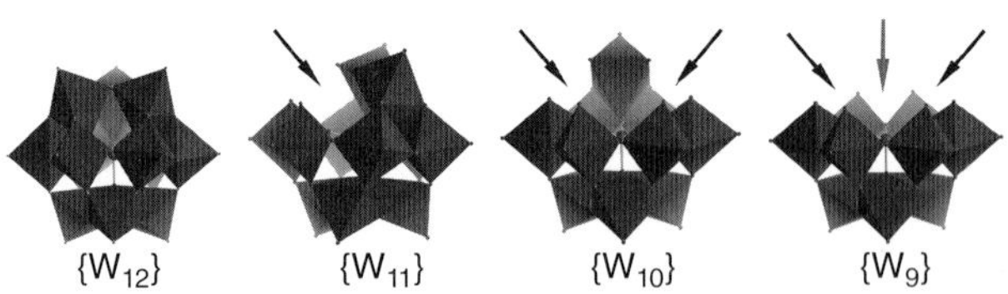

(a) (b) (c) (d)

Figure 9.2 Illustration of lacunary polyoxotungstate clusters. (a) The native Keggin anion $[SiW_{12}O_{40}]^{4-}$ ({W$_{12}$}), (b) the lacunary units $[SiW_{11}O_{39}]^{8-}$ ({W$_{11}$}), (c) $[SiW_{10}O_{36}]^{8-}$ ({W$_{10}$}), and (d) $[SiW_9O_{34}]^{10-}$ ({W$_9$}). Metal binding sites are highlighted by arrows.

$[SiW_{12}O_{40}]^{4-}$, hydrolytic removal of one, two, or three metal centers leads to the formation of so-called lacunary species, resulting in the formation of $[SiW_{11}O_{39}]^{8-}$, $[SiW_{10}O_{36}]^{8-}$, and $[SiW_9O_{34}]^{10-}$. Lacunary clusters feature vacant binding sites where additional heterometals can be coordinatively bound, resulting in modified cluster derivatives. This approach has been used extensively in the formation of functional clusters with promising redox, catalytic, and biological properties [2–5].

9.1.2 Principles of Polyoxometalate Redox Chemistry

Redox-activity is one of the most noted chemical properties of POMs because POMs are based on early, high-valent transition metals that can undergo multiple reductions and subsequent reoxidations. The stability of the reduced species depends mainly on the type of cluster and the metal centers present [8, 9]. Generally speaking, POM reduction requires more negative reduction potentials (i.e., stronger reducing conditions) in the order W > Mo > V. Further, group 6 POMs are more stable than group 5 POMs on reduction. In addition, heteropolyoxometalates are more stable on multielectron reduction than isopolyoxometalates, as the internal template anion stabilizes the cluster architecture. Reduction proceeds by one or two electron transfer; the electron transfers are often (quasi-)reversible. At more negative potentials, additional multielectron reductions are observed, leading to cluster decomposition. Because of the increasingly negative cluster charge, POM anions become more nucleophilic and more basic on reduction, so that reduction is often coupled with the protonation of cluster oxo ligands [8, 9].

9.1.3 Principles of Polyoxometalate Photochemistry

It has been known for several decades that POMs feature rich photochemical activity [10]. To understand why POMs have attracted much interest as photoactive materials [11], the electronic consequences of cluster irradiation need to be considered. As the clusters typically feature fully oxidized d^0 metal centers, light absorption is mainly controlled by $O \rightarrow M$ ligand-to-metal charge-transfer (LMCT) bands in the region of $\lambda = 200–500$ nm [12, 13]. As a result of photon absorption, an electron is promoted from a doubly occupied bonding orbital (highest occupied molecular orbital, HOMO) to an empty, antibonding orbital (lowest unoccupied molecular orbital, LUMO), resulting in the generation of an oxo-centered radical [14]. The photoexcited cluster species is, therefore, highly reactive and is a better oxidizing agent (higher electron affinity E_{ea}) and a better reducing agent (lower ionization energy E_I) than the ground state cluster species, see Figure 9.3 [13].

Besides the facile photoexcitation using near-visible or UV-light, POMs offer advantages as homogeneous or heterogeneous photocatalysts:

1) POMs show strong light absorption with high absorption coefficients ($\varepsilon > 1 \times 10^4\ M^{-1}\ cm^{-1}$), although the absorption maxima are often found in

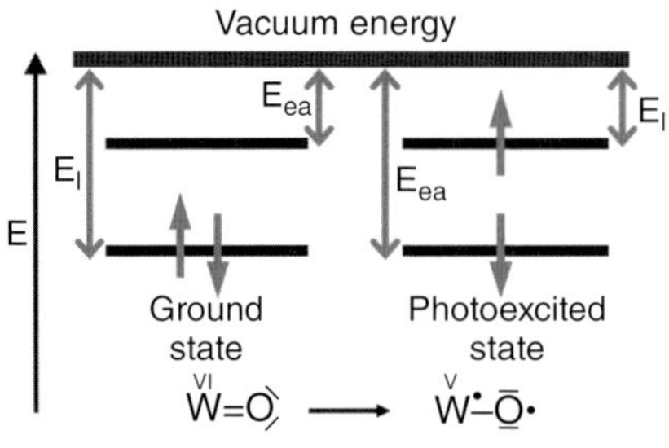

Figure 9.3 Top: Simplified orbital diagram illustrating the ground state and excited-state orbital occupation: photoexcitation results in the promotion of an electron from a bonding to an antibonding orbital. The resulting excited species features lower ionization energy (E_I) and higher electron affinity (E_{ea}) than the ground state species and is, therefore, both a better oxidizing agent and a better reducing agent than the ground state species. Bottom: Schematic illustration of the type of ligand-to-metal charge-transfer (LMCT) observed on cluster irradiation. (Reproduced from Ref. [13] with permission from The Royal Society of Chemistry.)

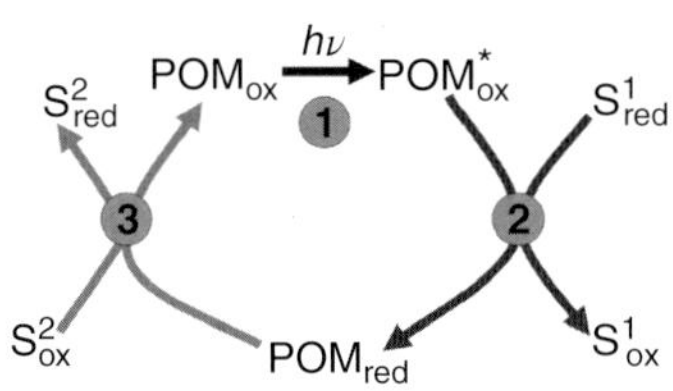

Figure 9.4 General scheme of a polyoxometalate-based photoredox-cycle: (1) photoexcitation of the oxidized cluster, POM_{ox} by photon absorption (irradiation of the cluster LMCT band); (2) oxidation of substrate (S^1) and reduction of the cluster giving POM_{red}; (3) reduction of a second substrate (S^2) and reoxidation of the cluster. Substrates can be a wide range of oxidizable/reducible organic-compounds. For energy conversion and storage, S^1 can be water (oxidized to O_2), S^2 can be protons (reduced to H_2) or CO_2 (reduced to lower-valent carbon species), for details see Chapter 15.

UV region. Tailing of these LMCT bands into the visible region can be used to tune light absorption, particularly for molybdate and vanadate clusters.

2) POMs show interesting redox-activity and can undergo photoredox reactions to catalyze substrate oxidation or substrate reduction and can be used with a wide range of substrate molecules, see Figure 9.4.
3) The structural integrity of the cluster shell is often maintained during the photoredox processes, thereby allowing the application of the POM as a catalytic species.
4) Because of the number and type of metal centers, POMs can undergo multielectron redox-events, making them interesting compounds for multistep photoredox-systems.
5) Reoxidation of the reduced species is often possible using molecular oxygen or hydrogen peroxide without the degradation of the cluster compounds even under harsh reaction conditions.

9.1.4
POMs for Energy Applications

The electrochemical and photochemical activity of POMs outlined above, combined with their wide structural and chemical variability, makes POMs promising molecular materials for the development of energy conversion and energy storage systems. The following sections outline areas that offer high prospects for the development of POM-based functional systems for sustainable energy systems. Each section outlines the current state of the art and provides the authors' subjective view on future developments as well as bottlenecks and challenges in POM development, which require attention from researchers in the field.

9.2
POM Photocatalysis

Photocatalysis is the science of converting light into chemical reactivity. When sunlight is used to drive a useful chemical reaction, photocatalysts can be employed instead of chemical or thermal processes, thus reducing energy usage and lowering CO_2 emissions. POMs have been studied as promising photocatalysts for several decades. Their applications as well as limitations are discussed in the following sections.

9.2.1
The Roots of POM-Photocatalysis Using UV-light

The POM-catalyzed photooxidation of organic substrates follows a general photoredox-cycle as outlined in Figure 9.4 with S^1 = organic substrate (i.e., electron donor) and S^2 = electron acceptor, for example, O_2, H_2O_2, H^+ [10–18]. In general, two possible oxidation mechanism are discussed, depending on whether the reaction proceeds under strictly inert conditions, in the absence of water, or in the presence of water molecules [19]:

1) In the presence of water, it is generally accepted that water molecules are preassociated with the cluster shell via hydrogen-bonds. After photoexcitation of the cluster M–O bonds, this preassociation allows the facile hydrogen-atom abstraction by homolytic H–OH bond cleavage, resulting in the formation of a hydroxyl radical $OH^\bullet$ and a protonated, reduced M–OH species [12, 13, 18].
2) In the absence of water, the substrate preassociates with the cluster, and on photoexcitation, hydrogen-atom abstraction from the substrate is achieved directly by a cluster-based oxo-radical [13, 20].

The mechanistic versatility of POM photocatalysis can be used to optimize cluster reactivity and selectivity to target a specific desired reaction product by controlling the reaction conditions.

9.2.2
Sunlight-Driven POM Photocatalysts

9.2.2.1 Structurally Adaptive Systems for Sunlight Conversion

Traditional POM photocatalysis has focused mainly on the reactivity of molybdate and tungstate clusters whereas vanadium-based clusters were often considered to be too labile or too unreactive to show useful photochemistry [12]. Recently, it was demonstrated that purely vanadium oxide based cluster systems can be employed as photoactive systems for the selective oxidation of organic substrates [21]. The systems illustrate some of the key advantages of vanadates over molybdate- or tungstate-based systems: in the study it was shown that visible-light photoexcitation can be achieved by the thermal activation of an inactive precursor, $[V_4O_{12}]^{4-}$ (= $\{V_4\}$). At elevated temperatures, $\{V_4\}$ undergoes a reversible rearrangement to the active cluster, $[V_5O_{14}]^{3-}$ (=$\{V_5\}$), see Figure 9.5. Structural change is accompanied by a significant bathochromic shift of the LMCT absorption band, resulting in visible-light absorption by $\{V_5\}$. In contrast, $\{V_4\}$ absorbs light only in the UV region [22].

Irradiation of a solution of $\{V_5\}$ in the presence of primary or secondary alcohols using visible light ($\lambda > 380$ nm) leads to the photooxidation of the alcohol to the corresponding carbonyl compound. It was demonstrated that in the case of methanol, this 2-electron oxidation is selective and results in the formation of formaldehyde only. No higher oxidation products (formic acid, CO_x) were observed, see Figure 9.5. The current working hypothesis for the observed selectivity suggests that methanol can preassociate to the cluster shell via hydrogen-bonding, thus enabling efficient substrate oxidation. In contrast, formaldehyde is not able to form a similar hydrogen-bonded species.

On reduction, $\{V_5\}$ undergoes an aggregation reaction as the $\{V_5\}$ unit is not stable under reductive conditions. The resulting aggregation can formally be described as a dimerization where two 1-electron reduced $\{V_5\}$ clusters aggregate and form a 2-electron reduced decavanadate species, $[V(IV)_2V(V)_8O_{26}]^{4-}$ (= $\{V_{10}\}$). The reduced $\{V_{10}\}$ species can be reoxidized using molecular oxygen or aqueous

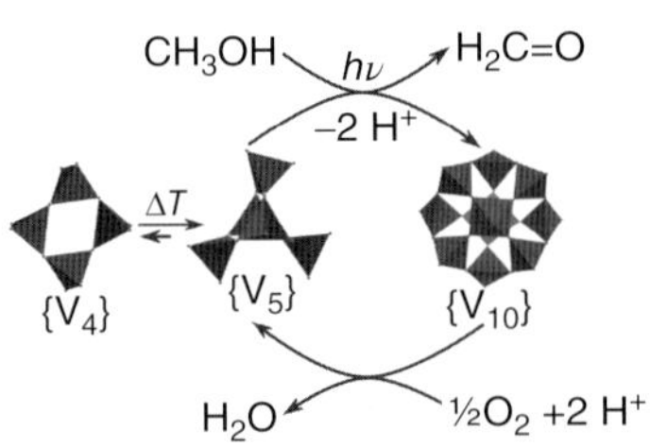

Figure 9.5 Vanadate-based photoredox-cycle [16]. Left: Thermally activated conversion of the inactive $\{V_4\}$ to the active $\{V_5\}$ species. Center: Selective photooxidation of methanol to formaldehyde and reductive cluster aggregation yielding the 2-electron reduced $\{V_{10}\}$ species. Subsequent reoxidation can be achieved by molecular oxygen (slow) or by H_2O_2 (fast). $\{V_4\} = [V_4O_{12}]^{4-}$; $\{V_5\} = [V_5O_{14}]^{3-}$; $\{V_{10}\} = [V(IV)_2V(V)_8O_{26}]^{4-}$.

hydrogen peroxide. It should be noted that the reoxidation using O_2 is slow and requires a prolonged period of time for complete reoxidation.

9.2.2.2 **Optimized Sunlight Harvesting by Metal Substitution**

Recently, it was put forward that functionalization of VIS-inactive metal oxo clusters (e.g., molybdates, tungstates) with VIS-active metals such as vanadium. This route can provide access to highly stable, homogeneous visible-light photocatalysts. In an initial study, the well-known Lindqvist cluster family $[M_6O_{19}]^{n-}$ (M = Mo, W (n = 2); Ta, Nb (n = 8)) was used as a particularly promising model system. The chemical modification of Lindqvist clusters by metal substitution has been studied intensely over the past decades, leading to the discovery of a range of heterometallic Lindqvist clusters $[M'(L)M_5O_{18}]^{n-}$ (M = Mo, W; M′ = transition metal, e.g., Co, V, Zr, Hf, L = ligand, e.g., RO^-, O^{2-}, etc.), see Figure 9.6 [23–27].

For the proposed photochemical study, the mono-vanadium-substituted Lindqvist cluster $[VMo_5O_{19}]^{3-}$ was compared with the native $[Mo_6O_{19}]^{2-}$ compound [28]. UV–vis-spectroscopy showed that the vanadium-substituted unit features several low-energy LMCT transitions in the near-visible range around 400 nm, whereas the molybdate cluster absorbs light in the UV-region only. This was confirmed by time-dependent density functional theory (TD-DFT) calculations, which showed that $[VMo_5O_{19}]^{3-}$ features a number of V–O-based LMCT transitions in the region around 400 nm. Vanadium substitution can, therefore, in principle be used to photosensitize polyoxomolybdate-based cluster

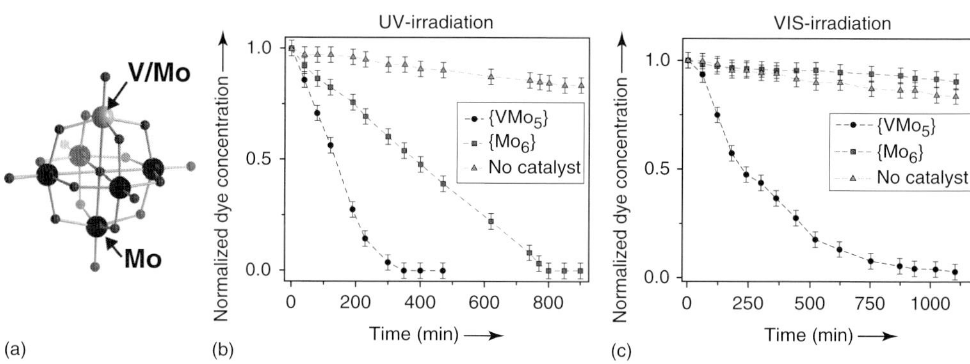

Figure 9.6 (a) Ball-and-stick representation of the general structure of the Lindqvist anion $[M'Mo_5O_{19}]^{n-}$ (M′ = V (n = 3) or Mo (n = 2)). The cluster is templated by a central μ^6-oxo ligand around which six $[MO_6]$ octahedra are arranged in an octahedral fashion. (b) Photocatalytic decomposition of a model pollutant (patent blue V dye) under broadband UV-irradiation (medium-pressure mercury lamp, P = 150 W) catalyzed by $[VMo_5O_{19}]^{3-}$ ({VMo_5}) and by $[Mo_6O_{19}]^{2-}$ ({Mo_6}). {Mo_6} shows significantly lower reactivity compared with {VMo_5}. (c) Photocatalytic decomposition of the model pollutant (patent blue V dye) under VIS-irradiation (λ = 470 nm, LED light source, P = 3 W) catalyzed by {VMo_5} and {Mo_6}. {Mo_6} shows no reactivity (because of negligible light absorption) whereas residual reactivity is observed for {VMo_5}. (Reproduced from Ref. [28] with permission from The Royal Society of Chemistry.)

systems. A number of photochemical dye degradation reactions were carried out to understand the photochemical activity of both clusters and it was demonstrated that $[VMo_5O_{19}]^{3-}$ (TON = 1606, TOF = 5.35 min^{-1}) shows significantly higher turnover numbers (TONs) and turnover frequencies (TOFs) for the degradation of a triphenylmethane dye (patent blue V) as compared with $[Mo_6O_{19}]^{2-}$ (TON = 1308, TOF = 1.92 min^{-1}) when irradiated with UV-light. In addition, only $[VMo_5O_{19}]^{3-}$ showed photoactivity when irradiated with monochromatic visible light ($\lambda = 470$ nm) whereas $[Mo_6O_{19}]^{2-}$ did not show any residual photoactivity. Metal substitution can, therefore, be used to improve general photoactivity as well as photoactivity in the visible region of molecular metal oxide clusters, see Figure 9.6.

9.2.2.3 **Visible-Light Photocatalysis – Inspiration from the Solid-State World**

It is noticeable that solid-state vanadium oxides have attracted considerably more interest as visible-light photocatalysts as compared with their molecular counterparts, the polyoxovanadates. On the basis of the well-known solid-state vanadium oxide photocatalysts, the synthesis of their molecular analogs might be a promising way for the development of homogeneous, molecular photocatalysts. Bismuth vanadium oxide, $BiVO_4$ is a prime example as the solid-state oxide has been known to be a promising VIS-active photocatalyst [29–31] whereas no molecular analogs have been known until recently. Access to molecular bismuth vanadium oxide compounds was obtained using a fragmentation-and-reassembly route where vanadium oxide precursors were fragmented in organic solution and reassembled in the presence of Bi(III) to give the first molecular bismuth vanadium oxide cluster, $[H_3(Bi(dmso)_3)_4V_{13}O_{40}]$ (= $\{Bi_4V_{13}\}$), see Figure 9.7. The cluster is based on the so-called ε-Keggin architecture $[\varepsilon\text{-}V_{12}O_{36}(VO_4)]^{15-}$ and is stabilized by four Bi(III) centers arranged on the cluster shell in a tetrahedral fashion. The compound shows visible-light absorption up to about 560 nm, and photooxidative dye decomposition

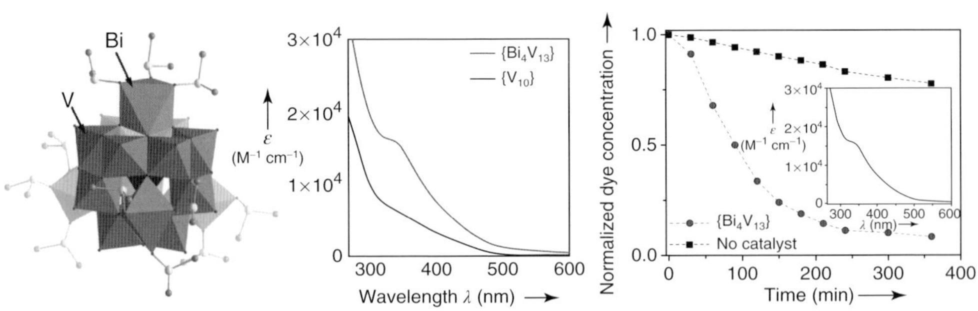

Figure 9.7 (a) Illustration of the $\{Bi_4V_{13}\}$ cluster $[H_3(Bi(dmso)_3)_4V_{13}O_{40}]$. The cluster is templated by a central $[VO_4]$ unit around which an ε-Keggin-cluster shell is formed, which is stabilized by four Bi(III) centers. (b) UV–vis absorption spectrum of $\{Bi_4V_{13}\}$ in comparison with a nonfunctionalized prototype vanadium oxide cluster, $\{V_{10}\}$ (= $[H_3V_{10}O_{28}]^{3-}$). (c) Photocatalytic decomposition of a model pollutant (patent blue V dye) under VIS-irradiation ($\lambda = 470$ nm, LED light source, $P = 3$ W) catalyzed by $\{Bi_4V_{13}\}$.

test reactions show that the cluster is stable as a homogeneous photocatalyst with TON = 1200 and TOF = 1.29 min^{-1} ($\lambda_{irradiation}$ = 470 nm). In addition, the cluster features three acidic protons that might make it interesting as a bifunctional acid and photooxidation catalyst [32].

9.2.3
Future Development Perspectives for POM Photocatalysts

Over the past decade, much research in POM photocatalyst science has been devoted to the coupling of POMs as photoactive materials to solid reactive supports such as semiconductors. The concept is aimed at harvesting the synergetic activity of the semiconductor and the POM in order to access more reactive or more selective photocatalysts. Much of this research was focused on well-known, commercially available POMs, typically Keggin and Dawson anions [33–47]. As these clusters are only UV-active, only few studies have focused on direct sunlight harvesting, where POMs could be employed as inorganic, highly stable visible-light active photosensitizers. However, this concept, if successful, would be a major advantage for several fields of research, notably for photocatalysis, photovoltaics, dye-sensitized solar cells (Grätzel-cells) and for photoelectrochemical applications, where the POM clusters could potentially replace the standard ruthenium-based noble-metal dyes used to date. To achieve this goal, basic research into the cluster-support interactions, including charge separation, lifetimes, quantum efficiencies, and synthetic routes for anchoring the cluster to the support are required to better understand and optimize these systems. Promising clusters for this type of study would be, besides pure vanadates, vanadium-substituted Keggin (e.g., $[SiV_xW_{12-x}O_{40}]^{n-}$) and Dawson (e.g., $[P_2V_xW_{18-x}O_{62}]^{m-}$) clusters, whose fundamental photoreactivities have already been established [13].

9.3
Energy Conversion

9.3.1
Water Splitting

Sunlight-driven, photochemical splitting of water into oxygen and hydrogen is a promising scheme for the conversion of solar energy into storable energy [48, 49]. It would allow the decoupling of energy-intense processes such as heating, transportation, electricity generation from fossil fuels. Further, it might give sustainable access to chemical feedstocks, for example, by means of syngas-to-methanol conversion. However, water splitting and particularly water oxidation remain challenging reactions requiring highly stable, complex catalysts. To compete with fossil-fuel production, water splitting catalysts need to be based on earth-abundant materials such as 3d-transition metals. Further, the resulting catalytic compounds need high stability, reactivity, and recyclability in order to be deployed on a global scale [50].

9.3.2
Water Oxidation by Molecular Catalysts

Water oxidation by molecular catalysts can be subdivided into two sections: catalysts with one redox-active metal site. These compounds are typically based on noble metals (particularly Ru); they are ideal for studying mechanistic details of the water oxidation process [51]. However, their utility in large-scale industrial operations is rather limited because of economic factors. They will not be considered further at this point. In contrast, much research is focused on mimicking the biological water oxidation catalyst known as the *oxygen evolving complex* (*OEC*) a calcium manganese oxo core [$CaMn_4O_5$] located within Photosystem II, a large protein assembly vital to photosynthesis in plants [52]. This unit is often used as a blueprint to develop similar synthetic systems comprising several redox-active metal centers to mimic the catalytic activity of the natural system. However, this development is still a major synthetic challenge [11, 50, 53–56].

One of the main challenges in catalyst development is the design of oxidatively stable catalysts. This is a challenge for traditional coordination compounds, as the organic ligands employed are thermodynamically unstable under oxidative conditions and often undergo decomposition reactions, thus destroying the catalyst [54, 57, 58]. To overcome this challenge, POMs have been used as purely inorganic molecular ligands, which are oxidatively stable under typical water oxidation conditions. The following section reports some of the main findings in POM-based water oxidation catalysis (WOC) and demonstrates the challenges faced to date.

9.3.2.1 Water Oxidation by Ru- and Co-Polyoxometalates

Over the past years, it has been demonstrated that transition metal-substituted POMs can be used as inorganic catalysts or precatalysts for the oxidation of water to molecular oxygen and protons [59–66]. The 4-electron water oxidation represents the critical step of the water splitting reaction, which can be used to generate hydrogen from water for use as a secondary energy carrier. Importantly, this system needs to be driven by sunlight to achieve direct conversion of solar to storable chemical energy. The development of stable and economically viable catalysts that can perform the sunlight-driven water oxidation would be a major progress toward carbon-neutral energy systems. Recently, POM-based systems have been reported where the active water oxidation catalyst (e.g., $[Ru_4(O)_4(OH)_2(H_2O)_4(\gamma\text{-}SiW_{10}O_{36})_2]^{10-}$ [62, 66] or $[Co_4(H_2O)_2(PW_9O_{34})_2]^{10-}$ (=Co_4-WOC)) has been coupled with ruthenium-based photosensitizers, which allowed the development of water oxidation complexes [59–61, 65]. Arguably, the most prominent example is the cobalt-POM-based system reported by Hill *et al.* [60, 64] where a $[Co(II)_4(H_2O)_2(PW_9O_{34})_2]^{10-}$ cluster is reported as the water oxidation catalyst and a visible-light-driven $[Ru(bpy)_3]^{2+/3+}$ redox couple (bpy = 2,2′-bipyridine) acts as electron shuttle that transfers electrons from the water-oxidizing Co_4-WOC to the stoichiometric electron acceptor, see Figure 9.8. Under optimized

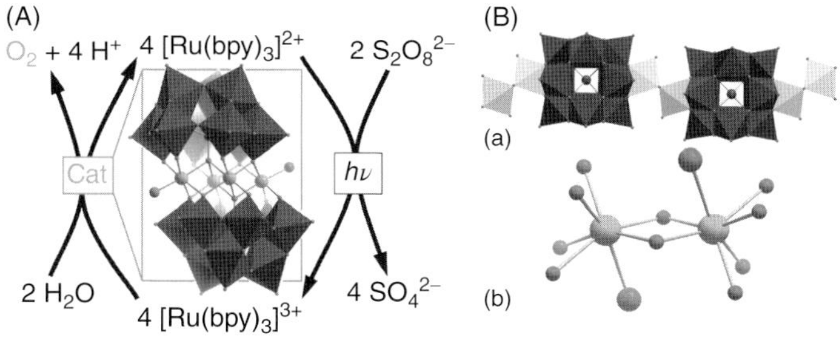

Figure 9.8 (A) Light-driven water oxidation catalyzed by the Cosubstituted POM cluster $[Co_4(H_2O)_2(PW_9O_{34})_2]^{10-}$ [26]. Step 1: Light-induced oxidation of $[Ru(bpy)_3]^{2+}$ in the presence of peroxodisulfate $S_2O_8^{2-}$. Step 2: Water oxidation catalyzed by the Co-POM and electron transfer to the oxidized $[Ru(bpy)_3]^{3+}$. (B) Polyhedral illustration of the heterogeneous niobate water oxidation catalyst $[\{Nb_2(O)_2(H_2O)_2\}\{SiNb_{12}O_{40}\}]^{10-}$ (a) and ball-and stick representation of the proposed catalytic site $\{Nb_2(O)_2(H_2O)_2\}$ featuring two bound water ligands shown as large ligand spheres (b) [67].

conditions, the system features TON > 220 and high quantum yields of up to $\phi = 0.30$ [60].

In addition to these prototype systems, several other POM-based systems with Co [68–70], Ru [71], Ir [72], and Ni [73] have recently been reported as POM-WOCs; however, little is understood about the catalytic mechanism or about the actual catalytic species, and future work needs to establish whether truly homogeneous or heterogeneous WOC is observed [58, 74, 75].

9.3.2.2 Polyoxoniobate Water Splitting

Recently, Peng *et al.* put forward a different concept for heterogeneous water splitting, based purely on niobium-based POMs, such as $[\{Nb_2(O)_2(H_2O)_2\}\{SiNb_{12}O_{40}\}]^{10-}$. The compound features Niobium-based Keggin-clusters linked into infinite 1D chains by dinuclear $\{Nb_2(O)_2(H_2O)_2\}$ moieties, see Figure 9.8 [67]. The $\{Nb_2(O)_2(H_2O)_2\}$ units possess binding sites that allow the coordination of water to the cluster and represent a vital prerequisite for efficient water oxidation. Initial studies show that the Nb-POM can be combined with a hydrogen-evolving catalyst, thus forming a heterogeneous system for the splitting of water into oxygen and hydrogen. When a suspension of the Nb-POM in pure water is irradiated with UV-light in the presence of the cocatalyst NiO, the formation of oxygen and hydrogen is detected, albeit at relatively low TONs. The authors suggest that the system can be combined with Ru-based photosensitizers to allow photoactivity in the visible range. In addition, higher TONs need to be obtained to achieve truly photocatalytic water splitting.

9.3.2.3 Water Oxidation by Dawson Anions in Ionic Liquids

A completely different concept for POM-based water oxidation has recently been reported by Bond *et al.* In their work, they demonstrated that UV-irradiation of classical Dawson clusters $[P_2W_{18}O_{62}]^{6-}$ in water-containing ionic liquids (ILs) such as 1-butyl-3-methylimidazolium tetrafluoroborate or diethanolamine hydrogen sulfate results in the oxidation of water, yielding molecular oxygen and protons [76, 77]. This reactivity is remarkable as the Dawson anion $[P_2W_{18}O_{62}]^{6-}$ does not show this type of reactivity in purely aqueous phase or in organic solvents such as acetonitrile or dichloromethane.

The authors attribute this unique reactivity to two effects, which are both related to the use of ILs. First, the formal reversible redox-potentials $E_F{}^0$ of the cluster are much more positive in ILs as compared with other organic solvents, and shifts of more than 500 mV are observed. In addition it has been reported that the structure of water aggregates in ILs is very different from the well-known hydrogen-bonded structure of bulk water and might, therefore, facilitate water oxidation. A drawback of the reaction is the use of UV-light instead of visible light, and photosensitization might be employed as one strategy to allow the use of visible light. Further, it would be of great interest to employ well-known POM-WOCs such as Co- or Ru-based systems [55, 56] in ILs to better understand their performance at altered redox-potentials.

9.3.2.4 **On the Stability of Molecular POM-WOCs**

One of the main concerns when investigating homogeneous catalysts is the question of "what is the catalytic species." In water oxidation particularly, this has become a major issue, because the highly oxidizing conditions of WOC often lead to oxidative degradation or conversion of the initial species (the so-called precatalyst) into the truly active species. This, of course, leads to major consequences, as the understanding of catalytic activity, selectivity, and stability all depend on the knowledge of what system is being investigated. Cases in point are POM-based WOCs, which are thought to be stable, molecular units under water oxidation conditions. Particularly, the Co_4-WOC reported recently demonstrates that utmost attention is required when determining whether the molecular POMs are truly the catalytic species or merely a precatalyst: in 2011, Finke *et al.* investigated Co_4-WOC (see above), and found that under electrocatalytic conditions, cluster decomposition is observed and the dominant catalyst is a heterogeneous cobalt oxide, CoO_x [58]. Further, photochemical studies by Bonchio *et al.* on Co_4-WOC found that the electron-transfer rate between Co_4-WOC and the oxidant $[Ru(bpy)_3]^{3+}$ increased with time, suggesting that a new, more active WOC species might be formed *in situ* [78]. However, it should be noted that both, the Bonchio and the Finke studies, used other reaction conditions than the original Hill report, so that different (homogeneous and/or heterogeneous) species might be the true catalyst in each study. Therefore, it is unclear which species (or combination of species) contributes to the homogeneous WOC activity of Co_4-WOC.

These findings exemplify the importance of comparing molecular versus heterogeneous catalysts [50, 53, 79], but also emphasize that the reaction conditions strongly affect catalyst stability. The controversies have recently led to a wider discussion on the stability of WOCs, and several publications and reviews on this topic have been published in 2012–2013 [53, 57, 78, 80].

9.3.3
Photoreductive H_2-Generation

One remarkable feature that has often been observed during the POM-catalyzed photooxidation of organic substrates was the ability of the photoreduced cluster species to spontaneously reduce protons to molecular hydrogen under anaerobic conditions where reoxidation of the reduced cluster species could occur by electron transfer to protons (facilitated in the presence of colloidal Pt or other hydrogen formation catalysts). It was shown by several groups that it is crucial to use a cluster that can undergo multiple electron reductions while preventing cluster decomposition [15, 81–83].

Typical examples for this behavior are the Keggin and Dawson anions as well as the decatungstate cluster $[W_{10}O_{32}]^{4-}$ [15, 81–83]. It should be noted though that to date this scheme requires the use of a sacrificial electron (and proton) donor such as primary or secondary alcohols. Therefore, the system can work economically only as a hydrogen generating catalyst if it is further developed to use readily available substrates such as water as the electron/proton donor.

Recently, it has been demonstrated that POM-based hydrogen production systems can be up-scaled from the laboratory scale toward small, portable pilot-plant dimensions [84]. The authors developed a flow-reactor where a reaction mixture consisting of the photocatalyst ($[W_{10}O_{32}]^{4-}$ or $[SiW_{12}O_{40}]^{4-}$), 2-propanol, water, and colloidal Pt was pumped through fused-silica tubing equipped with solar concentrators and exposed to sunlight to test the ability of hydrogen generation on a preparative scale. Under typical operative conditions (radiation flux 800–1000 $W\,m^{-2}$), the authors were able to obtain hydrogen production rates of up to 180 $ml\,h^{-1}$, and the system remained operational forseveral days.

9.3.4
Photoreductive CO_2-Activation

In order to activate small molecules such as CO_2 on POM clusters, a binding site needs to be incorporated into the cluster shell, as interactions between CO_2 and the negatively polarized oxo-shell of the cluster anion are electrostatically nonfavored. It was suggested previously that metal-substituted Keggin anions could be used for this purpose [30, 85]. These units of the type $[TM(L)XW_{11}O_{39}]^{n-}$ (X = heteroatom, e.g., P, S, Si) are based on the Keggin anion $[XW_{12}O_{40}]^{n-}$ and can formally be obtained by substituting one [W=O] unit with a [TM(L)] fragment (TM = transition metal, L = ligand, e.g., H_2O, solvent, etc.), see Figure 9.9.

The proposed scheme involved the use of functionalized cluster shells where a metal-center such as Co or Ru featuring a weakly bound ligand is present. Under nonaqueous conditions in a nonpolar solvent, it was hypothesized that the labile ligand can be exchanged for a CO_2 molecule; the metal-center would, therefore, act as a CO_2 binding site. To reductively activate the carbon dioxide, subsequent irradiation of the photoactive LMCT bands of the cluster ion in the presence of a sacrificial electron donor was proposed as a viable route.

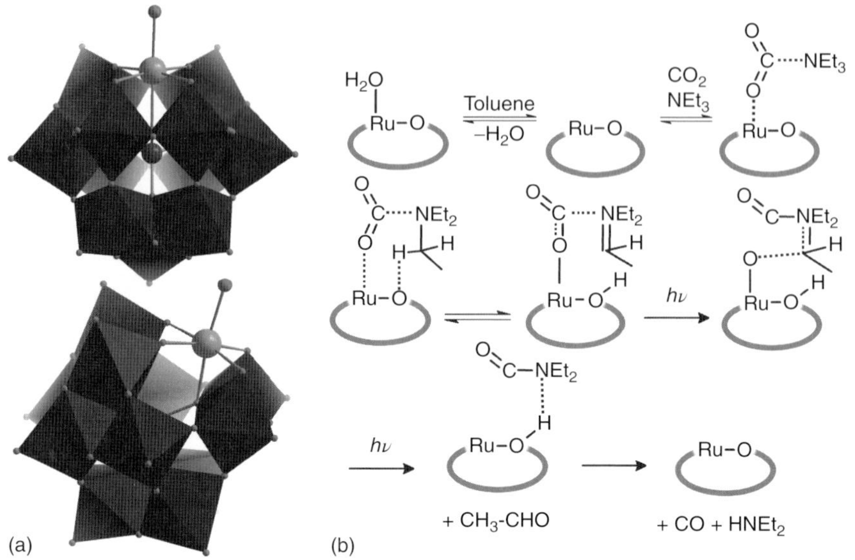

Figure 9.9 (a) Polyhedral representation of the $[Ru(III)(H_2O)SiW_{11}O_{39}]^{5-}$ cluster from front and side view. (b) Geometry-optimized model of the amine-assisted binding of CO_2 to the Ru center, illustrating the first steps of CO_2 activation on $[Ru(III)(H_2O)SiW_{11}O_{39}]^{5-}$. Proposed scheme of the triethyl amine-assisted activation of CO_2 on the Ru-substituted cluster anion $[Ru(CO_2)SiW_{11}O_{39}]^{5-}$.

Recently, this concept was employed by Neumann *et al.* [86] and resulted in the development of a homogeneous system for the photochemical reduction of CO_2 by C=O bond activation. Experimentally, a Ru-substituted lacunary Keggin anion, $[Ru(III)(H_2O)SiW_{11}O_{39}]^{5-}$ (Figure 9.9), was employed in a toluene solution in the presence of CO_2 using triethyl amine as the sacrificial electron donor.

The reaction mixture was irradiated with UV-light, and various analytical methods showed the formation of CO as the only CO_2-based reaction product. The computational study of the proposed reaction intermediates showed that triethyl amine acts not only as a sacrificial electron donor (reducing agent) but also as a supramolecular stabilizer in the cluster-based CO_2 activation. Computations at DFT-level suggest that the CO_2 molecule binds to the Ru center in an end-on-fashion. The amine aligns with the Ru-CO_2 moiety so as to form attractive $Et_3N\cdots CO_2$ interactions between the amine nitrogen and the CO_2 carbon as well as attractive N–C–H⋯O interactions between a C–H group of the amine alkyl substituents and cluster based, bridging oxo ligands located in proximity to the Ru center, see Figure 9.9a. On the basis of a series of experimental and computational studies, Neumann *et al.* proposed a reaction mechanism for the cluster-induced, amine-assisted photoreduction of CO_2 to CO as illustrated in Figure 9.9b.

Recently, Neumann *et al.* have reported a modified approach with the aim of substituting the original sacrificial electron donor, NEt_3, by molecular hydrogen, H_2.

The supramolecular system employed consists of a photoactive rhenium(I)-complex, $[Re(I)(L)(CO)_3Solv]^+$ to which a Keggin anion $[MHPW_{12}O_{40}]^-$ ($M = Na^+$, H_3O^+) is coupled via a crown ether-functionalized phenanthroline ligand L (L = 5,6-(15-crown-5)-1,10-phenanthroline, Solv = acetonitrile). It was demonstrated that the system can activate CO_2 by a cluster-mediated electron transfer. The initial step of the proposed reaction mechanism involves 2-electron-reduction of the cluster anion by H_2 in the presence of colloidal Pt. Photoexcitation of the reduced supramolecular species results in an intramolecular one-electron transfer from the cluster to the Re(I) center, resulting in the formation of an one-electron reduced cluster and a Re(0) species. CO_2 then undergoes an oxidative addition to the Re(0) center with simultaneous electron transfer of the remaining cluster-based electron, resulting in the formation of a $[Re(I)(CO_2H)]$ species. Subsequently, C–O bond cleavage is achieved and CO is released as the only CO_2-based reaction product. Note that in this process, the rhenium complex represents the photoactive site whereas the Keggin anion acts as the redox shuttle that provides two electrons for the CO_2 to CO reduction. However, it was demonstrated by the authors that the supramolecular coupling is a vital prerequisite for efficient CO_2 reduction: reaction of the nonfunctionalized complex $[Re(I)(phenanthroline)(CO)_3]^+$ with Keggin anions under identical experimental conditions did not lead to the formation of CO [87].

9.4 Promising Developments for POMs in Energy Conversion and Storage

9.4.1 Ionic Liquids for Catalysis and Energy Storage

Traditionally, ILs are formed when bulky organic cations such as alkylpyridinium or imidazolium cations are combined with a range of organic or inorganic anions such as Cl^-, BF_4^-, or PF_6^-. As a result, a compound is obtained, which is typically liquid at room temperature, has negligible vapor pressure, and offers a wide range of useful applications in synthesis [88, 89], catalysis [88–93], and electrochemistry [94, 95]. Some key features of ILs are summarized in Table 9.1 [88, 89].

It is important to note at this point that a truly IL must, at least in part, dissociate into ions, so that ILs are formed from dissociated ions and ion pairs and show a significant electrical conductivity. An important subclass of ILs are room-temperature ionic liquids (RTILs) [88, 89] that combine the usefulness of classical organic solvents with the benefits of ILs. Since the discovery of imidazolium-based RTILs, a wide range of technical applications has been developed, most notably in chemical synthesis [97], electrochemistry [94, 95, 98, 99], catalysis [88, 89, 91–93, 100], and sensor systems [101–103].

9.4.1.1 Polyoxometalate Ionic Liquids (POM-ILs)

POM-ILs are obtained when anionic POM clusters are combined with a range of organic bulky cations to give versatile materials that are liquid at or around room

Table 9.1 Key properties of modern, organocation-based ionic liquids.

Melting point	Typically <100 °C
Liquidus range	Typically >100–200 °C
Viscosity	Typically <100–50 cP
Thermal stability	High
Vapor pressure	Negligible
Dielectric constant	<30–40
Specific conductivity	<10–15 mS cm^{-1}
Electrochemical window	Up to 6 V [96]

temperature. True POM-ILs need to be distinguished from a simple POM cluster salt, which has been dissolved in an IL; firstly, the dissolution of a POM salt in ILs is limited by solubility; in contrast, true POM-ILs can achieve much higher cluster loadings (as the POM cluster acts as the anionic component of the IL). Further, dissolution of POM salts in ILs always results in the introduction of two additional ionic species (i.e., the cationic part of the POM salt and the anionic part of the IL), which will change the physical and chemical behavior of the mixture, thus leading to inadvertent effects, particularly at higher POM loadings. In this chapter, the syntheses, properties, and applications of true POM-ILs are discussed.

Over the past decade, several groups have addressed the purposeful synthesis of true POM-ILs. These materials have recently gained interest because of their unique physicochemical properties and wide applicability in various fields of research. Key findings and future applications are discussed in the following section.

In 2004, the first report of a true POM-IL was published. The material is based on the combination of the archetypal Keggin anion $[PW_{12}O_{40}]^{3-}$ with the polyethylene glycol (PEG)-based quaternary ammonium cation Ethoquad 18/25, $(CH_3)(C_{18}H_{37})N[(CH_2CH_2O)_nH][(CH_2CH_2O)_mH]^+Cl^-$ (=QCl) [104]. A viscous material was obtained where the protons of the cluster were partially exchanged with the Q^+ cations, giving $Q_2H[PW_{12}O_{40}]$. The compound is highly viscous at room temperature (75 000 mPa s), stable on heating up to about 160 °C, and features an ionic conductivity of about 6×10^{-4} S cm^{-1} at 140 °C. The conductivity mechanism is assumed to be based on proton conductivity arising from the partial protonation of the cluster anion, making the material interesting for applications such as proton transfer in fuel cells. It should be noted, however, that the IL properties are determined at least partially by the bulky PEG-based cation [98].

A different approach to true POM-ILs was reported in 2007 when a range of tetraalkylphosphonium POM-based POM-ILs were presented. Here, the choice of organic cation was rather straightforward as tetraalkylphosphonium cations of POMs are renowned for their facile preparation as well as their high thermal stability [105, 106]. The authors screened a range of imidazolium and phosphonium cations, several of which led only to high melting-point (>200 °C) solids, which were not further investigated. However, the combination of trihexyltetradecylphosphonium, $(C_{14}H_{29})(C_6H_{13})_3P^+$ cations with Keggin $[PW_{12}O_{40}]^{3-}$ and Lindqvist

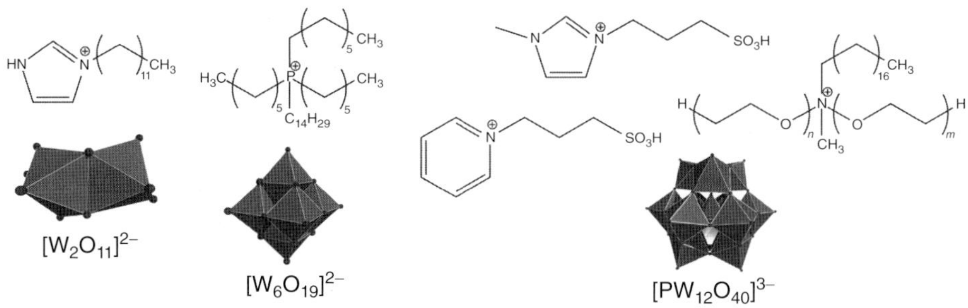

Figure 9.10 Representative examples of POM-ILs reported in the literature for catalysis and materials science [98, 107–109].

$[W_6O_{19}]^{2-}$ anions gave true POM-ILs with melting points of 65 °C (Keggin) and −48 °C (Lindqvist), respectively. Further work was focused on the characterization of the Lindqvist-based room-temperature POM-IL $((C_{14}H_{29})(C_6H_{13})_3P)_2[W_6O_{19}]$. Thermal stability measurements showed short-term stability to about 480 °C; long-term heating experiments gave long-term stability up to 360 °C. In the wake of these groundbreaking initial studies, POM-IL synthesis was expanded to involve a range of cationic and anionic components. The organic cations were based mostly on functionalized imidazolium salts whereas POM clusters were focused mainly on the well-known Keggin-based systems that are renowned for high catalytic reactivity [15].

In a landmark paper, the use of POM-ILs as highly active and selective, self-separating epoxidation catalysts was reported [107]. The authors synthesized three POM-ILs based on three organic cations and the plenary Keggin-cluster anion $[PW_{12}O_{40}]^{3-}$ (see Figure 9.10) and demonstrated facile separation of catalyst and reaction products after the reaction was complete. It should be noted that these compounds are not ILs truly but should rather be considered as molten salts, because their melting points were higher than 100 °C [107–109].

In addition to catalysis applications, POM-ILs are in demand as redox-active electrolyte or electrode materials. This has been demonstrated in several recent reports, which focused mainly on the design of redox-active electrodes based on Keggin-type POM clusters [110–115]. The reports demonstrated that POM-IL modified electrodes are not only highly redox-active but can also be employed as stable electrocatalysts for oxidative [110, 113] and reductive [110] catalysis. Further, electrochemical sensor systems capable of detecting several chemical species have been developed based on POM-IL modified carbonelectrodes [113, 114].

9.4.1.2 **Outlook: Future Applications of POM-ILs**

Thus far, true POM-ILs have mainly been used in catalysis and organic synthesis. However, POM-ILs also hold great promise in two energy-related fields as outlined below:

1) *Photocatalysis*: POMs are well-known photocatalysts and can be used to catalyze a wide variety of complex photochemical reactions [11, 22, 28, 32]. Recently,

it has been demonstrated that POMs dissolved in ILs can even be used as photochemical water oxidation catalysts, as IL solvents have dramatic effects on the redox-potentials of substrates [107]. Thus, photochemistry in POM-ILs remains a largely unexplored and highly promising field for future studies.

2) *Electrochemistry*: As POMs are inherently highly redox-active, they are ideal candidates to develop redox-active electrodes for sensing and electrocatalysis [99, 110–120]. This has been shown in several reports recently where electrode surfaces were modified by ILs containing a range of POMs; thus far the main focus was set on Keggin anion derivatives; however, future work can employ a wide variety of redox-active POM anions.

9.4.2 POM-Based Photovoltaics

To date, photovoltaics are the state-of-the-art technology for the conversion of direct sunlight into electricity. Besides classical solid-state silicon semiconductor-based systems, photoelectrochemical solar cells (e.g., dye-sensitized solar cells) receive much attention as their components can be designed on the molecular level for optimized stability and performance.

Recently, POM anions have been employed as redox-active components for the assembly of photoelectrical cells for sunlight to electricity conversion. In one recent example, Dawson anions $[P_2FeW_{17}O_{61}]^{7-}$ were electrostatically assembled with porphyrin cations using the so-called layer-by-layer approach. The resulting device was tested under visible-light-irradiation and stable photocurrent generation was observed on irradiation [121].

Further, several recent reports demonstrate that POMs, particularly Keggin anions, can be used to optimize the performance of solid-state titania or zinc oxide photoanodes in photoelectrochemical cells [38, 44, 45, 122]. Studies show that both photocurrent and energy conversion are increased on incorporation of the POM clusters, and this is attributed to a POM-induced enhancement of electron-hole separation, resulting in higher photoelectrochemical performance [38, 44, 45, 122].

Thus far, the clusters applied in this field have mainly been the prototype Keggin and Dawson anions and no chemical optimization of cluster structure and reactivity have been performed. This demonstrates that the field is wide open, and effective interdisciplinary projects between chemists, material scientists, and engineers can lead to the development of a new class of stable, economically viable solar cells based partly on POMs.

9.4.3 POM-Based Molecular Cluster Batteries

Electricity generation is growing more diversified day by day. Particularly, fossil-fuel independent electricity production is dependent on natural phenomena (wind, sun, tides, etc.) and maximum electricity production does not necessarily coincide with maximum electricity demand. For this reason, efficient electrochemical energy storage systems are vital for domestic, industrial, and transportation use.

A straightforward concept for the development of high capacity battery systems is the incorporation of highly redox-active metal oxides as electron storage units. Recently, this concept has been expanded into the realm of molecular metal oxide clusters. This offers several advantages, as the clusters can be designed and optimized at the molecular level; further, contacting between cluster and electrode is also possible at the nanoscale, making charging/discharging processes faster than for conventional solid-state oxide materials.

Recently, Awaga and coworkers presented a manganese-based concept study where a dodecanuclear cluster compound, $[Mn_{12}O_{12}(CH_3COO)_{16}(H_2O)_4]$, was employed as redox-active electron storage site [123, 124]. The resulting battery prototype showed high specific capacity (up to 200–300 A h kg^{-1}) on the first discharging and moderate capacity of about 70–90 A h kg^{-1} on further charging–discharging cycles. The drop in the capacity was assigned to cluster degradation. To overcome this, the authors employed Keggin anions, $[PMo_{12}O_{40}]^{3-}$, as electron storage site in an improved prototype system. Here, the cluster anions were deposited on carbon nanotubes (CNTs) so that each cluster is in contact with the conductive electrode material, thus providing optimum conditions for electron (and Li^+-counterion) transport. The material obtained was compared with a bulk mixture of POM and CNT without depositing the clusters on the CNTs on the molecular level. Comparison of both materials showed that the molecular deposition of the clusters results in a significantly enhanced capacity (320 A h kg^{-1} compared with 240 A h kg^{-1}) showing that the concept holds great promise for the design of molecular cluster batteries [125, 126].

9.5 Summary

In this chapter, an overview of POM materials for energy conversion and storage is given. Several areas are highlighted, ranging from visible-light photocatalysis and synthetic strategies for optimized light absorption to WOC and reductive hydrogen generation and CO_2 activation. An outlook is also provided, highlighting several key areas where current and future research can make significant contributions to the development of new molecular materials for clean energy systems.

References

1. Jackson, S.D. and Hargreaves, J.S.J. (eds) (2008) *Metal Oxide Catalysis*, Wiley-VCH Verlag GmbH, Weinheim.
2. Pope, M.T. (1983) *Heteropoly and Isopoly Oxometalates*, Springer-Verlag, Heidelberg.
3. Long, D.L., Burkholder, E., and Cronin, L. (2007) Polyoxometalate clusters, nanostructures and materials: from self assembly to designer materials and devices. *Chem. Soc. Rev.*, **36**, 105–121.
4. Long, D.L., Tsunashima, R., and Cronin, L. (2010) Polyoxometalates: building blocks for functional nanoscale systems. *Angew. Chem. Int. Ed.*, **49**, 1736–1758.
5. Pope, M.T. and Müller, A. (1991) Polyoxometalate chemistry – an old field

with new dimensions in several disciplines. *Angew. Chem., Int. Ed. Engl.*, **30**, 34–48.
6. Müller, A., Beckmann, E., Bögge, H., Schmidtmann, M., and Dress, A. (2002) Inorganic chemistry goes protein size: A Mo-368 nano-hedgehog initiating nanochemistry by symmetry breaking. *Angew. Chem. Int. Ed.*, **41**, 1162–1167.
7. Hayashi, Y. (2011) Hetero and lacunary polyoxovanadate chemistry: synthesis, reactivity and structural aspects. *Coord. Chem. Rev.*, **255**, 2270–2280.
8. Weinstock, I.A. (1998) Homogeneous-phase electron-transfer reactions of polyoxometalates. *Chem. Rev.*, **98**, 113–170.
9. Sadakane, M. and Steckhan, E. (1998) Electrochemical properties of polyoxometalates as electrocatalysts. *Chem. Rev.*, **98**, 219–237.
10. Yamase, T. (1998) Photo- and electrochromism of polyoxometalates. *Chem. Rev.*, **98**, 307–325.
11. Streb, C. (2012) New trends in polyoxometalate photoredox chemistry: from photosensitisation to water oxidation catalysis. *Dalton Trans.*, **41**, 1651–1659.
12. Hill, C.L. and Prosser-McCartha, C.M. (1993) in *Photosensitization and Photocatalysis Using Inorganic and Organometallic Compounds* (eds M. Grätzel and K. Kalyanasundaram), Kluwer Academic Publishers, Dordrecht, pp. 307–326.
13. Papaconstantinou, E. (1989) Photochemistry of polyoxometallates of molybdenum and tungsten and-or vanadium. *Chem. Soc. Rev.*, **18**, 1–31.
14. Duncan, D.C., Netzel, T.L., and Hill, C.L. (1995) Early-time dynamics and reactivity of polyoxometalate excited-states – identification of a short-lived LMCT excited-state and a reactive long-lived charge-transfer intermediate following picosecond flash excitation of $[W_{10}O_{32}]^{4-}$ in acetonitrile. *Inorg. Chem.*, **34**, 4640–4646.
15. Hill, C.L. (2003) POM reactivity. *Compre. Coord. Chem. II*, **4**, 679–759.
16. Jaynes, B.S. and Hill, C.L. (1995) Radical carbonylation of alkanes via polyoxotungstate photocatalysis. *J. Am. Chem. Soc.*, **117**, 4704–4705.
17. Yamase, T. and Usami, T. (1988) Photocatalytic dimerization of olefins by decatungstate(VI), $[W_{10}O_{32}]^{4-}$, in acetonitrile and magnetic-resonance studies of photoreduced species. *J. Chem. Soc., Dalton Trans.*, 183–190.
18. Mylonas, A. and Papaconstantinou, E. (1996) On the mechanism of photocatalytic degradation of chlorinated phenols to CO_2 and HCl by polyoxometalates. *J. Photochem. Photobiol., A*, **94**, 77–82.
19. Fox, M.A., Cardona, R., and Gaillard, E. (1987) Photoactivation of metal-oxide surfaces – photocatalyzed oxidation of alcohols by heteropolytungstates. *J. Am. Chem. Soc.*, **109**, 6347–6354.
20. Tanielian, C., Duffy, K., and Jones, A. (1997) Kinetic and mechanistic aspects of photocatalysis by polyoxotungstates: a laser flash photolysis, pulse radiolysis, and continuous photolysis study. *J. Phys. Chem. B*, **101**, 4276–4282.
21. Bouchard, D.A. and Hill, C.L. (1985) Catalytic photochemical dehydrogenation of organic substrates by polyoxometalates. *J. Am. Chem. Soc.*, **107**, 5148–5157.
22. Forster, J., Rosner, B., Khusniyarov, M.M., and Streb, C. (2011) Tuning the light absorption of a molecular vanadium oxide system for enhanced photooxidation performance. *Chem. Commun.*, **47**, 3114–3116.
23. Errington, R.J., Coyle, L., Middleton, P.S., Murphy, C.J., Clegg, W., and Harrington, R.W. (2010) Synthesis and structure of the alkoxido-titanium pentamolybdate $(^{n}Bu_4N)_3[(^{i}PrO)TiMo_5O_{18}]$: an entry into systematic $TiMo_5$ reactivity. *J. Cluster Sci.*, **21**, 503–514.
24. Errington, R.J., Harle, G., Clegg, W., and Harrington, R.W. (2009) Extending the Lindqvist family to late 3d transition metals: a rational entry to CoW_5 hexametalate chemistry. *Eur. J. Inorg. Chem.*, **2009**, 5240–5246.
25. Errington, R.J., Petkar, S.S., Horrocks, B.R., Houlton, A., Lie, L.H., and Patole, S.N. (2005) Covalent immobilization of a TiW_5 polyoxometalate on derivatized

silicon surfaces. *Angew. Chem. Int. Ed.*, **44**, 1254–1257.

26. Errington, R.J., Petkar, S.S., Middleton, P.S., and McFarlane, W. (2007) Synthesis and reactivity of the methoxozirconium pentatungstate $(^{n}Bu_4N)_6[\{(\mu\text{-MeO})ZrW_5O_{18}\}_2]$: insights into proton-transfer reactions, solution dynamics, and assembly of $\{ZrW_5O_{18}\}^{2-}$ building blocks. *J. Am. Chem. Soc.*, **129**, 12181–12196.
27. Filowitz, M., Ho, R.K.C., Klemperer, W.G., and Shum, W. (1979) O-17 nuclear magnetic-resonance spectroscopy of polyoxometalates. 1. Sensitivity and resolution. *Inorg. Chem.*, **18**, 93–103.
28. Tucher, J., Wu, Y.L., Nye, L.C., Ivanovic-Burmazovic, I., Khusniyarov, M.M., and Streb, C. (2012) Metal substitution in a Lindqvist polyoxometalate leads to improved photocatalytic performance. *Dalton Trans.*, **41**, 9938–9943.
29. Kudo, A., Omori, K., and Kato, H. (1999) A novel aqueous process for preparation of crystal form-controlled and highly crystalline $BiVO_4$ powder from layered vanadates at room temperature and its photocatalytic and photophysical properties. *J. Am. Chem. Soc.*, **121**, 11459–11467.
30. Xu, L., Gao, G.G., Li, F.Y., Liu, X.Z., and Yang, Y.Y. (2008) CO_2 coordination by inorganic polyoxoanion in water. *J. Am. Chem. Soc.*, **130**, 10838–10839.
31. Xi, G.C. and Ye, J.H. (2010) Synthesis of bismuth vanadate nanoplates with exposed {001} facets and enhanced visible-light photocatalytic properties. *Chem. Commun.*, **46**, 1893–1895.
32. Tucher, J., Nye, L.C., Ivanovic-Burmazovic, I., Notarnicola, A., and Streb, C. (2012) Chemical and photochemical functionality of the first molecular bismuth vanadium oxide. *Chem. Eur. J.*, **18**, 10949–10953.
33. Antonaraki, S., Triantis, T.M., Papaconstantinou, E., and Hiskia, A. (2010) Photocatalytic degradation of lindane by polyoxometalates: intermediates and mechanistic aspects. *Catal. Today*, **151**, 119–124.
34. Gu, C. and Shannon, C. (2007) Investigation of the photocatalytic activity of TiO_2-polyoxometalate systems for the oxidation of methanol. *J. Mol. Catal. A*, **262**, 185–189.
35. Guo, Y.H. and Hu, C.W. (2007) Heterogeneous photocatalysis by solid polyoxometalates. *J. Mol. Catal. A*, **262**, 136–148.
36. Li, J.H., Kang, W.L., Yang, X., Yu, X.D., Xu, L.L., Guo, Y.H., Fang, L.B., and Zhang, S.D. (2010) Mesoporous titania-based $H_3PW_{12}O_{40}$ composite by a block copolymer surfactant-assisted templating route: preparation, characterization, and heterogeneous photocatalytic properties. *Desalination*, **255**, 107–116.
37. Li, K.X., Guo, Y.N., Ma, F.Y., Li, H.C., Chen, L., and Guo, Y.H. (2010) Design of ordered mesoporous $H_3PW_{12}O_{40}$-titania materials and their photocatalytic activity to dye methyl orange degradation. *Catal. Commun.*, **11**, 839–843.
38. Luo, X.Z., Li, F.Y., Xu, B.B., Sun, Z.X., and Xu, L. (2012) Enhanced photovoltaic response of the first polyoxometalate-modified zinc oxide photoanode for solar cell application. *J. Mater. Chem.*, **22**, 15050–15055.
39. Marci, G., Garcia-Lopez, E., and Palmisano, L. (2009) Photo-assisted degradation of 2-propanol in gas-solid regime by using TiO_2 impregnated with heteropolyacid $H_3PW_{12}O_{40}$. *Catal. Today*, **144**, 42–47.
40. Nakajima, A., Koike, T., Yanagida, S., Isobe, T., Kameshima, Y., and Okada, K. (2010) Preparation and photocatalytic activity of $[PW_xMo_{12-x}O_{40}]^{3-}/TiO_2$ hybrid film composites. *Appl. Catal., A*, **385**, 130–135.
41. Sun, Z.X., Li, F.Y., Xu, L., Liu, S.P., Zhao, M.L., and Xu, B.B. (2012) Effects of Dawson-type tungstophosphate on photoelectrochemical responses of cadmium sulfide composite film. *J. Phys. Chem. C*, **116**, 6420–6426.
42. Tachikawa, T., Tojo, S., Fujitsuka, M., and Majima, T. (2006) One-electron redox processes during

polyoxometalate-mediated photocatalytic reactions of TiO_2 studied by two-color two-laser flash photolysis. *Chem. Eur. J.*, **12**, 3124–3131.
43. Tangestaninejad, S., Moghadam, M., Mirkhani, V., Mohammadpoor-Baltork, I., and Salavati, H. (2010) Vanadium-containing polyphosphomolybdate immobilized on TiO_2 nanoparticles: a recoverable and efficient catalyst for photochemical, sonochemical and photosonochemical degradation of dyes under irradiation of UV light. *J. Iran. Chem. Soc.*, **7**, 161–S174.
44. Wang, L.H., Xu, L., Mu, Z.C., Wang, C.A., and Sun, Z.X. (2012) Synergistic enhancement of photovoltaic performance of TiO_2 photoanodes by incorporation of Dawson-type polyoxometalate and gold nanoparticles. *J. Mater. Chem.*, **22**, 23627–23632.
45. Wang, S.M., Liu, L., Chen, W.L., Wang, E.B., and Su, Z.M. (2013) Polyoxometalate-anatase TiO_2 composites are introduced into the photoanode of dye-sensitized solar cells to retard the recombination and increase the electron lifetime. *Dalton Trans.*, **42**, 2691–2695.
46. Yanagida, S., Nakajima, A., Sasaki, T., Isobe, T., Kameshima, Y., and Okada, K. (2009) Preparation and photocatalytic activity of Keggin-ion tungstate and TiO_2 hybrid layer-by-layer film composites. *Appl. Catal., A*, **366**, 148–153.
47. Yang, H.X., Liu, T.F., Cao, M.N., Li, H.F., Gao, S.Y., and Cao, R. (2010) A water-insoluble and visible light induced polyoxometalate-based photocatalyst. *Chem. Commun.*, **46**, 2429–2431.
48. Lewis, N.S. and Nocera, D.G. (2006) Powering the planet. *Proc. Natl. Acad. Sci. U.S.A.*, **103**, 15729–15735.
49. Armaroli, N. and Balzani, V. (2007) The future of energy supply: challenges and opportunities. *Angew. Chem. Int. Ed.*, **46**, 52–66.
50. McAlpin, J.G., Stich, T.A., Casey, W.H., and Britt, R.D. (2012) Comparison of cobalt and manganese in the chemistry of water oxidation. *Coord. Chem. Rev.*, **256**, 2445–2452.
51. Cao, R., Lai, W., and Du, P. (2012) Catalytic water oxidation at single metal sites. *Energy Environ. Sci.*, **5**, 8134–8157.
52. Umena, Y., Kawakami, K., Shen, J.R., and Kamiya, N. (2011) Crystal structure of oxygen-evolving photosystem II at a resolution of 1.9 angstrom. *Nature*, **473**, 55–65.
53. Fontecave, M. and Artero, V. (2013) Solar fuel generation and molecular catalysis: is it homogeneous or heterogeneous catalysis. *Chem. Soc. Rev.*, **42**, 2338–2356.
54. Crabtree, R.H. (2012) Resolving heterogeneity problems and impurity artifacts in operationally homogeneous transition metal catalysts. *Chem. Rev.*, **112**, 1536–1554.
55. Lv, H.J., Geletii, Y.V., Zhao, C.C., Vickers, J.W., Zhu, G.B., Luo, Z., Song, J., Lian, T.Q., Musaev, D.G., and Hill, C.L. (2012) Polyoxometalate water oxidation catalysts and the production of green fuel. *Chem. Soc. Rev.*, **41**, 7572–7589.
56. Sartorel, A., Bonchio, M., Campagna, S., and Scandola, F. (2013) Tetrametallic molecular catalysts for photochemical water oxidation. *Chem. Soc. Rev.*, **42**, 2262–2280.
57. Limburg, B., Bouwman, E., and Bonnet, S. (2012) Molecular water oxidation catalysts based on transition metals and their decomposition pathways. *Coord. Chem. Rev.*, **256**, 1451–1467.
58. Stracke, J.J. and Finke, R.G. (2011) Electrocatalytic water oxidation beginning with the cobalt polyoxometalate $Co_4(H_2O)_2(PW_9O_{34})_2^{10-}$: identification of heterogeneous CoO_x as the dominant catalyst. *J. Am. Chem. Soc.*, **133**, 14872–14875.
59. Geletii, Y.V., Huang, Z.Q., Hou, Y., Musaev, D.G., Lian, T.Q., and Hill, C.L. (2009) Homogeneous light-driven water oxidation catalyzed by a tetraruthenium complex with all inorganic ligands. *J. Am. Chem. Soc.*, **131**, 7522–7523.
60. Huang, Z.Q., Luo, Z., Geletii, Y.V., Vickers, J.W., Yin, Q.S., Wu, D., Hou, Y., Ding, Y., Song, J., Musaev, D.G.,

Hill, C.L., and Lian, T.Q. (2011) Efficient light-driven carbon-free cobalt-based molecular catalyst for water oxidation. *J. Am. Chem. Soc.*, **133**, 2068–2071.

61. Puntoriero, F., La Ganga, G., Sartorel, A., Carraro, M., Scorrano, G., Bonchio, M., and Campagna, S. (2010) Photo-induced water oxidation with tetranuclear ruthenium sensitizer and catalyst: a unique 4 × 4 ruthenium interplay triggering high efficiency with low-energy visible light. *Chem. Commun.*, **46**, 4725–4727.

62. Botar, B., Geletii, Y.V., Kögerler, P., Hillesheim, D.A., Musaev, D.G., and Hill, C.L. (2008) An all-inorganic, stable, and highly active tetraruthenium homogeneous catalyst for water oxidation. *Angew. Chem. Int. Ed.*, **47**, 3896–3899.

63. Kuznetsov, A.E., Geletii, Y.V., Hill, C.L., Morokuma, K., and Musaev, D.G. (2009) Dioxygen and water activation processes on multi-Ru-substituted polyoxometalates: comparison with the "blue-dimer" water oxidation catalyst. *J. Am. Chem. Soc.*, **131**, 6844–6854.

64. Yin, Q.S., Tan, J.M., Besson, C., Geletii, Y.V., Musaev, D.G., Kuznetsov, A.E., Luo, Z., Hardcastle, K.I., and Hill, C.L. (2010) A fast soluble carbon-free molecular water oxidation catalyst based on abundant metals. *Science*, **328**, 342–345.

65. Orlandi, M., Argazzi, R., Sartorel, A., Carraro, M., Scorrano, G., Bonchio, M., and Scandola, F. (2010) Ruthenium polyoxometalate water splitting catalyst: very fast hole scavenging from photogenerated oxidants. *Chem. Commun.*, **46**, 3152–3154.

66. Sartorel, A., Carraro, M., Scorrano, G., De Zorzi, R., Geremia, S., McDaniel, N.D., Bernhard, S., and Bonchio, M. (2008) Polyoxometalate embedding of a tetraruthenium(IV)-oxo-core by template-directed metalation of $[\gamma\text{-SiW}_{10}O_{36}]^{8-}$: a totally inorganic oxygen-evolving catalyst. *J. Am. Chem. Soc.*, **130**, 5006–5007.

67. Zhang, Z.Y., Lin, Q.P., Kurunthu, D., Wu, T., Zuo, F., Zheng, S.T., Bardeen, C.J., Bu, X.H., and Feng, P.Y. (2011) Synthesis and photocatalytic properties of a new heteropolyoxoniobate compound: $K_{10}[Nb_2O_2(H_2O)_2][SiNb_{12}O_{40}] \times 12H_2O$. *J. Am. Chem. Soc.*, **133**, 6934–6937.

68. Goberna-Ferrón, S., Vigara, L., Soriano-López, J., and Galán-Mascarós, J.R. (2012) Identification of a nonanuclear $\{Co^{II}{}_9\}$ polyoxometalate cluster as a homogeneous catalyst for water oxidation. *Inorg. Chem.*, **51**, 11707–11715.

69. Zhu, G., Geletii, Y.V., Kögerler, P., Schilder, H., Song, J., Lense, S., Zhao, C., Hardcastle, K.I., Musaev, D.G., and Hill, C.L. (2012) Water oxidation catalyzed by a new tetracobalt-substituted polyoxometalate complex: $[\{Co_4(\mu\text{-}OH)(H_2O)_3\}(Si_2W_{19}O_{70})]^{11-}$. *Dalton Trans.*, **41**, 2084–2090.

70. Tanaka, S., Annaka, M., and Sakai, K. (2012) Visible light-induced water oxidation catalyzed by molybdenum-based polyoxometalates with mono- and dicobalt(III) cores as oxygen-evolving centers. *Chem. Commun.*, **48**, 1653–1655.

71. Besson, C., Huang, Z., Geletii, Y.V., Lense, S., Hardcastle, K.I., Musaev, D.G., Lian, T., Proust, A., and Hill, C.L. (2010) $Cs_9[\gamma\text{-}PW_{10}O_{36})_2Ru_4O_5(OH)(H_2O)_4]$, a new all-inorganic, soluble catalyst for the efficient visible-light-driven oxidation of water. *Chem. Commun.*, **46**, 2784–2786.

72. Cao, R., Ma, H., Geletii, Y.V., Hardcastle, K.I., and Hill, C.L. (2009) Structurally characterized iridium(III)-containing polytungstate and catalytic water oxidation activity. *Inorg. Chem.*, **48**, 5596–5598.

73. Zhu, G., Glass, E.N., Zhao, C., Lv, H., Vickers, J.W., Geletii, Y.V., Musaev, D.G., Song, J., and Hill, C.L. (2012) A nickel containing polyoxometalate water oxidation catalyst. *Dalton Trans.*, **41**, 13043–13049.

74. Lin, Y. and Finke, R.G. (1994) A more general-approach to distinguishing homogeneous from heterogeneous catalysis - discovery of polyoxoanion-stabilized and Bu_4N^+-stabilized, isolable and redissolvable, high-reactivity Ir-approximate-to-190-450

nanocluster catalysts. *Inorg. Chem.*, **33**, 4891–4910.
75. Widegren, J.A. and Finke, R.G. (2003) A review of the problem of distinguishing true homogeneous catalysis from soluble or other metal-particle heterogeneous catalysis under reducing conditions. *J. Mol. Catal. A*, **198**, 317–341.
76. Bernardini, G., Wedd, A.G., Zhao, C., and Bond, A.M. (2012) Photochemical oxidation of water and reduction of polyoxometalate anions at interfaces of water with ionic liquids or diethylether. *Proc. Natl. Acad. Sci. U.S.A.*, **109**, 11552–11557.
77. Zhao, C. and Bond, A.M. (2009) Photoinduced oxidation of water to oxygen in the ionic liquid $BMIMBF_4$ as the counter reaction in the fabrication of exceptionally long semiconducting silver-tetracyanoquinodimethane nanowires. *J. Am. Chem. Soc.*, **131**, 4279–4287.
78. Natali, M., Berardi, S., Sartorel, A., Bonchio, M., Campagna, S., and Scandola, F. (2012) Is $[Co_4(H_2O)_2(\alpha\text{-}PW_9O_{34})_2]^{10-}$ a genuine molecular catalyst in photochemical water oxidation? Answers from time-resolved hole scavenging experiments. *Chem. Commun.*, **48**, 8808–8810.
79. Cronin, L. and Müller, A. (2012) From serendipity to design of polyoxometalates at the nanoscale, aesthetic beauty and applications. *Chem. Soc. Rev.*, **41**, 7333–7334.
80. Hocking, R.K., Brimblecombe, R., Chang, L.Y., Singh, A., Cheah, M.H., Glover, C., Casey, W.H., and Spiccia, L. (2011) Water-oxidation catalysis by manganese in a geochemical-like cycle. *Nat. Chem.*, **3**, 461–466.
81. Argitis, P. and Papaconstantinou, E. (1985) Photocatalytic multielectron photoreduction of 18-tungstodiphosphate in the presence of organic-compounds – production of hydrogen. *J. Photochem.*, **30**, 445–451.
82. Ioannidis, A. and Papaconstantinou, E. (1985) Photocatalytic generation of hydrogen by 1:12 heteropolytungstates with concomitant oxidation of organic-compounds. *Inorg. Chem.*, **24**, 439–441.
83. Papaconstantinou, E. (1982) Photocatalytic oxidation of organic-compounds using heteropoly electrolytes of molybdenum and tungsten. *J. Chem. Soc., Chem. Commun.*, 12–13.
84. Muradov, N. and T-Raissi, A. (2006) Solar production of hydrogen using "self-assembled" polyoxometalate photocatalysts. *J. Sol. Energy*, **128**, 326–330.
85. Szczepankiewicz, S.H., Ippolito, C.M., Santora, B.P., Van de Ven, T.J., Ippolito, G.A., Fronckowiak, L., Wiatrowski, F., Power, T., and Kozik, M. (1998) Interaction of carbon dioxide with transition-metal-substituted heteropolyanions in nonpolar solvents. Spectroscopic evidence for complex formation. *Inorg. Chem.*, **37**, 4344–4352.
86. Khenkin, A.M., Efremenko, I., Weiner, L., Martin, J.M.L., and Neumann, R. (2010) Photochemical reduction of carbon dioxide catalyzed by a ruthenium-substituted polyoxometalate. *Chem. Eur. J.*, **16**, 1356–1364.
87. Ettedgui, J., Diskin-Posner, Y., Weiner, L., and Neumann, R. (2011) Photoreduction of carbon dioxide to carbon monoxide with hydrogen catalyzed by a rhenium(I) phenanthroline-polyoxometalate hybrid complex. *J. Am. Chem. Soc.*, **133**, 188–190.
88. Hallett, J.P. and Welton, T. (2011) Room-temperature ionic liquids: solvents for synthesis and catalysis. 2. *Chem. Rev.*, **111**, 3508–3576.
89. Welton, T. (1999) Room-temperature ionic liquids. Solvents for synthesis and catalysis. *Chem. Rev.*, **99**, 2071–2083.
90. Zhao, D., Wu, M., Kou, Y., and Min, E. (2002) Ionic liquids: applications in catalysis. *Catal. Today*, **74**, 157–189.
91. Dupont, J., de Souza, R.F., and Suarez, P.A. (2002) Ionic liquid (molten salt) phase organometallic catalysis. *Chem. Rev.*, **102**, 3667–3692.
92. Pârvulescu, V.I. and Hardacre, C. (2007) Catalysis in ionic liquids. *Chem. Rev.*, **107**, 2615–2665.
93. Welton, T. (2004) Ionic liquids in catalysis. *Coord. Chem. Rev.*, **248**, 2459–2477.

94. Buzzeo, M.C., Evans, R.G., and Compton, R.G. (2004) Non-haloaluminate room-temperature ionic liquids in electrochemistry - A review. *ChemPhysChem*, **5**, 1106–1120.
95. Galiński, M., Lewandowski, A., and Stępniak, I. (2006) Ionic liquids as electrolytes. *Electrochim. Acta*, **51**, 5567–5580.
96. Barrosse-Antle, L.E., Bond, A.M., Compton, R.G., O'Mahony, A.M., Rogers, E.I., and Silvester, D.S. (2010) Voltammetry in room temperature ionic liquids: comparisons and contrasts with conventional electrochemical solvents. *Chem. Asian J.*, **5**, 202–230.
97. Wasserscheid, P. and Welton, T. (2003) *Ionic Liquids in Synthesis*, vol. 1, Wiley Online Library.
98. Smarsly, B. and Kaper, H. (2005) Liquid inorganic-organic nanocomposites: novel electrolytes and ferrofluids. *Angew. Chem. Int. Ed.*, **44**, 3809–3811.
99. Bhatt, A.I., Bond, A.M., MacFarlane, D.R., Zhang, J., Scott, J.L., Strauss, C.R., Iotov, P.I., and Kalcheva, S.V. (2006) A critical assessment of electrochemistry in a distillable room temperature ionic liquid, DIMCARB. *Green Chem.*, **8**, 161–171.
100. Wasserscheid, P. and Keim, W. (2000) Ionic liquids-new "solutions" for transition metal catalysis. *Angew. Chem.*, **39**, 3772–3789.
101. Pandey, S. (2006) Analytical applications of room-temperature ionic liquids: a review of recent efforts. *Anal. Chim. Acta*, **556**, 38–45.
102. Wei, D. and Ivaska, A. (2008) Applications of ionic liquids in electrochemical sensors. *Anal. Chim. Acta*, **607**, 126–135.
103. Buzzeo, M.C., Hardacre, C., and Richard, G. (2004) Use of room temperature ionic liquids in gas sensor design. *Anal. Chem.*, **76**, 4583–4588.
104. Bourlinos, A.B., Raman, K., Herrera, R., Zhang, Q., Archer, L.A., and Giannelis, E.P. (2004) A liquid derivative of 12-tungstophosphoric acid with unusually high conductivity. *J. Am. Chem. Soc.*, **126**, 15358–15359.
105. Rickert, P.G., Antonio, M.R., Firestone, M.A., Kubatko, K.A., Szreder, T., Wishart, J.F., and Dietz, M.L. (2007) Tetraalkylphosphonium polyoxometalate ionic liquids: novel, organic-inorganic hybrid materials. *J. Phys. Chem. B*, **111**, 4685–4692.
106. Rickert, P.G., Antonio, M.R., Firestone, M.A., Kubatko, K.A., Szreder, T., Wishart, J.F., and Dietz, M.L. (2007) Tetraalkylphosphonium polyoxometalates: electroactive, "task-specific" ionic liquids. *Dalton Trans.*, 529–531.
107. Leng, Y., Wang, J., Zhu, D.R., Ren, X.Q., Ge, H.Q., and Shen, L. (2009) Heteropolyanion-based ionic liquids: reaction-induced self-separation catalysts for esterification. *Angew. Chem. Int. Ed.*, **48**, 168–171.
108. Li, K.X., Chen, L., Wang, H.L., Lin, W.B., and Yan, Z.C. (2011) Heteropolyacid salts as self-separation and recyclable catalysts for transesterification of trimethylolpropane. *Appl. Catal., A*, **392**, 233–237.
109. Li, H., Hou, Z.S., Qiao, Y.X., Feng, B., Hu, Y., Wang, X.R., and Zhao, X.G. (2010) Peroxopolyoxometalate-based room temperature ionic liquid as a self-separation catalyst for epoxidation of olefins. *Catal. Commun.*, **11**, 470–475.
110. Huang, B.Q., Wang, L., Shi, K., Xie, Z.X., and Zheng, L.S. (2008) A new strategy for the fabrication of the phosphor polyoxomolybdate modified electrode from ionic liquid solutions and its electrocatalytic activities. *J. Electroanal. Chem.*, **615**, 19–24.
111. Fernandes, D.M., Simoes, S.M.N., Carapuca, H.M., Brett, C.M.A., and Cavaleiro, A.M.V. (2010) Novel poly(hexylmethacrylate) composite carbon electrodes modified with Keggin-type tungstophosphate-tetrabutylammonium salts. *J. Electroanal. Chem.*, **639**, 83–87.
112. Chiang, M.H., Dzielawa, J.A., Dietz, M.L., and Antonio, M.R. (2004) Redox chemistry of the Keggin heteropolyoxotungstate anion in ionic liquids. *J. Electroanal. Chem.*, **567**, 77–84.
113. Ji, H.M., Zhu, L.D., Liang, D.D., Liu, Y., Cai, L., Zhang, S.W., and Liu, S.X. (2009) Use of a 12-molybdovanadate(V) modified ionic liquid carbon paste

electrode as a bifunctional electrochemical sensor. *Electrochim. Acta*, **54**, 7429–7434.
114. Haghighi, B., Hamidi, H., and Gorton, L. (2010) Formation of a robust and stable film comprising ionic liquid and polyoxometalate on glassy carbon electrode modified with multiwalled carbon nanotubes: toward sensitive and fast detection of hydrogen peroxide and iodate. *Electrochim. Acta*, **55**, 4750–4757.
115. Haghighi, B. and Hamidi, H. (2009) Electrochemical characterization and application of carbon ionic liquid electrodes containing 1:12 phosphomolybdic acid. *Electroanalysis*, **21**, 1057–1065.
116. Wang, R.Y., Jia, D.Z., and Cao, Y.L. (2012) Facile synthesis and enhanced electrocatalytic activities of organic-inorganic hybrid ionic liquid polyoxometalate nanomaterials by solid-state chemical reaction. *Electrochim. Acta*, **72**, 101–107.
117. Dong, T., Du, J.B., Cao, M.H., and Hu, C.W. (2010) The electrochemical properties of 12-molybdophosphoric acid modified ionic liquid carbon paste electrode. *J. Cluster Sci.*, **21**, 155–162.
118. Wang, L., Feng, Z.G., and Cai, H.N. (2009) A general strategy for the preparation of polyoxometalate coordination polymer modified electrodes via an ionic liquid route and their electrocatalytic activities. *J. Electroanal. Chem.*, **636**, 36–39.
119. Zhang, J., Bhatt, A.I., Bond, A.M., Wedd, A.G., Scott, J.L., and Strauss, C.R. (2005) Voltammetric studies of polyoxometalate microparticles in contact with the reactive distillable ionic liquid DIMCARB. *Electrochem. Commun.*, **7**, 1283–1290.
120. Liu, H.T., He, P., Li, Z.Y., Sun, C.Y., Shi, L.H., Liu, Y., Zhu, G.Y., and Li, J.H. (2005) An ionic liquid-type carbon paste polyoxometalate-modified electrode and its properties. *ElectroChem. Commun.*, **7**, 1357–1363.
121. Ahmed, I., Farha, R., Goldmann, M., and Ruhlmann, L. (2013) A molecular photovoltaic system based on Dawson type polyoxometalate and porphyrin formed by layer-by-layer self assembly. *Chem. Commun.*, **49**, 496–498.
122. Xie, Y., Zhou, L., and Huang, H. (2007) Enhanced photoelectrocatalytic performance of polyoxometalate-titania nanocomposite photoanode. *Appl. Catal., B*, **76**, 15–23.
123. Yoshikawa, H., Hamanaka, S., Miyoshi, Y., Kondo, Y., Shigematsu, S., Akutagawa, N., Sato, M., Yokoyama, T., and Awaga, K. (2009) Rechargeable batteries driven by redox reactions of Mn12 clusters with structural changes: XAFS analyses of the charging/discharging processes in molecular cluster batteries. *Inorg. Chem.*, **48**, 9057–9059.
124. Yoshikawa, H., Kazama, C., Awaga, K., Satoh, M., and Wada, J. (2007) Rechargeable molecular cluster batteries. *Chem. Commun.*, 3169–3170.
125. Wang, H., Hamanaka, S., Nishimoto, Y., Irle, S., Yokoyama, T., Yoshikawa, H., and Awaga, K. (2012) In operand X-ray absorption fine structure studies of polyoxometalate molecular cluster batteries: polyoxometalates as electron sponges. *J. Am. Chem. Soc.*, **134**, 4918–4924.
126. Kawasaki, N., Wang, H., Nakanishi, R., Hamanaka, S., Kitaura, R., Shinohara, H., Yokoyama, T., Yoshikawa, H., and Awaga, K. (2011) Nanohybridization of polyoxometalate clusters and single-wall carbon nanotubes: applications in molecular cluster batteries. *Angew. Chem. Int. Ed.*, **50**, 3471–3474.
127. Zhang, J., Bond, A.M., MacFarlane, D.R., Forsyth, S.A., Pringle, J.M., Mariotti, A.W.A., Glowinski, A.F., and Wedd, A.G. (2005) Voltammetric studies on the reduction of polyoxometalate anions in ionic liquids. *Inorg. Chem.*, **44**, 5123–5132.

10
The Next Generation of Silylene Ligands for Better Catalysts

Shigeyoshi Inoue

10.1
General Introduction

10.1.1
Silylenes

Silylenes, divalent neutral silicon species and heavier analogs of carbenes, represent a highly challenging topic for several decades and play an integral role with their interesting properties, reactivities, and potential applications in transition metal catalysis (Scheme 10.1) [1–6]. Strikingly, in 1986, Jutzi and coworkers reported the isolation of decamethylsilicocene, which was the first stable monomeric divalent silicon(II) compound [7, 8]. The chemistry of silylenes has been developed vastly since the original synthesis of N-heterocyclic silylenes (NHSi) in 1994 [9]. The advent of donor-stabilized silylenes as well as donor-free silylenes [10–12] has obviously made the study of the reactivity and particularly applications in organic synthesis much more practical [13–17]. Yet, even though stable silylenes are no longer just transient intermediates, they must be taken to synthesize stable species through either thermodynamic or kinetic stabilization. Usually a combination of both is used, as illustrated in many examples where the inclusion of π-donor heteroatoms such as nitrogen and the introduction of bulky substituents work in concert to stabilize the silylene (Scheme 10.1). Nowadays, two-coordinated acyclic silylenes can also be isolated and structurally characterized [18–23]. Silylenes now enjoy indefatigable research attention with many research groups in academia and industry worldwide feverishly pursuing their potential uses in catalysis, synthesis, and stoichiometric transformations. It must also be mentioned that recent studies have expanded to focus attention on their utility as building blocks for the synthesis of novel functional silicon compounds and their role as strikingly versatile coordination ligands toward transition metals. In particular, the chemistry of N-hetrocyclic silylene has been developed dramatically over the past 20 years.

Discovering the Future of Molecular Sciences, First Edition. Edited by Bruno Pignataro.

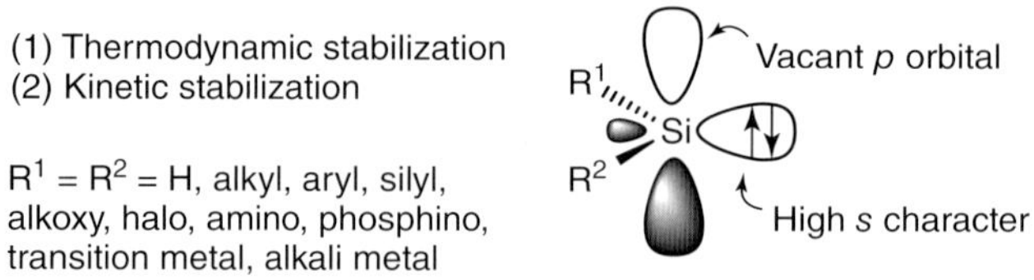

Scheme 10.1 Silylenes.

10.1.2 Bissilylenes

Bissilylene can be categorized as compounds with two divalent silicon atoms (silylene moiety) in a single molecule. While the chemistry of stable silylenes has enjoyed impressive advances [1–6, 13–17], the chemistry of bissilylenes is much less developed [24, 25]. This is probably because of a suitable synthetic method. In general, bissilylenes can be divided into two types of compounds: interconnected bissilylenes **A–C** where the two divalent silicon atoms are directly connected by a central silicon-silicon single bond (Scheme 10.2) [26–29] and spacer-separated bissilylenes (**D**) in which the divalent silicon atoms are separated by a various spacer (Scheme 10.2) [30, 31]. Recently, interconnected bissilylenes have attracted much attention owing to their isoelectronic structure of disilynes having silicon-silicon triple bond. The syntheses of interconnected bissilylenes **A–C** have been accomplished by employing chelate ligand or NHC (N-heterocyclic carbenes). The reactions of interconnected bissilylenes toward transition metal complexes and small organic molecules have also been reported [24, 25]. On the other hand, the

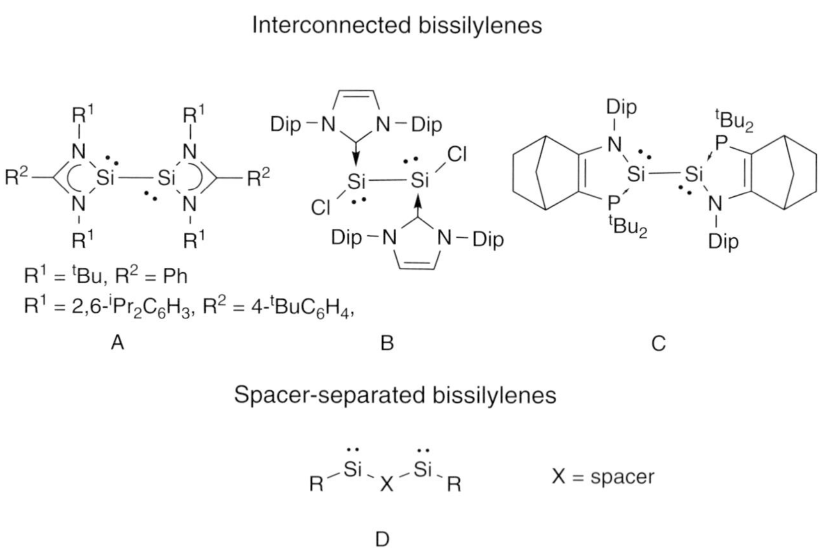

Scheme 10.2 Bissilylenes.

chemistry of spacer-separated bissilylenes is still in its infancy because of the lack of a suitable synthetic method. Although only a few examples of spacer-separated bissilylenes have been synthesized so far, these species would be excellent chelate ligand system featuring two strong σ-donor Si(II) moieties toward transition metal complexes. In other words, the development of new bissilylene ligand system could provide fruitful insights into organometallic chemistry, coordination chemistry, organic chemistry, and main group chemistry.

10.1.3
Silylene Transition Metal Complexes

During the past several decades, complexes between transition metals and low-coordinated carbon have gained interest. For example, a rich chemistry of transition metal carbene complexes ($R_2C{=}ML_n$) **I** has evolved in the quarter century since the original synthesis of $(CO)_5W{=}C(OMe)Me$ by Fischer and Maasböl [32]. The preparation, characterization, and reactivity of a wide variety of such complexes have been documented including, for example, heteroatom-stabilized electrophilic carbene complexes, electrophilic species that lack heteroatom stabilization, early transition metal alkylidene complexes that display nucleophilic character, several examples of simple methylene complexes, and numerous types of binuclear carbene transfers using diazo compounds [33, 34]. Recently, stoichiometric reactions of carbene complexes directed toward useful synthetic transformations have come under careful scrutiny. Fischer and Schrock carbene complexes are certainly among the most versatile organometallic reagents for organic synthesis that have ever been developed. These complexes have been utilized in a large variety of important reaction, the full scope of which is too vast to be listed here, and the discovery process continues unabated. For example, these two types of carbene complexes are known as catalysts for olefin metathesis reactions such as Ring Opening Metathesis Polymerization (ROMP) or Ring Closing Metathesis (RCM) [35–37]. Indeed, a Nobel Prize in this field was awarded to Grubbs, Schrock, and Chauvin in 2005, indicating that these complexes have had a tremendous impact on modern chemistry. Meanwhile, during the past two decades, transition metal silylene complexes have attracted interest as silicon analogs of carbene complexes and as potential intermediates in various metal-catalyzed syntheses and transformations of organosilicon compounds (Scheme 10.3) [38–42]. Silylene complexes ($R_2Si{=}ML_n$) **II** are postulated intermediates in a number of transition

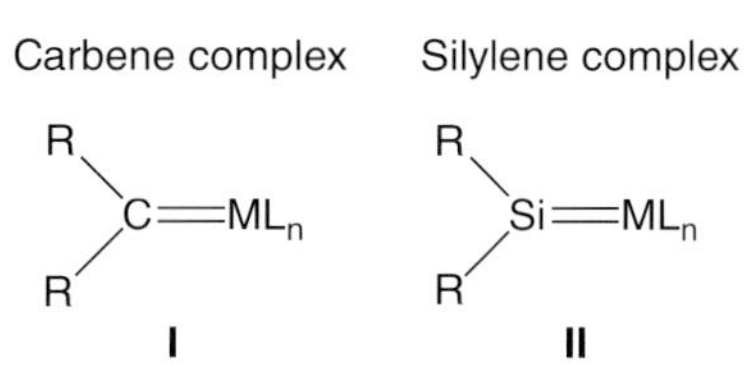

Scheme 10.3 Carbene complex and silylene complex.

metal-mediated transformations, including Rochow's direct process [43], catalytic redistribution of silanes [44–47], and various silylene-transfer reactions [48–51].

Since two base-stabilized silylene complexes, $(CO)_4Fe{=}Si(OtBu)_2{\cdot}HMPA$ (hexamethylphosphoramide) and $[Cp^*(PMe_3)_2Ru{=}SiPh_2{\cdot}NCMe]^+[BPh_4]^-$ (Cp* = pentamethylcyclopentadienyl) [52, 53], were independently isolated in 1987, several studies of base-stabilized or base-free transition metal silylene complexes have been published to date [38–42]. The ability of stable silylene to form transition metal complexes has been well investigated with a wide variety of transition metals ranging from late to early transition metals as well as several lanthanoides. In general, the respective silylene complexes are formed by ligand substitution reaction of hemilabile ligands such as phosphines, carbon monoxide, and 1,5-cyclooctadiene (cod). Alternatively, there are insertion and reduction reactions of stable silylene with transition metals and ligation reactions of silylene to lanthanides. However, the chemistry of silylene complexes, particularly their use as catalysts, has not seen the same explosion in interest as carbene complexes. In fact, only a few examples of catalytic applications have been reported: Suzuki cross-coupling of aryl boronic acids with bromoarenes [54] and Heck coupling of styrene and bromoacetophenone [55], clearly demonstrating their potential and spurring on further development. Because of the low-stability of silylene complexes, a larger variety of these catalytic results is lacking, but these catalytic results have proven their possiblity.

10.2 Synthesis and Catalytic Applications of Silylene Transition Metal Complexes

10.2.1 Bis(silylene)titanium Complexes

The synthesis of Sila-Schrock-type complexes **III** is a particularly challenging area. Since the first two base-stabilized silylene complexes were independently isolated, several studies of base-stabilized and unsupported transition metal Fischer-type silylene complexes **IV**, which typically have electrophilic silicon and nucleophilic metal centers (Scheme 10.4), have been undertaken [52, 53, 56–60]. Although Tilley *et al.* have reported the tungsten-silylene complex $[Cp^*(dmpeH_2W{=}SiR_2)][B(C_6H_5)_4]$ (dmpe = 1,2-bis(dimethyphosphino)ethane)

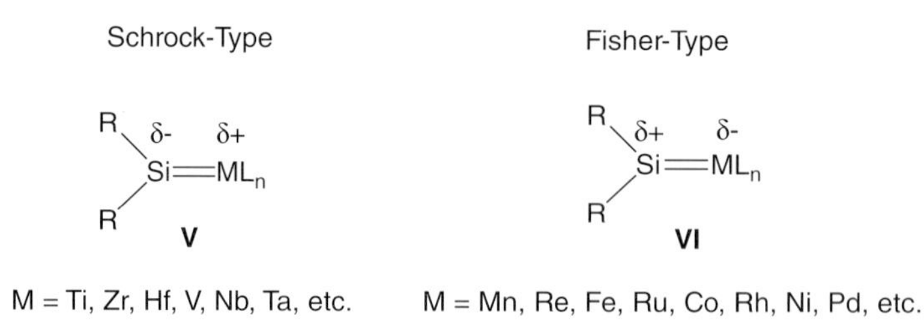

Scheme 10.4 Fisher- and Schrock-type silylene complexes.

[56], little is known about early transition metal silylene complexes, especially from group 4 metals such as titanium, zirconium, and hafnium, owing to a lack of suitable synthetic methods.

In 2006, Sekiguchi and coworkers reported on the synthesis and characterization of a stable Schrock-type hafnium-silylene complex using bulky di-*tert*-butyl(methyl)silyl groups $(\eta^5\text{-}C_5H_4Et)_2Hf(PMe_3){=}Si(SiMe^tBu_2)_2$ **E** (Scheme 10.5) [61]. In addition, very recently, the synthesis and isolation of Schrock-type titanium-silylene complexes **F** featuring Lewis base ligands has been successfully accomplished by Sekiguchi and coworkers (Scheme 10.5) [62]. Interestingly, the reaction of this 18-electron titanium-silylene complexes with terminal alkynes gives silatitanacyclobutanes. In contrast to sila-Fischer-type complexes, sila-Schrock-type complexes are expected to be useful as catalysts for olefin and alkane metathesis reactions because of the higher reactivity of this type of metal-silicon bond.

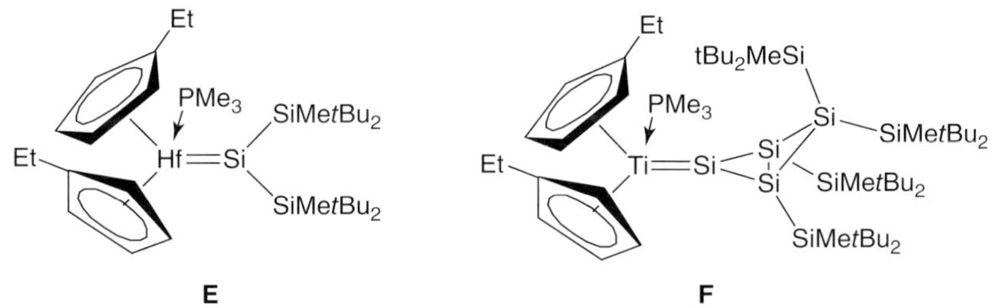

Scheme 10.5 Schrock-type silylene complexes.

On the other hand, the synthesis of bis(silylene) titanium complex is not reported to date. The reaction of *N*-heterocyclic chlorosilylene LSiCl (**1**) [63] [L = PhC(N*t*Bu)$_2$] with the titanium trimethyl phosphine complex $[(\eta^5\text{-}C_5H_5)_2Ti(PMe_3)_2]$ [64] in the molar ration of 2 : 1 in hexane results in the formation of bis(silylene)titanium complex $[(\eta^5\text{-}C_5H_5)_2Ti(LSiCl)_2]$ (**2**), which could be isolated as yellow crystals in 67% yield (Scheme 10.6). Furthermore, reaction of titanium complex **2** with 2 M equiv. of methyllithium MeLi in toluene led to the formation of the new bis(silylene)titanium complex $[(\eta^5\text{-}C_5H_5)_2Ti(LSiMe)_2]$ (**3**) in 57% yield. Equally important, reaction of **2** with 2 M equiv. of $LiBHEt_3$ afforded the first example of a bis(hydridosilylene)titanium complex, $[(\eta^5\text{-}C_5H_5)_2Ti(LSiH)_2]$ (**4**), in 40% yield (Scheme 10.6) [65].

Molecular structures of bis(silylene)titanium complexes **2** and **3** were determined by X-ray crystallography (Figure 10.1). Compounds **2** and **3** possess distorted-tetrahedral silicon centers with bonding angles around silicon ranging from 98.26° to 133.59° in **2** and 94.42° to 124.77° in **3**. The Ti-Si bond length in complexes **2** (2.486(1) Å) and **3** (2.515(1) Å) are significantly shorter than the known Ti–Si single bonds in silyltitanium complexes of 2.59–2.70 Å [66], suggesting some possible multiple bonding character. Multiple bonding character of the Ti-Si bonds in complexes **2**, **3**, and **4** has clearly been elucidated by DFT (density functional theory) calculation including the WBI (Wiberg Bond Index) values (0.999 for **2**; 1.007 for **3**; 1.045 for **5**) and the corresponding bonding orbitals. For example, a

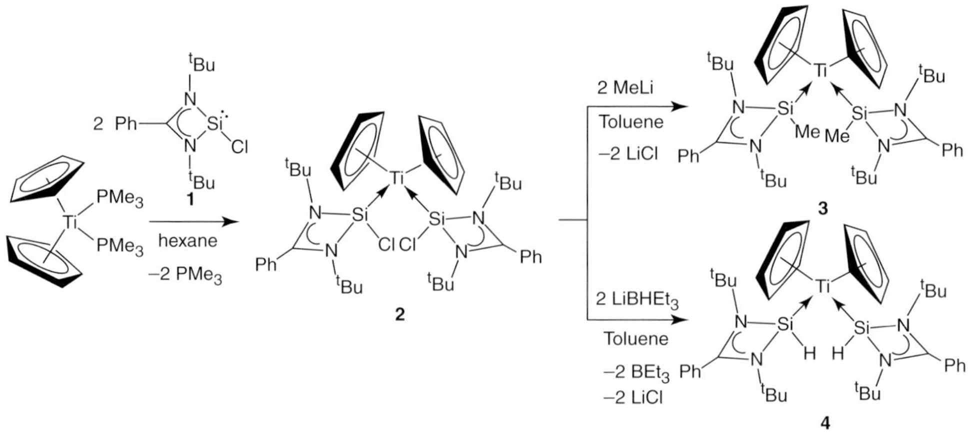

Scheme 10.6 Syntheses of bis(silylene)titanium(II) complexes **2, 3**, and **4**.

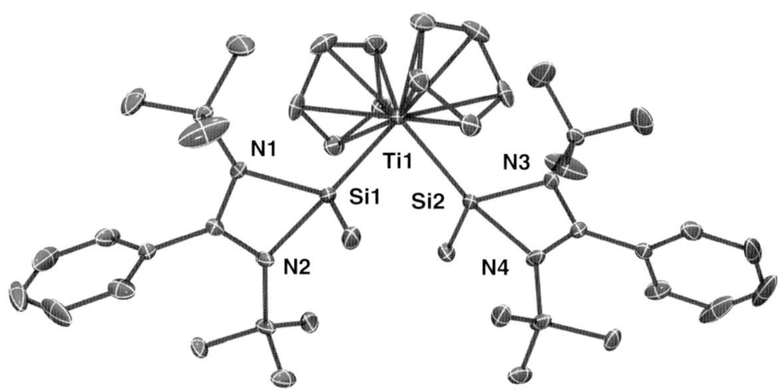

Figure 10.1 Molecular structure of bis(silylene)titanium(II) complex **3**.

good agreement exists for the Ti-Si bond lengths found in the optimized structures (**2**: 2.548 Å, **3**: 2.545 Å, **4**: 2.527 Å). Moreover, two σ-type orbitals and one π-type orbital appear to be delocalized over the Si-Ti-Si framework in compound **2**. This bonding nature is similar to that of the reported cyclic stannylene and germylene transition metal complexes $Cp_2M\{ER_2\}_2$ E = Sn, Pb; M = Ti, Zr, Hf (Cp = cyclopentadienyl) [67]. The bis(silylene) titanium(II) complexes **2–4** represent a compound with Ti-Si multiple bond and promising potential for catalytic reactions such as alkene and alkyne metathesis reactions.

10.2.2
Bis(silylene)nickel Complex

A large number of transition metal silylene complexes have been isolated and characterized since the isolation of two base-stabilized silylene complexes in 1987 [52, 53]. However, bis(silylene) transition metal complexes are still limited to the cyclic diaminosilylene complexes [68–73], cyclic dialkylsilylene complexes [74], base-stabilized silyl(silylene) complexes [59, 75–81], and NHC-stabilized bissilylene complexes [82–84]. In general, it is well known that the properties of the ligands influence greatly the reactivity, stability, and, in consequence, the catalytic performance of transition metal complexes. In order to increase the stability of silylenes and the electron density of the central transition metal, the spacer-separated bissilylene ligands have been employed. Again, bissilylene ligands with two silylene moieties would be ideal for the synthesis of electron rich transition metal complexes due to their strong σ-donor character.

The oxygen-bridged bissilylene **6**, which has two lone pairs, one at each of the silicon centers, would be a suitable target compound as a new bissilylene ligand. Reaction of 1,1,3,3-tetrachlorodisiloxane [$Cl_2SiH{-}O{-}SiHCl_2$] with 2 M equiv. of lithium amidinate LLi [L = PhC(N*t*Bu)$_2$] in diethylether gives the expected disiloxane [LSiH(Cl)–O–SiH(Cl)L] **5** in 53% yield (Scheme 10.7) [85]. The oxygen-bridged bissilylene [LSi–O–SiL] **6** was readily obtained by the dehydrochloration of **5** using 2 M equiv. of $LiN(SiMe_3)_2$ in toluene (Scheme 10.7). Recrystallization in toluene afforded pale yellow crystals of **6** in 76% yield. The molecular structure of **6** was characterized by means of spectroscopy and X-ray crystallography. The Si–O distances in **6** of 1.641(2) and 1.652(2) Å are typical value for silicon-oxygen single bonds of other disiloxanes [86]. Similar to a disiloxane-like system, the Si1–O–Si2 angle of **6** is bent (159.88(15)°) and larger than what was observed for a C–O–C moiety of ethers.

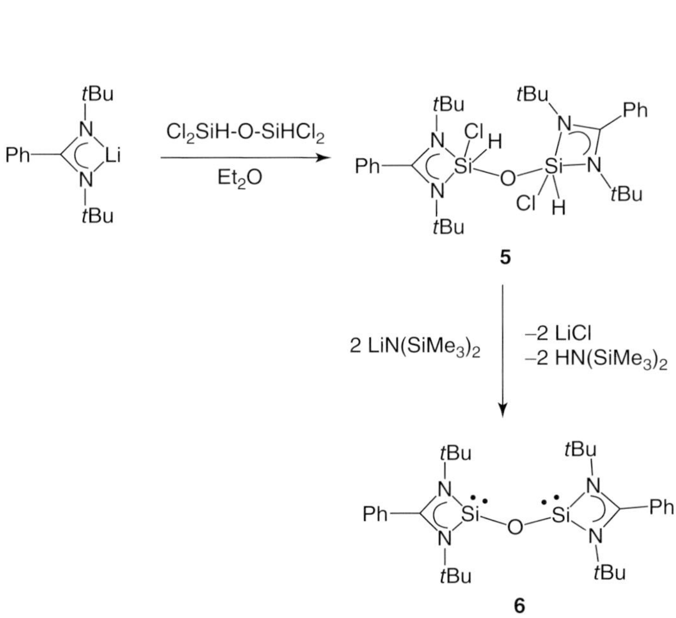

Scheme 10.7 Synthesis of oxygen-bridged bissilylene **7**.

In order to test the chelate coordination ability of oxygen-bridged bissilylene **6**, its reactivity toward [Ni(COD)$_2$] has been investigated. The reaction of the bissilylene **6** with 1 M equiv. of bis(cyclooctadiene)nickel(0) [Ni(COD)$_2$] in toluene at room temperature leads to the oxygen-bridged bis(silylene)nickel complex **7** in 91% yield (Scheme 10.8).

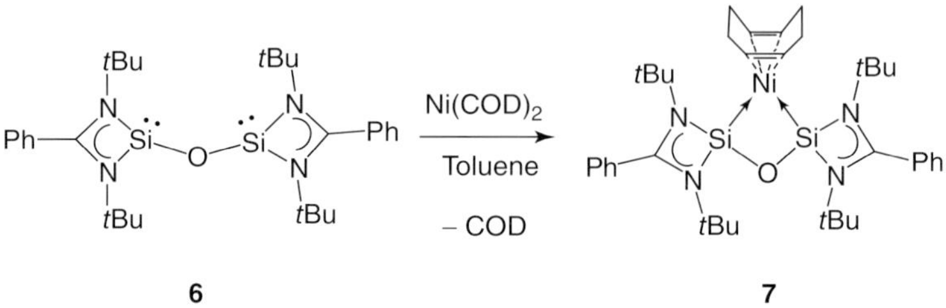

Scheme 10.8 Synthesis of oxygen-bridged bis(silylene) nickel complex **8**.

Molecular structure of **7** is confirmed by X-ray diffraction analysis (Figure 10.2). The two silylene moieties and one cyclooctadiene ligand are coordinated to the nickel atom with the Si1–Ni1–Si2 angle of 68.90(3)°. The Si–O bond lengths in **7** (1.7011 (15) and 1.7081(17) Å) are longer than that in **6** (1.641(2) and 1.652(2) Å), while the Si1–O–Si2 angle of **7** (93.44(8)°) is significantly smaller than that in **6**. The Ni–Si bond lengths (2.1908(7) and 2.1969(7) Å) lie in between the value of Ni–Si bond lengths for literature known compounds [82, 87–90].

The catalytic abilities of oxygen-bridged bissilylene nickel complex **7** were studied in the coupling reaction of aryl halides and benzylzinc bromide (Table 10.1) [91–94].

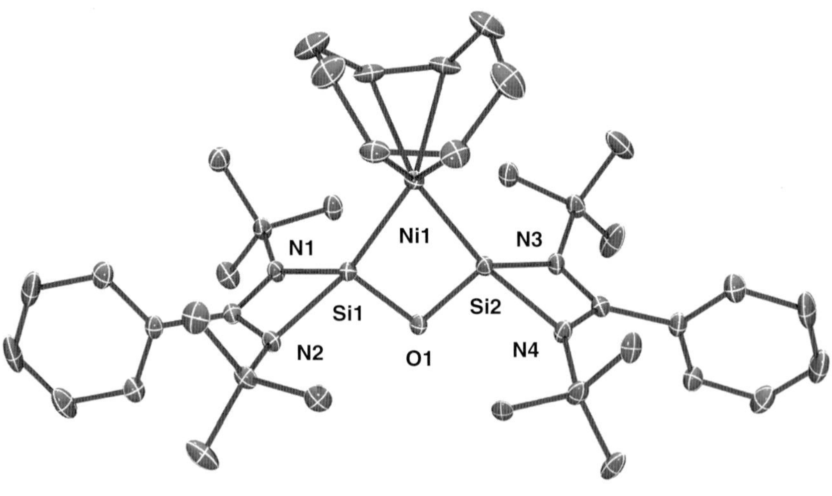

Figure 10.2 Molecular structure of oxygen-bridged bis(silylene) nickel complex **7**.

Table 10.1 Nickel-catalyzed C–C bond formation of benzylzinc bromides with aryl halides.

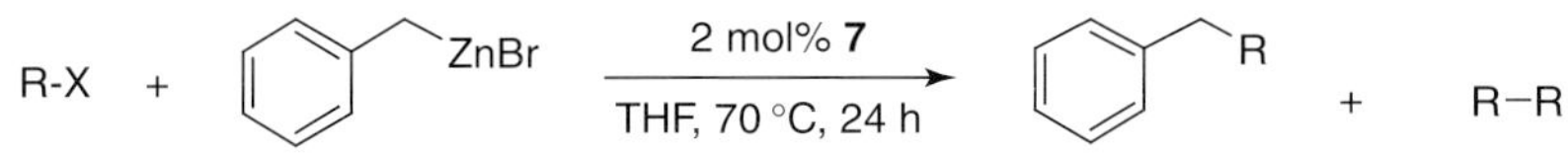

8, 10–18 **9** (1.5 equiv) **8a, 10a–18a** side product

Entry	Substrate	Product	Yield (%)[a]
1	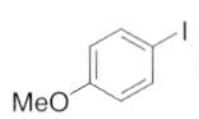 **8**	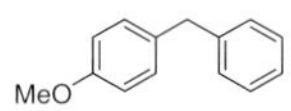 **8a**	98 (2)
2	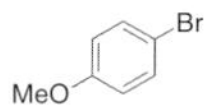 **10**	8a	29 (11)[b]
3	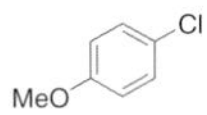 **11**	8a	8 (<1)
4	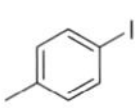 **12**	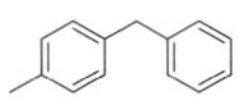 **12a**	97 (<1)
5	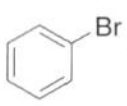 **13**	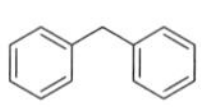 **13a**	74 (19)
6	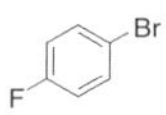 **14**	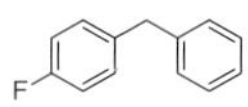 **14a**	49 (11)
7	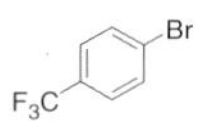 **15**	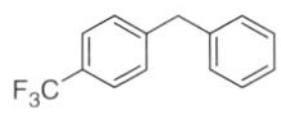 **15a**	85 (15)
8	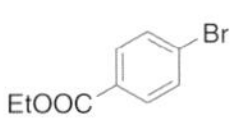 **16**	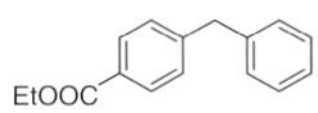 **16a**	>99 (<1)
9	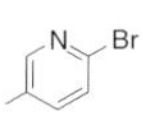 **17**	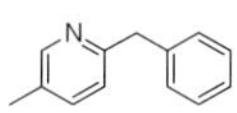 **17a**	>99 (<1)
10	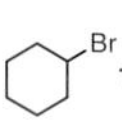 **18**	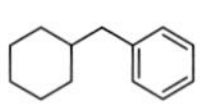 **18a**	<1

[a] Reaction conditions: substrate (0.72 mmol), benzyl zinc bromide (1.08 mmol, 0.5 M in THF), precatalyst **7** (2.0 mol%), 70 °C, 24 h. The yield was determined by GC-MS and ^{1}H NMR. The yield of the homocoupling of the aryl halide is stated in brackets.
[b] Dehalogenation of **8**: 4%.

As a model reaction, the cross-coupling of 1-iodo-4-methoxybenzene (**8**) with benzylzinc bromide (**9**) was performed in the presence of nickel complex **7** (Table 10.1). The reaction outcome was investigated at different temperatures. Only at 70 °C, the desired coupling product was observed in excellent yield. It should be noted that no reaction was observed in the absence of nickel complex **7**. After the optimization of reaction condition, 2.0 mol% of complex **7** in tetrahydrofuran (THF) performed the $C(sp^2)-C(sp^3)$ bond formation at 70 °C in excellent yield (Table 10.1, entry 1). In this reaction, the homocoupling product of **8** was observed in very low yield (2%). The application of bromo- and chloro-derivatives of **8** (substrates **10** and **11**) showed a decrease of reactivity (Table 10.1, entries 2–3) [95]. On the other hand, for aryl bromides containing electron-withdrawing groups, a better performance was observed (Table 10.1, entries 6–7). Moreover, the cross-coupling reaction could be realized in the presence of sensitive functional groups, such as esters, in excellent yield (Table 10.1, entry 8). In addition, the heteroaryl bromide **17** was converted into the coupling product **17a** with excellent yield and selectivity (Table 10.1, entry 9). Attempts to broaden the scope to the challenging $C(sp^3)-C(sp^3)$ cross-coupling have failed so far (Table 10.1, entry 10). Importantly, in several cases, as side reaction, the homocoupling of the aryl halides were observed. Probably the formation of R^1ZnX via metal exchange from **9** to R^1X and subsequent nickel-catalyzed coupling with R^1X resulted in the formation of the homocoupling product.

Besides organometallic zinc reagents, Grignard reagents were reacted with aryl halides in the presence of nickel complex **7** with a catalyst loading of 2.0 mol% in accordance with the reaction conditions described in Table 10.1. First 1-iodo-4-methoxybenzene (**8**) was reacted with *p*-tolyl magnesium bromide (**19**) to obtain the product **8b** in 63% yield (Table 10.2, entry 1). Along with the desired coupling product, the homocoupling and dehalogenation products were observed in 5 and 31% yield, respectively. In contrast, with aryl bromides and chlorides, a higher yield and a higher selectivity was observed for the $C(sp^2)-C(sp^2)$ bond formation (Table 10.2, entries 2–3). Moreover, various substituted aryl as well as hetero aryl halides were converted in good to excellent yields to the desired biaryls (Table 10.2, entries 4–9). On the other hand, the reaction of Grignard reagents with alkyl halides showed no coupling product formation (Table 10.2, entry 10). As monitored for the Negishi type coupling reaction with benzylzinc bromide, in the Kumada type reaction, as side product the homocoupling product was observed as well. Besides, the dehalogenation of the aryl halide was detected. The observation found that the dehalogenated product supports the formation of R^1MgX via metal exchange of the Grignard reagents with the starting material R^1X. During the aqueous work-up, the formed R^1MgX is hydrolyzed to R^1H.

On the basis of the previously reported mechanistic assumptions for other coupling reactions with the nickel catalyst, plausible mechanism is shown in Scheme 10.9. The precatalyst **7** generated the active species **A** by oxidative addition of the aryl halide to form a nickel(II) complex along with the elimination of cyclooctadiene [96–99]. Subsequently, a transmetallation reaction with the organometallic zinc or magnesium reagent produced the nickel(II) intermediate **B** accompanied with ZnX_2 or MgX_2. Reductive elimination generated a C–C bond and the desired organic products. On the other hand, the nickel(0) complex **C** stabilized by the bissilylene ligand could run through the catalytic cycle again.

Table 10.2 Nickel-catalyzed C–C bond formation of Grignard reagents with aryl halides.

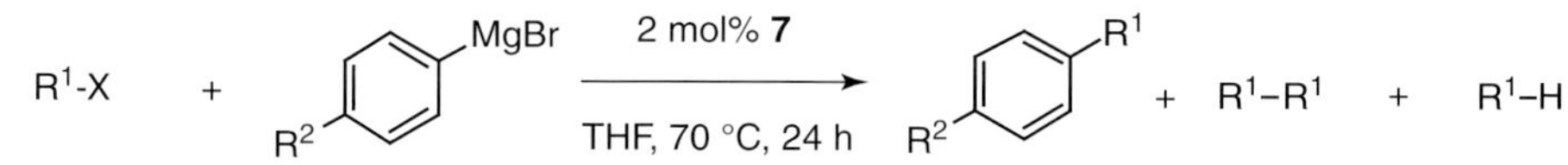

8,12,13,15,21–25 **19:** R^2 = Me; **20:** R^2 = OMe → **8b,15b,21b, 22b,24b,25b** + **8c,15c,21c, 22c,24c,25c** + **8d,15d,21d, 22d,24d,25d**

Entry[a]	Substrate	Product	Yield b (%)[b]	Yield c (%)[c]	Yield d (%)[d]
1	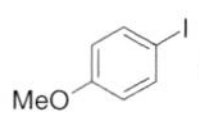 **8**	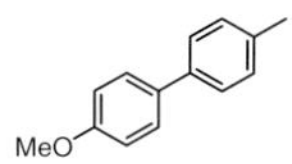 **8b**	63	5	31
2	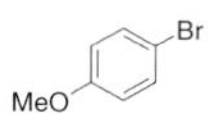 **12**	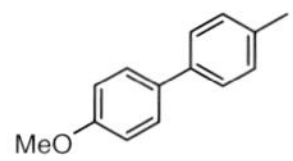 **8b**	74	6	19
3	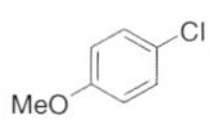 **13**	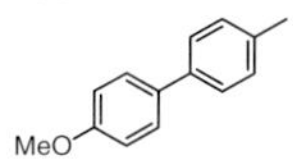 **8b**	79	9	8
4	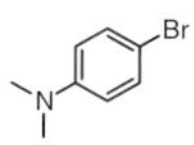 **21**	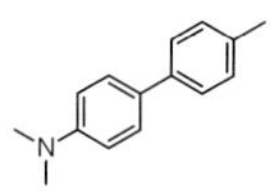 **21b**	72	<1	28
5	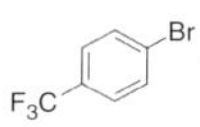 **15**	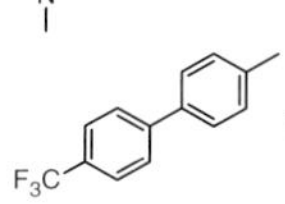 **15b**	93	<1	7
6	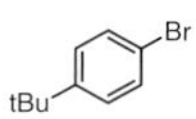 **22**	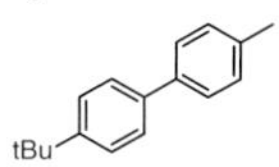 **22b**	61	7	32
7[e]	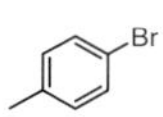 **23**	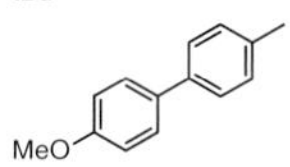 **8b**	66	4	30
8[f]	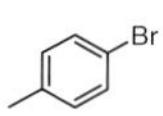 **23**	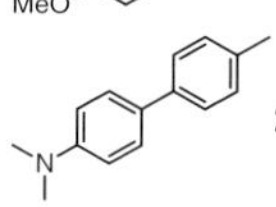 **21b**	55	<1	<1
9	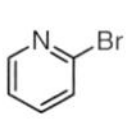 **24**	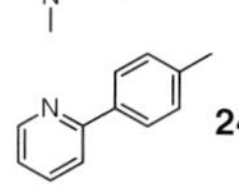 **24b**	92	<1	<1
10	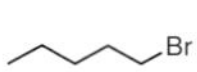 **25**	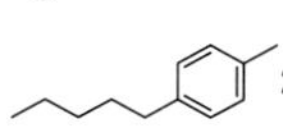 **25b**	<1	<1	<1

[a] Reaction conditions: substrate (0.72 mmol), Grignard reagent **18** or **19** (1.08 mmol, 1.0 M in THF), precatalyst **7** (2.0 mol%), 70 °C, 24 h. The yield was determined by Gas chromatography-mass spectrometry (GC-MS) and ^{1}H NMR.
[b] Yield of the cross-coupling product.
[c] Yield of the homocoupling product.
[d] Yield of the dehalogenated product.
[e] Grignard reagent **20**.
[f] Grignard reagent: 4-(*N,N*-dimethylaminophenyl)magnesium bromide/LiCl was generated by direct Mg insertion into the bromide in the presence of LiCl [179].

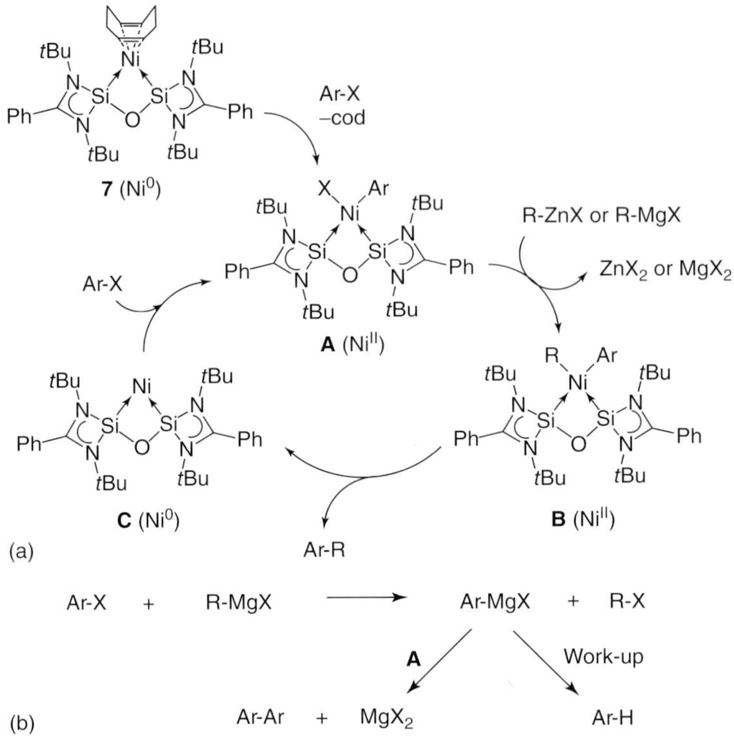

Scheme 10.9 (a,b) Proposed reaction mechanism for the cross-coupling reactions.

10.2.3 Pincer-Type Bis(silylene) Complexes (Pd, Ir, Rh)

Recently, the chemistry of pincer-ligated transition metal complexes underwent marvelous developments and has been utilized for various fascinating catalytic transformations [100–106]. In particular, catalytic applications mediated by palladium pincer complexes have been studied deeply in which the Pd atom is usually in the +II oxidation state. These complexes have a terdentate coordination mode with an arene carbon-palladium σ bond and two Lewis-donor atoms on side arms, which are coordinated to the palladium center. The coordination chemistry of the most common pincer ligands such as nitrogen-carbon-nitrogen (NCN-), phosphorus-carbon-phosphorus (PCP-), and sulfur-carbon-sulfur (SCS)-type ligands has been explored extensively, in particular with respect to their Pd complexes. It is generally well known that the properties of the donor group greatly influence the reactivity, stability, and catalytic performances of pincer aryl-metal complexes. Pincer arene ligands with stronger σ-donor atoms than those provided by group 15 and 16 atoms appear very attractive for the synthesis of more electron rich transition metal catalysts for chemical transformations of particularly unreactive small molecules (e.g., alkanes). Promising targets are pincer systems having divalent heavier group

14 elements (Si, Ge, Sn, and Pb). However, pincer ligands bearing low-valent electropositive group 14 donor atoms such as divalent silicon are unknown so far, despite the fact that isolable silicon(II) species (unsupported and donor-stabilized silylenes) are no longer laboratory curiosities and the Si(II) center can serve as stronger σ-donor and π-acceptor ligands toward transition metals than P atoms of phosphines [1–6]. Therefore, it can be expected that bis-silylene-like SiCSi pincer transition metal complexes could benefit from the advantages of Si(II) donors and may lead to a unique reactivity (Scheme 10.10).

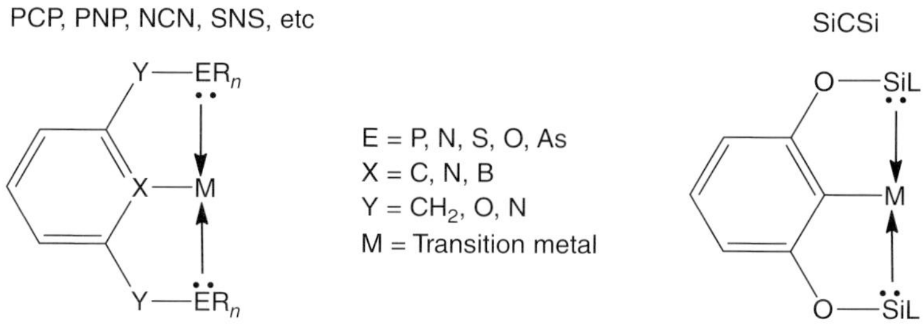

Scheme 10.10 ECE pincer complexes versus bis-silylene SiCSi pincer arene complexes.

As a new pincer ligand, SiCSi-type (silicon-carbon-silicon) bissilylene **26** is easily synthesized following the reactions shown in Scheme 10.10 [107]. The deprotonation of 4,6-di-*tert*-butylresorcinol **27** with n-BuLi gives to the corresponding resorcinolate dianion **28**. The reaction of **28** with *N*-heterocyclic chlorosilylene LSiCl [L = PhC(N*t*Bu)$_2$] **1** [63] in the molar ratio of 1 : 2 leads to the first SiCSi pincer compound **26**, which could be isolated as yellow crystals in 79% yield (Scheme 10.11). In the ^{29}Si NMR spectrum of bissilylene **26**, a sharp singlet signal was observed at δ = −24.0 ppm, which is comparable to that observed for alkoxy substituted silylenes LSiOR (R = *t*Bu, *i*Pr, Me) reported by Roesky and coworkers [108].

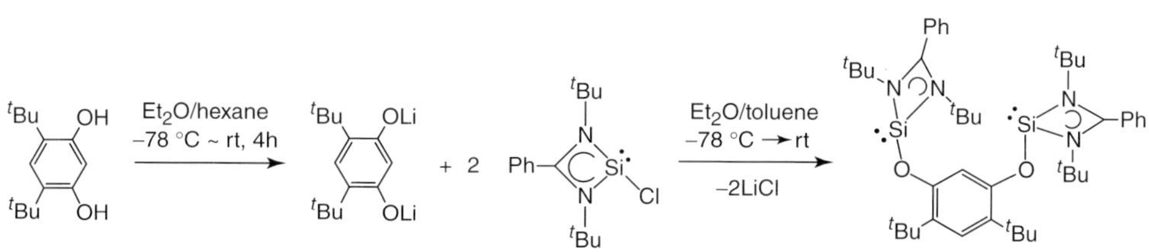

27 **28** **1** **26**

Scheme 10.11 Synthesis of the SiCSi pincer ligand **26**.

The treatment of **26** with 0.5 equiv. of tetrakis(triphenylphosphine)palladium(0) Pd(PPh$_3$)$_4$ in hexane at room temperature for one day afforded unexpected silyl(silylene)palladium complex **29**, which could be isolated as colorless crystals in 81% yield (Scheme 10.12) [107]. When a 1 : 1 molar ratio of starting materials

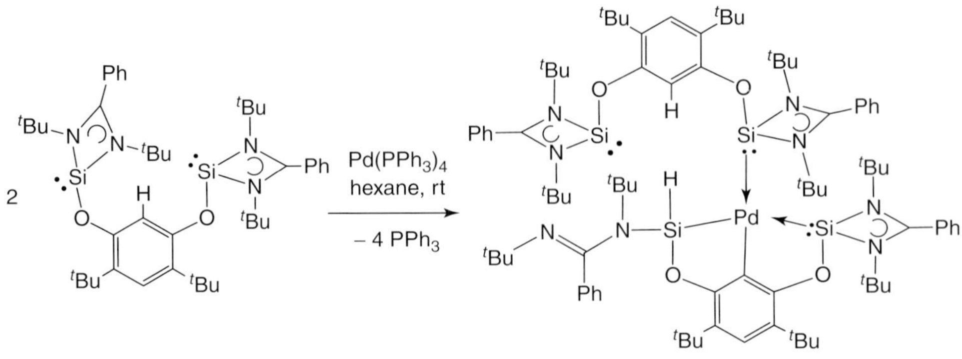

26 **29**

Scheme 10.12 Synthesis of silyl(silylene)palladium complex **29**.

was used, smaller yield was obtained. It should be highlighted that the palladium complex **29** was obtained in a 1 : 2 molar ratio of palladium : bis-silylene ligand. When a 1 : 1 molar ratio of starting materials was applied, a smaller yield of **29** (<40%) was obtained, but no other product or intermediate could be detected by ^{1}H NMR. The ^{29}Si NMR spectrum of **29** in C_6D_6 reveals four signals at $\delta = -8.7$ (L*Si*:), 39.7 (*Si*H), 62.3 and 65.8 ppm (L*Si*:Pd), respectively. The chemical shift of silyl hydride silicon atom is upfield shifted in comparison with those of the corresponding HSi → Pd complex reported by Kira and coworkers due to the nitrogen and oxygen atoms bound to the Si1 atom [74]. The ^{29}Si resonances of the two Si(II) atoms coordinated to the palladium center are significantly shifted to a higher field in comparison with those reported for donor-free silylene-palladium complexes owing to the additional N-donor coordination of the amidinate ligand to the Si(II) atom.

The molecular structure of compound **29** was determined by X-ray crystallography (Figure 10.3). Interestingly, during the complexation, one of the Si–N bond of amidinate ligand is opened, which attributed the 1,2-hydride migration to Si atom from Pd atom. Owing to the coordinative saturation of the Pd center, one silylene moiety remains "free." The Si–Pd bond lengths of silylene complex moieties (2.3271(12) and 2.3038(11) Å) in **29** are shorter than that of the Si–Pd distance of silyl complex moiety (2.3561(12) Å), reflecting the difference between a silyl ligand with a Si(IV)–Pd single bond versus silylene ligand with a Si(II) → Pd dative bond.

A schematic potential energy surface of the proposed pathway for the formation of bis(silylene) palladium complex is given in Scheme 10.13. The reaction starts with the complexation of bis-silylene ligand **26** to give SiCSi-type bis(silylene) palladium complex **30**. A 1,2-hydride migration from palladium to silicon in **30** yield 14-electron palladium complex **32** as an intermediate. This process has a barrier of +21.0 kcal mol^{-1} via transition state **31-TS**, and intermediate **32** is thermally more stable than **31-TS**, but only slightly (1.4 kcal mol^{-1}). Electron deficient-palladium silylene complex **32** could easily undergo the donor coordination with

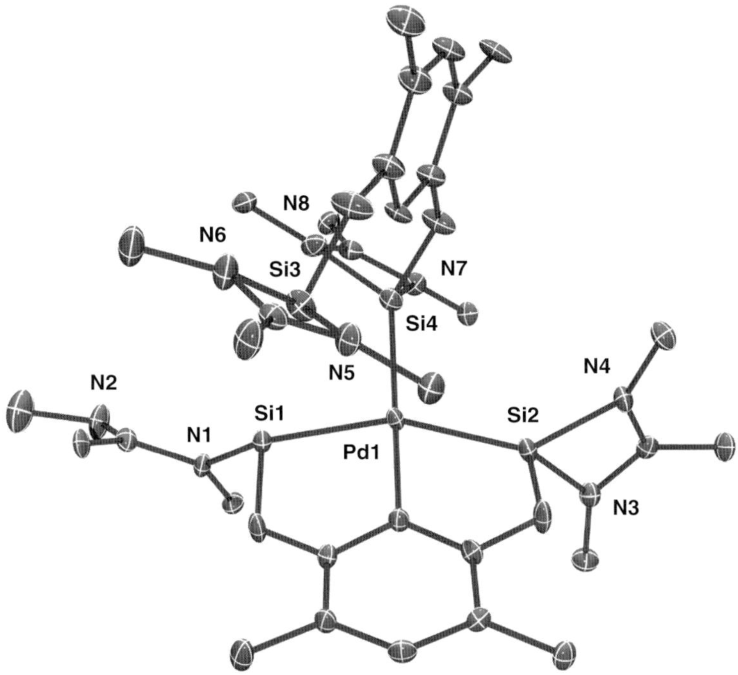

Figure 10.3 Molecular structure of silyl(silylene)palladium complex **29**. Hydrogen atoms and *tBu* groups were omitted for clarity.

phosphine or silylene, resulting in the formation of more stable 16-electron complexes. For instance, calculated relative energies of complex **33** stabilized by triphenylphosphine PPh_3 and complex **34** stabilized by of methoxysilylene LSiOMe as a simplified silylene donor were +9.1 and −4.1 kcal mol^{-1}, respectively. As the silylene is a strong σ-donor ligand than phosphine, complex **34** is 13.3 kcal mol^{-1} more stable than complex **33**. This computation is in good agreement with the experimental results. Accordingly, it is reasonable to expect that the complicated compound **29** is produced via 1,2-hydride shift and silylene donor coordination. In fact, this pathway from **30** to **34** is exothermic by 4.2 kcal mol^{-1}.

In addition, further coordination ability of SiCSi bissilylene **26** toward iridium and rohdium was also investigated. Reaction of pincer SiCSi ligand **26** with $[\{IrCl(coe)_2\}_2]$ (coe = cyclooctene) led to the formation of bis(silylene) iridium complex **35** in 92% yield (Scheme 10.14) [109]. In the ^1H NMR of **35**, a characteristic hydride signal was observed at δ = −25.6 ppm. The ^{29}Si NMR spectrum of pincer-type bis(silylene) iridium complex **35** has a singlet signal at δ = −54.9 ppm, which is downfield of that of the starting material **26**. X-ray crystallographic analysis revealed that the Ir–Si bond lengths in **35** (2.305(1) and 2.301(1) Å) are similar to those in the palladium complex **29** (Figure 10.4). Likewise, SiCSi ligand **26** reacts with $[IrCl(CO)(PPh_3)_2]$ in C_6D_6 for 1 h at 100 °C to yield bis(silylene) iridium complex **36** in 88% yield (Scheme 10.14). The IR spectrum of complex **36**

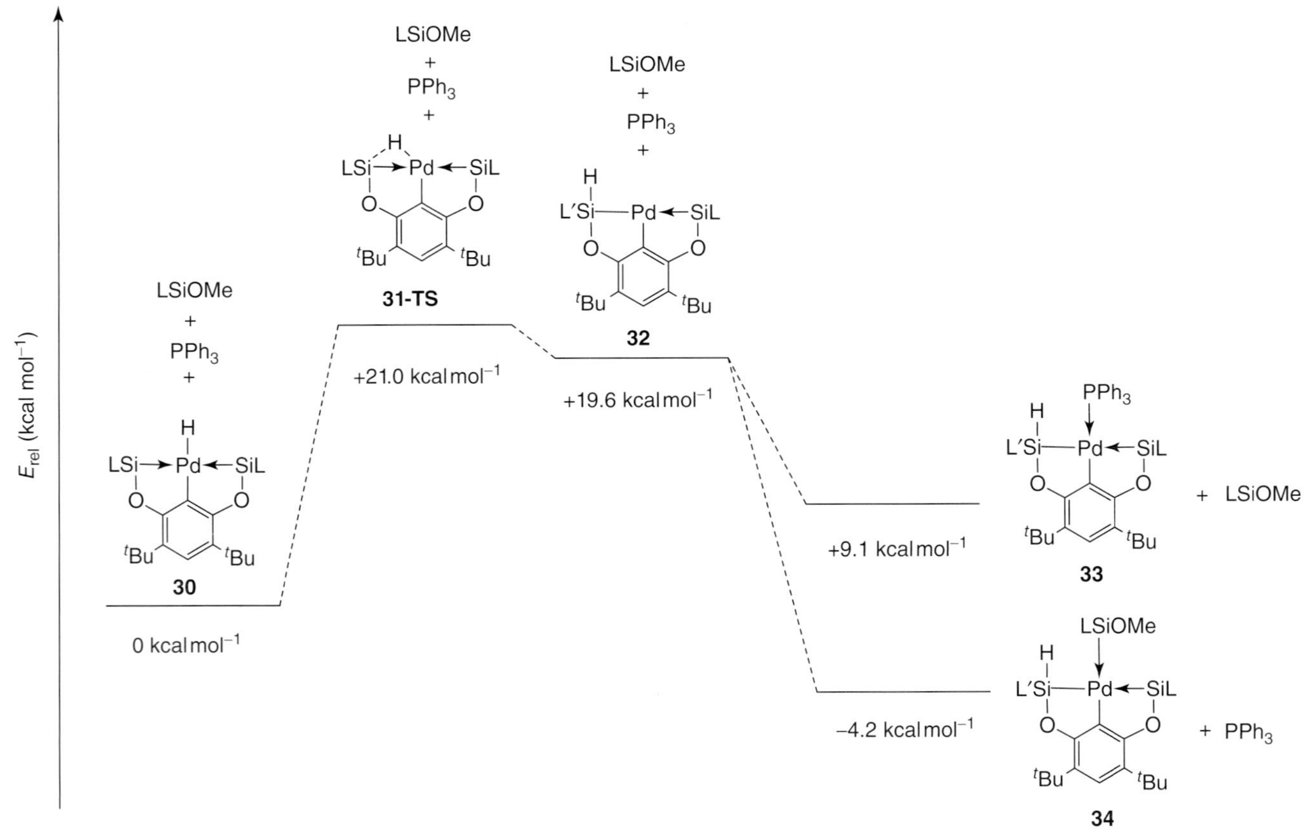

Scheme 10.13 Relative energies of model compounds **30–34** (L = 1,3-μ^2-PhC(N*t*Bu)^2, L′ = μ_1-(N*t*Bu)C(Ph)=N*t*Bu).

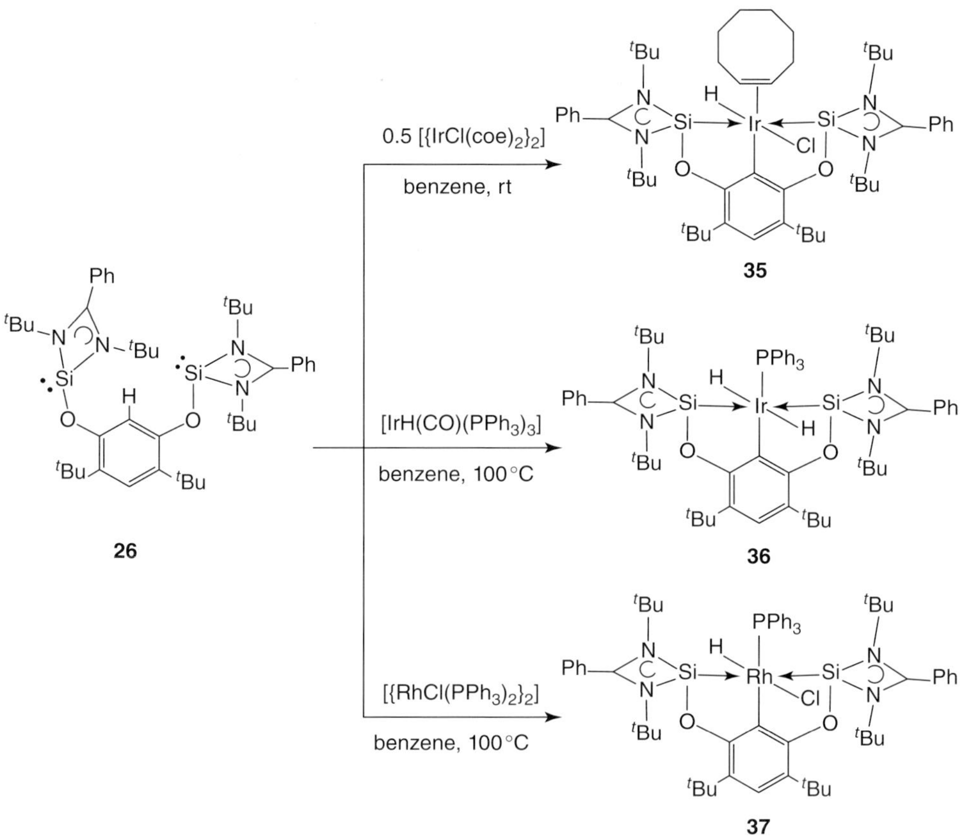

Scheme 10.14 Syntheses of bis(silylene) iridium and rhodium complexes **35–37**.

revealed one CO stretching vibration at $\nu = 1968\,\text{cm}^{-1}$ and one weak band for the hydrides at $\nu = 2251\,\text{cm}^{-1}$. In addition, bis(silylene) rhodium complex **37** can also be synthesized in 84% yield from the reaction of SiCSi ligand **26** with [{RhCl(PPh$_3$)$_2$}$_2$] (Scheme 10.14). While a doublet at $\delta = 36.6$ ppm ($^1J_{\text{PRh}} = 97.6$ Hz) was observed in the ^{31}P NMR spectrum, a doublet of doublets at $\delta = 66.4$ ppm ($^2J_{\text{SiP}} = 59.4$ Hz) appears in the ^{29}Si NMR spectrum.

Catalytic C–H borylation of arenes with pinacolborane using bis(silylene) iridium complex **35** was tested (Scheme 10.15). Using a pinacolborane (20 equiv.) in C_6D_6 led to the formation of small amounts of borylated benzene and hydrogenated coe after heating to 100 °C for 2 h. On the other hand, when coe was added, the fast formation of the product in high yield was observed. Thus, the reaction proceeds significantly faster with SiCSi pincer iridium complex **35** and is higher yielding in the presence of coe after 24 h (90% vs. 53%). However, the catalytic performance of SiCSipincer iridium complex **35** is significantly lower in comparison with benchmark systems employing bidentate nitrogen ligand [110]. It is probably attributed to the steric repulsion around the iridium.

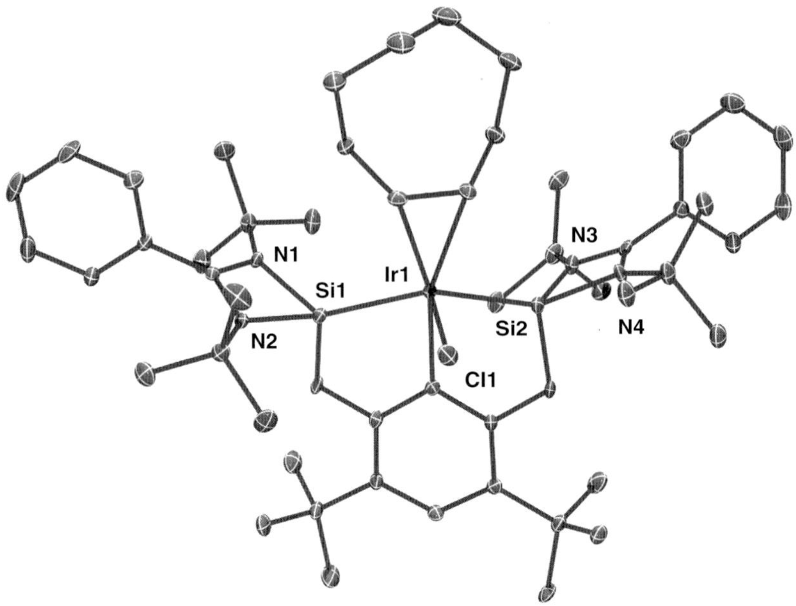

Figure 10.4 Molecular structure of bis(silylene) iridium complex **35**.

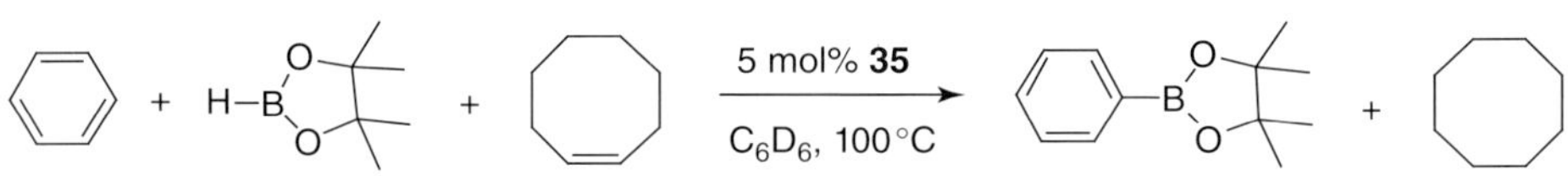

Scheme 10.15 C-H borylation of benzene using pinacolborane with 5 mol% precatalyst bis(silylene) iridium **35**.

10.2.4
Bis(silylenyl)-Substituted Ferrocene Cobalt Complex

As mentioned above, silylene transition metal complexes have received much attention because they can play a key role as intermediates in transition metal-catalyzed transformation of silicon compounds. Bissilylene ligands have been synthesized as promising chelate ligands for the electron rich transition metal complexes. In order to gain access to other new bis(silylene) ligands as potential bidentate σ-donor ligands, a novel type of bis(silylene) with ferrocenyl spacer have been investigated, which show unique electronic properties.

Treatment of chlorosilylene **1** [63] with the generated 1,1′-dilithioferrocene led to the formation of the desired bis(silylene) ligand **38** in 70% yield (Scheme 10.16) [111]. Bis(silylene) ligand **38** shows a singlet signal at $\delta = 43.3$ ppm in the ^{29}Si NMR spectrum. Compound **38** is the first example of an isolable ferrocene-substituted silylene.

The coordination behavior of bis(silylene) ligand **38** has been studied. Cobalt(I) complexes can be a useful tool, on the one hand, for studying the intrinsic properties of the new ligands in coordination chemistry and, on the other hand,

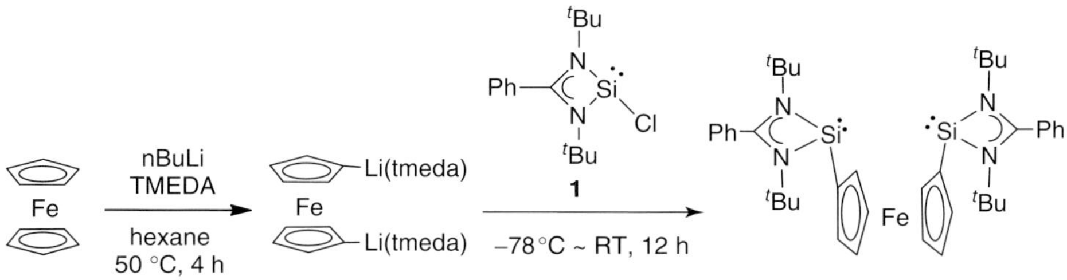

38

Scheme 10.16 Synthesis of bissilylene **38**.

for the evaluation of the catalytic potential of the complexes. Treatment of sodium cyclopentadienide and cobalt(II) dibromide with potassium graphite resulted in the generation of CpCo(I)L_n (L = toluene or THF) in solution, followed by in-situ reaction with bis(silylene) ligand **38** to give the corresponding bis(silylene) Co(I) complex **39** in 30% yield (Scheme 10.17) [112–114]. The coordination of bis(silylene) moieties to Co(I) was clearly confirmed by the ^{29}Si NMR spectra, in which the characteristic signal was observed at $\delta = 82.0$ ppm. Owing to the coordination to cobalt, its δ value is shifted significantly downfield from the starting material, bis-silylene **38** (43.3 ppm).

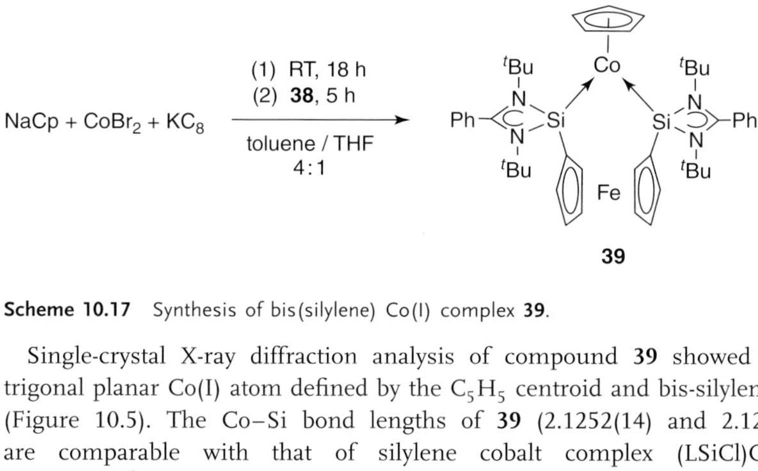

39

Scheme 10.17 Synthesis of bis(silylene) Co(I) complex **39**.

Single-crystal X-ray diffraction analysis of compound **39** showed distorted trigonal planar Co(I) atom defined by the C_5H_5 centroid and bis-silylene ligands (Figure 10.5). The Co–Si bond lengths of **39** (2.1252(14) and 2.1200(14) Å) are comparable with that of silylene cobalt complex (LSiCl)CoCp(CO) (2.1143(4) Å) [115]. Furthermore, the Co–Si distance of **39** are shorter than those of other reported silylene cobalt complexes, [(NHC)$SiCl_2$]CoCp (CO) (NHC = 1,3-bis(2,6-diisopropylphenyl)imidazole-2-ylidene) (2.1348(5) Å) [84], $[\{(NHC)Cl_2Si\}_2Co(CO)_3]^+[CoCl_3(thf)]^-$ (2.228(2) Å) [116], $[(LSiCl)_2Co(CO)_3]^+$ $[Co(CO)_4]^-$ (2.2060(6) and 2.2017(6) Å) [115], and PNP[HSi=Co$(H)_3(SiH_2Ph)_2$] (PNP = $\{Ph_2PCH_2SiMe_2\}N^-$) (2.3990(7) Å) [117]. The relative short Si–Co distances in **39** indicate that **38** is one of the strongest σ-donor ligands in the series of divalent Si donors.

When the bis(silylene) **38** was allowed to react with 2 M equiv. of $CpCo(CO)_2$, the bis(silylene) cobalt complex $[LSiCo(CO)Cp]_2Fc$ **40** could be obtained in 87%

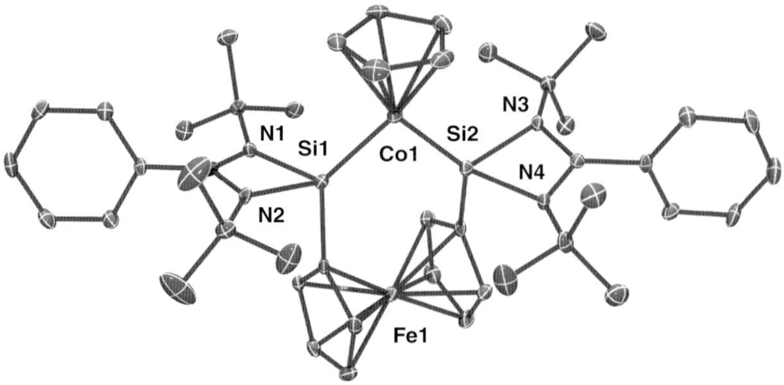

Figure 10.5 Molecular structure of bis(silylene) Co(I) complex **39**.

yield, accompanied by the elimination of CO (Scheme 10.18). The ^{29}Si NMR spectrum of **40** shows a singlet resonance at $\delta = 85.7$ ppm, similar to that of **39**. In the ^{13}C NMR spectrum of **40**, one characteristic signal for the CO ligands appears at $\delta = 207.8$ ppm. Moreover, the IR spectrum of **40** exhibits one strong stretching band at $\nu = 1888\,\text{cm}^{-1}$ attributed to the carbonyl groups on the cobalt(I) atoms. The observed CO stretching frequency of bis(silylene) cobalt complex **40** is much lower than those of (LSiCl)Co(CO)Cp ($\nu = 1968\,\text{cm}^{-1}$) [112], indicating that the bis(silylene)-substituted ferrocene **38** is a much strong σ-donor than chlorosilylene **1**.

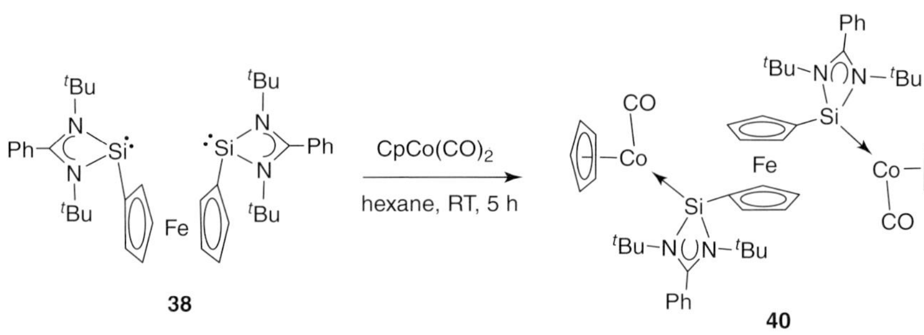

Scheme 10.18 Synthesis of bis(silylene) Co(I) complex **40**.

A large number of cobalt complexes have been successfully applied as precatalysts in [2+2+2] cycloaddition reactions, which are a powerful methodology in organic chemistry to create arene and heteroarene moieties [118–120]. On the basis of this, the catalytic abilities of complex **39** in [2+2+2] cycloaddition reactions was investigated (Scheme 10.19) [121–128]. The bis(silylene)-Co complex **39** is a precatalyst for the cyclotrimerization of phenylacetylene to give triphenyl-substituted benzenes **42** and for the [2+2+2] cycloaddition reaction of phenylacetylene PhC≡CH and acetonitrile $CH_3C{\equiv}N$, affording substituted pyridines **43**. The first set of experiments was dedicated to the benchmark trimerization of phenyl acetylene **41**, resulting in the quantitative formation of the isomers **42a** and **42b** in the

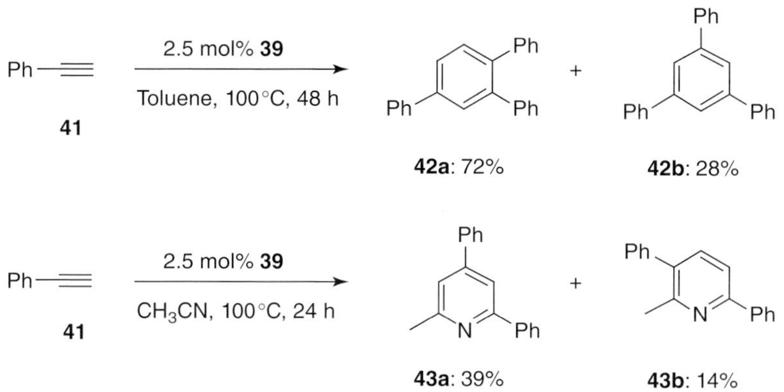

Scheme 10.19 Application of bis(silylene) Co(I) complex **39** as precatalyst in [2+2+2] cycloaddition reactions.

presence of complex **39** (Scheme 10.19). Moreover, the formation of substituted pyridines **43** by [2+2+2] cycloaddition of phenyl acetylene and an excess of acetonitrile was examined [129–133]. Once more, the catalytic activity was exhibited by Si(II)-Co complex **39**, while no product formation could be observed using Ge(II)-Co derivative [111]. Interestingly, the application of $CpCo(CO)_2$ as precatalyst showed to some extent lower yields of the substituted pyridines **43a** (3%) and **43b** (11%).

10.2.5 Silylene Iron Complexes

Since the original report of the first NHSi transition metal complex in 1977 [134], numerous NHSi transition metal complexes have been synthesized and isolated [38–42]. Furthermore, their application in a variety of activation of small molecules [88, 89, 135], stoichiometric transformations, for example, hydrosilylation of alkynes [136] or silane activation [137] have been highlighted by several recent publications. Although quite a number of synthetic routes for NHSi transition metal complexes exist now, these synthetic methodology are mainly based on ligand elimination reactions [69, 82, 83, 115, 138–143] from suitable transition metal precursors upon reaction with NHSi's. Interestingly, only a few examples of iron-NHSi transition metal complexes have been reported to date and they possess the $\{Fe(CO)_4\}$ fragment [69, 134, 144]. It seemed reasonable to think that this is because of the lack of suitable iron precursors to access other NHSi iron complexes. Accordingly, the iron complex $[(dmpe)_2Fe(PMe_3)]$ (**44**) has been employed as a starting material. This iron ligand would enhance the π-back bonding donation from iron to the silicon center by increasing the electron density at the iron center with electron pushing alkyl phosphane ligands. In the light of the potentially increasing reactivity of the complex, these target complexes having electron rich iron complex **44** may lead to better catalytic properties in comparison with the known silylene complexes.

To enhance the stability of Si(II) center of silylene subunit, amidinate ligand L would be an ideal due to intramolecular N-donor coordination into Si and bulky *t*Bu substituents on the N atom. Thus, chlorosilylene LSiCl **1** [63] was employed as the starting material. The reaction of the chlorosilylene **1** with the electron rich Fe complex [(dmpe)$_2$Fe(PMe$_3$)] **44** in diethylether leads to the formation of the expected silylene iron complex **45** in 81% yield (Scheme 10.20) [145]. The reactivity of complex **45** can also be carried out. For example, complex **45** reacts with MeLi to form the complex **46** by metathetical exchange at the silicon center, in 67% yield (Scheme 10.21). Furthermore, the reaction of complex **45** with lithium triethylborohydride Li[HBEt$_3$] furnishes the desired complex **47** by halide-hydride exchange at the silicon center, in 89% yield (Scheme 10.20). In general, silylene hydrides require stabilization by a transition metal complex or intramolecular donor-acceptor stabilization. Structurally characterized silylene hydrides are relatively rare because of their high reactivity [136, 146–149]. Silylene

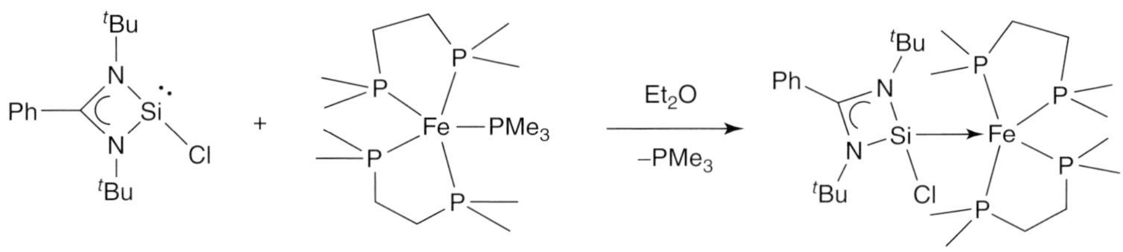

1 **44** **45** (81%)

Scheme 10.20 Synthesis of silylene iron complex **45**.

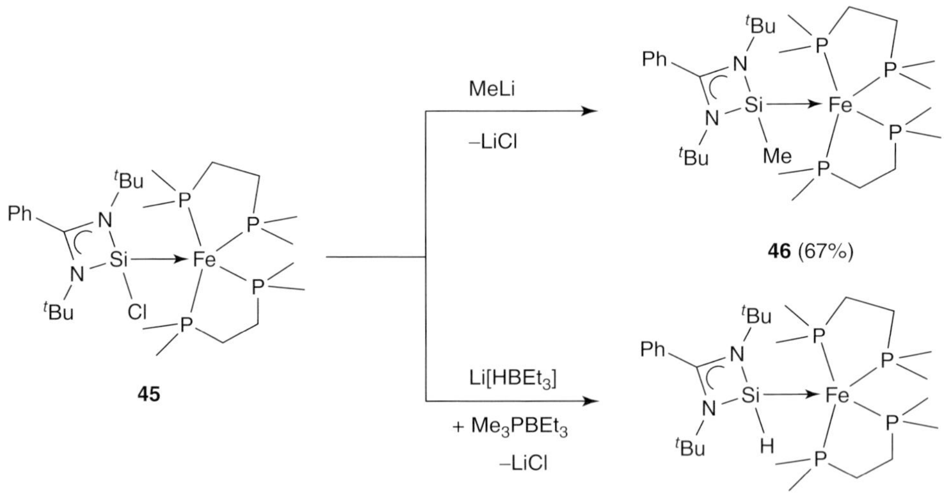

47 (89%)

Scheme 10.21 Syntheses of silylene iron complexes **46** and **47**.

iron complexes **45–47** are all highly thermally stable and their suitability is a desirable property as precatalysts for organic transformations.

The ^{29}Si NMR chemical shifts of silylene iron complexes (**45** ($\delta = 43.1$ ppm, $^2J_{SiP} = 12.1$ Hz); **46** ($\delta = 102.5$ ppm, $^2J_{SiP} = 16.1$ Hz); **47** ($\delta = 63.6$ ppm, $^2J_{SiP} = 12.1$ Hz)) show a negative linear relationship between the chemical shift and the Hammett constant (σ_p) of the substituents [150]. The same trend was observed for an analogous series of silylene titanium complexes [65], bearing the same substituents on silicon and can be explained on the basis of the Bent rule [151]. The shielding of the silicon nucleus in **45** results from a concentration of p-character in the Si–Cl bond, which concomitantly results in increased s-character in the contribution of the silicon atom of the iron silicon bond. This increase in the s-character results in a shielding effect and the lower chemical shift value. On the contrary, in the case of **46** the exact opposite can be expected on the basis of the Bent rule, and a lower s-character in the contribution of the silicon atom in the silicon iron bond is predicted, resulting in deshielding and a larger chemical shift. This was confirmed on the basis of the bonding analysis of the Si–Fe bond in these complexes using DFT calculations. The molecular structures of compounds **45–47** are elucidated by X-ray diffraction analysis and reveal slightly distorted trigonal-bipyramidal coordination geometries around the iron centers with the NHSi ligand occupying one of the equatorial coordination sites in each case. This is the typically preferred position for π-accepting ligands in a trigonal-bipyramidal geometry, which provides some evidence for the π-accepting ability of the NHSi ligand. The Si–Fe bond lengths for the three complexes range from 2.1634(9) Å in **45** to 2.200(2) Å in **46**, with complex **47** exhibiting a value of 2.184(2) Å. The observed Fe–Si lengths correlate perfectly with the Hammett constants (σ_p) associated with the substituents on the silicon centers in analogy to the trend observed with the respective ^{29}Si NMR shifts of compounds **45–47**. This is an additional indication of the influence of the substituent on the nature of the Fe-Si bond. The Fe–Si distances are in fact comparable with that of the donor-free silylene complex [$Cp_2Fe(CO)(SiMe_3)(=SiMes_2)$] (Mes = 2,4,6-$Me_3C_6H_2$) by Tobita and coworkers, where a formal double bonding interaction between iron and silicon is observed [152]. This suggests that the Si–Fe bond in compounds **45–47** can be considered as Si=Fe double bond.

Recently, much has been reported about iron catalysis because of the precious and toxic metals (e.g., Rh, Ir, Ru) and their abundance and low cost [153–163]. What needs to be emphasized is that excellent performances using iron precatalysts have been demonstrated in redox chemistry [164–166]. Equally important, it has been shown that the ligand sphere is of crucial importance for achieving excellent catalytic activities and selectivities [160, 167–175]. With this in mind, the catalytic abilities of complex **45** with ketone in the hydrosilylation have been evaluated. In the presence of the catalytic amounts of **45** (5 mol%), a variety of ketones bearing different steric and electronic properties (**48–54**) were converted in excellent yields to the corresponding alcohols (**48a–54a**) applying $(EtO)_3SiH$ as a hydridosilane source (Table 10.3). Furthermore, the catalytic ability of **45** in the reaction of

Table 10.3 Probing the catalytic ability of complex **45** for hydrosilylation of ketones.

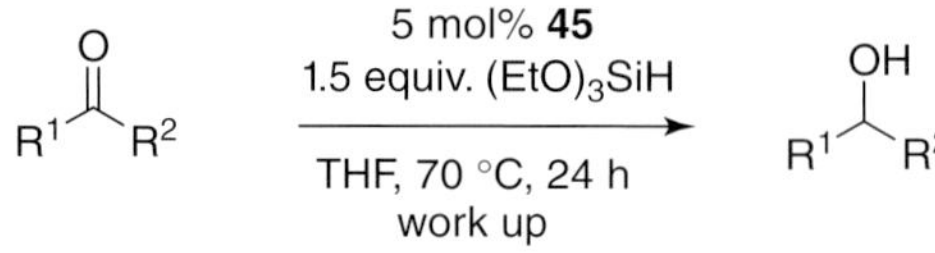

[a]

Entry	Substrate	Yield (%)[b]
1	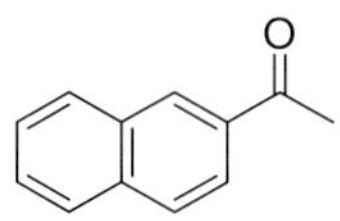 **48**	48a: 92
2	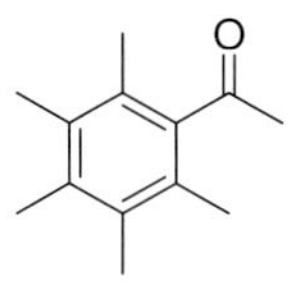 **49**	49a: 73
3	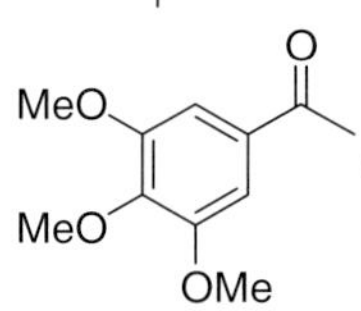 **50**	50a: 97 50a: >99[c]
4	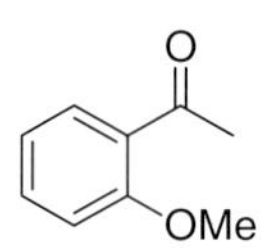 **51**	51a: >99
5	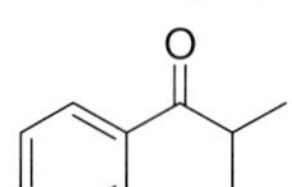 **52**	52a: 98
6	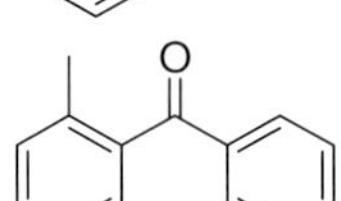 **53**	53a: 98
7	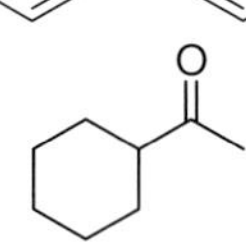 **54**	54a: 95

[a] Reaction conditions: precatalyst (5.0 mol%), substrate (0.16 mmol), $(EtO)_3SiH$ (0.24 mmol), 70 °C, 24 h.
[b] Yield determined by GC-MS using *n*-dodecane as internal standard.
[c] Catalytic reaction repeated using precatalyst (5.0 mol%), substrate (0.16 mmol), $(EtO)_3SiH$ (0.24 mmol), 25 °C, 24 h.

hydrosilylation with ketone **50** at room temperatures was studied, and quantitative conversion into the corresponding alcohols was observed without heating.

To get more insights into the reaction mechanism for these hydrosilylation reactions with silylene iron complex, the reaction of **45** with ketone **48**, in the absence of $(EtO)_3SiH$, was carried out on preparative scale in the hope of isolating the corresponding ketone adduct **55**, which is a likely first step in the catalytic process. The selective formation of a new iron hydride species was observed, when a sample of **45** with a stoichiometric amount of ketone **48** in toluene-d_8 was heated. The thermal stability of **45** was independently checked in the absence of ketone. On the basis of the observation of no H-migration from silicon to iron even upon prolonged heating, it can be concluded that the hydride transfer process is induced by ketone coordination to the silicon(II) center. This data suggest nonrigid stereochemistry, or fluxionality of the formed iron hydride species **55** in accordance with previous observations for some cis-configured Fe-hydrido complexes [176]. The initial formation of a ketone adduct with concomitant 1,2-hydride migration from silicon to iron was observed in the ^{29}Si NMR spectrum. The formation of the ketone adduct is probably the activation step in the catalytic process. Similar observations for silylene ruthenium complexes had been described by Tilley and coworkers [177] and Tobita and coworkers [178].

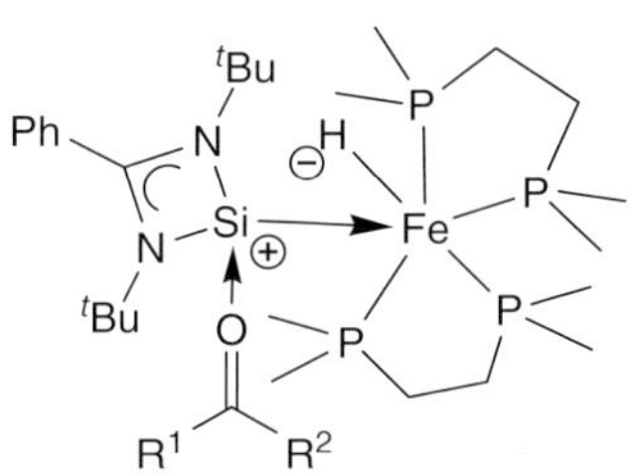

55

10.3
Conclusion and Outlook

The chemistry of compounds featuring multiple bonding between silicon and transition metal has been expanding in the past decade and the future promises exciting developments. The utilization of bulky ligands, with the appropriate steric and electronic effects, is a critical factor in the stabilization of such compounds. Beginning as laboratory curiosities, stable silylene transition metal complexes were proposed as silicon analogs of carbene complexes and as potential intermediates in various metal-catalyzed reactions and transformations of organosilicon compounds. The isolation and characterization of novel silylene transition metal complexes represented the new area of organometallic and coordination chemistry. The increased number of these complexes in the literature suggests that a better understanding of nature and property of these species is continually being attained. Moreover, the development of novel NHSi ligands and bis(silylene) ligands with various spacer underlines that their transition metal complexes are no longer laboratory curiosities and provide doorways to catalytic reactions. Thanks to intramolecular donor stabilization such as amidinate ligand, NHSi has likely been recognized as promising ligands for catalytic reaction in the next generation. In particular, these novel bis(silylene) species will find many more uses as ligands for transition metals and in other chemical applications. The potential for silylene transition metal complexes can be demonstrated in several recent publications. While the investigation of highly efficient catalysts employing these complexes is still in its infancy, the discovery of catalytic reactions using isolable silylene transition metal complexes shows that these complexes can engage in the application for organic transformations. For example, Suzuki-Miyaura cross-coupling, Heck coupling, Negishi and Kumada type C-C bond formation reactions, [2+2+2] cycloaddition reactions, and hydrosilylation reactions using silylene transition metal complexes as catalysts have strikingly been achieved. Although the importance of the development of highly efficient, environmentally friendly, and energy efficient catalysts is well known, the main stress falls on the investigation of catalytic activity of electron rich transition metal complexes stabilized by bis(silylene) ligands and their potential applications. Arguably, the development of novel complexes utilizing strong σ-donor silylene ligand system and demonstration of their catalytic activity could lead to a new field in this research area. Although a broad range of fascinating achievements has been disclosed recently, this research area is still unexplored, and more fascinating advances will be made in the near future.

References

1. Inoue, S. and Driess, M. (2012) *Science of Synthesis: Silylenes*, vol. 4, Georg Thieme Vorlag KG, Stuttgartt,·New York, pp. 213–295, and references therein.
2. Asay, M., Jones, C., and Driess, M. (2011) *Chem. Rev.*, **111**, 354.
3. Yao, S., Xiong, Y., and Driess, M. (2011) *Organometallics*, **30**, 1748.
4. Mandal, S.K. and Roesky, H.W. (2010) *Chem. Commun.*, **46**, 6016.
5. Mizuhata, Y., Sasamori, T., and Tokitoh, N. (2009) *Chem. Rev.*, **109**, 3479.
6. Lee, Y. and Sekiguchi, A. (2010) *Organometallic Compounds of Low-Coordinate Si, Ge, Sn and Pb*, Chapter 4, Wiley-VCH Verlag GmbH, London, pp. 139–188.
7. Jutzi, P., Kanne, D., and Krüger, C. (1986) *Angew. Chem.*, **98**, 163; *Angew. Chem., Int. Ed. Engl.*, 1986, **25**, 164.
8. Jutzi, P., Holtmann, U., Kanne, D., Krüger, C., Blom, R., Gleiter, R., and Hyla-Krispin, I. (1989) *Chem. Ber.*, **122**, 1629.
9. Denk, M., Lennon, R., Hayashi, R., West, R., Belyakov, A.V., Verne, H.P., Haaland, A., Wagner, M., and Metzler, N. (1994) *J. Am. Chem. Soc.*, **116**, 2691.
10. Kira, M., Ishida, S., Iwamoto, T., and Kabuto, C. (1999) *J. Am. Chem. Soc.*, **121**, 9722.
11. Asay, M., Inoue, S., and Driess, M. (2011) *Angew. Chem.*, **123**, 9763; *Angew. Chem. Int. Ed.*, 2011, **50**, 9589.
12. Abe, T., Tanaka, R., Ishida, S., Kira, M., and Iwamoto, T. (2012) *J. Am. Chem. Soc.*, **134**, 20029.

13. Haaf, M., Schmedake, T.A., and West, R. (2000) *Acc. Chem. Res.*, **33**, 704.
14. Kira, M., Iwamoto, T., and Ishida, S. (2007) *Bull. Chem. Soc. Jpn.*, **80**, 258.
15. Ottosson, H. and Steel, P.G. (2006) *Chem. Eur. J.*, **12**, 1576.
16. Gehrhus, B. and Lappert, M.F. (2001) *J. Organomet. Chem.*, **617**, 209.
17. Takeda, N. and Tokitoh, N. (2007) *Synlett*, **16**, 2483.
18. Rekken, B.D., Brown, T.M., Fettinger, J.C., Tuononen, H.M., and Power, P.P. (2012) *J. Am. Chem. Soc.*, **134**, 6504.
19. Protchenko, A.V., Birjkumar, K.H., Dange, D., Schwarz, A.D., Vidovic, D., Jones, C., Kaltsoyannis, N., Mountford, P., Aldridge, S., and Ingold, C. (2012) *J. Am. Chem. Soc.*, **134**, 6500.
20. Protchenko, A.V., Schwarz, A.D., Blake, M.P., Jones, C., Kaltsoyannis, N., Mountford, P., and Aldridge, S. (2013) *Angew. Chem.*, **125**, 596; *Angew. Chem. Int. Ed.* 2013, **52**, 568.
21. Driess, M. (2012) *Nat. Chem.*, **4**, 1.
22. West, R. (2012) *Nature*, **485**, 49.
23. Inoue, S. and Leszczyńska, K. (2012) *Angew. Chem.*, **124**, 8717; *Angew. Chem. Int. Ed.*, 2012, **51**, 8589.
24. Sen, S.S., Khan, S., Samuel, P.P., and Roesky, H.W. (2012) *Chem. Sci.*, **3**, 659.
25. Sen, S.S., Khan, S., Nagendran, S., and Roesky, H.W. (2012) *Acc. Chem. Res.*, **45**, 578.
26. Sen, S.S., Jana, A., Roesky, H.W., and Schulzke, C. (2009) *Angew. Chem.*, **121**, 8688; *Angew. Chem. Int. Ed.*, 2009, **48**, 8536.
27. Gau, D., Rodriguez, R., Kato, T., Saffon-Merceron, N., De Cózar, A., and Baceiredo, A. (2011) *Angew. Chem.*, **123**, 1124; *Angew. Chem. Int. Ed.*2011, **50**, 1092.
28. Wang, Y., Xie, Y., Wei, P., King, R.B., Schaefer, H.F. III, Schleyer, P.v.R., and Robinson, G.H. (2008) *Science*, **321**, 1069.
29. Jones, C., Bonyhady, S.J., Holzman, N., Frenking, G., and Stasch, A. (2011) *Inorg. Chem.*, **50**, 12315.
30. Gehrhus, B., Hitchcock, P.B., and Lappert, M.F. (2005) *Z. Anorg. Allg. Chem.*, **631**, 1383.
31. Yeong, H.-X., Xi, H.-W., Lim, K.H., and So, C.-W. (2010) *Chem. Eur. J.*, **16**, 12956.
32. Fischer, E.O. and Massböl, A. (1964) *Angew. Chem.*, **76**, 645, *Angew. Chem., Int. Ed. Engl.*, **3**, 580.
33. Brookhart, M. and Studabaker, W.B. (1987) *Chem. Rev.*, **87**, 411.
34. Stille, J.R. (1995) in *Comprehensive Organometalic Chemistry II*, vol. 12 (eds E.W. Abel, F.G.A. Stone, and G. Wilkinson), Pergamon, Oxford, pp. 577–600.
35. Schrock, R.R. (1990) *Acc. Chem. Res.*, **23**, 158.
36. Belderrain, T.R. and Grubbs, R.H. (1997) *Organometallics*, **16**, 400.
37. Wilhelm, T.E., Belderrain, T.R., Brown, S.N., and Grubbs, R.H. (1997) *Organometallics*, **16**, 3867.
38. Ogino, H. (2002) *Chem. Rev.*, **2**, 291.
39. Okazaki, M., Tobita, H., and Ogino, H. (2003) *J. Chem. Soc., Dalton Trans.*, 493.
40. Waterman, R., Hayes, P.G., and Tilley, T.D. (2007) *Acc. Chem. Res.*, **40**, 712.
41. Inoue, S. and Driess, M. (2011) *Angew. Chem.*, **123**, 5728; *Angew. Chem. Int. Ed.*, 2011, **50**, 5614.
42. Blom, B., Stolzel, M., and Driess, M. (2013) *Chem. Eur. J.*, **19**, 40, and references therein.
43. Buechner, W. (1980) *J. Organomet. Chem. Libr.*, **9**, 409.
44. Curtis, M.D. and Epstein, P.S. (1981) *Adv. Organomet. Chem.*, **19**, 213.
45. Yamamoto, K., Okinoshima, H., and Kumada, M. (1971) *J. Organomet. Chem.*, **27**, C31.
46. Ojima, I., Inaba, S.-I., Kogure, T., and Nagai, Y. (1973) *J. Organomet. Chem.*, **55**, C7.
47. Kumada, M. (1975) *J. Organomet. Chem.*, **100**, 127.
48. Seyferth, D., Shannon, M.L., Vick, S.C., and Lim, T.F.O. (1985) *Organometallics*, **4**, 57.
49. Okinoshima, H., Yamamoto, K., and Kumada, M. (1972) *J. Am. Chem. Soc.*, **94**, 9263.
50. Sakurai, H., Kamiyama, Y., and Nakadaira, Y. (1977) *J. Am. Chem. Soc.*, **99**, 3879.
51. Pannell, K.H., Cervantes, J., Hernandez, C., Cassias, J., and

Vincenti, S. (1986) *Organometallics*, **5**, 1056.

52. Zybill, T. and Müller, G. (1987) *Angew. Chem.*, **99**, 683; *Angew. Chem., Int. Ed. Engl.*, 1987, **26**, 669.
53. Straus, D.A., Tilley, T.D., Rheingold, A.L., and Geib, S.J. (1987) *J. Am. Chem. Soc.*, **109**, 5872.
54. Fürstner, A., Krause, H., and Lehmann, C.W. (2001) *Chem. Commun.*, 2372.
55. Zhang, M., Liu, X., Shi, C., Ren, C., Ding, Y., and Roesky, H.W. (2008) *Z. Anorg. Allg. Chem.*, **634**, 1755.
56. Mork, B.V. and Tilley, T.D. (2001) *J. Am. Chem. Soc.*, **123**, 9702.
57. Watanabe, T., Hashimoto, H., and Tobita, H. (2004) *Angew. Chem. Int. Ed.*, **43**, 218.
58. Ochiai, M., Hashimoto, H., and Tobita, H. (1987) *Angew. Chem. Int. Ed.*, **46**, 8192.
59. Koshikawa, H., Okazaki, M., Matsumoto, S., Ueno, K., Tobita, H., and Ogino, H. (2005) *Chem. Lett.*, **34**, 1412.
60. Takanashi, K., Ya, V., Lee, T., Yokoyama, A., and Sekiguchi, A. (2009) *J. Am. Chem. Soc.*, **131**, 916.
61. Nakata, N., Fujita, T., and Sekiguchi, A. (2006) *J. Am. Chem. Soc.*, **128**, 16024.
62. Lee, V.Y., Aoki, S., Yokoyama, T., Horiguchi, S., Sekiguchi, A., Gornitzka, H., Guo, J.-D., and Nagase, S. (2013) *J. Am. Chem. Soc.*, **135**, 2987.
63. Sen, S.S., Roesky, H.W., Stern, D., Henn, J., and Stalke, D. (2010) *J. Am. Chem. Soc.*, **132**, 1123.
64. Kool, L.B., Rausch, M.D., Alt, H.G., Herberhold, M., Thewalt, U., and Wolf, B. (1985) *Angew. Chem.*, **97**, 425; *Angew. Chem., Int. Ed. Engl.*, 1985,**24**, 394..
65. Blom, B., Driess, M., Gallego, D., and Inoue, S. (2012) *Chem. Eur. J.*, **18**, 13355.
66. (2011 release) *Cambridge Structural Database*, Cambridge Crystallographic Data Centre, Cambridge.
67. Arp, H., Baumgartner, J., Marschner, C., Zark, P., and Müller, T. (2012) *J. Am. Chem. Soc.*, **134**, 10864.
68. Denk, M., Hayashi, R.K., and West, R. (1994) *J. Chem. Soc. Chem. Commun.*, 33.
69. Schmedake, T.A., Haaf, M., Paradise, B.J., Millevolte, A.J., Powell, D.R., and West, R. (2001) *J. Organomet. Chem.*, **636**, 17.
70. Gehrhus, B. and Lappert, M.F. (2001) *J. Organomet. Chem.*, **617–618**, 209.
71. Clendenning, S.B., Gehrhus, B., Hitchcock, P.B., Moser, D.F., Nixon, J.F., and West, R. (2002) *J. Chem. Soc. Dalton Trans.*, 484.
72. Hills, N.J. and West, R. (2004) *J. Organomet. Chem.*, **689**, 4165.
73. Kong, J., Zhang, J., Song, H., and Cui, C. (2009) *Dalton Trans.*, 5444.
74. Watanabe, C., Iwamoto, T., Kabuto, C., and Kira, M. (2008) *Angew. Chem.*, **120**, 5466; *Angew. Chem. Int. Ed.* (2008), **47**, 5386.
75. Ueno, K., Tobita, H., Shimoi, M., and Ogino, H. (1988) *J. Am. Chem. Soc.*, **110**, 4092.
76. Tobita, H., Ueno, K., Shimoi, M., and Ogino, H. (1990) *J. Am. Chem. Soc.*, **112**, 3415.
77. Takeuchi, T., Tobita, H., and Ogino, H. (1991) *Organometallics*, **10**, 835.
78. Tobita, H., Wada, H., Ueno, K., and Ogino, H. (1994) *Organometallics*, **13**, 2545.
79. Ueno, K., Masuko, A., and Ogino, H. (1997) *Organometallics*, **16**, 5023.
80. Okazaki, M., Tobita, H., and Ogino, H. (1997) *Chem. Lett.*, 437.
81. Tobita, H., Kurita, H., and Ogino, H. (1998) *Organometallics*, **17**, 2844.
82. Tavčar, G., Sen, S.S., Azhakar, R., Thorn, A., and Roesky, H.W. (2010) *Inorg. Chem.*, **49**, 10199.
83. Azhakar, R., Sarish, S.P., Roesky, H.W., Hey, J., and Stalke, D. (2011) *Inorg. Chem.*, **50**, 5039.
84. Li, J., Merkel, S., Henn, J., Meindl, K., Döring, A., Roesky, H.W., Ghadwal, R.S., and Stalke, D. (2010) *Inorg. Chem.*, **49**, 775.
85. Wang, W., Inoue, S., Yao, S., and Driess, M. (2010) *J. Am. Chem. Soc.*, **132**, 15890.
86. Sheldrick, W.S. (1989) in *The Chemistry of Organic Silicon Compounds*, vol. 1, Chapter 3 (eds S. Patai and Z.

Rappoport), John Wiley & Sons, Inc., New York, and references therein.

87. Meltzer, A., Präsang, C., Milsmann, C., and Driess, M. (2009) *Angew. Chem.*, **121**, 3216; *Angew. Chem., Int. Ed.*, 2009, **48**, 3170..

88. Meltzer, A., Präsang, C., and Driess, M. (2009) *J. Am. Chem. Soc.*, **131**, 7232.

89. Meltzer, A., Inoue, S., Präsang, C., and Driess, M. (2010) *J. Am. Chem. Soc.*, **132**, 3038.

90. Watanabe, C., Inagawa, Y., Iwamoto, T., and Kira, M. (2010) *Dalton Trans.*, **39**, 9414.

91. Someya, C.I., Haberberger, M., Wang, W., Enthaler, S., and Inoue, S. (2013) *Chem. Lett.*, **42**, 286.

92. Someya, C.I., Irran, E., and Enthaler, S. (2012) *Asian J. Org. Chem.*, **1**, 322.

93. Haberberger, M., Someya, C.I., Company, A., Irran, E., and Enthaler, S. (2012) *Catal. Lett.*, **142**, 557.

94. Someya, C.I., Inoue, S., Krackl, S., Irran, E., and Enthaler, S. (2012) *Eur. J. Inorg. Chem.*, 1269.

95. Lide, D.R. (ed.) (2005) *CRC Handbook of Chemistry and Physics*, CRC Press, Boca Raton, FL.

96. Netherton, M.R. and Fu, G.C. (2004) *Adv. Synth. Catal.*, **346**, 1525.

97. Phapale, V.B. and Cárdenas, D.J. (2009) *Chem. Soc. Rev.*, **38**, 1598.

98. Jana, R., Pathak, T.P., and Sigman, M.S. (2011) *Chem. Rev.*, **111**, 1417.

99. Hu, X. (2011) *Chem. Sci.*, **2**, 1867.

100. Pugh, D. and Danopoulos, A.A. (2007) *Coord. Chem. Rev.*, **251**, 610.

101. Benito-Garagorri, D. and Kirchner, K. (2008) *Acc. Chem. Res.*, **41**, 201.

102. Choi, J., MacArthur, A.H.R., Brookhart, M., and Goldman, A.S. (2011) *Chem. Rev.*, **111**, 1761.

103. Gunanathan, C. and Milstein, D. (2011) *Acc. Chem. Res.*, **44**, 588, and references therein.

104. Selander, N. and Szabó, K. (2009) *J. Dalton Trans*, 6267.

105. Selander, N. and Szabó, K.J. (2011) *Chem. Rev.*, **111**, 2048, and references therein.

106. Zhang, H. and Lei, A. (2011) *Dalton Trans.*, **40**, 8745, and references therein.

107. Wang, W., Inoue, S., Irran, E., and Driess, M. (2012) *Angew. Chem.*, **124**, 3571; *Angew. Chem. Int. Ed.*, 2012, **51**, 3691.

108. So, C.-W., Roesky, H.W., Gurubasavaraj, P.M., Oswald, R.B., Gamer, M.T., Jones, P.G., and Blaurock, S. (2007) *J. Am. Chem. Soc.*, **129**, 12049.

109. Brück, A., Gallego, D., Wang, W., Irran, E., Driess, M., and Hartwig, J.F. (2012) *Angew. Chem.*, **124**, 11645; *Angew. Chem. Int. Ed.*, 2012, **51**, 11478.

110. Hartwig, J.F. (2010) *Organotransition Metal Chemistry. From Bonding To Catalysis*, University Science Book, Sausalito, CA, p. 47.

111. Wang, W., Inoue, S., Enthaler, S., and Driess, M. (2012) *Angew. Chem.*, **124**, 6271; *Angew. Chem. Int. Ed.*, 2012, **51**, 6167.

112. Baumann, F., Dormann, E., Ehleiter, Y., Kaim, W., Krächer, J., Kelemen, M., Krammer, R., Saurenz, D., Stalke, D., Wachter, C., Wolmershäuser, G., and Sitzmann, H. (1999) *J. Organomet. Chem.*, **587**, 267.

113. Hung-Low, F. and Bradley, C.A. (2011) *Organometallics*, **30**, 2636.

114. Hung-Low, F., Krogman, J.P., Tye, J.W., and Bradley, C.A. (2012) *Chem. Commun.*, **48**, 368.

115. Ramachandran, A., Ghadwal, R.S., Roesky, H.W., Hey, J., and Stalke, D. (2012) *Chem. Asian. J.*, **7**, 528.

116. Ghadwal, S., Azhakar, R., Pröpper, K., Holstein, J.J., Dittrich, B., and Roesky, H.W. (2011) *Inorg. Chem.*, **50**, 8502.

117. Ingleson, M., Fan, H., Pink, M., Tomaszewski, J., and Caulton, K.G. (2006) *J. Am. Chem. Soc.*, **128**, 1804.

118. Eichman, C.C., Bragdon, J.P., and Stambuli, J.P. (2011) *Synlett*, 1109.

119. Geny, A., Agenet, N., Iannazzo, L., Malacria, M., Aubert, C., and Gandon, V. (2009) *Angew. Chem.*, **121**, 1842; *Angew. Chem. Int. Ed.*, 2009, **48**, 1810.

120. Hilt, G., Hengst, C., and Hess, W. (2008) *Eur. J. Org. Chem.*, 2293, and references therein.

121. Bönnemann, H. (1978) *Angew. Chem.*, **90**, 517; *Angew. Chem., Int. Ed. Engl.*, 1978, **17**, 505.

122. Peter, K. and Vollhardt, C. (1984) *Angew. Chem.*, **96**, 525; *Angew. Chem., Int. Ed. Engl.*, 1984, **23**, 539.
123. Bönnemann, H. (1985) *Angew. Chem.*, **97**, 264; *Angew. Chem., Int. Ed. Engl.*, 1985, **24**, 248.
124. Hess, W., Treutwein, J., and Hilt, G. (2008) *Synthesis*, **22**, 3537.
125. Varela, A. and Caá, C. (2008) *Synlett*, 2571.
126. Shinata, T. and Tsuchikama, K. (2008) *J. Org. Biomol. Chem.*, **6**, 1317.
127. Scheuermann née Taylor, C.J. and Ward, B.D. (2008) *New J. Chem.*, **32**, 1850.
128. Tanaka, K. (2009) *Chem. Asian. J.*, **4**, 508, and references therein.
129. Nehl, H. (1994) *Chem. Ber.*, **127**, 2535.
130. Taârit, Y.B., Diab, Y., Elleuch, B., Kerkani, M., and Chihaoui, M. (1986) *J. Chem. Soc., Chem. Commun.*, 402.
131. Brändli, C. and Ward, T.R. (2000) *J. Comb. Chem.*, **2**, 42.
132. Wakatsuki, Y. and Yamazaki, H. (1985) *Bull. Chem. Soc. Jpn.*, **58**, 2715.
133. Wang, C., Li, X., Wu, F., and Wan, B. (2011) *Angew. Chem.*, **123**, 7300; *Angew. Chem. Int. Ed.*, 2011, **50**, 7162.
134. Schmid, G. and Welz, E. (1977) *Angew. Chem.*, **89**, 823; *Angew. Chem., Int. Ed. Engl.*, 1977, **16**, 785.
135. Amoroso, D., Haaf, M., Yap, G.P.A., West, R., and Fogg, D.E. (2002) *Organometallics*, **21**, 534.
136. Stoelzel, M., Präsang, C., Inoue, S., Enthaler, S., and Driess, M. (2012) *Angew. Chem.*, **124**, 411; *Angew. Chem. Int. Ed.*, 2012, **51**, 399.
137. Campion, B.K., Heyn, R.H., and Tilley, T.D. (1988) *J. Chem. Soc., Chem. Commun.*, 278.
138. Azhakar, R., Ghadwal, R.S., Roesky, H.W., Wolf, H., and Stalke, D. (2012) *J. Am. Chem. Soc.*, **134**, 2423.
139. Neumann, E. and Pfaltz, A. (2005) *Organometallics*, **24**, 2008.
140. Gehrhus, B., Hitchcock, P.B., Lappert, M.F., and Maciejewski, H. (1988) *Organometallics*, **17**, 5599.
141. Perti, S.H.A., Eikenberg, D., Neumann, B., Stammler, H.-G., and Jutzi, P. (1999) *Organometallics*, **18**, 2615.
142. Herrmann, W.A., Härter, P., Gstöttmayr, C.W.K., Bielert, F., Seeboth, N., and Sirsch, P. (2002) *J. Organomet. Chem.*, **649**, 141.
143. Zeller, A., Bielert, F., Härter, P., Herrmann, W.A., and Strassner, T. (2005) *J. Organomet. Chem.*, **690**, 3292.
144. Yang, W., Fu, H., Wang, H., Chen, M., Ding, Y., Roesky, H.W., and Jana, A. (2009) *Inorg. Chem.*, **48**, 5058.
145. Blom, B., Enthaler, S., Inoue, S., Irran, E., and Driess, M. (2013) *J. Am. Chem. Soc.*, **135**, 6703–6713. doi: 10.1021/ja402480v
146. Jana, A., Leusser, D., Objartel, I., Roesky, H.W., and Stalke, D. (2011) *Dalton Trans.*, **40**, 5458.
147. Rodriguez, R., Gau, D., Contie, Y., Kato, T., Saffon-Merceron, N., and Baceiredo, A. (2011) *Angew. Chem.*, **123**, 11694; *Angew. Chem. Int. Ed.*, 2011, **50**, 11492.
148. Abraham, M.Y., Wang, Y., Xie, Y., Wei, P., Schaefer, H.F. III, Schleyer, P.v.R., and Robinson, G.H. (2011) *J. Am. Chem. Soc.*, **133**, 8874.
149. Al-Rafia, S.M.I., Malcolm, A.C., McDonald, R., Ferguson, M.J., and Rivard, E. (2012) *Chem. Commun.*, **48**, 1308, and references cited therein.
150. Hansch, C., Leo, A., and Taft, R.W. (1991) *Chem. Rev.*, **91**, 165.
151. Bent, H.A. (1961) *Chem. Rev.*, **61**, 275.
152. Tobita, H., Matsuda, A., Hashimoto, H., Ueno, K., and Ogino, H. (2004) *Angew. Chem.*, **116**, 223; *Angew. Chem. Int. Ed.*, 2004, **43**, 221.
153. Morris, R.H. (2009) *Chem. Soc. Rev.*, **38**, 2282.
154. Plietker, B. (ed.) (2008) *Iron Catalysis in Organic Chemistry*, Wiley-VCH Verlag GmbH, Weinheim.
155. Gaillard, S. and Renaud, J.-L. (2008) *ChemSusChem*, **1**, 505.
156. Correa, A., Mancheño, O.G., and Bolm, C. (2008) *Chem. Soc. Rev.*, **37**, 1108.
157. Sherry, B.D. and Fürstner, A. (2008) *Acc. Chem. Res.*, **41**, 1500.
158. Bullock, R.M. (2007) *Angew. Chem.*, **119**, 7504; *Angew. Chem. Int. Ed.*, 2007, **46**, 7360.
159. Bolm, C., Legros, J., Le Paih, J., and Zani, L. (2004) *Chem. Rev.*, **104**, 6217.

160. Junge, K., Schröder, K., and Beller, M. (2011) *Chem. Commun.*, **47**, 4849.

161. Bauer, E.B. (2008) *Curr. Org. Synth.*, **12**, 1341.

162. Czaplik, W.M., Mayer, M., Cvengros, J., and Jacobi von Wangelin, A. (2009) *ChemSusChem*, **2**, 396.

163. Enthaler, S., Junge, K., and Beller, M. (2008) *Angew. Chem.*, **120**, 3363; *Angew. Chem. Int. Ed.*, 2008, **47**, 3317.

164. Zhang, M. and Zhang, A. (2010) *Appl. Organomet. Chem.*, **24**, 751.

165. Darwish, M. and Wills, M. (2012) *Catal. Sci. Technol.*, **2**, 243.

166. Shaikh, N.S., Enthaler, S., Junge, K., and Beller, M. (2008) *Angew. Chem.*, **120**, 2531; *Angew. Chem. Int. Ed.*, 2008, **47**, 2497.

167. Bart, S.C., Lobkovsky, E., and Chirik, P.J. (2004) *J. Am. Chem. Soc.*, **126**, 13794.

168. Haberberger, M., Irran, E., and Enthaler, S. (2011) *Eur. J. Inorg. Chem.*, **2011**, 2797.

169. Enthaler, S., Haberberger, M., and Irran, E. (2011) *Chem. Asian. J.*, **6**, 1613.

170. Belger, C. and Plietker, B. (2012) *Chem. Commun.*, **48**, 5419.

171. Enthaler, S. (2011) *ChemCatChem*, **3**, 666.

172. Cardoso, J.N.S. and Royo, B. (2012) *Chem. Commun.*, **48**, 4944.

173. Tondreau, A.M., Lobkovsky, E., and Chiril, P.J. (2008) *Org. Lett.*, **10**, 2789.

174. Dieskau, A.P., Begouin, J.-M., and Plietker, B. (2011) *Eur. J. Org. Chem.*, 5291.

175. Addis, D., Shaikh, N., Zhou, S., Das, S., Junge, K., and Beller, M. (2010) *Chem. Asian. J.*, **5**, 1687.

176. Meakin, P., Muetterties, E.L., and Jesson, J.P. (1973) *J. Am. Chem. Soc.*, **95**, 75.

177. Fasulo, M.E. and Tilley, T.D. (2012) *Organometallics*, **31**, 5049.

178. Ochiai, M., Hashimoto, H., and Tobita, H. (2012) *Organometallics*, **31**, 527.

179. Krasovskiy, A. and Knochel, P. (2004) *Angew. Chem.*, **116**, 3396; *Angew. Chem. Int. Ed.*, 2004, **43**, 3333.

11
Halide Exchange Reactions Mediated by Transition Metals

Alicia Casitas Montero

11.1
Introduction

Aryl halides are commonly applied as versatile intermediates for many organic reactions as depicted in Scheme 11.1. They can be easily activated by many transition metals and participate in a wide range of C-C cross-couplings reactions such as Heck, Suzuki-Miyaura, Negishi, and Stille reactions [1, 2], as well as in several C-heteroatom bond forming reactions such as Buchwald-Hartwig [3, 4] and Ullmann-type reactions [5–7]. Moreover, they are widely used for the synthesis of organometallic compounds, for instance, organolithium and Grignard reagents, and as precursors for the generation of free-radical intermediates. Generally, the reactivity order of aryl halides increases significantly as the bond strength of the C-halogen decreases, being more reactive aryl iodides > bromides > chlorides. Among the different halides, the cheapest and most commercially available aryl chlorides are the least reactive and, therefore, there is a high interest in developing aromatic halogen exchange methods to produce aryl bromides and iodides from the corresponding aryl chlorides.

On the other hand, aryl halides are also important synthetic targets in their own, because they can be found in a large number of pharmaceuticals and agrochemicals (Scheme 11.2). Because aryl chlorides and fluorides are usually inert to chemical transformations they are much more commonly found in biological active molecules than aryl bromides and iodides. Currently, the presence of fluoroaromatic compounds in pharmaceutical industry is increasing due to the beneficial attributes related to fluorine atoms. The incorporation of fluorine into a drug increases metabolic chemical inertness, thermal stability, lipophilicity, solubility, and noncovalent interactions with biological targets. These attributes lead not only to an enhancement of drug efficacy, but also to lower dosing, which can reduce undesirable side effects [8]. Moreover, the synthesis of aryl fluoride compounds containing radioactive ^{18}F isotope is of high interest for medical applications, for example, for positron emission tomography (PET), an imaging technique used for cancer diagnose and disease staging, among others [9]. Current industrial methods used for the preparation of fluoroarenes, for instance,

Discovering the Future of Molecular Sciences, First Edition. Edited by Bruno Pignataro.

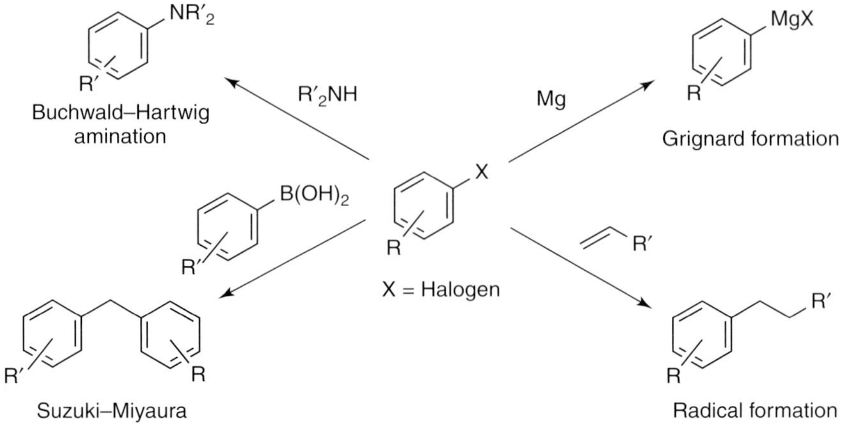

Scheme 11.1 Representative examples of the reactivity of aryl halides in organic chemistry.

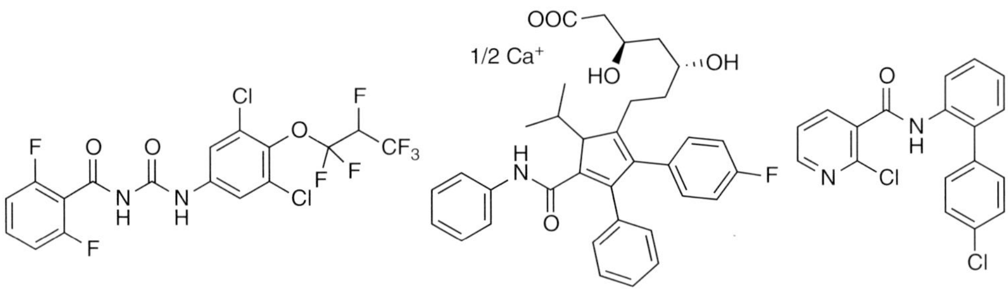

Scheme 11.2 Haloarene compounds as pharmaceuticals and agrochemicals.

the Balz–Schiemann reaction [10] or the Halex reaction [11], have several practical limitations such as modest scope, the requirement of potentially explosive reagents, low yields, and long reactions times, and, thus, the development of new synthetic methods are of great current interest.

Therefore, the ability to exchange a given halide in an aryl group for another halide, regarding the target of the haloarene, is a high useful transformation. The halide exchange in alkyl halides, known as Finkelstein reaction, is a useful methodology for obtaining alkyl iodides from the corresponding aryl chlorides or bromides in the presence of an alkali iodide salt [12]. This reaction occurs by means of a S_N2 nucleophilic substitution, where the driving force is the low solubility of the alkali chloride or bromide formed during the reaction. In contrast, the analogous aromatic and vinylic halogen exchange reaction cannot be readily achieved by direct nucleophilic substitution of the halogen atom due to the inaccessibility of the C–X antibonding orbital. Only electron-deficient haloarenes can participate in halogen exchange processes by nucleophilic aromatic substitution (S_NAr) mechanism, and this method is used for the preparation of fluoroarenes compounds. The use of transition metals, such as palladium, nickel, and copper, has allowed the development of aromatic halide exchange reactions of a wide range of aryl halides under relatively mild conditions [13].

From a mechanistic point of view, halide exchange reactions catalyzed by transition metals can be envisioned to occur through the typical catalytic cycle of most cross-coupling reactions (Scheme 11.3). Thus, the aryl halide is activated by oxidative addition, affording an arylmetal intermediate that is oxidized by two electrons. Then, ligand exchange followed by C-halogen reductive elimination renders the desired aryl halide and regenerates the metal catalyst. Whereas oxidative addition and ligand exchange steps are known to occur commonly in metal complexes, C-halogen reductive elimination is considered the bottleneck of the catalytic cycle [14]. Although the latter step is thermodynamically disfavored, by using appropriate ligands coordinated to the metal as well as modifying the reaction conditions, C-halogen reductive elimination can be promoted by kinetic control [15]. Alternatively, other mechanistic pathways, for instance, single-electron transfer processes initiated by transition metals, have also been proposed for halide exchange reactions [7].

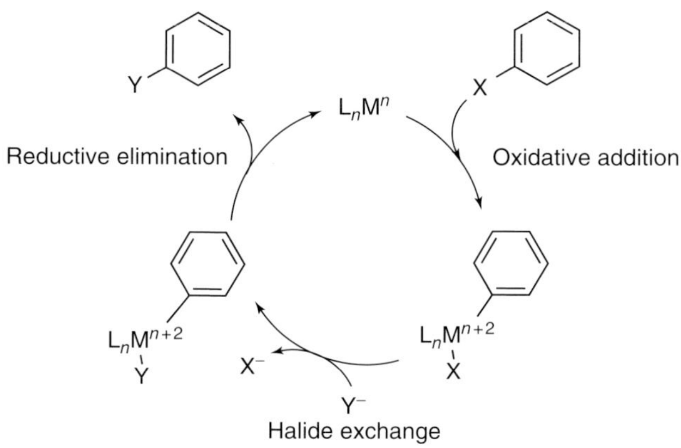

Scheme 11.3 Catalytic cycle based on two-electron redox steps for halide exchange reactions.

Examples of aromatic halide exchanges catalyzed by transition metals date back to more than 50 years, even though, the harsh conditions required as well as their narrow substrate scope limited their utility in organic synthesis. However, recent advances in halide exchange reactions mainly with palladium and copper catalysts have enabled to consider them as versatile tools for having access to a wide range of aryl halides. In this chapter, a general overview of transition metal-catalyzed aromatic halide exchange reactions are covered, with special emphasis on the most synthetic useful halide exchanges: (i) preparation of aryl iodides from aryl chlorides and bromides and (ii) synthesis of fluoroarenes compounds from different halogen derivatives. Moreover, sulfonate esters, often referred to as pseudohalides, can be readily accessed from a wide range of commercially available phenols, and their conversion to aryl halides is also discussed in this chapter.

11.2 Nickel-Based Methodologies for Halide Exchanges

Nickel catalyzed halide exchange reactions have been known since the 1970s, when Cramer and Coulson first published that nickel salts effectively catalyze halide exchange of nonactivated aryl halides. They reported the formation of chlorobenzene from bromobenzene using very low catalyst loading of $NiCl_2$ (0.001 mol%) and lithium chloride at very high temperatures (>150 °C) [16]. In a more detailed study, Tsou and Kochi observed that the use of nickel complexes containing phosphine ligands efficiently catalyzed the conversion of several aryl iodides to aryl bromides using quaternary ammonium bromide salts under milder reaction conditions [17]. In the presence of 3 mol% of nickel catalyst (o-$CH_3Ph)Ni(PEt_3)_2$ and 1 equiv of tetrabutylammonium bromide, iodobenzene was converted to bromobenzene in 74% yield in benzene at 80 °C (Scheme 11.4). The halide exchange process is dependent on the solvent, being more efficient in benzene and tetrahydrofuran in comparison with other tested solvents such as ethanol, DMSO, or acetone. Furthermore, the reaction of iodobenzene/Bu_4NBr or bromobenzene/Bu_4NI was found to reach equilibrium in the presence of the catalytic amounts of nickel complex (o-$CH_3Ph)Ni(PEt_3)_2$. Mechanistic insights suggest the implication of nickel(I) species as the active catalyst for the halide exchange reaction in aryl halides.

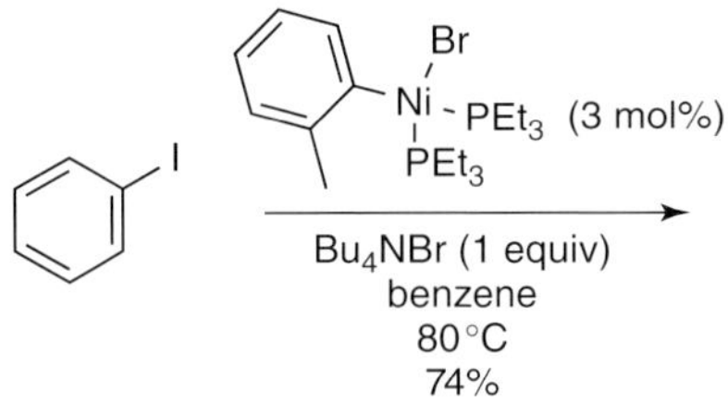

Scheme 11.4 Synthesis of bromoarenes catalyzed by nickel complexes and bromide salts.

More recently, Arvela and Leadbeater reported a more general methodology for halide exchange reactions, using stoichiometric amounts of nickel salts in dimethylformamide (DMF) at high temperatures (170 °C) (Scheme 11.5). The use of microwave heating allowed the desired transformation in very short reaction times (5 min), and, in addition, the exclusion of air and water was not required. However, this method is limited to the preparation of simple aryl chlorides from the corresponding aryl bromides and iodides, and to the synthesis of aryl bromides from aryl iodides.

Regarding reverse halide exchange reactions for the preparation of aryl iodides, several reports have demonstrated the feasibility of low valent nickel complexes to promote this transformation. By *in situ* reduction of nickel(II) salts with zinc powder in the presence of potassium iodide in polar solvent hexamethylphosphoramide (HMPA), aryl iodides were obtained in moderate yields from the corresponding aryl bromides [18, 19]. However, the requirement for a large excess of Zn powder

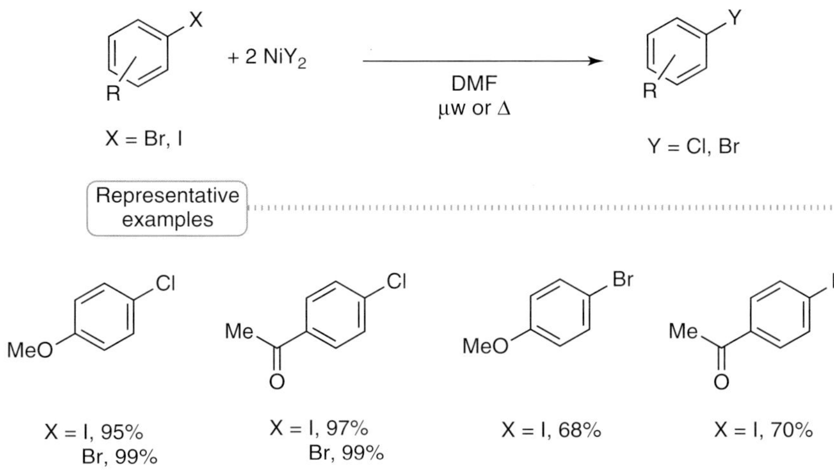

Scheme 11.5 Preparation of aryl chlorides and bromides using stoichiometric nickel salts.

caused the formation of homo-coupled biaryl side products that decreases the yield of the desired aryl iodide. The use of less reactive Ni(0) powder instead of *in situ* generated Ni(0) complexes was found to be more selective toward halide exchange process, even though higher temperatures were required [20]. Therefore, Ni powder with potassium iodide (KI) converted aryl bromides, and in less extent aryl chlorides, to the corresponding aryl iodides in DMF at 150 °C with moderate yields (Scheme 11.6).

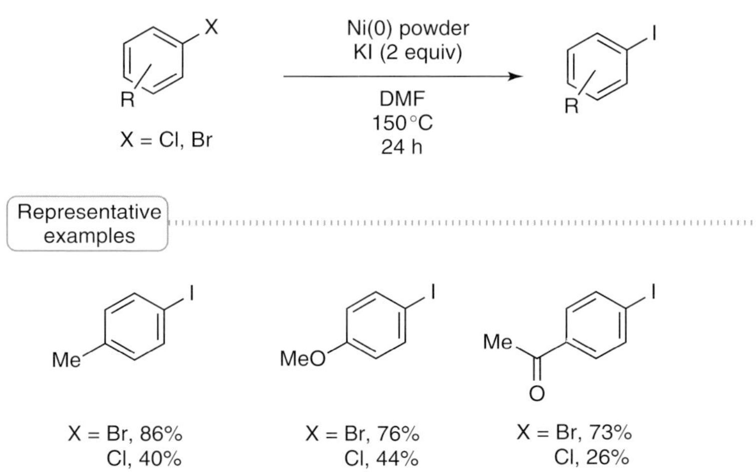

Scheme 11.6 Reverse halogen exchange toward aryl iodides catalyzed by nickel(0) powder.

These examples showed that nickel complexes are able to promote halide exchange reactions in simple haloarenes. However, some current limitations, for instance, the requirement of high temperatures, the use of stoichiometric nickel salts, or the formation of biaryl side products have to be improved for further implementation of nickel-based methodologies in the synthesis of more complex haloarenes.

11.3
Recent Advances in Palladium-Catalyzed Aryl Halide Exchange Reactions

Palladium-catalyzed cross-coupling reactions are well-established methodologies for the formation of C–C bonds as well as C-heteroatom bonds (essentially C–N and C–O) [21]. Nevertheless, the research directed toward the development of halide exchange reactions based on palladium catalysts has been very scarce until the appearance of very recent publications. In this context, palladium-catalyzed halogen exchange reactions could be envisaged through the standard catalytic cycle Pd(0)/Pd(II) proposed for cross-coupling reactions (Scheme 11.3). However, whereas the oxidative addition of an aryl halide to palladium(0) is a well-known step, the reversible C-halogen reductive elimination is rare because the free energy favors in most cases the oxidative addition. Nevertheless, few examples have shown the feasibility of palladium(II) complexes to undergo C-halogen reductive elimination [15].

In this regard, practical methods for halide exchange reactions based on palladium catalysts are still undeveloped, even though, some recent reports have demonstrated the capability of palladium to promote such reactions. The synthetically useful conversion of aryl chlorides to more reactive aryl iodides is limited to a single report, where halide exchange and Sonogashira reactions are performed in a multistep one-pot reaction [22]. Electron-deficient aryl chlorides are activated to aryl iodides, prior to the coupling reaction with alkynes, using sodium iodide, potassium fluoride as base and Pd/C as heterogeneous catalyst at 130 °C in DMF.

On the other hand, Buchwald and coworkers developed the first general method for the conversion of aryl triflates to aryl bromides or chlorides catalyzed by palladium (Scheme 11.7). The use of palladium catalyst that contains sterically hindered dialkylbiaryl monophosphine ligand was key for the success of the reaction. However, this method suffered from low applicability due to the large number of additives needed for achieving good conversions of the corresponding aryl halides. In this regard, a phase-transfer catalyst, such as PEG3400, was required to increase the solubility of potassium bromide or chloride. Moreover, the addition of $i\text{-Bu}_3\text{Al}$ was required to sequester the formed potassium triflate that inhibited the reaction. Therefore, the addition of 2-butanone to generate the dialkylaluminum alkoxide *in situ* with $i\text{-Bu}_3\text{Al}$ was needed in order to suppress the formation of undesired by-products (both C–C cross-coupling products and reduction products).

In a latter report, Buchwald and coworkers developed a more practical method for the conversion of aryl triflates to aryl bromides or chlorides [23]. The same

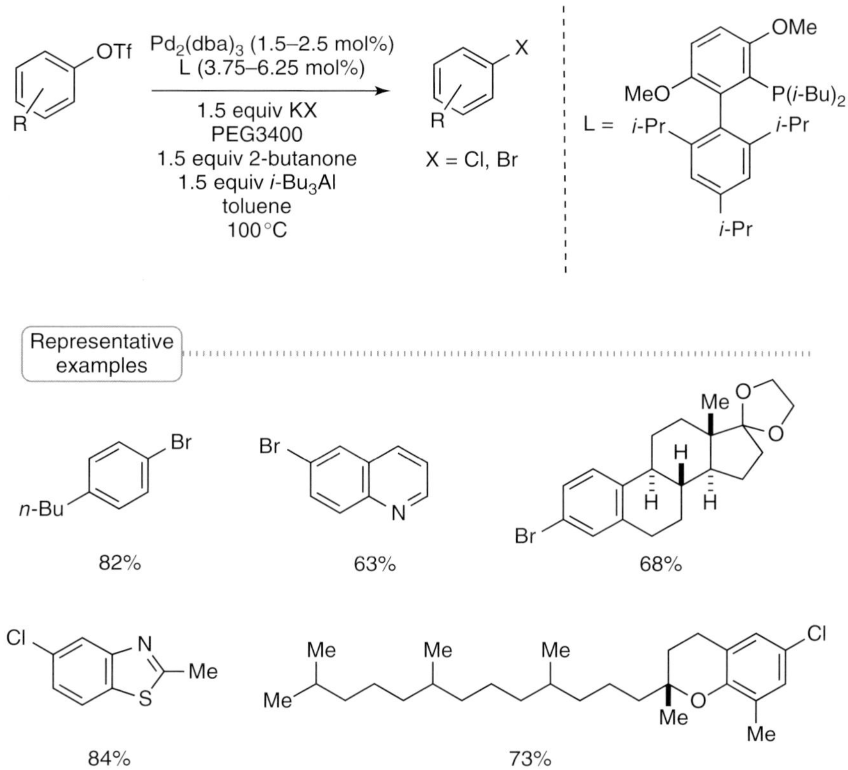

Scheme 11.7 Palladium-catalyzed conversion of aryl triflates to aryl bromides and chlorides.

palladium catalyst, the use of bromide or chloride salts in 1,4-dioxane as solvent, instead of toluene, and the use of potassium fluoride as a single additive allowed to synthesize aryl halides in moderate to good yields, even though higher temperatures were required (130 °C). However, no mechanistic studies were reported and the crucial role of potassium fluoride in producing the desired product was not understood.

Fluoroarene formation by means of C–F reductive elimination from well-defined arylpalladium(II)-fluoride complexes has proven to be extremely challenging, because the formation of strong metal-fluorine bonds causes a slower C–F reductive elimination in comparison with competing side reactions, such as P–F and C–C reductive elimination reactions [24, 25]. Nevertheless, Buchwald and coworkers reported a breakthrough in the field in 2009 [26]. The use again of a bulky monophosphine ligand was found key to promote C_{aryl}–F reductive elimination from an aryl-Pd(II) fluoride complex. Catalytic nucleophilic fluorination of aryl triflates or bromides was developed using cesium fluoride and palladium catalyst (Scheme 11.8). However, the selectivity of the method was sparse due to the formation of undesired regioisomeric fluoroarenes and the formation of reduced products under reaction conditions. Strikingly, this is the first example of aromatic fluorination based on Pd(0)/Pd(II) catalytic cycle operating under mild reaction conditions, even though a more complex mechanism was proposed in a later report due to *in situ* palladium catalyst modification [27].

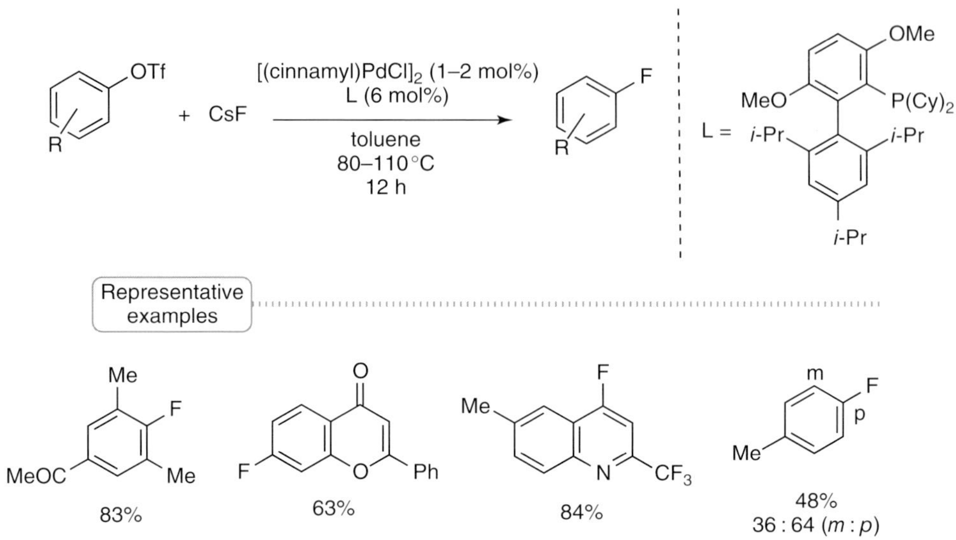

Scheme 11.8 Palladium-catalyzed aromatic fluorination of aryl triflates reported by Buchwald and coworkers.

Whereas C-halogen reductive elimination step from arylpalladium(II) complexes requires forcing conditions in most cases, the use of electrophilic halogenating reagents that gives access to high-valent arylpalladium complexes is a promising strategy for promoting C-halogen reductive elimination under milder reaction conditions [10, 28]. In this context, aryl halides can be synthesized by means of palladium-catalyzed ligand-directed C–H bond functionalization through a Pd(II)/Pd(IV) catalytic cycle (Scheme 11.9) [29, 30]. Coordination of the directing group to palladium(II) species facilitates selective C–H activation in *ortho* position to this directing group forming an arylpalladium(II) complex. Then, formation of an arylpalladium(IV) intermediate in the presence of electrophilic halogenating reagent and subsequent C-halogen reductive elimination affords the corresponding aryl halide and regenerates the active Pd(II) species. Nevertheless, the isolation of both monomeric Pd(IV) and bimetallic Pd(III) complexes that are able to undergo reductive elimination causes some mechanistic controversy about the nature of active palladium complex in catalytic C–H functionalization reactions [31]. It is important to highlight that even fluorine atoms can be inserted into the aromatic rings by means of this strategy (Scheme 11.9) [30].

In addition, C–F reductive elimination has been observed in several well-defined arylpalladium(IV) fluoride complexes [32–34]. Significantly, Ritter and

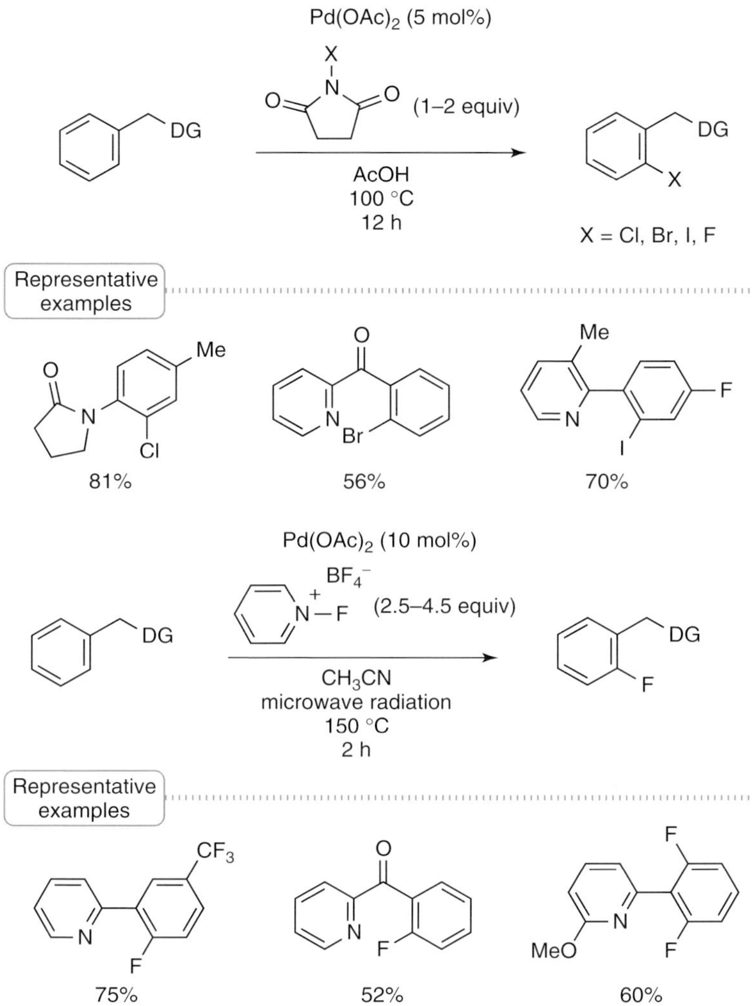

Scheme 11.9 Palladium-catalyzed ligand-directed C–H functionalization reactions for the preparation of haloarenes developed by Sanford and coworkers. DG = directing group.

coworkers [35] have developed late-state fluorination of complex molecules promoted by palladium(IV) complexes (Scheme 11.10). This strategy consisted of the generation of arylpalladium(II) complexes by transmetalation of boronic acids and subsequent oxidation to Pd(IV) intermediate, with an electrophilic fluorinating reagent, which undergoes C–F reductive elimination. The synthesis and application of an octahedral Pd(IV) fluoride complex as highly reactive fluorinating agent, instead of common F-TEDA (1-chloromethyl-4-fluoro-1,4-diazoniabicyclo[2.2.2]octane bis(tetra-fluoroborate)) [36], allowed to perform the fluorinating reaction at very mild conditions. Furthermore, the relevance of this

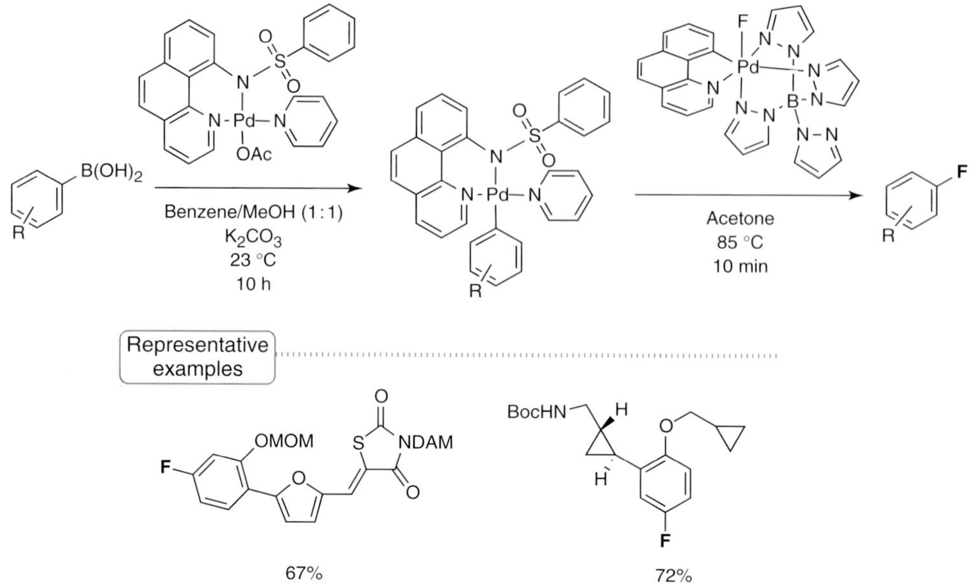

Scheme 11.10 Late-stage fluorination promoted by arylpalladium(IV) complexes. The indicated yield corresponds to the C–F reductive elimination step.

methodology is illustrated by their implementation in the synthesis ^{18}F-radiolabeled fluoroarenes, which can serve as ^{18}F-PET tracers, because very short reaction times, less than 10 min, are required for the C–F reductive elimination step.

The field of palladium-catalyzed halide exchange reactions have changed completely during the past decade following the recent advances summarized in this section. These reports have demonstrated for the first time that palladium catalysts, in specific ligand environments, such as bulky monophosphine ligands, catalyze halide exchange reactions, including fluorination of simple aryl halides or pseudohalides. Moreover, the synthesis of high-valent organometallic palladium complexes has allowed overcoming the difficulties associated with C-halogen reductive elimination. Current methodologies are still at a great distance from industrial applications and, therefore, further research points toward the development of more general and practical halide exchange reactions with palladium catalysts.

11.4 The Versatility of Copper-Catalyzed Aryl Halide Exchange Reactions

Early examples of copper-catalyzed halogen exchange reactions of aromatic halides were reported by Bacon and Hill in the 1960s [37]. Heating copper(I) halide salts, Cu(I)X (X = Cl, Br, I), in the presence of aryl halides in polar solvents afforded the exchanged aryl halides (Scheme 11.11). The reactivity order observed in copper salts was CuCl > CuBr > CuI, whereas regarding the several aryl halides, the trend found was Ar–I > Ar–Br > Ar–Cl ≫ Ar–F. This reactivity pattern can be explained by the energy of the formed C-halogen bond, which is favored toward the formation of a stronger C–X bond (Cl > Br > I). Therefore, the reaction is useful only synthetically for the preparation of aryl chlorides from aryl bromides or iodides and the preparation of aryl bromides from aryl iodides. Alternatively, Bacon and Hill [38] also developed halide exchange reactions catalyzed by copper(I) oxide in the presence of lithium, sodium, or quaternary ammonium halide salts. However, these procedures cannot be applied for the conversion of lighter halides to heavier halides. For example, the synthesis of more reactive aryl iodide, which will be a more preparatively useful halide exchange method, gives only very low yields (<10%).

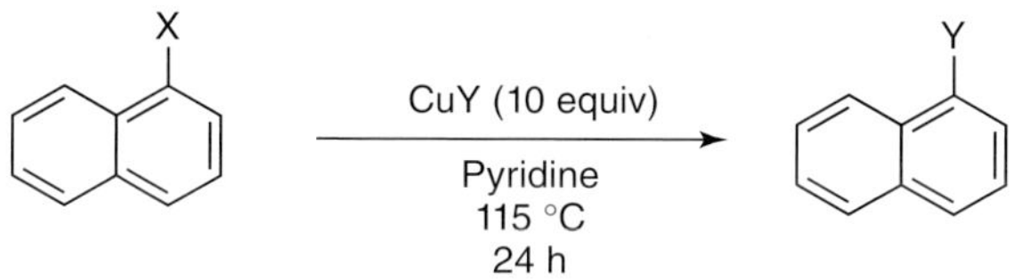

X	Y	Yield (%)
Br	Cl	84
I	Br	100
Cl	Br	5

Scheme 11.11 Early example of halide exchange reaction catalyzed by copper reported by Bacon and Hill.

An earlier example of reverse halogen exchange reaction was reported in 1985 by Ogawa and coworkers. Their work consisted of the synthesis of aryl iodides from aryl bromides, in moderate yields, using potassium iodide and copper(I) iodide in HMPA at 150 °C [39, 40]. Nevertheless, the most significant contribution to the copper-catalyzed halide exchange reactions appeared in 2002, when Buchwald and Klapar [41] reported a general and practical procedure to synthesize aryl iodides from the corresponding aryl bromides. They found that the addition of bidentate amines had an accelerating effect in copper-catalyzed halide exchange reactions, affording the desired transformation under mild conditions. Thus, the strategy consisted of using catalytic amounts of copper(I) iodide (5 mol%), *trans*-cyclohexane-1,2-diamine ligand (10 mol%), and 2 equiv of sodium iodide at 110 °C in 1,4-dioxane (Scheme 11.12). This methodology afforded high conversions of a wide range of aromatic and heteroaromatic iodides from the corresponding bromides, under relatively mild reaction conditions. Importantly, the presence of very versatile functional groups on the substrates, such as amides, esters, pyridines, and amines, was well tolerated. This procedure was an important improvement in comparison with the previous halogen exchange methodologies as the use of stoichiometric amount of copper, high polar solvents, and high temperatures were not required. In analogy to the Finkelstein reaction [12], the driving force for conversion to the aryl iodide is provided by the difference in the solubility of the sodium halides. The conversion to the aryl iodide is higher in solvents in which sodium bromide has relatively low solubility, such as dioxane or pentanol, because its precipitation displaces the equilibrium reaction.

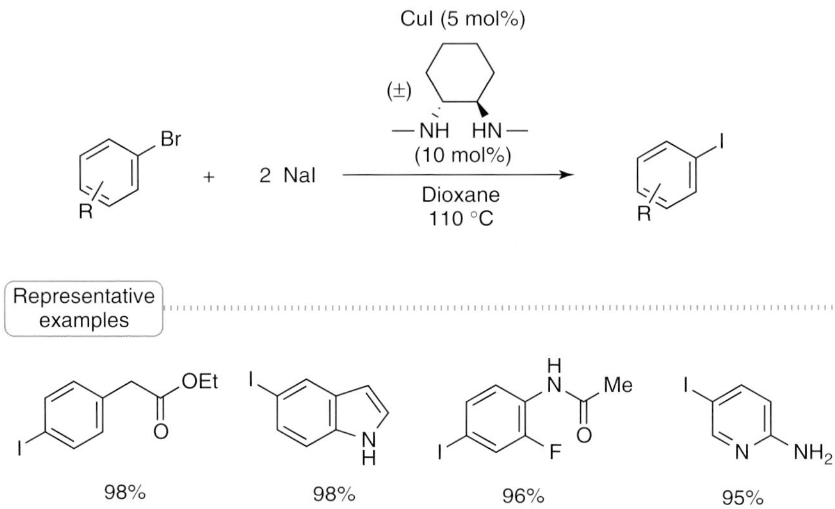

Scheme 11.12 Conversion of aryl bromides to aryl iodides using as catalyst copper salts and diamine ligands.

This methodology has found several synthetic applications, for example, in multiple consecutive cross-coupling reactions on functionalized substrates, which take advantage of the fact that aryl bromides are less reactive than aryl iodides in common cross-coupling reactions. In this context, Willand and coworkers developed a multicomponent reaction, with three consecutive copper-catalyzed reactions, that is, Ullmann/Halide-Exchange/Ullmann, that gives access to double dissymmetrical cross-coupling products in high yields from the reaction of 1-bromo-4-iodobenzene with a mixture of carbamate and amide nucleophiles (Scheme 11.13a) [42]. In the first step, the most acidic and less sterically hindered nitrogen nucleophile reacts with aryl iodide; then, halide exchange allows the conversion *in situ* of aryl bromide to more reactive aryl iodide, which reacts with the second nitrogen nucleophile, in the subsequent step.

Furthermore, copper-catalyzed halide exchange reactions have been applied in *in situ* copper-catalyzed conversion of aryl bromides to aryl iodides followed by N-arylation and S-arylation for the synthesis of anilides or aromatic thioethers [43, 44]. Thus, poorly reactive aryl bromides are activated by halide exchange toward the corresponding aryl iodides, prior to the cross-coupling reaction with

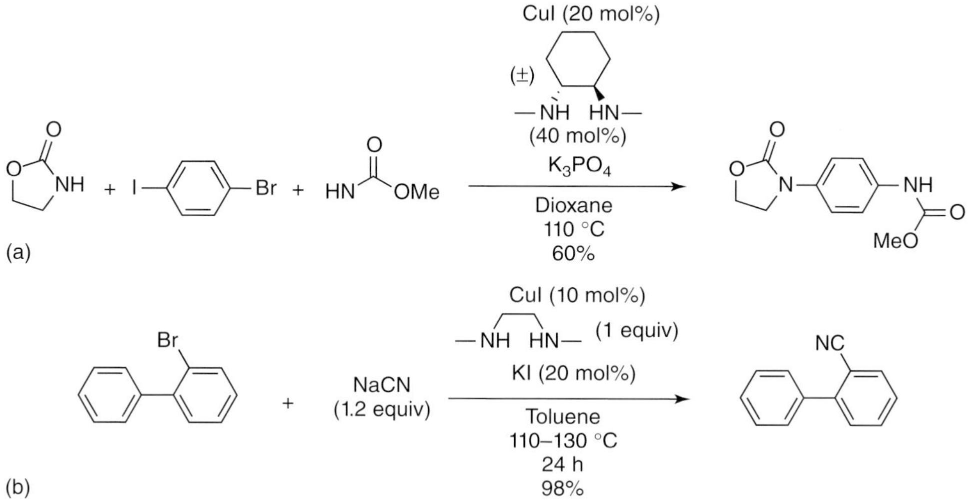

Scheme 11.13 (a, b) Representative applications of copper-catalyzed halide exchange reactions in domino reactions.

heteronucleophiles. Similarly, the groups of Buchwald and Taillefer, independently, incorporated this methodolgy in the synthesis of aromatic nitrile compounds through copper-catalyzed domino halogen exchange-cyanation of aryl bromides (Scheme 11.13b) [45, 46].

Significantly, copper-catalyzed halide exchange reactions have been applied in total synthesis of natural products recently. In this context, (±)-aspercyclide [47], an anti-inflammatory agent used in the prevention of allergic disorders, and (−)-kaitocephalin [48], a glutamate receptor antagonist studied as potential neuroprotectant to prevent neurodegenerative disorders, are some examples that illustrate the relevance of halide exchange methodologies as practical tools for organic synthesis.

Although copper-catalyzed aromatic fluorinations from aryl halides are not as developed as the rest of halide exchange reactions so far, recent reports have demonstrated that copper is a promising transition metal catalyst for such transformations. In an early report, Vigalok and coworkers showed that copper(I) complex $(Ph_3P)_2CuF$ reacts stoichiometrically with 2-bromonitrobenzene and potassium fluoride in refluxing DMF to afford quantitatively the corresponding fluorobenzene [49]. However, copper-mediated aromatic fluorination reactions from haloarenes were limited to this single example until 2007. Grushin patented a general methodology for preparing fluoroarenes from a wide range of nonactivated haloarenes mediated by copper difluoride, TMEDA (*N*,*N*,*N*′,*N*′-tetramethylethylenediamine) as auxiliar ligand and alkali metal fluorides under high temperature conditions (usually above 150 °C) and in high polar solvent HMPA (Scheme 11.14). This is a remarkable methodology because of the difficulty of introducing fluorine atoms in aromatic rings, even though the harsh conditions employed and the stoichiometric loadings of CuF_2 are clear drawbacks for the implementation of this reaction in late-stage fluorination of highly functionalizated compounds. It is worth mention that another CuF_2 mediated process provides the formation of aryl fluorides from benzene by using hydrofluoric acid and oxygen with only water as a by-product, but the low yields obtained (below 30%) and the very high temperatures required (>500 °C) preclude a wide implementation [50].

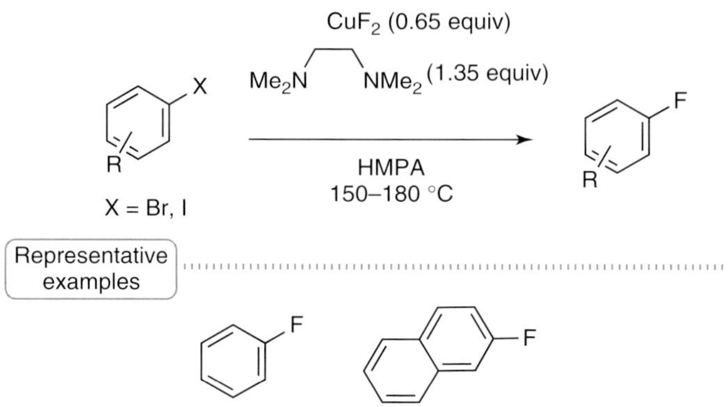

Scheme 11.14 Copper-mediated fluorination reaction mediated by CuF_2 patented by Grushin and coworkers.

Despite the fact that the formation of C–F bonds by reductive elimination is a challenging reaction, Ribas and coworkers have demonstrated the feasibility of high-valent arylcopper(III)-halide complexes (Cl, Br, I), which are stabilized in triazamacrocyclic ligands, to promote C–F bond forming reaction by reductive elimination [51]. By reacting arylcopper(III)-halide complex with common fluoride salts, such as silver fluoride or potassium fluoride, the corresponding fluoroarene and copper(I) salt were obtained at room temperature. However, the intermediacy of arylcopper(III)-fluoride complex is supported only by computational data, because the halide exchange step, prior to the C–F reductive elimination, is the rate-determining step. In addition, Ribas and coworkers developed copper-catalyzed aromatic fluorination in model aryl halide triazamacrocyclic substrates by using catalytic amounts of copper(I) salts and silver fluoride, achieving moderate to excellent yields of fluorination of aryl halides (chlorides and bromides) (Scheme 11.15). Under reaction conditions, arylcopper(III)-halide complex was detected by spectroscopic methods, which gives support to a oxidative addition/reductive elimination mechanistic pathway for the aromatic fluorination reaction. Furthermore, copper-catalyzed halide exchange reactions between chlorides, bromides, and iodides were also developed in these macrocyclic aryl halides using catalytic amounts of copper(I) salts and excess of halide salts. Similarly, Wang and coworkers reported another family of arylcopper(III) complexes, stabilized in azacalixarene ligands, that are able to participate in C-halogen bond forming reactions (F, Cl, Br, I) in the presence of several alkali halide salts [52]. Although the reactions of arylcopper(III) complexes are limited to intramolecular examples, these studies open the door to explore more general and efficient methodologies for copper-catalyzed aromatic halide exchange reactions including fluorinations based on Cu(I)/Cu(III) catalytic cycle.

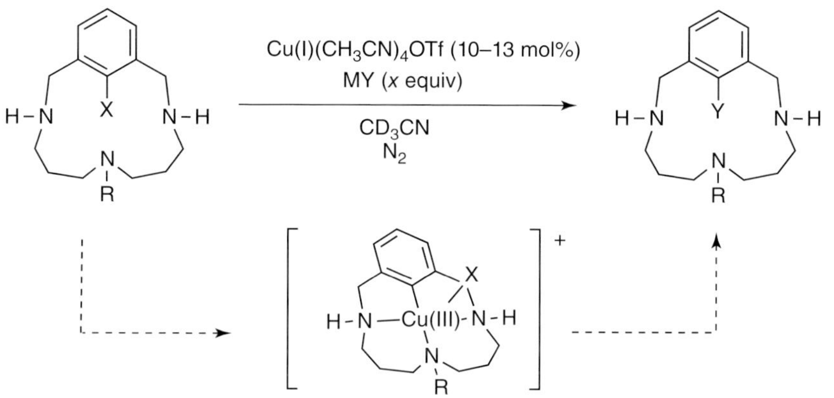

R	X	MY (equiv)	Yield (%)
H	Cl	AgF (2)	76
H	Br	AgF (2)	71
CH_3	Cl	AgF (2)	98
CH_3	Br	AgF (2)	97
H	I	Bu_4NBr (10)	87
H	I	Bu_4NCl (10)	99
H	Br	Bu_4NCl (10)	99
H	Br	NaI (10)	84
H	Cl	NaBr (10)	37
H	Cl	NaI (10)	62

Scheme 11.15 Intramolecular halide exchange reactions occurring by the intermediacy of arylcopper(III)-halide complexes.

In this context, very recently, Fier and Hartwig made an important contribution to intermolecular copper-mediated aromatic fluorinations [53]. Simple aryl iodides were converted to the corresponding fluoroarenes using silver fluoride and Cu(*t*-BuCN)$_2$OTf as copper(I) complex (Scheme 11.16). The authors gave support to mechanistic pathway initiated by oxidative addition of copper(I) to the aryl iodide to form a copper(III) intermediate and C–F reductive elimination from an arylcopper(III)fluoride complex. The use of nitriles as the weakly donating ligand was postulated to be key to promote the C–F reductive elimination from copper(III) intermediate. However, this methodology required the use of an excess of Cu(*t*-BuCN)$_2$OTf with respect to aryl iodide, as well as temperatures above 140 °C for the reaction to occur in high yields. In addition, the presence of adventitious water causes the formation of dehalohydrogenation products, which decreases the yield of the desired fluorinated compounds and makes difficult their purification.

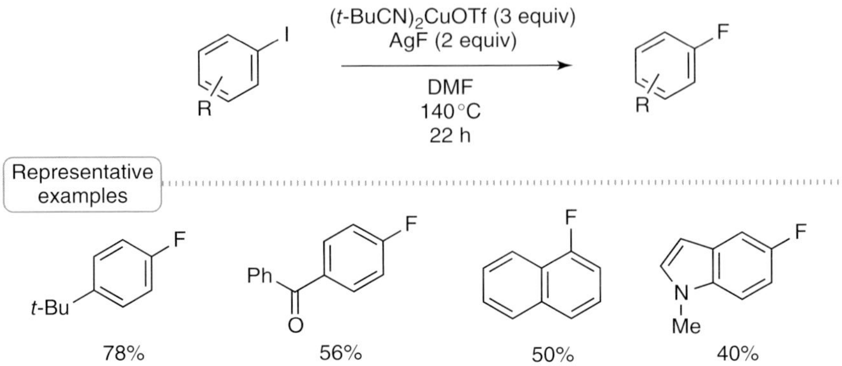

Scheme 11.16 Copper-mediated aromatic fluorination of simple aryl iodides reported by Fier and Hartwig.

Finally, it is worth mentioning that the use of copper(I) complexes and electrophilic fluorinating reagents has allowed the synthesis of fluorinated compounds in moderate yields from the corresponding aryl stannanes, aryl trifluoroborates, or arylboronate esters, which are synthesized directly from the corresponding aryl halides (Scheme 11.17) [54, 55]. Independently, the groups of Hartwig and Sanford have developed different fluorinating methodologies that are accommodated by the same mechanistic pathway. Initial oxidation of copper(I) by the electrophilic fluorinating agent generates a high-valent copper(III) fluoride complex. This intermediate reacts with an organometallic reagent by transmetalation and generates an arylcopper(III)-fluoride complex, which is prone to C–F reductive elimination. Experimental evidences for the formation of copper(III) species were provided by Hartwig and coworkers by means of ^{19}F-NMR spectroscopy. Remarkably, the reaction is carried out under mild reaction conditions, even though stoichiometric amounts of copper salts and excess of fluorinated oxidant are required. These reports are the starting point for further development of catalytic fluorinations based on a Cu(I)/Cu(III) catalytic cycle.

11.5 Conclusions and Perspectives

Recent advances in aromatic halogen exchange reactions catalyzed by transition metals have demonstrated the increased potential of nickel, palladium, and copper complexes for catalyzing such transformations. However, most current methodologies are not qualified for practical applications as they are still limited in scope or

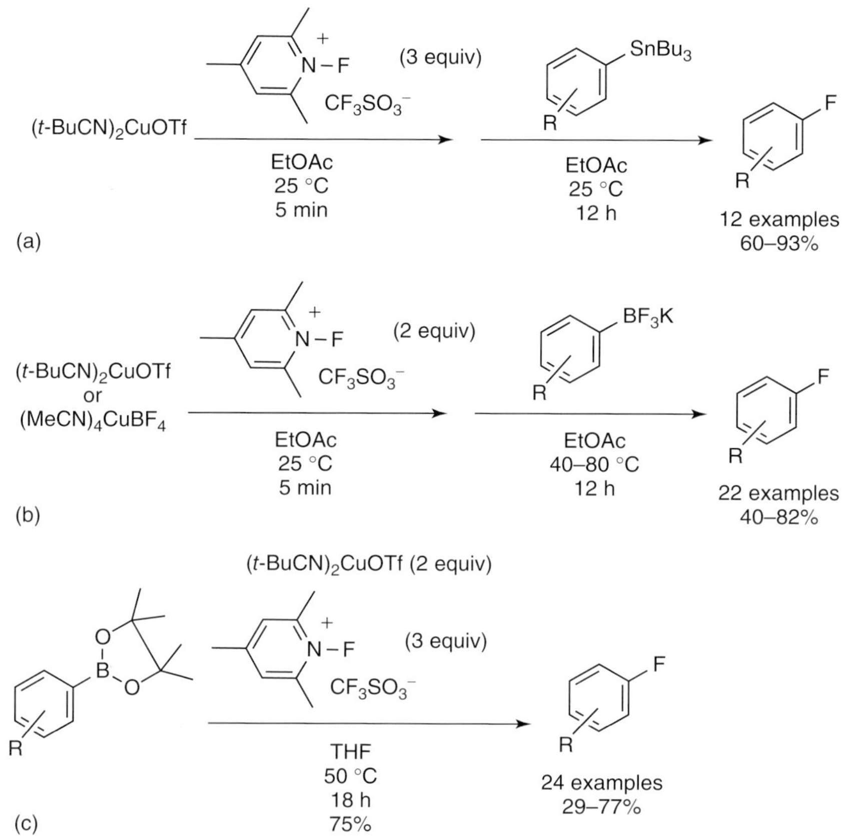

Scheme 11.17 Fluorine insertion methodologies mediated by copper salts in the presence of electrophilic fluorinating reagents from (a) aryl stannanes, (b) aryl trifluoroborates, and (c) arylboronate esters.

to certain experimental conditions, which are generally far from being considered mild and environmentally benign. Copper complexes have shown to be the most efficient catalysts to date for mediating halide exchange reactions, and their use in the synthesis of complex aryl iodides have found applications in the syntheses of natural products. Detailed mechanistic studies of halide exchange reactions need to be made because mechanistic understanding is key for developing more efficient and practical methods. It is clear that oxidative addition/reductive elimination pathway is a plausible mechanism for accommodating a wide range of halide exchange reactions. Some breakthroughs regarding aromatic fluorination methods have disclosed that both palladium and copper complexes can mediate the insertion of fluorine atoms by means of C−F reductive elimination. Nevertheless, these fluorinating methodologies are far from developed but further improvement of the existent methods and the research toward the development of more efficient catalysts will be made in due course.

References

1. Beletskaya, I.P. and Cheprakov, A.V. (2000) *Chem. Rev.*, **100**, 3009–3066.
2. Henry, P.M. (2002) in *Handbook of Organopalladium Chemistry for Organic Synthesis* (ed. E. Negishi), John Wiley & Sons, Inc., New York, p. 2119.
3. Hartwig, J.F. (1998) *Acc. Chem. Res.*, **31**, 852–860.
4. Hartwig, J.F. (2008) *Acc. Chem. Res.*, **41**, 1534–1544.
5. Beletskaya, I.P. and Cheprakov, A.V. (2004) *Coord. Chem. Rev.*, **248**, 2337–2364.
6. Ley, S.V. and Thomas, A.W. (2003) *Angew. Chem. Int. Ed.*, **42**, 5400–5449.
7. Sperotto, E., van Klink, G.P.M., van Koten, G., and de Vries, J.G. (2010) *Dalton Trans.*, **39**, 10338–10351.
8. Müller, K., Faeh, C., and Diederich, F. (2007) *Science*, **317**, 1881.
9. Gouverneur, V. (2012) *Nat. Chem.*, **4**, 152–154.
10. Powers, D.C. and Ritter, T. (2009) *Nat. Chem.*, **1**, 302–309.
11. Langlois, B., Gilbert, L., and Forat, G. (1996) *Ind. Chem. Lib.*, **8**, 244–292.
12. Finkelstein, H. (1910) *Ber. Dtsch. Chem. Ges.*, **43**, 1528–1532.
13. Sheppard, T.D. (2010) *Org. Biomol. Chem.*, **7**, 1043–1052.
14. Vigalok, A. (2008) *Chem. Eur. J.*, **14**, 5102–5108.
15. Roy, A.H. and Hartwig, J.F. (2003) *J. Am. Chem. Soc.*, **125**, 13944–13945.
16. Cramer, R. and Coulson, D.R. (1975) *J. Org. Chem.*, **40**, 2267–2273.
17. Tsou, T.T. and Kochi, J.K. (1980) *J. Org. Chem.*, **45**, 1930–1937.
18. Takagi, K., Hayama, N., and Inokawa, S. (1980) *Bull. Chem. Soc. Jpn.*, **53**, 3691–3695.
19. Takagi, K., Hayama, N., and Okamoto, T. (1978) *Chem. Lett.*, 191–192.
20. Yang, S.H., Li, C.S., and Cheng, C.H. (1987) *J. Org. Chem.*, **52**, 691–694.
21. de Meijere, A. and Diederich, F. (2004) *Metal-Catalyzed Cross-Coupling Reactions*, vol. **1**, Wiley-VCH Verlag GmbH, Weinheim.
22. Thathagar, M.B. and Rothenberg, G. (2006) *Org. Biomol. Chem.*, **4**, 111–115.
23. Pan, J., Wang, X., Zhang, Y., and Buchwald, S.L. (2011) *Org. Lett.*, **13**, 4974–4976.
24. Grushin, V.V. (2010) *Acc. Chem. Res.*, **43**, 160–171.
25. Ball, N.D., Kampf, J.W., and Sanford, M.S. (2010) *Dalton Trans.*, **39**, 632–640.
26. Watson, D.A., Su, M., Teverovskiy, G., Zhang, Y., García-Fortanet, J., Kinzel, T., and Buchwald, S.L. (2009) *Science*, **325**, 1661–1664.
27. Maimone, T.J., Milner, P.J., Kinzel, T., Zhang, Y., Takase, M.K., and Buchwald, S.L. (2011) *J. Am. Chem. Soc.*, **133**, 18106–18109.
28. Whitfield, S.R. and Sanford, M.S. (2007) *J. Am. Chem. Soc.*, **129**, 15142–15143.
29. Kalyani, D., Dick, A.R., Anani, W.Q., and Sanford, M.S. (2006) *Org. Lett.*, **8**, 2523–2526.
30. Hull, K.L., Anani, W.Q., and Sanford, M.S. (2006) *J. Am. Chem. Soc.*, **128**, 7134–7135.
31. Powers, D.C., Xiao, D.Y., Geibel, M.A.L., and Tobias Ritter, T. (2010) *J. Am. Chem. Soc.*, **132**, 14530–14536.
32. Furuya, T. and Ritter, T. (2008) *J. Am. Chem. Soc.*, **130**, 10060–10061.
33. Ball, N.D. and Sanford, M.S. (2009) *J. Am. Chem. Soc.*, **131**, 3796–3797.
34. Furuya, T., Benitez, D., Tkatchouk, E., Strom, A.E., Tang, P., Goddard-III, W.A., and Ritter, T. (2010) *J. Am. Chem. Soc.*, **132**, 3793–3807.
35. Lee, E., Kamlet, A.S., Powers, D.C., Neumann, C.N., Boursalian, G.B., Furuya, T., Choi, D.C., Hooker, J.M., and Ritter, T. (2011) *Science*, **334**, 639–642.
36. Furuya, T., Kaiser, H.M., and Ritter, T. (2008) *Angew. Chem. Int. Ed.*, **47**, 5993–5996.
37. Bacon, R.G.R. and Hill, H.A.O. (1964) *J. Chem. Soc.*, 1097–1107.
38. Bacon, R.G.R. and Hill, H.A.O. (1964) *J. Chem. Soc.*, 1108–1112.
39. Suzuki, H., Kondo, A., Inouye, M., and Ogawa, T. (1986) *Synthesis*, (2), 121–122.
40. Suzuki, H., Kondo, A., and Ogawa, T. (1985) *Chem. Lett.*, 411–412.

41. Klapars, A. and Buchwald, S.L. (2002) *J. Am. Chem. Soc.*, **124**, 14844–14845.
42. Toto, P., Gesquière, J.-C., Cousaert, N., Deprez, B., and Willand, N. (2006) *Tetrahedron Lett.*, **47**, 4973–4978.
43. Jones, C.P., Anderson, K.W., and Buchwald, S.L. (2007) *J. Org. Chem.*, **72**, 7968–7973.
44. Carril, M., SanMartin, R., Domínguez, E., and Tellitu, I. (2007) *Chem. Eur. J.*, **13**, 5100–5105.
45. Zanon, J., Klapars, A., and Buchwald, S.L. (2003) *J. Am. Chem. Soc.*, **125**, 2890–2891.
46. Cristau, H.-J., Ouali, A., Spindler, J.-F., and Taillefer, M. (2005) *Chem Eur. J.*, **11**, 2483–2492.
47. Carr, J.L., Offermann, D.A., Holdom, M.D., Dusart, P., White, A.J.P., Beavil, A.J., Leatherbarrow, R.J., Lindell, S.D., Sutton, B.J., and Spivey, A.C. (2010) *Chem. Commun.*, **46**, 1824–1826.
48. Vaswani, R.G. and Chamberlin, A.R. (2008) *J. Org. Chem.*, **73**, 1661–1681.
49. Antipin, I.S., Vigalok, A.I., and Konovalov, A.I. (1991) *Zh. Org. Khim.*, **27**, 1577–1577.
50. Subramanian, M.A. and Manzer, L.E. (2002) *Science*, **297**, 1665.
51. Casitas, A., Canta, M., Costas, M., Solà, M., and Ribas, X. (2011) *J. Am. Chem. Soc.*, **133**, 19386–19392.
52. Yao, B., Wang, Z.-L., Zhang, H., Wang, D.-X., Zhao, L., and Wang, M.-X. (2012) *J. Org. Chem.*, **77**, 3336–3340.
53. Fier, P.S. and Hartwig, J.F. (2012) *J. Am. Chem. Soc.*, **134**, 10795–10798.
54. Ye, Y. and Sanford, M.S. (2013) *J. Am. Chem. Soc.*, **135**, 4648–4651.
55. Fier, P.S., Luo, J., and Hartwig, J.F. (2013) *J. Am. Chem. Soc.*, **135**, 2552–2559.

12
Nanoparticle Assemblies from Molecular Mediator

Marie-Alexandra Neouze Gauthey

12.1
Introduction

The study of nanoparticles represents a breakthrough in material science not only from a fundamental point of view but also for numerous applications, such as materials reinforcement, doping agent for materials conductivity, for drug delivery purposes, and for cancer cell detection [1, 2].

For about 10–15 years, the progress in understanding nanoparticles drove material scientists at focusing on specific materials related to the assemblies of nanoparticles [3–7]. These new materials aim at making use of the nanoparticles' collective properties by bringing nanoparticles next to one another, at distances where they can interact with one another without aggregating.

Toward this purpose, the synthesis pathway to create nanoparticle assemblies has to be perfectly controlled in order to get fixed distances between the nanoparticles. This distance has to be long enough for the nanoparticles to not aggregate, in which case they would lose their nanometer-size-related properties; at the same time, the distance between the nanoparticles has to be short enough to get good "communication," interaction possibilities, between the individual nanoparticles.

This chapter, not being exhaustive, presents some of these new materials, the way they are prepared and the application they are aiming at. This review is constructed as follows (Figure 12.1): The first part of the chapter analyzes the main synthesis pathways proposed in the literature, such as template-assisted synthesis, layer-by-layer deposition, and the use of covalent molecular mediators. In this part, attention is given to associate various synthesis ways with the nature of the nanoparticle, such as metal oxide, organic, and metallic. Examples have been chosen to highlight the main aspects and the principle of these methods.

As the field of nanoparticle assemblies is already very broad, we focus only on the nanoparticle assemblies obtained from molecular mediators, or linker-assisted syntheses. Hence, some preparation methods, such as nanolithography [8], chemical vapor synthesis [9], and the use of microcavities, are not described [10].

Discovering the Future of Molecular Sciences, First Edition. Edited by Bruno Pignataro.

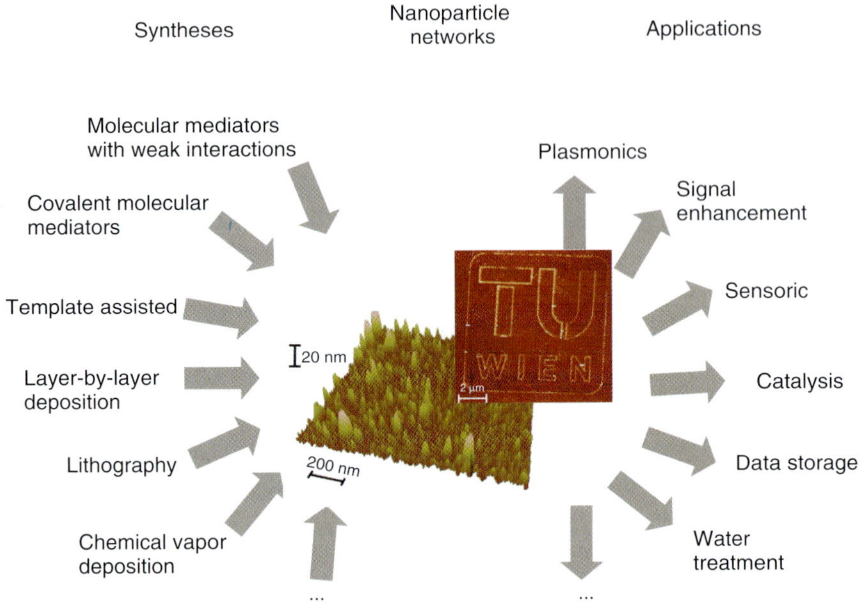

Figure 12.1 Overview of various synthesis pathways and applications for nanoparticle assemblies.

Linker-assisted syntheses are based on the use of molecules that are present between the nanoparticles. By controlling the length and rigidity of these molecules, one can control the distance between the nanoparticles.

The second part of the chapter presents some of the various applications targeted by the new materials. Among these applications, we can mention sensoric or catalysis as well as plasmonics, assemblies of metal nanoparticles that could drive to a great breakthrough in materials for communication.

In this chapter, if the deposition of films is treated in a separate paragraph, the formation of one-dimensional assemblies is not separately but briefly discussed for each concerned method. This choice was driven by the fact that, to produce 1D assemblies, most of the main approaches are similar as those for 2D and 3D architectures. This aspect has been highlighted in the review articles of Kotov and coworkers [7, 11].

12.2
Assembly or Self-assembly

The borderline between assembly and self-assembly was discussed by Yu and coworkers [12], on the basis of the definition given in 2002 by Whitesides *et al.* [13]: "self-assembly is an autonomous organization of components into patterns or structures without human intervention." This borderline was then "defined" by "human intervention" and thus can be very delicate and sometimes even difficult to distinguish. Hence, this chapter discusses the assembly and self-assembly of nanoparticles.

Also, a more difficult aspect concerns the "patterns or structures" formed by the assembly, as defined by Whitesides *et al.* [13]. As stressed by Yu and coworkers, if restricting to assemblies highly ordered over long distances, more than 90% of the articles referring to assembly or self-assembly would be rejected. Indeed, most of the works are reporting structures without pattern, with only short-range order, or even disorder between the nanoparticles. In this chapter, we keep in mind that the definition should be the formation of a pattern. Nevertheless, short-range ordered and disordered nanoparticle assemblies will also be considered as they help to progress in the material science of assemblies.

12.3 Nanoparticles and Their Protection against Aggregation or Agglomeration

12.3.1 Finite-Size Objects

Nanoparticles are defined as particles measuring less than 100 nm in, at least, one dimension. Nanoparticles can be spheres but they can also be nano-cubes, nano-rods, nano-pyramid even complex or irregular geometries, as long as the characteristic dimension is smaller than 100 nm. The specificities of nanoparticles are threefold:

- First, given the size of a nanoparticle, the surface energy of these nanoparticles is extremely high. This nanometric size develops a *considerable specific surface area* that is extremely advantageous for surface reactions, like in catalysis.
- Second, corollary to the previous point, owing to the limited size of the nanoparticle, the *ratio of surface atoms to bulk atoms is much higher* than in conventional materials. For example, in a particle with a diameter of 5 nm, 50% of the atoms are surface atoms, while in a 1 cm diameter particle only a few 10^{-4}% atoms are surface atoms. Considering surface controlled or surface limited reactions, such as catalysis, the number of bulk atoms, which are not involved in the reactions, is limited. "Few unreactive bulk atoms" is synonymous of reaction-efficiency as well as cost-efficiency. The cost-efficiency starts to be a non-negligible aspect when considering expensive raw materials.
- Third, the finite-size of the nanoparticle gives rise to an *unusual electronic structure*. The energy levels of a nanoparticle are distinct from the defined structure of a molecular compound, where the combination of a few atomic orbitals leads to a very define structure with well separated energy levels. The structure of a nanoparticle is also distinct from the one of larger-size solid-state materials where the molecular orbitals are so close to one another that they end in merging into energy bands. In the case of zinc oxide, for example, quantum confinement can be obtained for particles with a diameter lower than 7 nm.

Nanoparticles generally tend to aggregate in order to reduce their high surface energy. Depending on the type of interaction involved between the nanoparticles, one speaks of nanoparticle aggregation or nanoparticle agglomeration. The agglomerates, also described as "soft agglomerates," relate to physically interacting nanoparticles. While aggregates, also called "hard agglomerates", correspond to nanoparticles that are strongly bonded with one another by means of chemical bonds. In agglomerates, the bonds between the nanoparticles are relatively easy to break up as opposed to the chemical bonds in aggregates. Thus *aggregates are considered to be irreversible.*

12.3.2 Protection against Aggregation

As has been stressed many times in the previous paragraph, all the specific features of nanoparticles are directly linked to their finite, nanometric size, and to the limited number of atoms per particle. In consequence, all the features get lost when aggregates are formed.

Hence, scientists developed methods to hinder the aggregation of nanoparticles. The two main methods are the steric and the electrostatic hindrances:

- The steric hindrance is based on the modification of the nanoparticle surface with ligands [14]. Owing to the presence of molecules on the surface of the nanoparticles, the nanoparticles can no more interact directly with one another. This method was extensively used for protecting luminescent quantum dots [15].
- The electrostatic hindrance is based on charge repulsion whose repulsion forces the maintaining of a minimum distance between two nanoparticles, proportional to the inverse of the square radius of the nanoparticle.

In order to get a nanoparticle charged, two main routes can be used. The first route consists in using a ligand with a charged functional group, such as a triethyldodecyl quaternary ammonium ligand [14–16]. A second method is to get far away from the nanoparticle point of zero charge.

Nanoparticles tend naturally to aggregate in order to diminish their high surface energy. With aggregation, all the specificities of nanoparticle would get lost. Thus, many methods have been developed to hinder aggregation of nanoparticles.

When considering nanoparticle assemblies, the challenge is to bring the nanoparticles near enough for them to interact, but simultaneously far enough for them not to be considered as aggregated.

12.4 Nanoparticle Assemblies Synthesis Methods

In literature, some major methods for the preparation of nanoparticle assemblies have been reported. We concentrate on four main methods here: (i) interligand bonding, where a molecule is introduced between the nanoparticles, and will remain in the final material; (ii) template-assisted method, where the molecules will force the organization of the nanoparticles; (iii) deposition of 2D assemblies, where the interaction with the surface helps to organize the nanoparticle assembly; and (iv) pressure driven assemblies.

12.4.1 Interligand Bonding

12.4.1.1 Noncovalent Linker Interactions and Self-assembly

Electrostatic Interactions Futamata and coworkers prepared a gold nanoparticles assembly where the bifunctional cationic rhodamine 123 molecules linked the nanoparticles with one another [17]. The gold nanoparticles were obtained by the reduction of $HAuCl_4$ with citric acid in water. The authors measured an average diameter for the gold nanoparticles of 20 nm. The citrate anions on the gold nanoparticles were replaced with chloride ions from NaCl treatment. In this experiment, cationic aromatic rhodamine was used as the linker. Owing to the aromatic character of the molecule, the positive charge was delocalized and could be considered to be half on both amino groups (schematized with a gray bar on the rhodamine molecule in Figure 12.2). The cationic rhodamine linker interacted with the Cl^- treated gold nanoparticles (Figure 12.2) via electrostatic interactions.

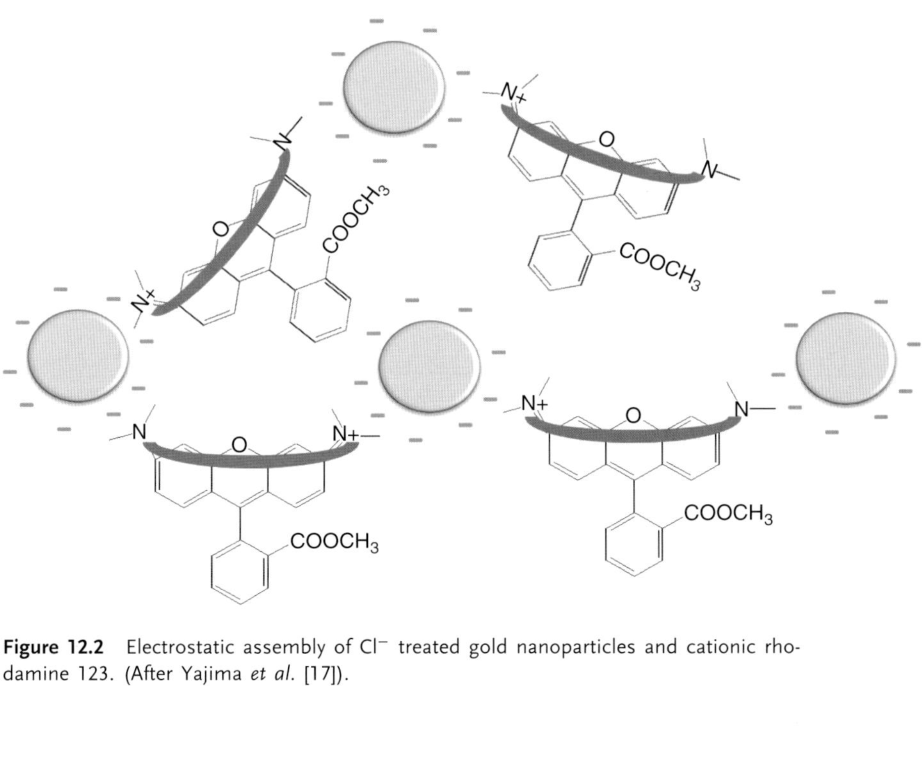

Figure 12.2 Electrostatic assembly of Cl^- treated gold nanoparticles and cationic rhodamine 123. (After Yajima *et al.* [17]).

Biomolecules are intensively studied in bio- or biomimetic-material sciences for their accessibility, no-toxicity. For example, DNA is one of the most studied biomolecule in material science. It offers a perfectly defined structure and multiple binding sites.

In the example reviewed by Gao and coworkers [18], the negatively charged phosphate groups of DNA chains in dimethylsulfoxide (DMSO) allowed the binding of Zn(II) cations from zinc acetate. The zinc atoms immobilized on the DNA chains reacted with the hydroxides added and conducted to the formation of ZnO nanoparticles by sol-gel reaction. The DNA promoted and controlled the ZnO nanoparticle formation. This reaction ended in one-dimensional chains of around 10 nm diameter zinc oxide nanoparticles. Photoluminescence spectra of the chains showed the typical excitation and emission wavelength awaited for 10 nm diameter zinc oxide nanoparticles.

Hydrogen Bonding For the formation of nanoparticle assemblies, H-bonding is very often involved. In order to control the H-bonding between nanoparticles, the key issue lies in the choice of the functional groups anchored on the surface of the nanoparticles.

Polyhedral oligomeric silsesquioxanes (POSS) are intensively studied among the targeted applications one can cite, for example, the encapsulation of biomolecules [19]. In a recent work, Kuo and coworkers made use of amino acids compatibilities to synthesis POSS assemblies [20].

Similarly, Herves and coworkers presented the reversible assembly of silver nanoparticles by means of interactions between the capping ligands [21]. The capping ligand was DL-penicillamine (Figure 12.3). This ligand possesses a thiol group used for the interaction with the metal surface, ammonium group (R-NH_3^+), and a carboxylic acid/carboxylate group depending on the pH. The authors have shown that electrostatic interactions (Figure 12.3a) at high pH can be neglected. Only at low pH, between 3 and 7, the ligands interact through H-bonding (Figure 12.3b). This is proven by UV-visible spectroscopy, where the assembly of the nanoparticles can be directly observed by a clear shift in the absorption maximum toward longer wavelengths, namely, dispersed capped silver nanoparticles absorb around 500 nm whereas assembled nanoparticles absorb around 650 nm.

Pelton and coworkers reported the assembly of polystyrene nanoparticles of various sizes [22]. The most important aspect in this work was that the interactions involved in the assembly were found to be reversible. When the pH is modified, the interactions can be "turned off."

To get back to biomolecules, as early as 1996, Mirkin and Alvisatos reported the first two examples of DNA-controlled self-assembly of nanoparticles [3, 23]. In both works, complementary DNA oligo-nucleotides anchored via thiol groups onto gold nanoparticles were used to build gold nanoparticle or nanocrystal assemblies.

Coradin and coworkers grafted silica particles on DNA to form three-dimensional networks. In this work, the complementary pairing of the DNA chains helped to get the high stability of the network [24].

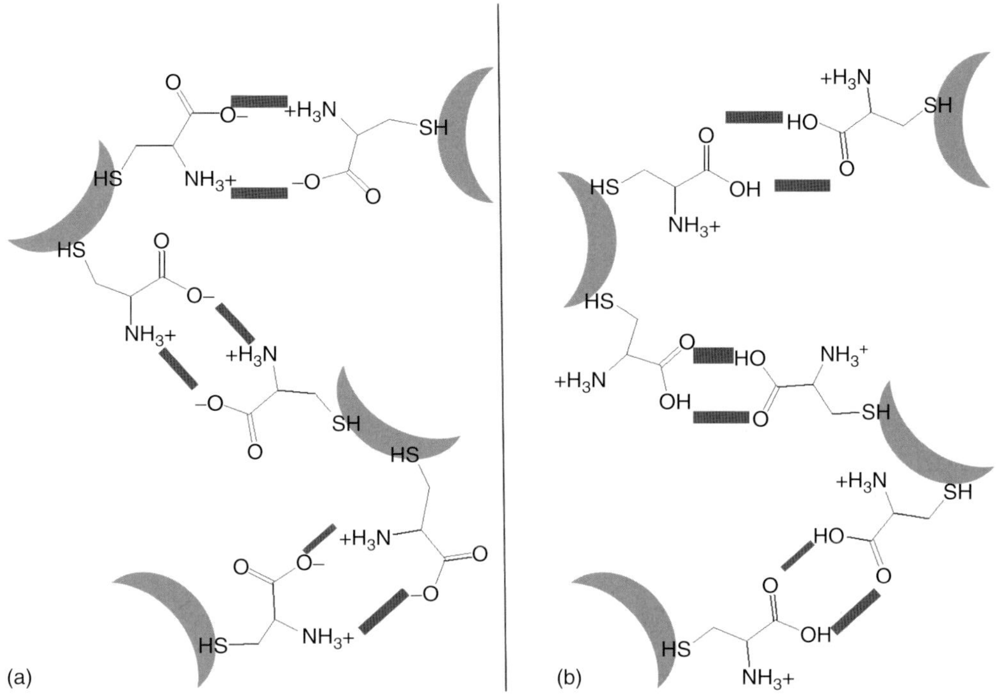

Figure 12.3 Assembly of silver nanoparticles by means of DL-penicillamine ligands interacting via H-bonding (b) or electrostatic interactions (a). (After Taladriz-Blanco *et al.* [21]).

Hydrophobic and π-π Stacking Interactions In another approach, physical self-assembly can be highly interesting for the networking of nanoparticles. This was exemplified in the work of Donato *et al.* [25], among others. Here, the authors functionalize the surface of silica nanoparticles by H-bonding between the anions of an imidazolium based ionic liquid, decylmethylimidazolium, and the OH groups on the surface of the nanoparticles. Then, the long alkyl chains are found to be pending on the surface of the silica nanoparticles, and those hydrophobic chains can interact with one another, maintaining the silica nanoparticle network. It should be noted that in addition to the self-assembly between the alkyl chains, epoxy chains are maintaining the structure.

Liz-Marzan and coworkers recently proposed a work where the assembly of gold nanoparticles is "controlled" by hydrophobic interactions [26]. The hydrophobic interactions occur between the gold nanoparticles with polystyrene shell when exposed to water. Indeed, the gold/polystyrene nanoparticles are well dispersed in tetra hydroflurane (THF), but when water is added, the hydrophobic groups come close to one another, forming a reversible assembly of gold nanoparticles. The authors proposed the polystyrene polymer chains as a proof of principle for hydrophobic driven gold nanoparticle assembling.

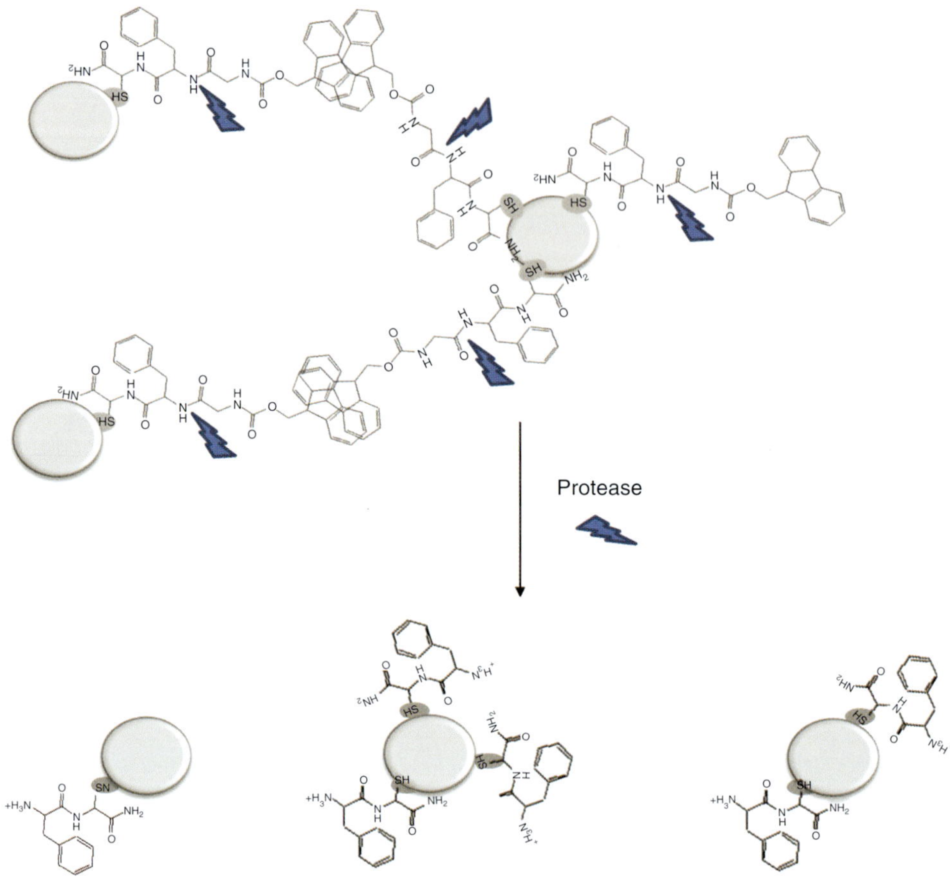

Figure 12.4 Gold nanoparticle assembly maintained by π-π stacking interactions Tripeptide ligand Fmoc-Gly-Ph-Cys-NH_2 and (below) dispersed nanoparticles after action of the protease. (After Laromaine *et al.* [27]).

Ulijn and coworkers proposed the preparation of bioresponsive nanoparticle assemblies [27]. Their assembly is based on modified gold nanoparticles. The surface modification of the gold is performed with ligands possessing thiol functional groups, for the interaction on the surface of the metal present in the cysteine group (Cys). In addition, the ligand has a part with large aromatic groups that could interact by means of π-π stacking; this part comes from the N-(fluorenyl-9-Methoxycarbonyl) (Fmoc). In addition, the ligand possesses a section that can be cleaved by a protease (Gly-Ph). The authors labeled the tripeptide ligand as follows: Fmoc-Gly-Ph-Cys-NH_2 (Figure 12.4). Without protease, the assembly is maintained by π-π stacking interactions (Figure 12.4, top). After the action of the protease (marked as a lightening in Figure 12.4), the ligand is cleaved, and a positively charged ligand remains on the surface of the gold nanoparticles. The electrostatic hindrance then forces the dispersion of the nanoparticles (Figure 12.4, bottom). Here, the color of the solution changes after the dispersion of the nanoparticles. This change indicates the presence of protease in solution.

To conclude, weak interactions such as electrostatic, H-bonding, and hydrophobic/hydrophilic interactions are used largely for assembling nano-objects [4]. The process is now well known and provides the possibility of using it to get reversible assemblies.

12.4.1.2 Covalent Molecular Mediators

In order to get stable nanoparticle assemblies with molecular mediators, one of the often chosen syntheses is to covalently link the nanoparticles.

Interligands Reactions One of the common ways to link nanoparticles with one another by means of covalent bonds is to induce interligand reactions. Click-chemistry reactions are well suited for such assemblies. Click-chemistry reactions are characterized by high reaction yield, at low temperature, in environmental friendly medium, such as water or ethanol. In addition, no side product is produced by a click-chemistry reaction. Owing to these features, click-chemistry reactions are considered as green chemistry reactions [28–30].

For example, organic particles were functionalized and were linked in a subsequent step by click-chemistry reaction, namely, alkyne-azide click-reaction in the presence of a copper (I) catalyst [31]. First, the authors prepared hyperbranched polyester nanoparticles containing functional azide groups, which was used as cargo. These nanoparticles possess a hydrodynamic average diameter of 88 nm. The authors had reacted these azide nanoparticles with various types of nanoparticles possessing alkyne groups on the surface, such as pure hyperbranched polyester nanoparticles or polymer nanoparticles containing iron oxide or cerium oxide core (Figure 12.5). The reaction was performed in the presence of the catalytic amount of copper (I) and bicarbonate buffer.

Quaternarization reactions can also be considered as click-chemistry reaction. Thus, the quaternarization of imidazole units into imidazolium groups was implemented by Neouze and coworkers to prepare nanoparticles assemblies of titania or silica nanoparticles [32–35]. The nanoparticle assemblies were prepared following two similar pathways. The first way was the reaction of two different batches of nanoparticles, the first functionalized with imidazole groups and the other one with chloroalkyl groups (Figure 12.6, left). The second way consisted of introducing difunctional ligands to the suspension of imidazole functionalized nanoparticles (Figure 12.6, right). The quaternization reactions end in the formation of imidazolium units.

Strongin and coworkers used N-homocysteinylation reactions to assemble gold nanoparticles [36]. Gold nanoparticles capped with citrate were prepared by them. Then the citrate ligands were exchanged for ε-Amino Lysine. The amino groups of the protein could interact with the surface of the gold nanoparticles. In addition, the amino groups reacted when a γ-thiolactone was added. The reaction between the protein and the lactone enabled the formation of a disulfide covalent bridge between two neighboring nanoparticles.

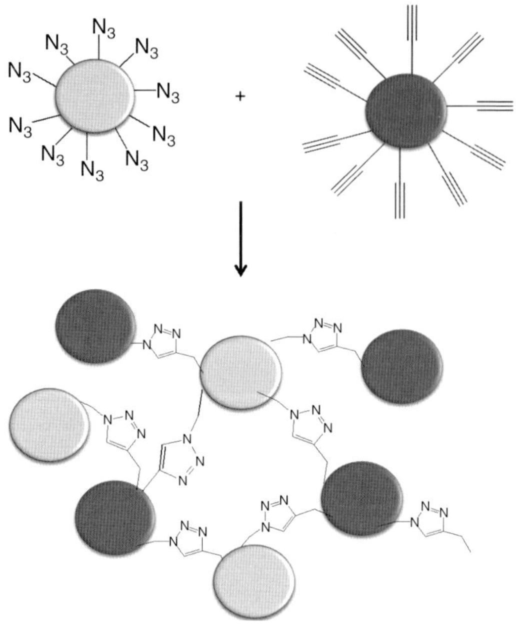

Figure 12.5 Reaction between hyperbranched polyester nanoparticles functionalized with azide groups and other particles functionalized with alkyne groups to form an assembly of various nanoparticles. (After Santra *et al.* [31]).

Di-functional Ligands To the number of reaction steps and thus, to simplify the overall synthesis, di-functional ligands can be used. With these di-functional ligands, a strong link between the nanoparticles is obtained while the ligand is prepared in a preliminary step.

Neouze and coworkers prepared platinum nanoparticle assemblies using platinum nanoparticles modified with mercaptopropionic acid on the one hand and platinum nanoparticles modified with mercaptoethanol on the other hand. The thiol functional groups anchored the ligand onto the platinum surface and the pending carboxylic acid or hydroxyl groups could interact with the surface of neighboring platinum nanoparticles, thus maintaining the assembly [37].

In a mini review, Boennemann and coworkers detailed the formation and characterization of platinum nanoparticle networks obtained with di-functional spacer molecules of different sizes [38]. In this review, the authors focused on diol spacer molecules. In the examples detailed in this review, the platinum nanoparticles were prepared by reducing platinum diacetylacetonate, $[Pt(acac)_2]$, with trimethylaluminum, $Al(CH_3)_3$, in THF. Once the nanoparticles were formed, reactive methyl ammonium groups, coming from the reducing agent, were present on the surface of the platinum nanoparticles. The methyl aluminum groups can react with di-alcohols in a protonolytic reaction.

The principle is also applicable to other multifunctional ligands, such as trifunctional ligands as, for example, reported by Sevov and coworkers in the

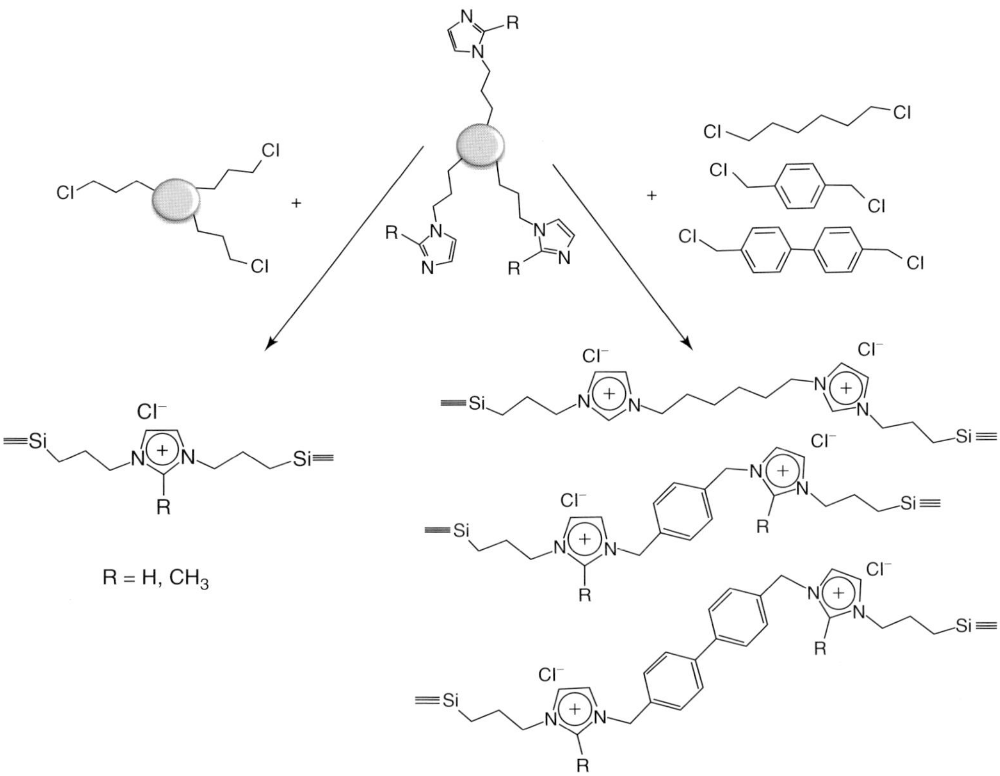

Figure 12.6 Quaternarization reaction to form nanoparticle assemblies. Left: Reaction of nanoparticles functionalized with imidazole and chloroalkyl. Right: Addition of a di-functional ligand to the imidazole functionalized nanoparticles. Note: The triple bound on the silica stands for the covalent link of the silicon atoms to the nanoparticles.

case of cobalt clusters [39]. The tri-functional ligand studied was a deprotonated N-(phosphonomethyl)iminoacetic acid $[N(CH_2\text{-}COO^-)_2(CH_2PO_3^{2-})]$. The cluster network was obtained under hydrothermal synthesis with the protonated ligand and cobalt acetate salt.

The use of di-functional ligands allows quite easy synthesis pathways and leads to strong linkers. However, in this section, it is important to stress the few potential disadvantages of such a reaction pathway. First, depending on the anchoring function, such as diols or di-carboxylic acids, exchange reactions can occur when other strongly coordinating functional groups are introduced [14]. For example, diol linkers can be replaced with thiols or di-thiols. Second, with this method, it is often better to start with naked, nonmodified, nanoparticles.

12.4.1.3 **Noncovalent versus Covalent Interaction**

For the synthesis of nanoparticle assemblies obtained via interacting molecular mediators, either with strong or weak bonding, the main step is always the

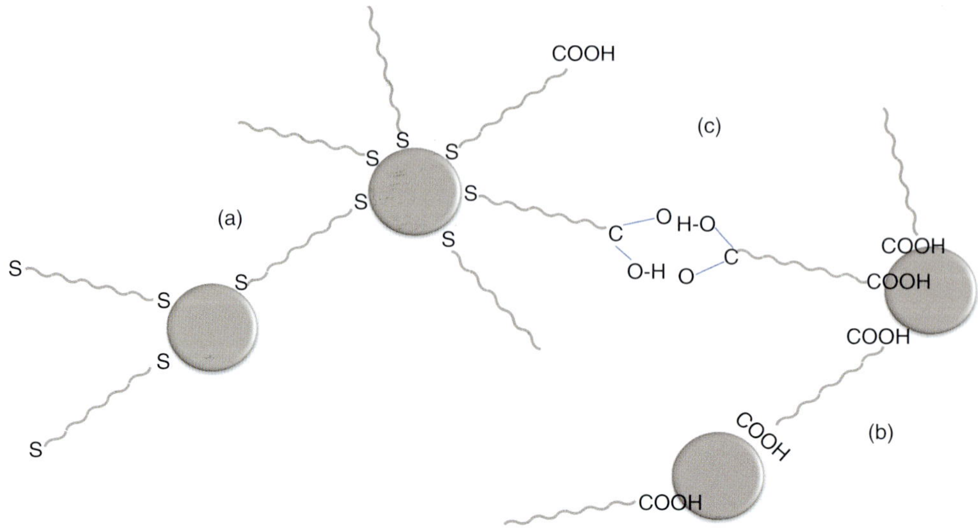

Figure 12.7 Nanoparticle networks based on (a and b) covalent interactions, via (a) di-thiol molecular mediators or (b) di-carboxylate molecular mediators, or (c) noncovalent interligand H-bonding. (After Wang *et al.* [40]).

modification of the nanoparticle surface. Once the particles are modified, they are stable and can easily be manipulated toward the formation of an assembly.

In a feature article, Zhong and coworkers compared covalent and noncovalent linking of nanoparticles in a network [40]. In their work they focused on specific cases: the covalent linking between nanoparticles, obtained from di-functional ligands (Figure 12.7a and b), and the noncovalent binding obtained from interligands H-bonding (Figure 12.7c).

It turned out that both pathways, with covalent or noncovalent linking, are largely reported to form nanoparticle assemblies. Such pathways allow adapting the interparticle distances, the ligands rigidity, or organization. Nevertheless, in general, for applications such as molecular recognition or sensoric, noncovalent linking will be preferred. Contrarily, for more stable nanostructures, with stability requirements toward solvents or temperature for example, covalent linking will be advantageous.

12.4.2
Template Assisted Synthesis

In a recent review, Mirkin and coworkers gave a very complete overview on various template assisted techniques for the preparation of assemblies of plasmonic nanostrucutres [41]. The authors insisted on the very broad variety of possible templates, from solution-phase colloidal templates, biological molecules, such as DNA, solid-phase templates, such as porous materials, to cite only a few. Moreover, they stressed the versatility of the method, where various types of templating methods could be implemented in the same procedure, leading to highly complex structures or even to organize different types of materials in a single assembly.

When considering most of the reported procedures using templates, it is striking that the nano-objects are assembled only as long as the template is present. To fill this gap, Mirkin and coworkers reported some methods where the organization is maintained after getting rid of the template, which is then called sacrificial template. In this review, we focus on nanoparticle assemblies obtained from molecular mediators. Thus, we do not discuss the use of sacrificial templates.

Interestingly, the use of a covalent mediator was considered by Mirkin and coworkers as a template assisted procedure, as the ligands structure the assembly.

Most of the examples of 1D nanoparticle assemblies obtained from template assisted deposition have been carried out with biomolecules.

As has been already mentioned, the DNA is one of the most, or even the most, prominent biomolecules used in material sciences. In one example described in the section dedicated to electrostatic interactions [18], the negatively charged phosphate groups of DNA chains in DMSO allowed the binding of Zn(II) cations from zinc acetate. The zinc atoms immobilized on the DNA chains then reacted with the hydroxides added and led to the formation of ZnO nanoparticles by sol-gel reaction. The DNA promoted and controlled the ZnO nanoparticle formation. This reaction ended in one-dimensional chains of around 10 nm diameter zinc oxide nanoparticles.

Anisotropic assemblies can also be of great interest. Kanaras and coworkers presented a "universal" method to prepare gold nanoparticle chains [42]. The formation of chains occurred following the electric dipole-dipole interactions induced by the ligand exchange between the charged molecules attached on the nanoparticle surface and the thiol molecules added.

Knecht and coworkers showed that amino acids, such as arginine, can drive the assembly of gold nanoparticles into nano-chains [43].

However, other functional biomolecules were used to prepare one-dimensional assemblies of nanoparticles, such as catechol or chitosan. Catechol based templates were developed by Park and coworkers as versatile preparation method for 1D assemblies. This template was used in the case of gold, iron oxide, or quantum dots [44]. Xia and coworkers used Cysteine-grafted chitosan templates to drive the assembly of gold nanoparticles [45].

12.4.3
Deposition of 2D Nanoparticle Assemblies: Monolayers, Multilayers, or Films

Another widely reported approach for realizing highly organized nano-objects assemblies is the deposition of nanoparticle monolayers or films. Such approaches induce mostly either two-dimensional organization or a vertically stratified organization. Various possibilities for the deposition of films such as lithography, layer-by-layer deposition, Langmuir-Blodgett films, and a more exotic method of bubble deposition are reported. However, in this section, lithography is not presented because for nanoparticle assemblies, the resolution obtained by such lithography techniques can be considered as poor due to the much too large interparticle distances of some tenth of nanometers [46].

12.4.3.1 Layer-by-Layer Deposition

The principle of layer-by-layer deposition lies in the alternate adsorption of complementary species. Layer-by-layer deposition has been performed first and largely to assemble polyelectrolytes using electrostatic interactions between oppositely charged objects [47]. These complementary species are then assembled by interactions such as electrostatic interactions, biorecognition, or hydrogen bonding. Some rare examples present layer-by-layer assemblies using covalent bonding [32, 48]. Layer-by-layer deposition is a simple and versatile approach to prepare multilayered assemblies of nanoparticles under controlling capillary and gravitational forces. Layer-by-layer deposition allows a good control on the deposition density and the thickness of the films.

Decher and coworkers described the preparation of iron oxide nanoparticle assemblies [49]. In their work, they insisted on the importance of the first step: the preparation of stable nanoparticle suspensions, by introducing a ligand on the nanoparticle surface. The 12 nm diameter iron oxide nanoparticles were first stabilized by oleic acid ligands. Then, due to their lack of surface charge, the oleic acid ligands were exchanged by citric acid ligands. The layer-by-layer deposition was performed by alternatively depositing a polyelectrolyte layer and negatively charged modified nanoparticles. The polyelectrolyte consisted of poly(allylamine hydrochloride) and poly(sodium 4-styrenesulfonate). The citric acid modified nanoparticles were absorbed on the poly(allylamine hydrochloride) (Figure 12.8).

With this layer-by-layer deposition approach, the authors obtained a strongly controlled multilayered assembly of nonaggregated iron oxide nanoparticles. The distance of the nanoparticles could be tuned by controlling the thickness of the polyelectrolyte layers. The layered nano-structure induced an antiparallel coupling of the nanoparticles, interacting not only in one layer but also in adjacent layers by means of dipolar interactions.

Click-chemistry was used by some groups as a simple method to prepare 2D assemblies of non-interacting nanoparticles. This method based on click-chemistry reaction is a derivatization of layer-by-layer assembly. The nanoparticles were covalently linked to a substrate by click-chemistry reaction between the ligands on the surface of the substrate and the ligands on the surface of the nanoparticles. Simultaneously, the ligands capping the nanoparticles hinder the direct interactions between the nanoparticles of the layer and maintain a given distance between the nanoparticles.

One example was proposed by Neouze and coworkers [32, 48]. The silicon wafer was modified with N-(3-propyltrimethoxysilane)imidazole (Figure 12.9a) while the titania nanoparticles were capped with 3-chloropropyl phosphonic acid (Figure 12.9b). Then the chloroalkyl functionalities on the nanoparticles reacted with the imidazole units attached on the substrate, without the need of a catalyst. This reaction could be considered as a click-chemistry-like reaction given that they were performed at low temperature (from room temperature to 70 °C) in

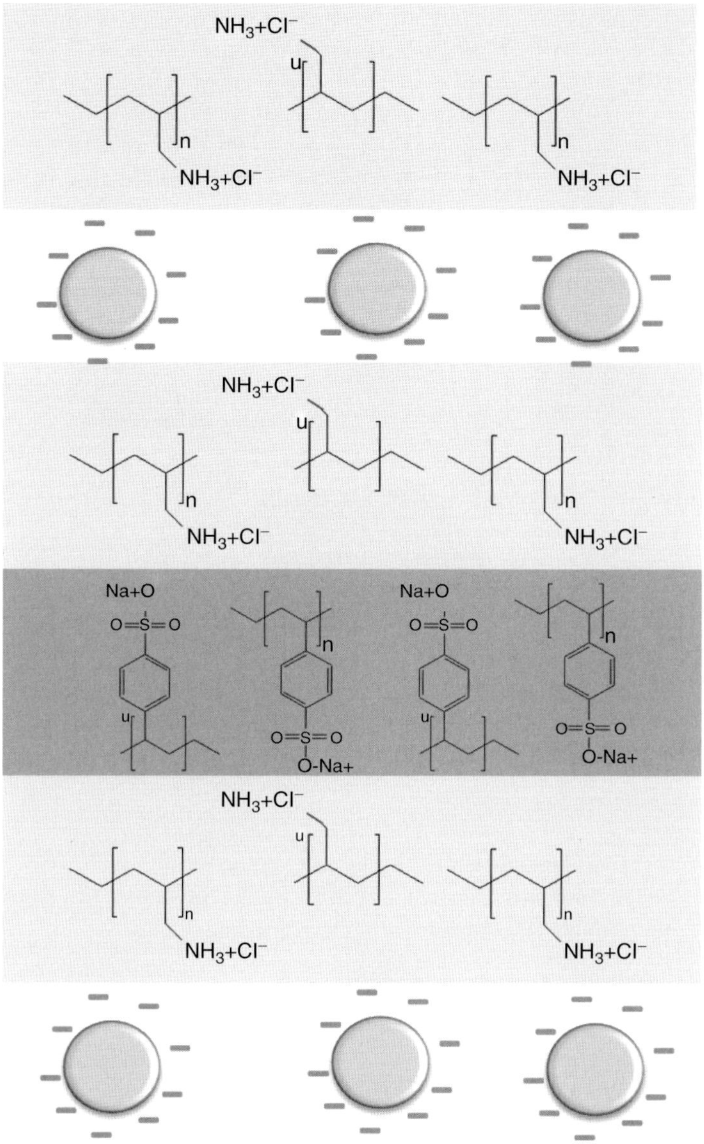

Figure 12.8 Layer-by-layer assembly of poly(allylamine hydrochloride), poly(sodium 4-styrenesulfonate), and citric acid modified iron oxide nanoaparticles on a silicon wafer surface. (After Pichon *et al.* [49]).

environmental friendly solvent (water or ethanol), with no formation of a side product. Under this reaction, an imidazolium unit was formed (Figure 12.9c). Thus, a monolayer of nanoparticles is relatively regularly deposited on the substrate (Figure 12.9d). In this monolayer, the nanoparticles were not interacting as the ligands were maintaining a constant distance by steric hindrance.

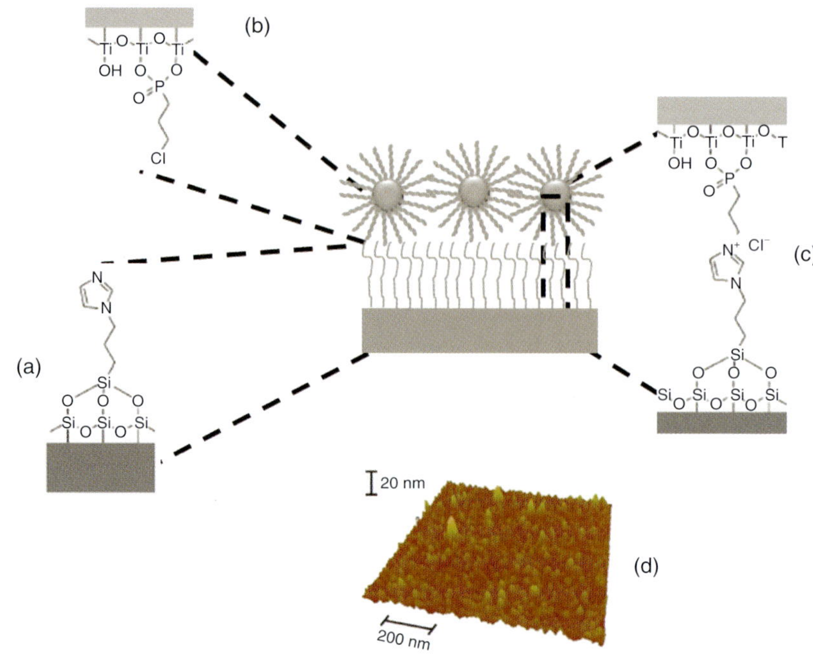

Figure 12.9 Click-chemistry reaction to form 2D assemblies of non-interacting nanoparticles. (a) Modification of the substrate; (b) modification of the nanoparticles; (c) click-chemistry reaction between the modified substrate and the modified nanoparticles; and (d) Atomic Force Microscopy (AFM) image of a monolayer of modified titania nanoparticles. (After Basnar *et al.* [48]).

Neouze and coworkers extended their results by performing further layer-by-layer depositions. Controlled multilayer deposition was obtained with one or two types of nanoparticles, either only titania or alternating layers titania–silica.

12.4.3.2 **Langmuir-Blodgett Deposition**

With the Langmuir-Blodgett deposition method, large-area monolayers of nanoparticles can be deposited. The driving forces of the assembly with this method are mainly mechanical and capillary forces.

The principle of the Langmuir-Blodgett deposition (Figure 12.10) consists in first spreading a surfactant at a water/air interface. Then the surfactant is compressed, which forces its self-assembling. Finally, a slide is pulled up through this monolayer forced at the air/water interface. The monolayer is thus transferred on to the surface of the slide. Initially, this technique was studied only for water/air interfaces but was enlarged to other solvent/gas interfaces, such as toluene/nitrogen interfaces. In addition, modified nanoparticles can successfully be used instead of the pure surfactants as described in this section.

The Langmuir-Blodgett deposition method was extended to multilayered assemblies.

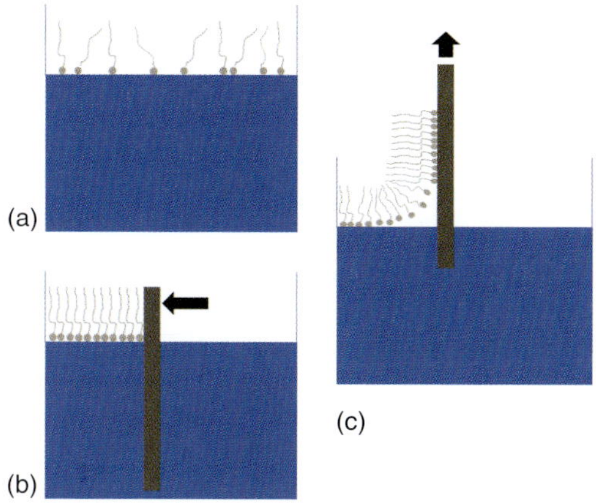

Figure 12.10 Principle of the Langmuir-Blodgett monolayer deposition. (a) Dispersion of a surfactant at the fluid/gas interface, (b) compression to self-assemble the surfactant, and (c) pull-up of a slide on with the surfactant will remain self-organized.

For example, multilayers of spherical iron oxide nanoparticles modified with oleic acid were deposited by successive Langmuir-Blodgett cycles [50]. Classically, for Langmuir-Blodgett deposition, the films were homogeneous and dense over large areas. The authors could improve the local order by "playing on" the substrate–nanoparticle as well as on the nanoparticle–nanoparticle interactions. Nevertheless, hexagonal packing of the nanoparticles was limited to local environment. There, layer-by-layer Langmuir-Blodgett deposition was obtained by successive dipping and pulling out of the film.

In conclusion, Langmuir Blodgett deposition is an easy method to get homogeneous and dense assemblies of nanoparticles over large areas. The organization of the nano-objects is formed by the solvent/air interface and later transferred to the substrate.

12.4.3.3 **Evaporation Induced Assembly**

The concept of Evaporation Induced Self-Assembly (EISA) consists of the controlled evaporation of the solvent of a solution to reach a concentration higher than the critical micellar concentration (cmc) of the dissolved surfactant. When the concentration oversteps the cmc, the surfactants self-organize into micelles. Depending on the concentration, various patterns can be obtained, such as lamellar or hexagonal ones. This principle was used to form porous films.

A solution is prepared containing dissolved matrix precursors, such as some alkoxy silanes for a silicon oxide matrix, and surfactants such as cetyltrimethylammonium bromide. A substrate, like a glass slab or a silicon wafer, is dipcoated with the solution. The solvent of the deposited solution evaporates slowly and micelles of the surfactant form and self-assemble inside the matrix precursors (Figure 12.11). After the reaction of the matrix precursors, condensation of the alkoxy silanes, the surfactant micelles can be extracted. The extraction is performed by washing or

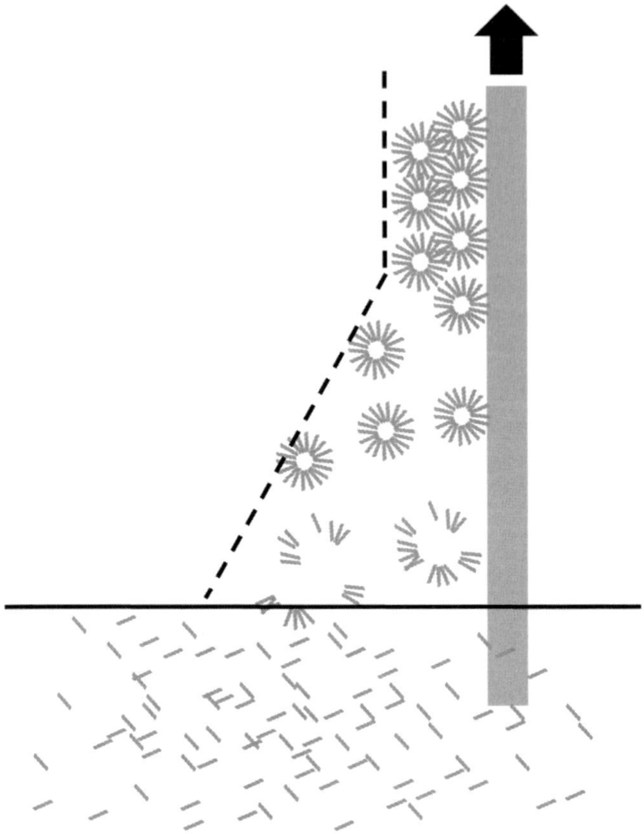

Figure 12.11 Principle of evaporation induced self-assembly of surfactant. The solution contains the dissolved surfactants as well as the matrix precursors, the former being omitted for clarity.

calcination. The matrix film, for example silicon oxide, is then porous, the pore sizes corresponding to the size of the micelles [51, 52]. The EISA method is often used for the preparation of porous inorganic films [53].

More recently, authors adapted the principle of EISA, micelles assembly, to the formation of nanoparticles assemblies.

Takeda and coworkers [54] prepared nano-fibers of silver, obtained from the anisotrope aggregation of silver nanoparticles on DNA. This anisotrope aggregation was driven by evaporation-induced self-assembly with DNA and drying front movement. The obtained silver nano-fibers are a few tenths of nanometers in diameter and a few millimeters long.

Both these examples of evaporation induced assemblies with either polystyrene balls or DNA combine EISA and template assisted deposition methods.

Zaho and coworkers described what they called Evaporation Induced Aggregating Assembly (EIAA) [55]. One of the main differences with "common" EISA is that

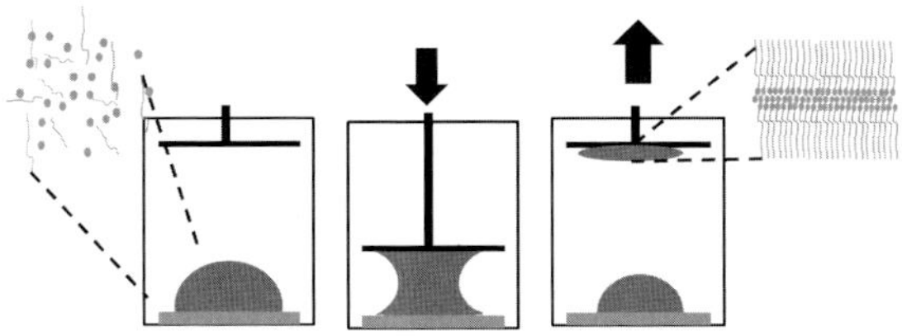

Figure 12.12 Bubble deposition method. (a) Deposition of a droplet on soaked filter paper in a closed chamber. (b) The pretreated substrate is driven down to the bubble surface. (c) The substrate is driven up back. (inset right) A film is transferred on the surface. (After Andreatta *et al.* [56]).

in EISA the organization is formed at the substrate/liquid interface while in EIAA the organization is formed at a liquid/liquid interface. In their work, the authors start with a mixture containing water, THF, silica precursors as well as a water-immiscible block-co-polymer that plays the role of a surfactant, namely, poly(ethylene oxide)-b-poly(methyl methacrylate). After mixing all the components, the THF is slowly evaporated, which forces the organization of the water-immiscible co-polymer into micelles with the silicate oligomers. Thus, composite micelles are formed in the solution. The micelles consist of regular nanoparticles of around 50 nm diameter. Then, continuing the evaporation, the micelles aggregate into larger structures: nanoparticle assemblies with a face centered cubic organization.

In conclusion, Evaporation Induced Assembly is an easy method to get homogeneous assemblies of nanoparticles over large areas. As opposed to Langmuir Blodgett deposition, the deposited assemblies are not necessarily dense, and the organization pattern is formed during the evaporation and is not transferred.

12.4.3.4 **Bubble Deposition**

In order to deposit hydrophilic nanoparticles onto hydrophobic substrates, Benattar and coworkers developed recently a method called Bubble deposition method [56]. The principle of the method is dispersing the hydrophilic nanoparticles, such as silica [22] or gold nanoparticles [57], in water containing the surfactant. In the case of the silica nanoparticles, the authors used sodium dodecyl benzensulfonate, while they used cetyltrimethylammonium bromide for the gold nanoparticles. A filter paper is soaked with this solution and a bubble is deposited on this filter paper with a pipet. The filter paper with the bubble are placed in a closed chamber, on top of which a pretreated silicon substrate is placed (Figure 12.12a). The silicon (111) is carefully pretreated with ammonium fluoride in order to get a hydrophobic surface. The substrate is driven down toward the bubble until getting into contact (Figure 12.12b). Afterward, the substrate is driven back up. On the surface of the substrate, a film is deposited. The deposited film consists of an assembly of nanoparticles, "sandwiched" between the surfactants. The hydrophobic head of the surfactants points toward the nanoparticles while the hydrophobic chain interacts with the surface of the substrate (Figure 12.12, inset right). As for the deposition method described in the previous paragraphs, the deposition procedure lasts only a few minutes.

It is shown that the morphology of the nanoparticle layer on the substrate can be controlled by the deposition conditions such as the concentration of the surfactant, the concentration of the nanoparticles, or the deposition time. By varying these parameters, morphologies ranging from isolated particles up to full surface coverage can be achieved.

The organization of the nanoparticles is obtained by solvent evaporation. However, in contrast to EISA deposition, the main force for the assembly of the silica nanoparticles seems to be the surfactant bilayer confinement force during the last stage of solvent drying.

12.4.4
Pressure-Driven Assembly

Fan and coworkers developed a high-pressure approach for the preparation of 3D or 1D nanoparticle assemblies [58, 59]. In this approach the mechanism underlying the assembly of the homo-dispersed gold nanoparticles is the sintering of neighboring nanoparticles.

Much lower pressures were used by Kundu and coworkers to prepare multilayered assemblies of gold nanoparticle [60]. This procedure is based on controlling the lateral compression force before pulling out the silicon substrate. Prior to the deposition, Si(001) substrates were made hydrophobic, by keeping it in a solution of hydrogen fluoride (10%) for 3 min at room temperature. The 6 nm dodecanethiol-capped gold nanoparticles were spread on water from a 0.95 mg/ml toluene solution. Films were deposited by the Langmuir-Blodgett method with six different lateral compression rates: 2.5, 8, 16, 21, 25, and 32 mN/m, at room temperature (26 °C). By carefully choosing the compression lateral force, Kundu could obtain controlled multilayers (Figure 12.13). For example, with the compression force of 8 mN/m, a monolayer was obtained, and with 16 and 21 mN/m, bi- and tri-layers were obtained. Kundu has also shown that after decompression, the multilayered structures were maintained. This example could also be classified as a kind of pressure induced assembly.

12.5
Applications of Nanoparticle Assemblies

This part presents briefly some of the main application fields of nanoparticle assemblies, with some representative examples.

12.5.1
Plasmonics

The interaction, coupling, between neighboring plasmon resonances can lead to extraordinary properties [61, 62].

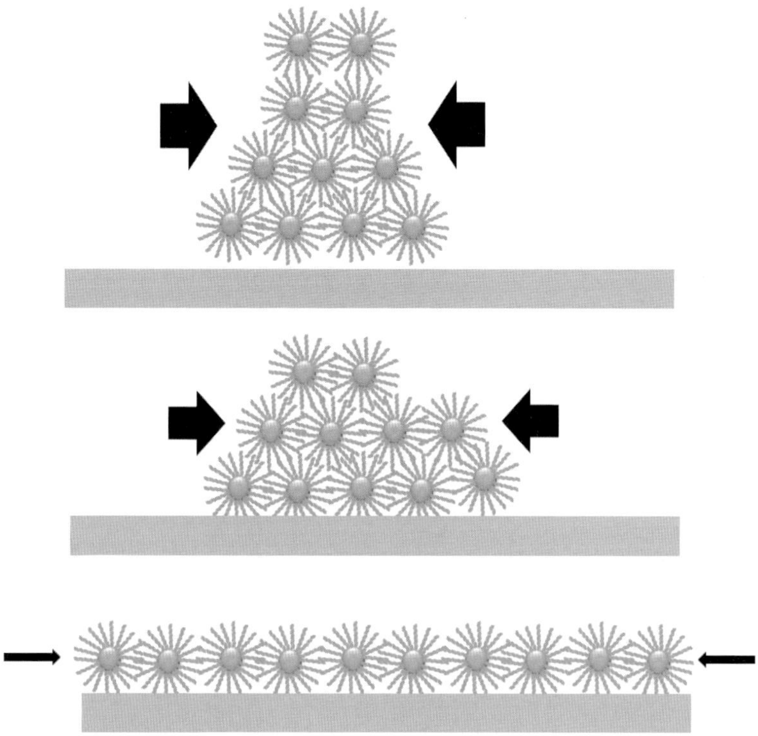

Figure 12.13 Langmuir-Blodgett deposition of mono-, di-, or trilayers (from bottom to top) of dodecanethiol capped gold nanoparticles using increasing lateral compression. (After Kundu [60]).

Localized surface plasmon resonance (SPR) refers to the collective motion of conduction band electrons (electron clouds or plasma) of metal nanoparticles, relative to the fixed metal atoms when excited by an electric field, an evanescent wave. In the case of localized SPR, the *dimension of the considered surface is smaller than the wavelength of the incident light,* which is typically the case for metal nanoparticles. In such case, the surface of the nanoparticle is too small to support the propagating wave, and the surface plasmon is confined to a small volume. The plasmon resonance then oscillates near the nanoparticle surface in the dielectric neighboring medium, over short distances, that is, a few nanometers.

The plasmon resonance wave decreases exponentially to the distance with the nanoparticle surface and thus propagates up to a few nanometers or tens of nanometers. When two nano-objects are near enough, in this propagation "region" meaning distances in the nanometer range, the plasmon resonance waves can interact with one another and lead to new resonances. When this interparticle distance is well controlled, for the nanoparticles not to aggregate, very high local fields can be reached. Such local fields are called "hot spots."

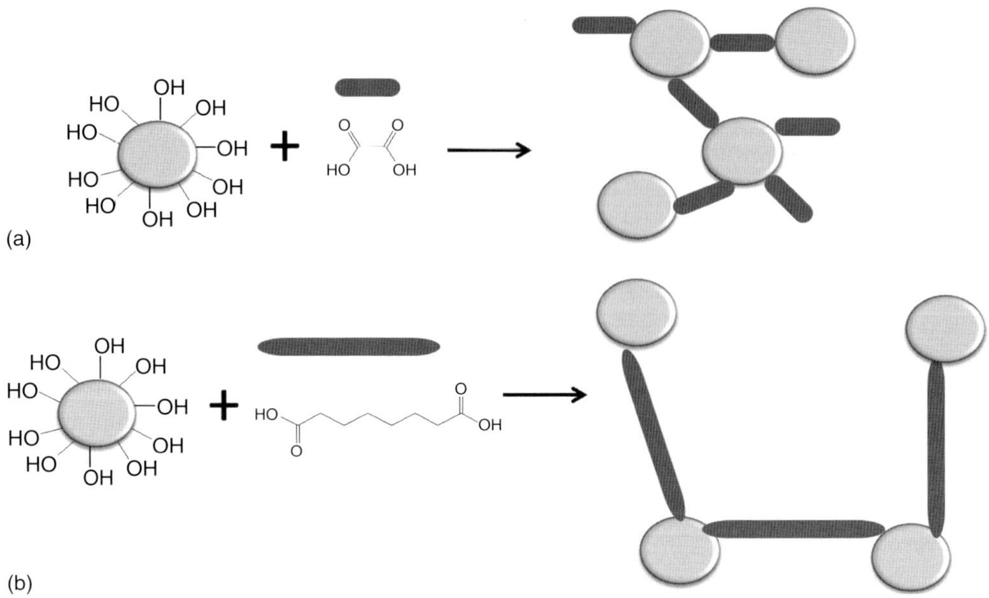

Figure 12.14 Hydroxyl capped silver nanoparticles assembles by reaction with (a) oxalic acid and (b) sebacic acid ligands. (After Abargues *et al.* [63]).

Controlling the assembly of metal nanoparticles is a way to control the generation of very intense electromagnetic fields on the nanometer scale. Direct applications of this phenomenon are presented in the following paragraphs.

Plasmonics is the most strongly investigated application for nanoparticle assemblies.

12.5.1.1 **Plasmonic Nanostructures**

The work of Abargues and colleagues of silver nanoparticle networks based on various dicarboxylic acid ligands was described earlier in the paragraph dedicated to difunctional ligands [63]. The hydroxyl functionalized silver nanoparticles were linked by various dicarboxylic acid ligands, namely, oxalic acid (Figure 12.14a), tartaric acid, glucaric acid, and sebacic acid (Figure 12.14b). The study of the shifts in the absorption spectra of the various nanoparticle assemblies allowed extracting that not only the distance between two nanoparticles influence the coupling of the plasmon resonances of neighboring particles but also the polarizability of these ligands. As a consequence, linkers with higher polarizability enabled better propagation of the plasmon coupling. For example, the propagation of the plasmon resonance in assemblies with tartaric acid was delocalized over up to three nanoparticles, and for the oxalic acid based material, only over up to two nanoparticles.

Wessels and coworkers also pointed out the issue of polarizability, in the gold nanoparticle assemblies based on dithiol ligands [64].

12.5.1.2 Sensoric

One of the first applications of SPR reflectivity was the detection of adsorbed molecules on metallic surfaces. Indeed, absorbed molecules, such as the DNA, polymers, antibody, induce a local change in the refractive index of the metallic surface. The change in local refractive index, and thus in the local dielectric constant, corresponds to a change in the resonance conditions of a surface plasmon.

This is because the plasmon is directly related to the dielectric constant of the metal, as discussed earlier. And, from Maxwell equations, the square of the refractive index is equal to the product of the dielectric constant and the magnetic permeability.

The sensitivity of SPR to absorbed molecules, or modification on adsorbed molecules, can be extended to localized SPR. This sensitivity makes nanoparticle and nanoparticle assemblies particularly suited for sensoric applications.

Ye and coworkers made use of this property for the detection of phtalates [65]. They prepared the sensor, gold nanoparticles modified with uridine-5′-triphosphate, and a cross-linker, Cu(II). The principle of the reaction was a complexation of the cupper (II) by the phtalates and the uridine functional group. In the system reported, the uridine functionality was introduced on the surface of the gold nanoparticles. To this purpose, the gold nanoparticles were modified with uridine-5′-triphosphate, where the triphosphate units interacted with the gold surface. In a solution containing the modified gold nanoparticles and copper (II) ions, the presence of di(2-ethyl-hexyl) phtalate induces complexation reactions, and, in consequence, the assembly of the gold nanoparticles (Figure 12.15). This assembly is characterized by a change in the maximum wavelength of the absorption band between "free" modified gold nanoparticles and assembles gold nanoparticles. Before complexation, the absorption maximum of the "free" modified gold nanoparticles is centered at 520 nm, corresponding to the plasmon resonance of the uridine-5′-triphosphate modified gold nanoparticles with a diameter of 15 nm. After complexation, a strong decrease of the plasmon absorption at 520 nm is observed, and the maximum of absorption is located to 600–650 nm. The surface

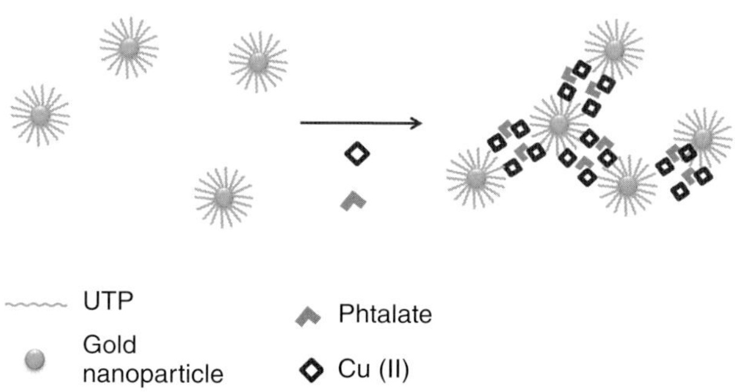

Figure 12.15 Assembly of gold nanoparticles through complexation reaction of copper (II) with uridine-5′-triphosphate modified gold nanoparticles and a phtalate. (After Zhang *et al.* [65]).

plasmon absorption at 600–650 nm comes from the interaction between neighboring plasmon absorption of single gold nanoparticles. This shift of the absorption maximum is observed as a color change, from red to blue or purple, as signature for the presence of phthalates. The authors could reach with this system a phthalate detection limit of around 0.5 ppm.

Wang and coworkers recently published a feature article describing many systems where gold nano-rods were used for sensoric applications [66]. In the specific example of nano-rods, various localized SPRs could be distinguished: namely, longitudinal and transverse localized SPRs. Using assembled nano-rods, up to 200 nm shift, could be observed in the maximum absorption.

Many of the presented applications are based on the assembly of the gold nano-rods by recognition between biologically active molecules.

Liu *et al.* published a feature article focusing on biological and chemical sensing applications for nanoparticle assemblies [67]. Among other examples, the review reports the use of organic cross-linked nanoparticles for electro-chemical sensing of specific molecules. Indeed the electrical properties of nanoparticle assemblies are strongly influenced by the distance as well as by the functionalities between the nanoparticles. Following this principle, the pi-donor molecules of adrenaline can be detected using an array of gold nanoparticles.

This review [67] reported examples based not only on gold nanoparticles but also on other nanoparticles. For example, a blood glucose sensor based on assembly of CdTe nanoparticles was discussed. The CdTe nanoparticles were assembled with glucose oxidase molecules by means of opposite charge layer-by-layer deposition method. In the presence of glucose, the glucose oxidase of the assembly produced H_2O_2, which quenched the photoluminescence of the CdTe nanoparticles. This nanoparticle assembly based sensor was successfully tested in real serum samples [68].

Recently, Neouze and coworkers also demonstrated the potential use of ionic nanoparticle networks for sensoric [69]. In this work, the chlorine counter ion of the imidazolium bridging ligand was coupled with copper chloride to form a $CuCl_4$ complex within the frame of the material. The nanoparticle network provided stability to the material, and the copper complex resulted in thermochromic behavior of the material with a clear color change from green to yellow at 180 °C.

12.5.1.3 Signal Amplification/Surface-Enhanced Raman Scattering

In this chapter, a paragraph is dedicated to Surface enhanced Raman Scattering (SERS), even though SERS is a type of sensor application. However, the use of nano-objects assemblies for SERS is so largely discussed in the literature that it is interesting to discuss it in a specific paragraph. More generally, in sensoric and detection devices, nanoparticles are broadly investigated for their signal amplification abilities [70].

SERS describes the increase of a Raman signal for a molecule absorbed on a surface. This increase comes from an increased electric field appearing on the surface. Such enhanced electric field can be obtained by SPR. However, to observe an enhanced Raman scattering for the absorbed molecule, the increased electric field must propagate perpendicular to the surface. For this reason, no SERS can be obtained for plane surfaces (with SPR), while it can be observed with rough surfaces of nanoparticle assemblies (with localized SPR).

Kim and coworkers investigated dimers of 55 nm gold nanoparticles with 1,4-phenylene diisocyanide molecules on the surface with AFM, Transmission Electron Microscopy (TEM), and Raman scattering, supported by calculations [71]. They verified that the "hot spot" for SERS is very precisely located between two nanoparticles. This gap has to be smaller than 10 nm, while for distances between 8 and 10 nm, the signal is already only very weakly enhanced.

Herves and coworkers presented the reversible assembly of silver or gold nanoparticles by means of interactions between DL-penicillamine (Figure 12.3) capping ligands [21]. As discussed before, the nanoparticles can reversibly assemble by means of H-bonding for pH in the range 3–6. The DL-penicillamine ligands present in the interparticle space are then located in the "hot-spots" where the Raman signal can be particularly enhanced. The authors have recorded the SERS spectra of the nanoparticle assemblies, with silver as well as with gold, for pH = 3 and pH = 7, with a laser excitation wavelength of 785 nm. Here, the enhanced Raman signal could be clearly highlighted. Indeed, at pH 7, where no nanoparticle interaction and thus assembly can be formed, no signal can be observed; whereas at pH 3, when the nanoparticle assemblies are formed, Raman signals could be observed. Among others, signals at 1117 cm^{-1}, characteristic of CC and CN stretching, and at 888 cm^{-1}, corresponding to the CH_3 rocking, could be clearly observed.

Li and coworkers published a review mainly dedicated to SERS applications of metallic nanostructures assembled by DNA [72], which can be used (i) for DNA detection, (ii) for the detection of proteins or smaller molecules, and (iii) for the detection of metallic ions.

The DNA bases (adenine, guanine, cytosine, and thymine) possess two secondary or tertiary amine groups that can interact with the surface of metal nanoparticles and hence serve as linker between two adjacent nanoparticles. In consequence, these small molecules can "easily" be detected in SERS even at low concentration. DNA-based nanoparticle assemblies are versatile supports for molecular recognition and enhanced Raman scattering. The above mentioned review reported such systems to detect cocaine at concentrations as low as 1 $\mu mol\,l^{-1}$, or lead ions at concentration as low as 20 $nmol\,l^{-1}$.

12.5.2 Interacting Super-Spins/Magnetic Materials

Small size magnetic nanoparticles are highly interesting for material scientists. And magnetic nanoparticle assemblies are highly interesting for the preparation of logic devices, sensors, magnetic data storage devices, or ultra-high density magnetic recording.

The research on magnetic nanoparticle assemblies was, and is, driven by the research of Pileni and coworkers. They reported the extraordinary properties of superlattices of magnetic nanocrystals or nanoparticles for more than 10 years [6, 73–77]. This section thus presents briefly some aspects of this huge research field (for more details, see the cited reviews).

Pileni and her colleagues have shown that the remanence as well as the coercivity of nanoparticle assemblies can be isotropically enhanced by preparing cobalt nano-chains under a magnetic field [7, 11, 74]. For 3D assemblies [78], the cobalt nanocrystals were obtained from the reduction of cobalt bis(2-ethyl-hexyl)-sulfosuccinate with sodium borohydrate in reverse micelles. The nanocrystals were extracted with lauric acid. Then the lauric acid surfactant was removed and the particles redispersed in toluene. The nanocrystals possess a main diameter of 8 nm and are stable for at least one week without the formation of an oxide shell. The deposition of the cobalt nanocrystal superlattices were performed with magnetic fields of 0, 0.26, and 0.57 T. They have shown a drastic increase in the coercivity for ordered nanocrystals as compared with isolated, nondeposited nano-crystals, 0.27 and 0.18 T, respectively. After deposition, in oriented nanocrystals superlattices, the coercivity remained relatively constant, from 0.27 to 0.30 T for increasing deposition fields. By increasing the deposition field, the remanence was strongly enhanced, from 0.50 to 0.70.

Magnetic nanoparticles, with diameters smaller than 10 nm, can be considered as single-domain structures as their diameter is lower than the exchange correlation length. These single-domain magnetic nanoparticles are also called super-spins.

For ultra-high density magnetic recording, each spin should be able to "react" individually, as a single bit, to a stimulus. Nanoparticle assemblies bring a high concentration in super-spins, for the high density aspect; however, if the nanoparticles are not far enough from one another, dipolar interactions will start to be prominent and disable individual response of the super-spins. Komatsu and coworkers have shown with $SiO_2@Fe_3O_4$ core-shell particles that the interparticle distance needs to be longer than 15–18 nm [79]. For this investigation, the authors have prepared core-shell nanoparticles with various silica shell thicknesses, inducing distances between neighboring magnetite cores ranging from 12 to 47 nm.

Complementarily, Margaris and coworkers have shown that the assembly magnetization, or the collective behavior in the assembly, decreases for constant interparticle distances when the core anisotropy increases [80].

Begin-Colin and coworkers prepared magnetite nanoparticles of various sizes, from 5 to 16 nm, from the thermal decomposition of iron stearate complex in the presence of oleic acid [81]. The authors compared the magnetic properties of the nanoparticles either dispersed, coated in poly-methylmethacrylate matrix, or in assemblies obtained by Langmuir-Blodgett deposition. They observed a much stronger dipolar interaction within the nanoparticle assemblies, in comparison with the other systems. This effect was characterized by an increase in the blocking temperature for these samples. They attributed this stronger dipolar nanoparticle interaction to the sample anisotropy.

12.5.3 Metamaterials

Metamaterials are materials with properties above those obtained from materials found in nature. Metamaterials are highly ordered periodic arrays with building blocks possessing subwavelength dimensions. The metamaterial affects the incident wavelength owing to its subwavelength structuration. In addition, metamaterials behave like homogeneous materials, which can be described by a negative refractive index.

To present a negative refractive index, the meta material possesses both a negative permittivity and a negative permeability. Metals have already been reported for presenting a negative permittivity over a wide frequency range. But to get a meta material, both the permittivity and the permeability have to be negative simultaneously in the same frequency region. As schematized below, a metamaterial, with a negative refractive index, will reverse the direction of the refracted light (Figure 12.16). Such metamaterials were predicted in 1968, [82] and first observed in the late 1990s [83–85].

The applications that are targeted by such materials are, for example, super-lenses with a subwavelength resolution for biomedical imaging or photolithography, electromagnetic absorbers of electrically small resonators, or electrically small antennas [85–87].

In their review concerning wire meta-materials, the authors stress the importance of a perfect structure in the material. This requirement limits strongly the applicable methods to prepare meta-materials [88]. Among the techniques that are suitable, lithography can be cited. Some examples of metamaterials based on silica, gold, and alumina nanoparticles obtained by means of interferrometric lithography have been reported by Xia and coworkers [89].

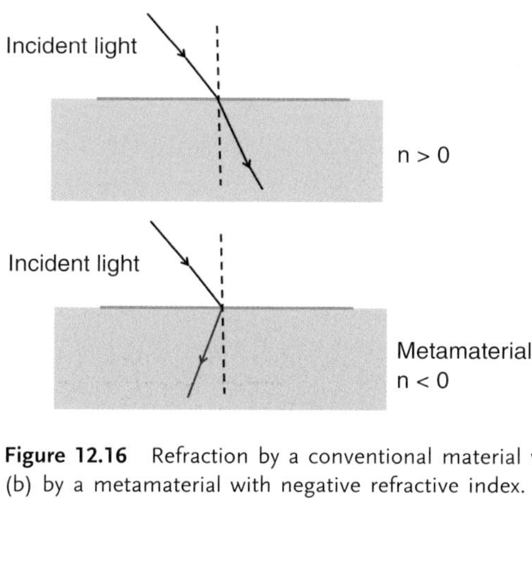

Figure 12.16 Refraction by a conventional material with positive refractive index (a) and (b) by a metamaterial with negative refractive index.

12.5.4
Catalysis/Electrocatalysis

The catalytic properties of stabilized nanoparticles were largely reported in the literature. Some examples could also be found concerning the catalytic properties of nanoparticle assemblies, even though much less investigated than the plasmonic based applications of nanoparticle assemblies.

In a prospective article, Mori and coworkers presented some synthesis aspects and systems based on metal nanoparticles for catalytic applications [90]. The authors highlighted various complex architectures based on nanoparticles that are highly promising. It turned out that two main architectures are stressed: first core-shell particles and second nanoparticle assemblies.

The reason for the interest given to nanoparticle assemblies comes from the stability of the materials, the nanoparticles are non-aggregated, as well as the high amount of nanoparticles.

Pt_3Ni nanoparticle networks were prepared by Zhang and coworkers by the co-reduction of platinum and nickel precursors at room temperature in a two-phase synthetic method [91]. The nanoparticles of the network were 3–8 nm in diameter with an irregular morphology. The material was tested for oxygen reduction reaction. The specific activity of the Pt_3Ni nanoparticle network was shown to be 8.4 times higher than that of commercial Pt/C catalysts, with a specific activity of 0.270 mA/cm^2 at 0.9 V. The electro-catalytic performances of the material were also evaluated for methanol oxidation reactions. For this reaction, the activity of Pt_3Ni nanoparticle network, with current densities of 2.3 mA/cm^2, was shown to be six times higher than that of commercial Pt/C catalysts and two times higher than the activity of pure platinum nanoparticle networks. With these results, the authors concluded that this type of alloy nanoparticle networks are good candidates for cathode and anode catalysts in fuel cells.

The activity of nanoparticle networks for oxygen reduction reaction was also investigated by Zhou and coworkers [92] or Sun and coworkers [93] in the case of platinum nanoparticles assembled into branched chains and FePt nanoparticles assembled on graphene, respectively.

Another example of the use of nanoparticle networks for catalysis application was reported by Neouze and coworkers [94]. In this work, silica nanoparticle networks, assembled by imidazolium covalent molecular mediators, were efficient heterogeneous catalysts for the conversion of CO_2 into organic cyclic carbonates. If imidazolium moieties were already investigated for this kind of reactions, the use of nanoparticle networks remains very interesting as it gives access to a highly functionalized material easy to recycle.

12.5.5
Water Treatment/Photodegradation

Titanium oxide nanoparticles are catalytically very active, relatively cheap, and easy to prepare; in consequence, they are attractive catalyst. Saha and coworkers

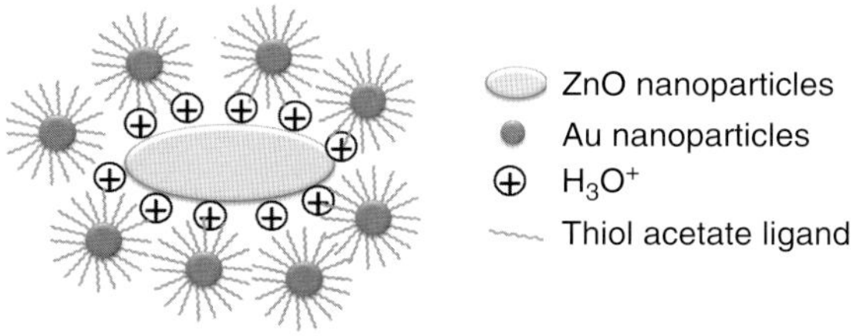

Figure 12.17 Assembly of negatively charged modified gold nanoparticles on positively charged ZnO particles.

investigated the activity of TiO_2 nanoparticle networks as catalyst for the synthesis of 5-hydroxymethylfurfural [95]. The morphology of TiO_2, nanoparticle networks instead of bulk TiO_2, was used in order to get a high specific surface area. The material was prepared by precipitation from titanium isopropoxyde in presence of DL-aspartic acid as a template. The activity of the TiO_2 nanoparticle network for catalytic conversion of fructose and glucose into 5-hydroxymethylfurfural under microwave activation was studied. The porous TiO_2 nanoparticle network in presence of butylmethylimidazolium chloride additive allowed reaching fructose dehydration yields of 82.3%. The catalysts could easily be removed and recycled.

Water treatment is nowadays another main issue in materials for catalysis. In this frame, Yu and co-workers studied the preparation and use of nano-chain assemblies of magnetite nanoparticles [96]. The magnetite nanoparticles were obtained by thermal decomposition of iron acetylacetonate. The authors used these nano-chains for the degradation under UV-irradiation of congo-red dye molecules.

The degradation of organic dye pollutants was also studied by Zheng and co-workers. To this purpose they used a gold/zinc oxide nanoparticle assemblies [97]. The nanocomposite material, based on gold nanoparticles assembles on the surface of ZnO nanoparticles was built by means of electrostatic interactions (Figure 12.17). The zinc oxide particles were stabilized in water at $pH = 7$, as the ξ potential of the particles was measured to be 9.4, the surface of the zinc oxide particles is loaded by acidic water molecules. The gold nanoparticles, prepared by reduction of $HAuCl_4$ precursors, were capped with dithiolated diethylendiaminepentaacetate derivatives. The thiol and/or disulfide groups can anchor the molecules onto the gold surface, while the overall modified nanoparticle is negatively charged owing to the presence of the acetate groups. Thus, negatively charged modified gold nanoparticles can assemble onto positively charged ZnO particles.

12.6 Conclusion

This chapter aims at presenting a brief overview on nanoparticle assemblies. It comes in the frame of the great development observed for nanoparticle assemblies based materials in the past years.

The main strategies reported for the synthesis of nano-objects assemblies are briefly presented. These preparation methods are ordered in four subsections: (i) interligand bonding, where the molecule between the nanoparticles can induce a covalent or a noncovalent linking of the nano-particles. If both procedures are interesting, it can be stressed that the preparation of assemblies through covalently linked nano-objects will lead to irreversible and thus highly stable assemblies, while non-covalent bonding may allow reversible assembly. (ii) The use of a template for the preparation of nano-objects assemblies which was largely reported. However, most of the examples in literature are using sacrificial templates. (iii) The preparation of two-dimensional assemblies was presented in a separated chapter as many papers are specially focusing on 2D assemblies, and the techniques presented are not always adapted for other materials dimensionalities. (iv) In addition some less frequent procedures based on the use of pressure to assemble the nano-objects were reported.

It turned out that a large panel of synthesis procedures are accessible to materials chemists to prepare assemblies of nano-objects. But in a very large majority of the reported synthesis the first step, and also the key step, consists in preparing stabilized nano-objects.

The focus on assemblies containing DNA can also be highlighted. In this review the versatile role of DNA was indirectly stressed. DNA can play the role of a template; In addition, DNA can be used to attach the nano-objects by means of electrostatic interactions or hydrogen bondings.

In the second part of the review, the main applications of nanoparticle assemblies, such as (i) plasmonic nanostructures for sensing, communication, and signal enhancement, (ii) magnetic nanostructures, (iii) metamaterials that are very promising but still in their infancy, (iv) catalysis, and (v) up to photodegradation, were briefly presented.

Even though extensive works have been dedicated to nano-objects assemblies, and most particularly to gold nanoparticle assemblies, some aspects, such as thermodynamic and kinetics of the assembly formation, remain to be elucidated.

References

1. Doane, T.L. and Burda, C. (2012) *Chem. Soc. Rev.*, **41**, 2885–2911.
2. Jiang, S., Win, K.Y., Liu, S., Teng, C.P., Zheng, Y., and Han, M.-Y. (2013) *Nanoscale*, **5**, 3127–3148.
3. Alivisatos, A.P., Johnsson, K.P., Peng, X., Wilson, T.E., Loweth, C.J., Bruchez, M.P. Jr., and Schultz, P.G. (1996) *Nature*, **382**, 609–611.
4. Depero, L.E. and Curri, M.L. (2004) *Curr. Opin. Solid State Mater. Sci.*, **8**, 103–109.
5. Jones, M.R. and Mirkin, C.A. (2012) *Nature*, **491**, 42–43.
6. Pileni, M.P. (2001) *J. Phys. Chem. B*, **105**, 3358–3371.
7. Wang, L., Xu, L., Kuang, H., Xu, C., and Kotov, N.A. (2012) *Acc. Chem. Res.*, **45**, 1916–1926.
8. Murray, W.A., Astilean, S., and Barnes, W.L. (2004) *Phys. Rev. B: Condens. Matter*, **69**, 165401–165407.
9. Siedl, N., Guegel, P., and Diwald, O. (2013) *Langmuir*, **29**, 6077–6083.
10. Ahn, W., Hong, Y., Boriskina, S.V., and Reinhard, B.M. (2013) *ACS Nano*, **7**, 4470–4478.

11. Tang, Z. and Kotov, N.A. (2005) *Adv. Mater.*, **17**, 951–962.
12. Jiang, X.C., Zeng, Q.H., Chen, C.Y., and Yu, A.B. (2011) *J. Mater. Chem.*, **21**, 16797–16805.
13. Grzybowski, A.A. and Whitesides, G.M. (2002) *Science*, **296**, 718–721.
14. Neouze, M.-A. and Schubert, U. (2008) *Monatsh. Chem.*, **139**, 183–195.
15. Schmid, G. (ed) (2010) *Nanoparticles: From Theory to Application*, 2nd edn, Wiley-VCH Verlag GmbH, Weinheim.
16. Atkins, T.M., Louie, A.Y., and Kauzlarich, S.M. (2012) *Nanotechnology*, **23**, 294001–294009.
17. Yajima, T., Yu, Y., and Futamata, M. (2011) *Phys. Chem. Chem. Phys.*, **13**, 12454–12462.
18. Li, N., Gao, Y., Hou, L., and Gao, F. (2011) *J. Phys. Chem. C*, **115**, 25266–25272.
19. Tanaka, K. and Chujo, Y. (2013) *Polym. J.*, **45**, 247–254.
20. Wu, Y.-C. and Kuo, S.-W. (2012) *J. Mater. Chem.*, **22**, 2982–2991.
21. Taladriz-Blanco, P., Buurma, N.J., Rodriguez-Lorenzo, L., Perez-Juste, J., Liz-Marzan, L.M., and Herves, P. (2011) *J. Mater. Chem.*, **21**, 16880–16887.
22. Zhang, D. and Pelton, R. (2012) *Langmuir*, **28**, 3112–3119.
23. Mirkin, C.A., Letsinger, R.L., Mucic, R.C., and Storhoff, J.J. (1996) *Nature*, **382**, 607–609.
24. Wu, J., Silvent, J., Coradin, T., and Aime, C. (2012) *Langmuir*, **28**, 2156–2165.
25. Donato, R.K., Matejka, L., Schrekker, H.S., Plestil, J., Jigounov, A., Brus, J., and Slouf, M. (2011) *J. Mater. Chem.*, **21**, 13801–13810.
26. Sanchez-Iglesias, A., Grzelczak, M., Altantzis, T., Goris, B., Perez-Juste, J., Bals, S., Van Tendeloo, G., Donaldson, S.H., Chmelka, B.F., Israelachvili, J.N., and Liz-Marzan, L.M. (2012) *ACS Nano*, **6**(12), 11059–11065.
27. Laromaine, A., Koh, L., Murugesan, M., Ulijn, R.V., and Stevens, M.M. (2007) *J. Am. Chem. Soc.*, **129**, 4156–4157.
28. Dervaux, B. and Du Prez, F.E. (2012) *Chem. Sci.*, **3**, 959–966.
29. Hawker, C.J. and Wooley, K.L. (2005) *Science*, **309**, 1200–1205.
30. Kolb, H.C., Finn, M.G., and Sharpless, K.B. (2001) *Angew. Chem. Int. Ed.*, **40**, 2004–2021.
31. Santra, S., Kaittanis, C., and Perez, J.M. (2010) *Langmuir*, **26**, 5364–5373.
32. Basnar, B., Litschauer, M., Strasser, G., and Neouze, M.-A. (2012) *J. Phys. Chem. C*, **116**, 9343–9350.
33. Litschauer, M. and Neouze, M.-A. (2008) *Monatsh. Chem.*, **139**, 1151–1156.
34. Litschauer, M. and Neouze, M.-A. (2008) *J. Mater. Chem.*, **18**, 640–646.
35. Litschauer, M., Peterlik, H., and Neouze, M.-A. (2009) *J. Phys. Chem. C*, **113**, 6547–6552.
36. Gates, A.T., Moore, L., Sylvain, M.R., Jones, C.M., Lowry, M., El-Zahab, B., Robinson, J.W., Strongin, R.M., and Warner, I.M. (2009) *Langmuir*, **25**, 9346–9351.
37. Neouze, M.-A., Malenovska, M., and Peterlik, H. (2010) *Polyhedron*, **29**, 569–573.
38. Wen, F., Waldoefner, N., Schmidt, W., Angermund, K., Boennemann, H., Modrow, S., Zinoveva, S., Modrow, H., Hormes, J., Beuermann, L., Rudenkiy, S., Maus-Friedrichs, W., Kempter, V., Vad, T., and Haubold, H.-G. (2005) *Eur. J. Inorg. Chem.*, **18**, 3625–3640.
39. Gu, Z.-G. and Sevov, S.C. (2009) *J. Mater. Chem.*, **19**, 8442–8447.
40. Wang, L., Luo, J., Schadt, M.J., and Zhong, C.-J. (2010) *Langmuir*, **26**, 618–632.
41. Jones, M.R., Osberg, K.D., MacFarlane, R.J., Langille, M.R., and Mirkin, C.A. (2011) *Chem. Rev.*, **111**, 3736–3827.
42. Fernandes, R., Li, M., Dujardin, E., Mann, S., and Kanaras, A.G. (2010) *Chem. Commun.*, **46**, 9270.
43. Sethi, M. and Knecht, M.R. (2010) *Langmuir*, **26**, 9860–9874.
44. Lee, Y., Lee, H., Messersmith, P.B., and Park, T.G. (2010) *Macromol. Rapid Commun.*, **31**, 2109–2114.
45. Ding, Y., Gu, G., Xia, X.-H., and Huo, Q. (2009) *J. Mater. Chem.*, **19**, 795–799.
46. Gacoin, T., Besson, S., and Boilot, J.P. (2006) *J. Phys. Condens. Matter*, **18**, S85–S95.
47. Decher, G. and Schlenoff, J.B. (eds) (2012) *Multilayer Thin Films: Sequential Assembly of Nanocomposite Materials*,

2nd edn, Wiley-VCH Verlag GmbH, Weinheim.

48. Basnar, B., Litschauer, M., Abermann, S., Bertagnolli, E., Strasser, G., and Neouze, M.-A. (2011) *Chem. Commun.*, **47**, 361–363.

49. Pichon, B.P., Louet, P., Felix, O., Drillon, M., Begin-Colin, S., and Decher, G. (2011) *Chem. Mater.*, **23**, 3668–3675.

50. Pauly, M., Pichon, B.P., Albouy, P.-A., Fleutot, S., Leuvrey, C., Trassin, M., Gallani, J.-L., and Begin-Colin, S. (2011) *J. Mater. Chem.*, **21**, 16018–16027.

51. Brinker, C.J., Lu, Y., Sellinger, A., and Fan, H. (1999) *Adv. Mater.*, **11**, 579–585.

52. Grosso, D., Cagnol, F., Soler-Illia, G.J.D.A.A., Crepaldi, E.L., Amenitsch, H., Brunet-Bruneau, A., Bourgeois, A., and Sanchez, C. (2004) *Adv. Funct. Mater.*, **14**, 309–322.

53. Smarsly, B. and Fattakhova-Rohlfing, D. (2009) *Solution Processing of Inorganic Materials*, John Wiley & Sons, Inc., Hoboken, NJ, pp. 283–312.

54. Nakao, H., Tokonami, S., Hamada, T., Shiigi, H., Nagaoka, T., Iwata, F., and Takeda, Y. (2012) *Nanoscale*, **4**, 6814–6822.

55. Wei, J., Wang, H., Deng, Y., Sun, Z., Shi, L., Tu, B., Luqman, M., and Zhao, D. (2011) *J. Am. Chem. Soc.*, **133**, 20369–20377.

56. Andreatta, G., Wang, Y.-J., Fuk, K.L., Polidori, A., Tong, P., Pucci, B., and Benattar, J.-J. (2008) *Langmuir*, **24**, 6072–6078.

57. Andreatta, G., Cousty, J., and Benattar, J.-J. (2011) *Chem. Commun.*, **47**, 3571–3573.

58. Wu, H., Bai, F., Sun, Z., Haddad, R.E., Boye, D.M., Wang, Z., Huang, J.Y., and Fan, H. (2010) *J. Am. Chem. Soc.*, **132**, 12826–12828.

59. Wu, H., Bai, F., Sun, Z., Haddad, R.E., Boye, D.M., Wang, Z., and Fan, H. (2010) *Angew. Chem. Int. Ed.*, **49**, 8431–8434.

60. Kundu, S. (2011) *Langmuir*, **27**, 3930–3936.

61. Murray, W.A. and Barnes, W.L. (2007) *Adv. Mater.*, **19**, 3771–3782.

62. Ozbay, E. (2006) *Science*, **311**, 189–193.

63. Abargues, R., Albert, S., Valdes, J.L., Abderrafi, K., and Martinez-Pastor, J.P. (2012) *J. Mater. Chem.*, **22**, 22204–22211.

64. Wessels, J.M., Nothofer, H.-G., Ford, W.E., von Wrochem, F., Scholz, F., Vossmeyer, T., Schroedter, A., Weller, H., and Yasuda, A. (2004) *J. Am. Chem. Soc.*, **126**, 3349–3356.

65. Zhang, M., Liu, Y.-Q., and Ye, B.-C. (2011) *Chem. Commun.*, **47**, 11849–11851.

66. Xu, L., Kuang, H., Wang, L., and Xu, C. (2011) *J. Mater. Chem.*, **21**, 16759–16782.

67. Liu, S. and Tang, Z. (2010) *J. Mater. Chem.*, **20**, 24–35.

68. Li, X., Zhou, Y., Zheng, Z., Yue, X., Dai, Z., Liu, S., and Tang, Z. (2009) *Langmuir*, **25**, 6580–6586.

69. Kronstein, M., Kriechbaum, K., Akbarzadeh, J., Peterlik, H., and Neouze, M.-A. (2013) *Phys. Chem. Chem. Phys.* doi: 10.1039/C1033CP50430A

70. Wang, J. (2005) *Small*, **1**, 1036–1043.

71. Kim, K., Shin, D., Kim, K.L., and Shin, K.S. (2010) *Phys. Chem. Chem. Phys.*, **12**, 3747–3752.

72. Sun, Y., Xu, F., Zhang, Y., Shi, Y., Wen, Z., and Li, Z. (2011) *J. Mater. Chem.*, **21**, 16675–16685.

73. Pileni, M.-P. (2002) *Nano-Surf. Chem.*, 315–331.

74. Pileni, M.-P. (2003) *Nat. Mater.*, **2**, 145–150.

75. Pileni, M.P. (2008) *Acc. Chem. Res.*, **41**, 1799–1809.

76. Pileni, M.P. (2011) *J. Mater. Chem.*, **21**, 16748–16758.

77. Pileni, M.P. (2012) *J. Colloid Interface Sci.*, **388**, 1–8.

78. Legrand, J., Petit, C., and Pileni, M.P. (2001) *J. Phys. Chem. B*, **105**, 5643–5646.

79. Hiroi, K., Komatsu, K., and Sato, T. (2011) *Phys. Rev. B: Condens. Matter*, **83**, 224421–224429.

80. Margaris, G., Trohidou, K., and Kachkachi, H. (2012) *Phys. Rev. B: Condens. Matter*, **85**, 024411–024419.

81. Pauly, M., Pichon, B.P., Panissod, P., Fleutot, S., Rodriguez, P., Drillon, M., and Begin-Colin, S. (2012) *J. Mater. Chem.*, **22**, 6343–6350.

82. Veselago, V.G. (1967) *Usp. Fiz. Nauk*, **92**, 517–526.

83. Grima, J.N. and Caruana-Gauci, R. (2012) *Nat. Mater.*, **11**, 565–566.
84. Hess, O., Pendry, J.B., Maier, S.A., Oulton, R.F., Hamm, J.M., and Tsakmakidis, K.L. (2012) *Nat. Mater.*, **11**, 573–584.
85. Ying, J.Y. (2012) *Nat. Chem.*, **4**, 159–160.
86. Ehrenberg, I.M., Sarma, S.E., and Wu, B.-I. (2012) *J. Appl. Phys.*, **112**, 073111–073114.
87. Watts, C.M., Liu, X., and Padilla, W.J. (2012) *Adv. Mater.*, **24**, OP98–OP120.
88. Simovski, C.R., Belov, P.A., Atrashchenko, A.V., and Kivshar, Y.S. (2012) *Adv. Mater.*, **24**, 4229–4248.
89. Xia, D., Ku, Z., Lee, S.C., and Brueck, S.R.J. (2011) *Adv. Mater.*, **23**, 147–179.
90. Mori, K. and Yamashita, H. (2010) *Phys. Chem. Chem. Phys.*, **12**, 14420–14432.
91. Xu, Y., Hou, S., Liu, Y., Zhang, Y., Wang, H., and Zhang, B. (2012) *Chem. Commun.*, **48**, 2665–2667.
92. Wang, H.-H., Zhou, Z.-Y., Yuan, Q., Tian, N., and Sun, S.-G. (2011) *Chem. Commun.*, **47**, 3407–3409.
93. Guo, S. and Sun, S. (2012) *J. Am. Chem. Soc.*, **134**, 2492–2495.
94. Roeser, J., Kronstein, M., Litschauer, M., Thomas, A., and Neouze, M.-A. (2012) *Eur. J. Inorg. Chem.*, **2012**, 5305–5311.
95. De, S., Dutta, S., Patra, A.K., Bhaumik, A., and Saha, B. (2011) *J. Mater. Chem.*, **21**, 17505–17510.
96. Gao, M.-R., Zhang, S.-R., Jiang, J., Zheng, Y.-R., Tao, D.-Q., and Yu, S.-H. (2011) *J. Mater. Chem.*, **21**, 16888–16892.
97. Xiao, F., Wang, F., Fu, X., and Zheng, Y. (2012) *J. Mater. Chem.*, **22**, 2868–2877.

13
Porous Molecular Solids

Shan Jiang, Abbie Trewin, and Andrew I. Cooper

13.1
Introduction

Microporous materials with pore sizes smaller than 2 nm are of strong current interest and have potential applications in separations, gas storage, catalysis, sensors, and drug delivery [1]. In general, microporous materials can be designed and synthesized via two main synthetic strategies: extended networks and discrete organic molecules [2]. Extended networks are constructed from strong covalent or coordination bonds to form one-dimensional to three-dimensional (1D to 3D) pore structures [3]. Well-known classes of porous networks are disordered amorphous porous solids (such as activated carbons [4] and porous polymers [5]), crystalline porous solids (such as zeolites [6], metal organic frameworks (MOFs)) [7], and covalent organic frameworks (COFs) [8]). These materials have exhibited high apparent Brunauer–Emmett–Teller (BET) surface areas (up to 6000 $m^2 g^{-1}$) with good chemical and thermal stability [9]. An alternative strategy is to use discrete molecular building blocks to assemble noncovalently to form either crystalline or amorphous porous molecular materials.

Recently, the design of porous molecular materials where packing is dictated by weak van der Waals forces has been attracting considerable attention [2]. The porosity in the molecular materials consists of an intrinsic and an extrinsic one [2a]. The intrinsic pores arise from the voids within a molecule, whereas extrinsic pores result from inefficient molecular packing. The challenge for this research field is that most organic molecules are packed in such a way to minimize free volume; therefore, they do not show porosity in the solid state. Crystalline solvates of organic molecules sometimes have cavities or channels in which the solvent molecules are situated; however, on desolvation, these voids usually collapse. As a result, permanent porous organic molecules are comparatively rare with respect to extended porous networks. However, porous molecular materials offer a number of potential advantages: for example, good solution processability, that is, porous membranes or thin films can be used for devices [10]; synthetic diversity by functional group modification [11]; molecular mobility to open and close pore

Discovering the Future of Molecular Sciences, First Edition. Edited by Bruno Pignataro.

channels in response to guest molecules [12]; and reversible switching of porosity by switching between different polymorphs [13].

There are two distinct classes of porous molecules: crystalline and amorphous. To date, there has been little focus in the area of amorphous porous molecules compared with crystalline ones, because permanent porosity in amorphous molecular solids is extremely rare and difficult to design and rationalize the properties because of the lack of the molecular structures. In this chapter, we discuss the classes of porous molecular crystals, in particular with respect to porous organic cage systems. We also discuss several synthetic approaches to produce amorphous molecular materials and review theoretical studies on generating representative structural models for amorphous molecular materials.

13.2 Porous Organic Molecular Crystals

13.2.1 Porous Organic Molecules

Early examples of discrete organic molecules that can form permanent porosity after desolvation include calix[n]arenes [14], cucurbit[n]urils [15], tris-*o*-phenylenedioxycyclotriphosphazene (TPP) [16], some dipeptides [17], and 3,3′,4,4′-tetrakis (trimethylsilylethynyl)biphenyl (4TMSEBP) [18]. Calix[*n*]arenes are a class of macrocyclic compounds, which possess bowl-shaped conformations with defined upper and lower rims [19]. The internal cavity of calixarenes can be used to encapsulate guest molecules. A low density polymorph of *p-tert*-butylcalix[4]arene has demonstrated a porosity "without pores" manner [12]. The material does not show channels or voids in the static structure; however, on incorporation of guest molecules, the guest can be situated in the cavity of *p-tert*-butylcalix[4]arene following a cooperative process between the host and the guest [12]. Some flexible MOFs also exhibit the dynamic gate opening in response to the guest [20]. This structural flexibility has shown a significant influence on the gas sorption properties. Furthermore, *p-tert*-butylcalix[4]arene crystals obtained by sublimation have revealed O_2, N_2, and CO_2 absorption properties under ambient conditions [21]. However, the H_2 uptake was not observed up to the pressure of 7 atm. Hence, this material can be used to separate H_2 from a mixture of these gases [21]. The porous organic molecule, 1,2-dimethoxy-*p-tert*-butylcalix[4]dihydroquinone, shows a type I N_2 adsorption isotherm with a BET surface area of 230 $m^2\,g^{-1}$ [14]. Cucurbit[*n*]urils are "pumpkin"-shaped molecules that can be easily prepared by condensing glycoluril with formaldehyde [22]. The cavity of the cucurbit[n]urils is nonpolar and is widely used for the binding of hydrophobic guests [23]. Kim and coworkers reported a porous cucurbit[6]uril (CB[6]) with a BET surface area of 210 $m^2\,g^{-1}$, and an extrinsic 1D channel in the crystal structure [15]. The material has high adsorption capacity of acetylene (C_2H_2), and two C_2H_2 molecules are found in the channel, which can offer specific adsorption sites to form hydrogen bonds with CB[6] [15]. A polymorph of CB[6] obtained by recrystallization gives a high CO_2 adsorption capacity at 298 K and 1 bar [24]. Structural analysis indicates that CO_2 molecules are located not only in the extrinsic 1D channels but also in the intrinsic cavities of CB[6] [24]. TPP is a "paddle-wheel" molecule. The TPP molecules can form 1D pore channels via a hexagonal packing in the crystalline solid state [16]. The Langmuir surface area of TPP crystals is 240 $m^2 g^{-1}$, and the total N_2 uptake is 1.1 mol N_2 per mol TPP (corresponding to 2.5 mmol g^{-1} of TPP) [25]. The TPP crystals are observed to selectively adsorb CH_4 (1.5 mmol g^{-1} at 195 K and 1 atm) and CO_2 (2.7 mmol g^{-1} at 195 K and 1 atm) [26]. Moreover, the crystalline dipeptides assemble in a hexagonal packing to form 1D hydrophobic channels by hydrogen bonds [17]. The gas sorption analysis shows the CO_2/CH_4 selectivity of 2–2.5 for L-valyl-L-alanine (VA) and L-alanyl-L-valine (AV) at 195 K and 1 atm [17]. In addition, the organic porous crystal of 4TMSEBP was discovered by searching crystal structures in Cambridge Structural Database (CSD) [18]. The crystal structure of 4TMSEBP shows 3D interconnectivity of pore voids and channels. N_2 sorption measurements give a type I isotherm at 77 K, with a BET surface area of 278 $m^2 g^{-1}$ [18]. The material also absorbs a large amount of H_2 at 77 K, with 3.9 mmol g^{-1} absorbed at 10 bar [18]. Recently, Mastalerz and coworkers reported a triptycenetrisbenzimidazolone (TTBI) porous crystal assembled by hydrogen bonds [27]. This material absorbs a significant amount of N_2 (33.7 mmol g^{-1}) at 77 K and $P/P_0 = 0.95$. The measured BET surface area is 2796 $m^2 g^{-1}$ and the Langmuir surface area is 3020 $m^2 g^{-1}$. The material adsorbs CO_2 (3.6 mmol g^{-1}) over CH_4 (0.9 mmol g^{-1}) at 273 K and 1 bar. The H_2 uptake is 11 mmol g^{-1} at 77 K and 1 bar. These values exceed all other porous organic molecules and are comparable with some MOFs, and this study represents a remarkable achievement in terms of crystal engineering for porous organic solids.

13.2.2
PorousOrganic Cages

Recently, a series of organic cage molecules has been designed in several research groups, and this research area is attracting increasing attention [11a, 28]. The porosity of these materials depends on the molecular packing motifs. Slow precipitation of cage molecules can form crystalline materials, whereas rapid desolvation results in the porous amorphous cage materials. It has been noted that the degree of crystallinity has a significant effect on the porosity of these materials [29].

Initially, Warmuth and coworkers synthesized imine cage molecules with tetrahedral, octahedral, or rhombicuboctahedral symmetry as molecular capsules [30]. These cage molecules possess large cavity volumes that can be used to encapsulate guest molecules; however, these cage molecules have not been investigated after gas uptakes following desolvation. They are likely to collapse after the removal of the solvents.

The cage molecules developed in the Cooper research group were prepared by the cycloimination condensation reaction of 1,3,5-triformylbenzene

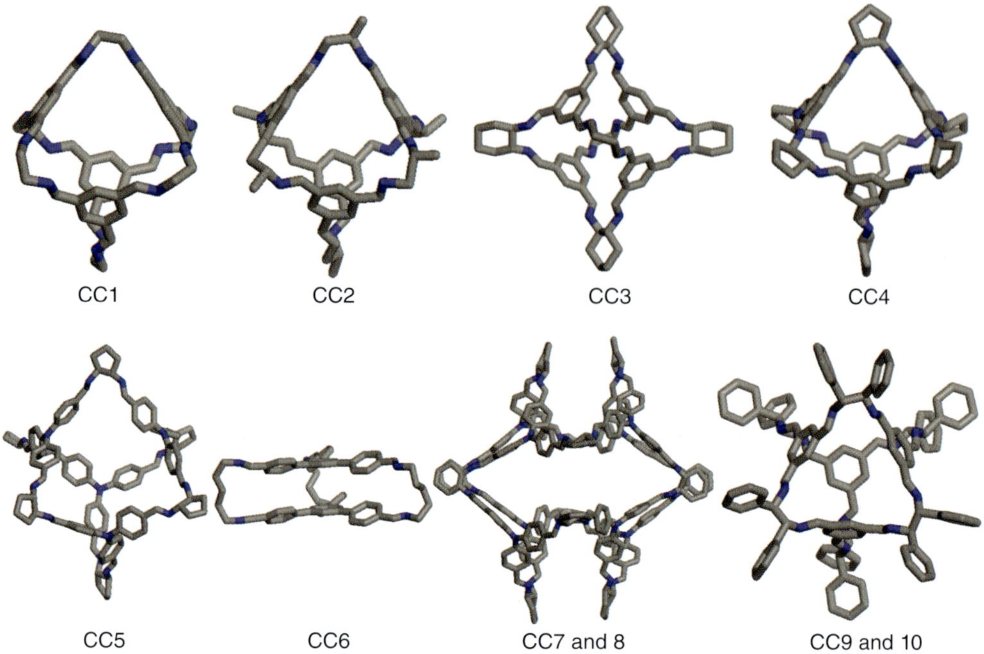

Figure 13.1 The molecular stick models of covalent cages (**CC1–CC10**) obtained from single crystal X-ray diffraction. Hydrogen atoms are omitted for clarity; carbon and nitrogen atoms are colored gray and black. (Figures reproduced from Refs [11a, 31]. Copyright 2009 and 2011 Nature publishing group, 2011 American Chemical Society and 2011 Wiley-VCH.)

(TFB), tri(4-formylphenyl)amine, or 1,3,5-tri-(4-formylphenyl)benzene with 1,2-ethylenediamine (EDA), 1,2-propylenediamine, 1,2-cyclohexanediamine (CHDA), 1,2-diaminocyclopentane, 1,5-pentanediamine, 1,2-diphenylethylenediamine, or 1,2-bis(4-fluorophenyl)ethane-1,2-diamine [11a, 31]. The products were isolated directly as crystals and characterized by single crystal X-ray diffraction. The molecular structures are shown in Figure 13.1.

The cage molecules possess various functional groups on the cage vertices. **CC1** has unfunctionalized ethylene linkers on the six vertices. **CC2** possesses six positional isomers due to the methyl group on each vertex. **CC3** has relatively bulky cyclohexyl groups on the vertices, whereas the vertices of **CC4** consists of cyclopentyl groups. By contrast, **CC5** is synthesized from tri(4-formylphenyl)amine; therefore, it exhibits a larger internal volume. A [2+3] cage molecule, **CC6**, shows a compact structure with little cage voids. **CC7** and **CC8** have a [8+12] topology with large internal voids; however, the voids for these cages collapse upon solvent removal. **CC9** and **CC10** cage molecules have bulky aryl groups attached the vertices.

Interestingly, pore structure and connectivity are directed by the functional groups that are attached to the cage vertices; therefore, it is possible to form interconnected or unconnected pore voids in the structures. The **CC1** crystal packing shows isolated cage voids from the Connolly surfaces generated using a N_2 probe of 1.84 Å in Figure 13.2a [13]. The **CC1** material adsorbs little N_2 at 77 K, 1 bar [13]. **CC1** can form three polymorphs in the solid state by recrystallization in different organic solvents and the gas porosity and selectivity in these polymorphs can be interconverted reversibly [13]. **CC1α** is nonporous to N_2 and H_2. **CC1β** is selectively porous to H_2. **CC1γ** is porous to both N_2 and H_2 [13]. Owing to the methyl groups on the vertices to direct the molecular packing, **CC2** gives 1D extrinsic pore channels between the cages, and also shows isolated intrinsic accessible voids (Figure 13.2a) [13]. **CC2** exhibits a type I N_2 sorption isotherm with a BET surface area of 533 $m^2\,g^{-1}$ [13]. The bulkier cyclohexyl groups in **CC3** direct the cages to pack window to window, leading to an interconnected 3D pore channel structure (Figure 13.2a) [13]. The BET surface area of **CC3** is 624 $m^2\,g^{-1}$ [13]. **CC4** crystal packing shows 2D pore connection with close intermolecular interaction [31d]. The N_2 sorption shows a step isotherm with a large degree of hysteresis (Figure 13.2b). **CC5** crystals exhibit a large pore volume and a higher BET surface area of 1333 $m^2\,g^{-1}$ [31c]. **CC6** has a compact molecular structure with little internal voids. The BET surface area of this material is 99 $m^2\,g^{-1}$. However, the material shows selective adsorption of H_2 and CO_2 over N_2 [31e]. The materials of **CC7** and **CC8** become amorphous on desolvation, and they are not found to adsorb N_2 at 77 K, 1 bar [31b]. The vertex functionality on **CC9** (diphenyl groups) and **CC10** (difluorophenyl groups) vertices enhances the gas uptake [31a]. Two polymorphs of **CC9** demonstrate BET surface areas of 854 and 501 $m^2\,g^{-1}$, respectively, while the BET surface area of **CC10** is 460 $m^2\,g^{-1}$ [31a].

The Mastalerz group has introduced a salicylbisimine cage molecule synthesized by the condensation reaction of trisaminotriptycene and salicylbisaldehyde in a one-step [4+6] cycloimination (Figure 13.3a,b) [28]. The cage molecule has tetrahedral symmetry with a large intrinsic volume. The packing of the cage molecules creates an interconnected 3D pore network, which is dictated by π–π stacking interactions. To evaluate its permanent porosity, gas sorption isotherms (N_2, CH_4, and CO_2) were measured. The calculated surface areas based on the N_2 isotherm are 1566 $m^2\,g^{-1}$ (Langmuir) and 1375 $m^2\,g^{-1}$ (BET). The porous cage molecule shows good selectivity for CO_2 over CH_4 (10/1 (mmol/mmol)). In addition, a series of periphery-substituted shape-persistent cage compounds were synthesized by Mastalerz and coworkers [32]. The cage compounds showed permanent porosity in both crystalline and amorphous solid states. The amorphous cages with various functionalities have surface areas in the range of 690–727 $m^2\,g^{-1}$. The bulkiness of the peripheral groups has a significant effect on the surface areas of the crystalline cage materials. The cage compound with methyl groups at the periphery has a BET surface area of 1291 $m^2\,g^{-1}$, while the value is 309 $m^2\,g^{-1}$ for the cage with 3-ethylpentyl groups, and 22 $m^2\,g^{-1}$ for the cage with trityl groups. One crystalline polymorph of the cage compound with *tert*-butyl groups at the periphery displays a remarkably high BET surface area of 2071 $m^2\,g^{-1}$.

Another study has been carried out to investigate the influence of the rigidity of the molecular structure of [2+3] cage compounds synthesized by an imine

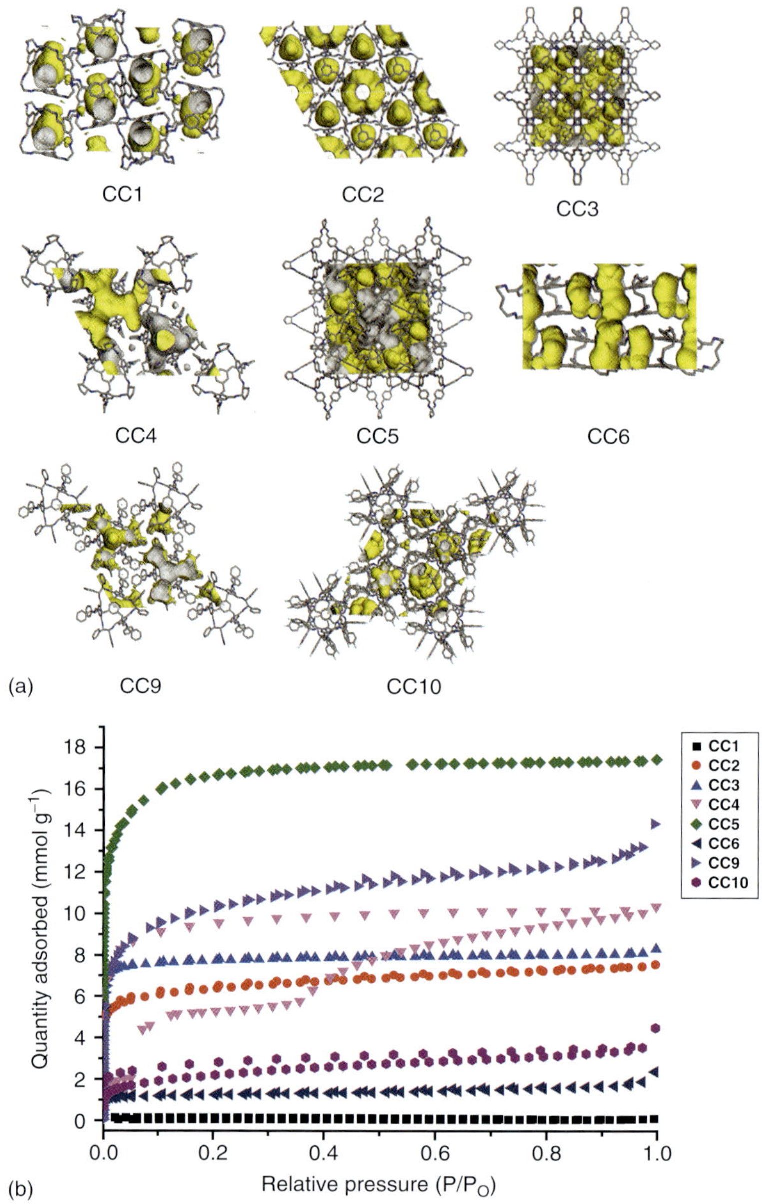

Figure 13.2 (a) Connolly surface area (shown in black) generated using a N_2 probe radius of 1.82 Å to show pore structures for **CC1–CC10** cages. (b) N_2 adsorption/desorption isotherms for these materials at 77 K. (Figure adapted with permission from Refs [11a, 31]. Copyright 2009 and 2011 Nature publishing group, 2011 American Chemical Society and 2011 Wiley-VCH.)

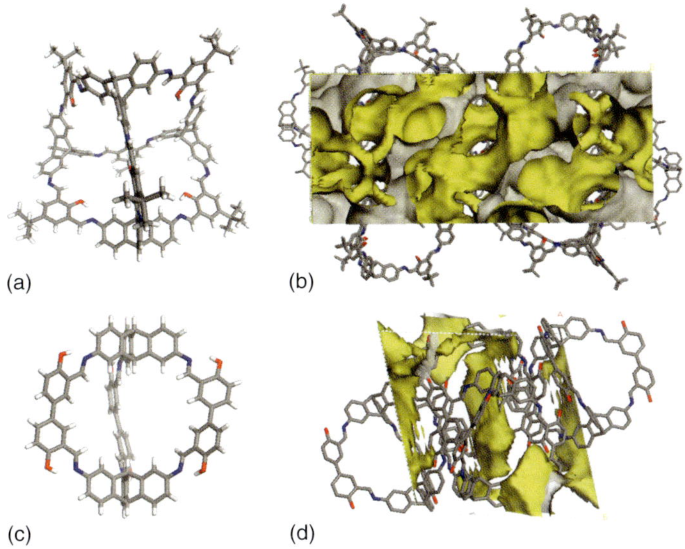

Figure 13.3 (a,c) Molecular structures of [4+6] and [2+3] cages that are synthesized in the Mastalerz group (hydrogen, carbon, nitrogen, and oxygen are colored by white, gray, blue, and red, respectively). (b,d) Solvent accessible surface area (shown in yellow) generated using a N_2 probe radius of 1.82 Å for [4+6] cage structure and [2+3] cage structure, respectively. (Figure adapted with permission from Refs [28, 33]. Copyright 2011 and 2012 Wiley-VCH.)

condensation of triptycenetriamine and bissalicylaldehydes [33]. The [2+3] cage compound with a short organic linker has a BET surface area of $744\,m^2\,g^{-1}$ (Figure 13.3c,d). In contrast, the [2+3] cage with a long organic linker shows a low level of porosity with a BET surface area of $30\,m^2\,g^{-1}$.

Zhang and his coworkers have synthesized a series of organic cage molecules in a [2+3] cycloimination reaction [34]. The cage molecules have very small internal voids, and the packing diagram shows that the molecules close pack to each other in a layered structure form. The materials hardly adsorbed CO_2 and N_2 at 20 °C and 1 bar, while the ideal selectivity of CO_2/N_2 for these [2+3] cage materials is in the range from 36/1 to 138/1 (mmol/mmol) [35]. However, in the practical application of gas separation, the materials should not only possess high selectivity, but also show a good gas sorption capacity.

Recently, Doonan and coworkers described a trigonal cage molecule that is constructed from carbon-carbon bonds rather than the imine formation [36]. One of the polymorphs is produced by rapid precipitation and shows a permanent porosity to N_2 with a high BET surface area of $1153\,m^2\,g^{-1}$, and pore channels are observed that connect the cage internal voids via the cage windows as shown in Figure 13.4. However, a polymorph synthesized by a slow crystallization method gives a gas selectivity behavior of H_2 over N_2 at 77 K. The work provides a synthetic control of polymorph formation for a cage molecule by a different synthetic method.

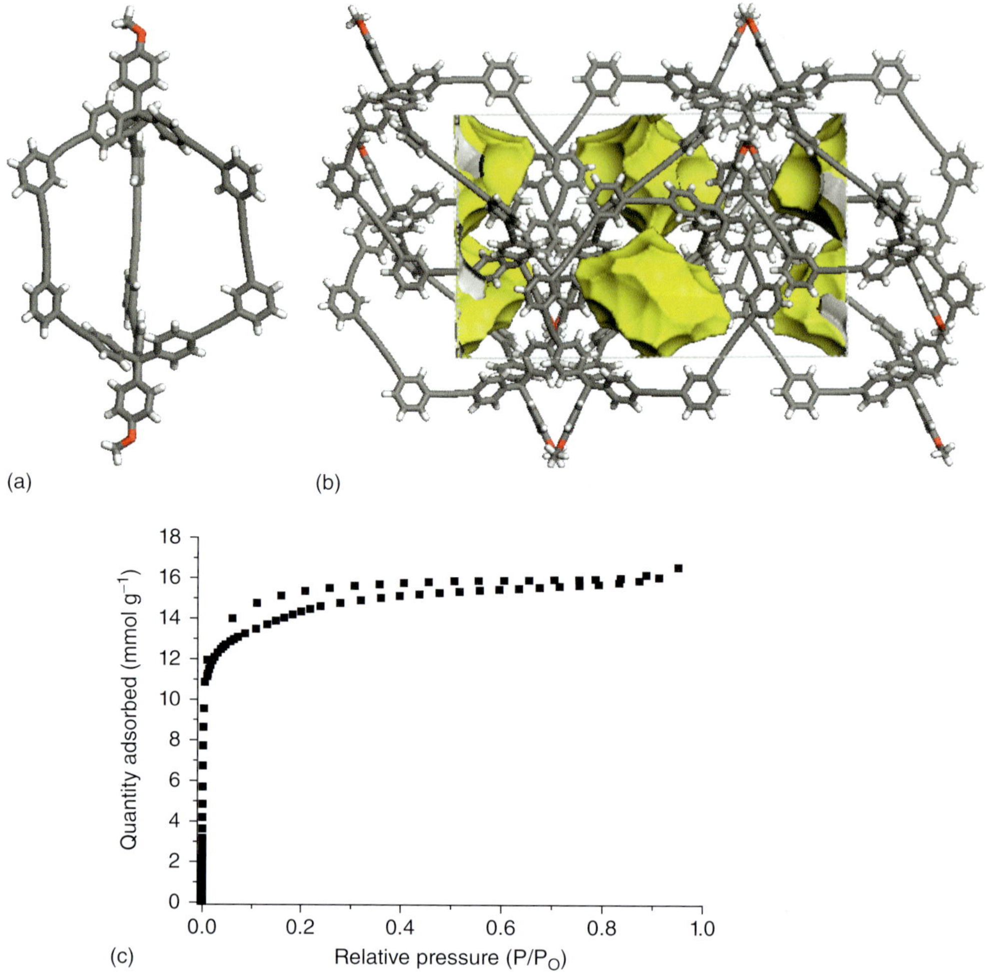

Figure 13.4 (a) Molecular structures of [2+3] cage that are synthesized in the Doonan group (hydrogen, carbon, nitrogen, and oxygen are colored by white, gray, blue, and red, respectively). (b) Solvent accessible surface area (shown in yellow) generated using a N_2 probe radius of 1.82 Å for the cage molecule. (c) N_2 adsorption/desorption isotherm for a polymorph of the material at 77 K. (Figure adapted with permission from Ref. [36]. Copyright 2013 Wiley-VCH.)

13.2.3
Simulation of Porous Organic Molecular Crystals

Molecular modeling and simulation are very useful tools to design and rationalize the properties of porous materials. For example, Density functional theory (DFT) calculations can be used to investigate the binding energy landscape of guest–host interactions. Grand canonical Monte Carlo (GCMC) simulations can provide information on gas uptakes and sorption sites in the host structure. GCMC simulations have been widely used for high throughput screening of crystalline pore networks for a specific application, such as storage of a desired guest molecule or separations [2c, 37]. In addition, the flexibility and dynamic motions of porous molecular materials have been shown to have a significant influence on the gas sorption properties and gas diffusion behavior [12, 38]. Therefore, it is essential to explore and design materials that are able to respond to guest molecules dynamically, for example, using molecular dynamic (MD) simulation. It should be noted that MD simulation results may depend critically on the force field (FF) used. It may be necessary to develop bespoke FFs to describe the hosts reliably. A better understanding of gas diffusion and gas sorption capacity features will aid the optimization and development of industrial applications of porous molecular materials in gas storage, separation, and catalytic processes.

Compared with simulations of porous networks, there are only a few reports simulating porous organic molecules [38]. Simulating gas uptakes of CO_2, CH_4, N_2, and Xe in TPP has been carried out using GCMC [39]. The FF was parameterized accurately for the TPP molecule in this study. However, the simulated gas uptakes overestimated the experimental values. This ascribed to the simulation based on a rigid and ideal structure, which did not take into account the molecular mobility and crystal imperfections in the experiment. On the other hand, Alavi *et al.* [40] have studied small guest molecule (Xe, N_2, H_2, CH_4, and SO_2) inclusion in *p-tert*-butylcalix[4]arene using MD simulations. The inclusion energy was calculated in a rigid calixarene with AMBER [41] FF. Furthermore, Ripmeester and coworkers [42] reported an experimental and molecular simulation study on hydrogen adsorption and diffusion in *p-tert*-butylcalix[4]arene. ^1H NMR and GCMC simulations were used to evaluate the H_2 sorption capacity and adsorption sites. MD simulations suggested that H_2 molecules diffused between calixarene bowls, but the diffusivity was low for this material. Rawashdeh *et al.* [43] indicated the inclusion complexation of methyl viologen with cucurbit[*n*]uril using MD simulation. The energy barriers for the host and guest inclusion process were obtained from the potential of mean force (PMF). We have used MD simulations to rationalize H_2/N_2 gas selectivity for a crystalline porous cage, **CC1β**. A Cage Specific Force Field (CSFF) was developed, which was parameterized based on PCFF [44] and fitted for imine tetrahedral cage molecules. A H_2 molecule was observed to diffuse between internal cage volumes and external channel volumes via a hopping mechanism, although a N_2 molecule could not diffuse. We have also applied MD simulations to explain the collapse of large voids of [8+12] cage molecule (**CC7**) on desolvation and hence loss of porosity, which means that the rigidity of the organic molecules is important in maintaining the structure. Moreover, MD simulations have been used to investigate larger guests (such as *para*-xylene, 4-ethyltoluene, and mesitylene) diffusion in crystalline **CC3** in order to rationalize molecular selectivity [45].

Recently, crystal structure prediction (CSP) has become a useful tool to predict the structures of porous molecular solids for rational design strategy. Day *et al.* used CSP to successfully predict porous organic cage packing [31c]. A racemic (*R,S*) packing of **CC3** was found as the lowest energy structure from CSP, which is consistent with the crystal structure observed in the experiment. However, the cocrystal of **CC5** was not observed experimentally, as homochiral crystals with window to window packing were strongly preferred compared with the racemic crystals from calculations [31c]. The lowest energy structure for a cocrystal of **CC1** and **CC3** (heterochiral **CC1**-*S* and **CC3**-*R*) was also found in the experiment.

13.2.4
Applications for Porous Molecular Crystals

Porous molecular materials have been widely used in various fields such as gas storage, separation, shape/size selectivity, and sensors [10, 11, 45]. To date, porous molecular materials have demonstrated high porosity to gas molecules, for example, TTBI crystals demonstrate high BET and Langmuir surface areas of 2796 and 3020 $m^2\,g^{-1}$, and high values of H_2, N_2, CO_2, and CH_4 uptakes have been achieved for this material [27]. Moreover, recently, there has been significant increased interest in molecular organic crystals that are capable of being select guest molecules. A series of trigonal [2+3] imine cage molecules exhibited a high ideal CO_2/N_2 selectivity of up to 138/1 at 293 K and 1 bar [35]. Organic cage molecules **CC1** and **CC3** have been designed for separation of structural isomers of 4-ethyltoluene, 3-ethyltoluene, 2-ethyltoluene over mesitylene, and positional isomers of *para*-xylene over *meta*- and *ortho*-xylene [45]. A covalent organic polyhedron (COP) synthesized by alkyne metathesis can selectively encapsulate C70 over C60 [46]. Owing to the good solubility of these materials, molecular porous materials have been solution produced to nanoparticles where particle sizes and porosity can be tuned [29], and porous composites by mixing porous organic cage with soluble porous polymers, PIMs (polymers of intrinsic microporosity) [47], and molecular sensors to organic vapors due to high binding affinity [10]. A reversible switching of porosity can be achieved by switching between different polymorphs of organic molecules by recrystallizations [13]. Furthermore, it has been shown that cocrystals can be produced by mixing different organic cage molecules in the Cooper group, heterochiral cocrystal of **CC1**-*S* and **CC3**-*R*, the cocrystal of enantiomers of **CC3**-*S* and **CC3**-*R*, and a racemic cocrystal of **CC4**-*S* and **CC3**-*R* [31c]. The crystal packing arrangement of mixtures of cage can also be predicted by molecular simulations.

13.3
Porous Amorphous Molecular Materials

Compared with crystalline porous molecular materials, the number of porous amorphous molecular materials is relatively small [2c]. In addition, due to the lack of molecular structures, understanding and design of the amorphous molecular materials is more challenging. However, these materials are a very promising class of porous solids and have a number of potential benefits [2a,c]. For example, desolvating molecular crystals often leads to a loss of crystallinity and porosity, which may not happen in the amorphous solid. The porosity of the materials can be further enhanced by increasing molecular bad packing (extrinsic porosity). Control and tuning of the properties of amorphous molecular solids can be achieved via functional group modification of discrete molecules. Moreover, the materials are able to form membranes or thin porous films because of their good solubility.

13.3.1
Synthesis of Porous Amorphous Molecular Materials

This is a brief overview of amorphous molecular solids in the literature. PIMs are prepared either as insoluble networks or soluble polymers [48]. The soluble linear PIMs with rigid and contorted molecular structures exhibit high surface areas (700–900 $m^2 g^{-1}$). The molecular weight of these liner polymers is in the range from 5×10^3 to 270×10^3 $g mol^{-1}$, which can be considered as porous molecular materials [48]. The "intrinsic" porosity for PIMs results from the bad packing of macromolecular chain building blocks, which is defined as extrinsic voids in this chapter. It is noted that PIM-1 with small chains has a low surface area, and that the porosity increases with the increasing chain length [49]. A significant advantage of these materials is the solution processability. PIM-1 has been processed into porous thin films or membranes for gas separation [49]. On the other hand, there are a few examples of amorphous porous molecular solids comprising small organic molecules. The Noria molecule and analogs show a "waterwheel" structure with a large hydrophobic cavity. These materials exhibit gas adsorption properties in the amorphous state as reported by Tian *et al.* [50]. The Noria molecule shows a low N_2 and H_2 uptake with BET surface area of 40 $m^2 g^{-1}$ calculated from the N_2 isotherm at 77 K. However, it displays a high absorption capacity to CO_2. The amorphous Noria material exhibits a type I CO_2 isotherm from which the surface area is calculated, ranging from 280 to 350 $m^2 g^{-1}$ and a pore volume of 0.13 $cm^3 g^{-1}$. The Noria and analog is the first amorphous porous organic molecule to show selective uptake of CO_2 over H_2 and N_2. Furthermore, Tian *et al.* reported amorphous cucurbit[7]uril (CB[7]) material for a high CO_2 sorption capacity (50 $cm^3 g^{-1}$ at 1 bar, 297 K) among known amorphous molecular materials [51]. In contrast, the N_2 and CH_4 uptakes of amorphous CB[7] are very low with only 5.5 and 6.0 $cm^3 g^{-1}$ under the same conditions, respectively. The high selectivity of CO_2 over CH_4 and N_2 can be ascribed to the high affinity of CB[7] and CO_2 [51]. As mentioned earlier, Jin *et al.* synthesized a series of amorphous [2+3] cage molecules [35]. Although the actual CO_2 adsorption capacity of these [2+3] cage molecules at 1 bar and 20 °C is much lower than other organic molecules published, the selectivity is extraordinarily high (ideal CO_2/N_2 selectivity in the range from 36/1 to 138/1 (mmol/mmol)). Schneider *et al.* [52] reported a procedure to control the degree of crystallinity of periphery-substituted cage compounds during the synthesis. They mentioned a BET surface areas comparison between bulk amorphous, nanospheric amorphous, and bulk crystalline materials for these cages compounds. They found that the amorphous cage materials exhibit BET surface areas of approximately 700 $m^2 g^{-1}$. When these cage compounds are in the crystalline or nanospheric solid state, the surface areas are much lower [52]. In addition, there is an exo-functionalized [4+6] imine cage compound in the amorphous solid state with a BET surface area of 919 $m^2 g^{-1}$ [53].

13.3.1.1 Synthesis of Amorphous Cage Materials by Scrambling Reactions and Freeze-Drying

We have demonstrated scrambled cage molecules that are synthesized by three novel methods. They are based on cage–diamine exchange (**CC1** and CHDA), cage–cage interchange (**CC1** and **CC3**), and coreaction of TFB with a mixture of both EDA and CHDA, respectively as shown in Figure 13.5 [54]. The thermodynamically stable products consist of a series of cage molecules incorporating both EDA- and CHDA-linked vertices in a single cage molecules, which can be confirmed by high performance liquid chromatography (HPLC) combined with mass spectrometry (MS). The equilibration of the scrambling reactions is affected by adjusting the reaction conditions such as reaction ratios and temperature in order to control the distribution of products.

The distribution of scrambled cage molecules packs together ineffectively and creates permanent porosity in the amorphous solid. The gas sorption analysis showed type I N_2 isotherms for samples by co-reactions with a various reaction ratio, and the sample by cage–cage interchange. The gas sorption properties can be controlled as a function of EDA:CHDA reaction ratio, and H_2/N_2 gas selectivity behavior can be tuned. The amorphous cage materials synthesized from a high EDA ratio can adsorb much more H_2 than N_2. For example, a sample with a

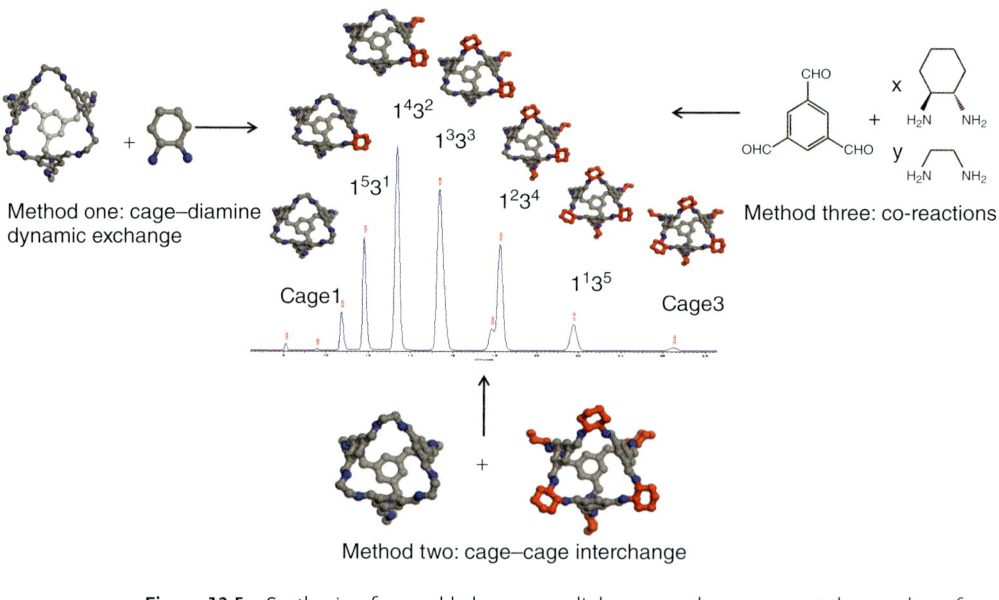

Figure 13.5 Synthesis of scrambled cage molecules by cage–diamine dynamic exchange, cage–cage interchange, and coreactions. The products are represented as $\mathbf{1}^n\mathbf{3}^m$ where **1** and **3** are EDA and CHDA linkers. n and m represent the number of EDA and CHDA linkers, respectively. (Figure adapted from the Ref. [54]. Copyright 2011 Nature publishing group.)

ratio of EDA and CHDA at 5 : 1 adsorbs around five times as much H_2 as N_2. However, the materials synthesized from a low EDA ratio are highly porous to both H_2 and N_2. The BET surface area of one of these amorphous scrambled cage materials is up to 818 $m^2\,g^{-1}$, which has exceeded crystalline porous cage molecules and other porous amorphous molecules. These amorphous materials are not only porous to N_2, but also adsorb H_2 and CO_2. They adsorb amounts of CO_2 in the range 1.60–1.93 $mmol\,g^{-1}$ (7.0–8.7 wt%) at 293 K and 1 bar. The materials are able to adsorb a large amount of H_2, 3.70–6.07 $mmol\,g^{-1}$ at 77 K and 1 bar. The H_2 uptakes for the samples (a ratio of EDA and CHDA at 5 : 1 and 4 : 2) were 4.81 and 3.71 $mmol\,g^{-1}$, respectively, but they had a very low N_2 uptake. Therefore, high H_2/N_2 gas selectivity was observed for these two samples.

We also synthesized amorphous **CC1** and amorphous **CC3** by the freeze-drying method in Figure 13.6 [29]. The method was used to rapidly precipitate cage

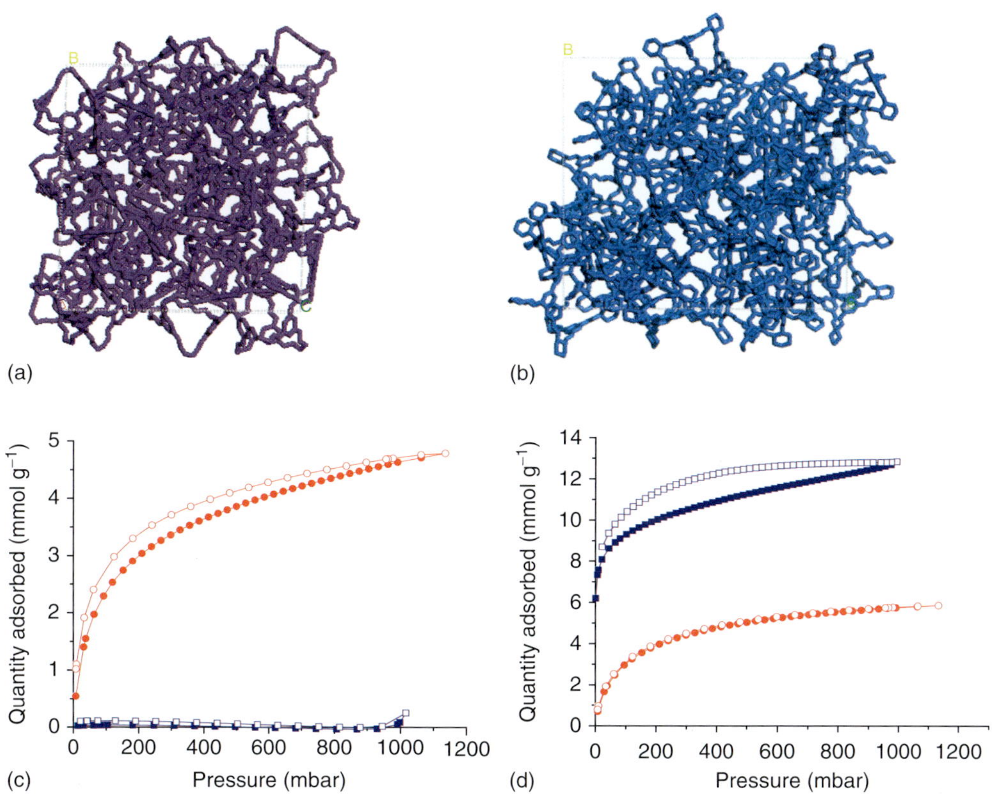

Figure 13.6 (a) A structural model for amorphous **CC1** with a simulation density of 0.92 $g\,cm^{-3}$. (b) A structural model for amorphous **CC3** with a simulation density of 0.82 $g\,cm^{-3}$. H_2 (red circles) and N_2 (blue squares) adsorption/desorption isotherms for (c) amorphous **CC1** and (d) amorphous **CC3** at 77 K. Filled and open represent adsorption and desorption isotherms. (The figure adapted with permission form Ref. [29]. Copyright (2012) American Chemical Society.)

molecules from the solution and produce the **CC1** and **CC3** in the amorphous solid state. The amorphous phase was confirmed on the basis of powder X-ray diffraction (PXRD) and scanning electron microscope (SEM) images. These amorphous cage materials showed a high level of porosity and tunable gas selectivity. The BET surface area of amorphous **CC3** is up to 898 $m^2 g^{-1}$, which exceeds comparable amorphous molecular solids and is twice the surface area of crystalline **CC3** [29]. H_2/N_2 gas selectivity was observed in these materials. Amorphous **CC1**, with fully EDA vertices, displays a selectively porous of H_2 and gives a H_2/N_2 selectivity of 19 at 77 K, 1 bar. However, amorphous **CC3** with fully CHDA vertices adsorbs both N_2 and H_2. It is noted that there is a direct correlation between the degree of crystallinity and gas uptakes for porous molecular materials [29]. The rate of precipitation of cage molecules from solution has a significant influence on the crystallinity and porosity of the materials [29].

13.3.2
Simulation of Porous Amorphous Molecular Materials

Crystalline porous materials are well understood in terms of pore topologies and sizes, because the crystal structures can be obtained from single crystal X-ray diffraction. By contrast, amorphous materials have no long-range order, meaning that the molecular structures are not available from the experimental techniques. Approaches such as measuring the surface area, pore volume, and pore size distribution can be used to characterize the properties of amorphous porous materials in the macroscopic scope. Furthermore, molecular simulation becomes a useful tool for generating structural models, designing, and rationalizing the properties in the microscopic scope. In order to simulate the properties of amorphous porous materials, a representative structural model should be built. In general, for the generation of a realistic model for amorphous materials, a large simulation cell is required and different models should be constructed for averaging in order to give a better representation of bulk materials.

A number of computational studies have been carried out for generating porous amorphous polymers structures such as hyper-cross-linked polymers (HCPs) [55], conjugated microporous polymers (CMPs) [56], and PIMs [57]. The molecular building blocks for amorphous polymers are uncertain, and hence a range of molecular chains with different molecular masses need to be constructed as a first step. These polymer chains can be packed either at the experimental density or a low density by Monte Carlo (MC), and then followed by either geometry energy minimization or several steps compression and relaxation with MD simulations to achieve the final structural model [58]. The physical properties obtained from the structural models such as density, surface areas, pore volumes can be compared with experimental data in order to validate the simulation methodology. However, some simulation methods are dependent on the experimental target density, which cannot be determined accurately in the experiments. This might lead to a poor estimation or prediction of physical properties [58b, 59]. Recently, Colina and coworkers reported a simulation procedure for generating HCPs and PIMs models that followed the synthesis route of simultaneous growth of chains and cross-linkers [55a]. The final step of structure generation was followed by a 21-step compression and slow decompression protocol. The simulated sorption isotherms agreed well with the experimental data without scaling factors [55a]. The method does not need any experimental inputs; therefore, it is a useful one to predict the structural properties before synthesis in the lab.

There are fewer studies of simulating amorphous molecular materials. Colina and coworkers developed simulation protocol for generating structural models of amorphous organic molecules of intrinsic microporosity (OMIMs) [60]. OMIMs have irregular and concave shapes of organic molecules, and the porosity results from the badly packing moleculars. Four simulation methods were described: (i) the conformational search; (ii) packing the molecular conformers at a specified final density or a low density; (iii) performing a MC annealing procedure; and (iv) a 21-step NVT (moles (N), volume (V) and temperature (T) are conserved)/NPT (moles (N), pressure (P) and temperature (T) are conserved) MD compression scheme to obtain final structural models. Structural analysis was carried out by the comparison of simulated and experimental densities, surface areas, pore volumes, and Wide-angle X-ray scattering (WAXS). Overall, one method that included 21-step compression and decompression procedures was proposed at the best simulation methodology for simulating OMIMs, as the experimental density was not required as a target density for simulations. Furthermore, a series of OMIMs structural modelswere generated in order to investigate the effect of the terminal functional groups on the packing behavior and gas sorption capacity [61].

We generated a structural model for one of the scrambled materials [54]. A simulation cell was loaded by the numerical distribution of scrambled cage species observed by HPLC. The model was geometry optimized to reach the experimental target density. The accessible surface area and pore volume had to have a good agreement with the experimental values. The pore structure was identified in the model in terms of the intrinsic pores in the cage molecules and the extrinsic pores that result from the ineffective packing of the distribution of scrambled cages. It has been suggested that 80% of the total free volume arises from the voids of inefficient packing for this amorphous scrambled cage material. As a result, the porosity in these materials is enhanced by the extra extrinsic volume. Gas uptake simulation was also carried out for the structural model.

In addition, we now describe the simulation methodology for packing amorphous organic cages without the consideration of experimental target density. The approach is implemented in the following four steps: (i) randomly loading cages at a low density; (ii) stabilization of the initial configuration using a microcanonical NVE ensemble (moles (N), volume (V) and energy (E) are conserved); (iii) compression of the system using NPT ensemble; and (iv) force field (CSFF) geometry optimization for the final configuration [62]. In order to ensure that the simulation models are representative of experimental bulk materials, the structural models were characterized in terms of surface areas, pore volumes, which can then be compared with the experimental data. When the models were constructed, MD simulations were performed for gas diffusion in amorphous **CC1** and **CC3**. Gas diffusion trajectories, gas hopping analysis, and self-diffusivities were carried out to rationalize gas selectivity in these materials.

13.4 Summary

Discrete molecules can pack inefficiently in the solid state, which results in permanent porosity for gas storage and separation applications. The porosity in these materials is generated from both intrinsic volumes in the molecules and extrinsic voids created from inefficient molecular packing. To date, the surface area in porous molecular solids (SA_{BET} up to 2796 $m^2 g^{-1}$) is lower than that in porous frameworks (SA_{BET} up to 6000 $m^2 g^{-1}$), but porous molecular solids possess potential advantages in the solubility, molecular mobility, and synthetic diversity in comparison with porous frameworks, and this research area has great potential for expanding.

The packing of the organic molecules can be directed by the functionalities in the molecules, thus giving synthetic control over the pore connectivity. For example, in crystalline cage packing, the degree of porosity depends upon the cage packing motifs, which is determined by the cage vertex functionality [11a]. In addition, the pore surface can be modified by the functionalities to increase the affinity of porous materials for a particular target adsorbate molecule and improve gas selectivity. A higher level of porosity has been observed for amorphous cage materials due to the enhanced extrinsic porosity that results from the distribution of molecular shapes. Further enhancing porosity could be done by introducing more bulky functionality on cage vertices, such as biphenyl or naphthyl groups.

To further understand and rationalize the structure and properties of these porous molecular materials, molecular simulations can offer essential insight and understanding of the molecular nature of the materials and assist with the synthesis and development of new porous cage structures. We have proposed a simulation strategy to predict the molecular structure from the precursors, and then the structure and physical properties of both crystalline and amorphous porous molecular solids [38]. For molecular materials, MD motions are particularly important for understanding guest–host interactions, as it has been suggested that the rearrangement of host occurs when the guest diffuses via a cooperative diffusion. Therefore, MD simulations have an important role for understanding microscopic structural properties including diffusion mechanism and predicting gas diffusion coefficients.

References

1. (a) Schüth, F., Sing, K.S.W., and Weitkamp, J. (2002) *Handbook of Porous Solids*, Wiley-VCH Verlag GmbH, Heidelberg; (b) Willems, T.F., Rycroft, C.H., Kazi, M., Meza, J.C., and Haranczyk, M. (2012) *Microporous Mesoporous Mater.*, **149**, 134–141.
2. (a) Holst, J.R., Trewin, A., and Cooper, A.I. (2010) *Nat. Chem.*, **2**, 915–920; (b) McKeown, N.B. (2010)

J. Mater. Chem., **20**, 10588–10597; (c) Tian, J., Thallapally, P.K., and McGrail, B.P. (2012) *CrystEngComm*, **14**, 1909–1919.
3. Janiak, C. and Vieth, J.K. (2010) *New J. Chem.*, **34**, 2366–2388.
4. Lee, J., Kim, J., and Hyeon, T. (2006) *Adv. Mater.*, **18**, 2073–2094.
5. Jiang, J.-X. and Cooper, A.I. (2010) *Top. Curr. Chem.*, **293**, 1–33.
6. Broach, R.W. (2010) *Zeolites in Industrial Separation and Catalysis*, Wiley-VCH Verlag GmbH, pp. 27–59.
7. Ferey, G. (2008) *Chem. Soc. Rev.*, **37**, 191–214.
8. Côté, A.P., Benin, A.I., Ockwig, N.W., O'Keeffe, M., Matzger, A.J., and Yaghi, O.M. (2005) *Science*, **310**, 1166–1170.
9. (a) Furukawa, H., Ko, N., Go, Y.B., Aratani, N., Choi, S.B., Choi, E., Yazaydin, A.Ö., Snurr, R.Q., O'Keeffe, M., Kim, J., and Yaghi, O.M. (2010) *Science*, **329**, 424–428; (b) Ben, T., Ren, H., Ma, S., Cao, D., Lan, J., Jing, X., Wang, W., Xu, J., Deng, F., Simmons, J.M., Qiu, S., and Zhu, G. (2009) *Angew. Chem. Int. Ed.*, **48**, 9457–9460.
10. Brutschy, M., Schneider, M.W., Mastalerz, M., and Waldvogel, S.R. (2012) *Adv. Mater.*, **24**, 6049–6052.
11. (a) Tozawa, T., Jones, J.T.A., Swamy, S.I., Jiang, S., Adams, D.J., Shakespeare, S., Clowes, R., Bradshaw, D., Hasell, T., Chong, S.Y., Tang, C., Thompson, S., Parker, J., Trewin, A., Bacsa, J., Slawin, A.M.Z., Steiner, A., and Cooper, A.I. (2009) *Nat. Mater.*, **8**, 973–978; (b) Culshaw, J.L., Cheng, G., Schmidtmann, M., Hasell, T., Liu, M., Adams, D.J., and Cooper, A.I. (2013) *J. Am. Chem. Soc.*, **135**, 10007–10010; (c) Schneider, M.W., Oppel, I.M., Griffin, A., and Mastalerz, M. (2013) *Angew. Chem. Int. Ed.*, **52**, 3611–3615; (d) Brutschy, M., Schneider, M.W., Mastalerz, M., and Waldvogel, S.R. (2013) *Chem. Commun.*, **49**, 8398–8400.
12. Atwood, J.L., Barbour, L.J., Jerga, A., and Schottel, B.L. (2002) *Science*, **298**, 1000–1002.
13. Jones, J.T.A., Holden, D., Mitra, T., Hasell, T., Adams, D.J., Jelfs, K.E., Trewin, A., Willock, D.J., Day, G.M., Bacsa, J., Steiner, A., and Cooper, A.I. (2011) *Angew. Chem. Int. Ed.*, **50**, 749–753.
14. Thallapally, P.K., McGrail, B.P., Atwood, J.L., Gaeta, C., Tedesco, C., and Neri, P. (2007) *Chem. Mater.*, **19**, 3355–3357.
15. Lim, S., Kim, H., Selvapalam, N., Kim, K.-J., Cho, S.J., Seo, G., and Kim, K. (2008) *Angew. Chem. Int. Ed.*, **47**, 3352–3355.
16. Sozzani, P., Comotti, A., Bracco, S., and Simonutti, R. (2004) *Angew. Chem. Int. Ed.*, **43**, 2792–2797.
17. Comotti, A., Bracco, S., Distefano, G., and Sozzani, P. (2009) *Chem. Commun.*, 284–286.
18. Msayib, K.J., Book, D., Budd, P.M., Chaukura, N., Harris, K.D.M., Helliwell, M., Tedds, S., Walton, A., Warren, J.E., Xu, M., and McKeown, N.B. (2009) *Angew. Chem. Int. Ed.*, **48**, 3273–3277.
19. Gutsche, C.D. (2008) *Calixarenes: An Introduction*, 2nd edn, RSC Publishing.
20. (a) Yang, W., Davies, A.J., Lin, X., Suyetin, M., Matsuda, R., Blake, A.J., Wilson, C., Lewis, W., Parker, J.E., Tang, C.C., George, M.W., Hubberstey, P., Kitagawa, S., Sakamoto, H., Bichoutskaia, E., Champness, N.R., Yang, S., and Schroder, M. (2012) *Chem. Sci.*, **3**, 2993–2999; (b) Hamon, L., Llewellyn, P.L., Devic, T., Ghoufi, A., Clet, G., Guillerm, V., van Beek, W., Jolimaitre, E., Vimont, A., Daturi, M., and Ferey, G. (2009) *J. Am. Chem. Soc.*, **131**, 17490–17499.
21. Atwood, J.L., Barbour, L.J., and Jerga, A. (2004) *Angew. Chem. Int. Ed.*, **43**, 2948–2950.
22. Lagona, J., Mukhopadhyay, P., Chakrabarti, S., and Isaacs, L. (2005) *Angew. Chem. Int. Ed.*, **44**, 4844–4870.
23. Isaacs, L. (2009) *Chem. Commun.*, 619–629.
24. Kim, H., Kim, Y., Yoon, M., Lim, S., Park, S.M., Seo, G., and Kim, K. (2010) *J. Am. Chem. Soc.*, **132**, 12200–12202.
25. Couderc, G., Hertzsch, T., Behrnd, N.R., Krämer, K., and Hulliger, J. (2006) *Microporous Mesoporous Mater.*, **88**, 170–175.

26. Sozzani, P., Bracco, S., Comotti, A., Ferretti, L., and Simonutti, R. (2005) *Angew. Chem. Int. Ed.*, **44**, 1816–1820.
27. Mastalerz, M. and Oppel, I.M. (2012) *Angew. Chem. Int. Ed.*, **51**, 5252–5255.
28. Mastalerz, M., Schneider, M.W., Oppel, I.M., and Presly, O. (2011) *Angew. Chem. Int. Ed.*, **50**, 1046–1051.
29. Hasell, T., Chong, S.Y., Jelfs, K.E., Adams, D.J., and Cooper, A.I. (2011) *J. Am. Chem. Soc.*, **134**, 588–598.
30. (a) Liu, Y., Liu, X., and Warmuth, R. (2007) *Chem. Eur. J.*, **13**, 8953–8959. (b) Liu, X., Liu, Y., Li, G., and Warmuth, R. (2006) *Angew. Chem. Int. Ed.*, **45**, 901–904. (c) Liu, X., Liu, Y., and Warmuth, R. (2008) *Supramol. Chem.*, **20**, 41–50.
31. (a) Bojdys, M.J., Briggs, M.E., Jones, J.T.A., Adams, D.J., Chong, S.Y., Schmidtmann, M., and Cooper, A.I. (2011) *J. Am. Chem. Soc.*, **133**, 16566–16571; (b) Jelfs, K.E., Wu, X., Schmidtmann, M., Jones, J.T.A., Warren, J.E., Adams, D.J., and Cooper, A.I. (2011) *Angew. Chem. Int. Ed.*, **50**, 10653–10656; (c) Jones, J.T.A., Hasell, T., Wu, X., Bacsa, J., Jelfs, K.E., Schmidtmann, M., Chong, S.Y., Adams, D.J., Trewin, A., Schiffman, F., Cora, F., Slater, B., Steiner, A., Day, G.M., and Cooper, A.I. (2011) *Nature*, **474**, 367–371; (d) Mitra, T., Wu, X., Clowes, R., Jones, J.T.A., Jelfs, K.E., Adams, D.J., Trewin, A., Bacsa, J., Steiner, A., and Cooper, A.I. (2011) *Chem. Eur. J.*, **17**, 10235–10240; (e) Jiang, S., Bacsa, J., Wu, X., Jones, J.T.A., Dawson, R., Trewin, A., Adams, D.J., and Cooper, A.I. (2011) *Chem. Commun.*, **47**, 8919–8921.
32. Schneider, M.W., Oppel, I.M., Ott, H., Lechner, L.G., Hauswald, H.-J.S., Stoll, R., and Mastalerz, M. (2012) *Chem. Eur. J.*, **18**, 836–847.
33. Schneider, M.W., Oppel, I.M., and Mastalerz, M. (2012) *Chem. Eur. J.*, **18**, 4156–4160.
34. Jin, Y., Voss, B.A., Noble, R.D., and Zhang, W. (2010) *Angew. Chem. Int. Ed.*, **49**, 6348–6351.
35. Jin, Y., Voss, B.A., Jin, A., Long, H., Noble, R.D., and Zhang, W. (2011) *J. Am. Chem. Soc.*, **133**, 6650–6658.
36. Avellaneda, A., Valente, P., Burgun, A., Evans, J.D., Markwell-Heys, A.W., Rankine, D., Nielsen, D.J., Hill, M.R., Sumby, C.J., and Doonan, C.J. (2013) *Angew. Chem. Int. Ed.*, **52**, 3746–3749.
37. Krishna, R. (2012) *Chem. Soc. Rev.*, **41**, 3099–3118.
38. Jelfs, K.E. and Cooper, A.I. (2013) *Curr. Opin. Solid State Mater. Sci.*, **17**, 19–30.
39. Li, W., Grimme, S., Krieg, H., Möllmann, J., and Zhang, J. (2012) *J. Phys. Chem. C*, **116**, 8865–8871.
40. Alavi, S., Afagh, N.A., Ripmeester, J.A., and Thompson, D.L. (2006) *Chem. Eur. J.*, **12**, 5231–5237.
41. Cornell, W. D., Cieplak, P., Bayly, C. I., Gould, I. R., Merz, K. M., Ferguson, D. M., Spellmeyer, D. C., Fox, T., Caldwell, J. W., and Kollman, P. A. (1995) *J. Am. Chem. Soc.*, **117**, 5179.
42. Alavi, S., Woo, T.K., Sirjoosingh, A., Lang, S., Moudrakovski, I., and Ripmeester, J.A. (2010) *Chem. Eur. J.*, **16**, 11689–11696.
43. El-Barghouthi, M.I., Assaf, K.I., and Rawashdeh, A.M.M. (2010) *J. Chem. Theory Comput.*, **6**, 984–992.
44. Sun, H. (1995) *Macromolecules*, **28**, 701–712
45. Mitra, T., Jelfs, K.E., Schmidtmann, M., Ahmed, A., Chong, S.Y., Adams, D.J., and Cooper, A.I. (2013) *Nat. Chem.*, **5**, 276–281.
46. Zhang, C., Wang, Q., Long, H., and Zhang, W. (2011) *J. Am. Chem. Soc.*, **133**, 20995–21001.
47. Skowronek, P., Warzajtis, B., Rychlewska, U., and Gawronski, J. (2013) *J. Chem. Commun.*, **49**, 2524.
48. Budd, P.M., Ghanem, B.S., Makhseed, S., McKeown, N.B., Msayib, K.J., and Tattershall, C.E. (2004) *Chem. Commun.*, 230–231.
49. McKeown, N.B. and Budd, P.M. (2006) *Chem. Soc. Rev.*, **35**, 675–683.
50. Tian, J., Thallapally, P.K., Dalgarno, S.J., McGrail, P.B., and Atwood, J.L. (2009) *Angew. Chem. Int. Ed.*, **48**, 5492–5495.
51. Tian, J., Ma, S., Thallapally, P.K., Fowler, D., McGrail, B.P., and Atwood, J.L. (2011) *Chem. Commun.*, **47**, 7626–7628.

52. Schneider, M.W., Lechner, L.G., and Mastalerz, M. (2012) *J. Mater. Chem.*, **22**, 7113–7116.

53. Schneider, M.W., Siegfried Hauswald, H.-J., Stoll, R., and Mastalerz, M. (2012) *Chem. Commun.*, **48**, 9861–9863.

54. Jiang, S., Jones, J.T.A., Hasell, T., Blythe, C.E., Adams, D.J., Trewin, A., and Cooper, A.I. (2011) *Nat. Commun.*, **2**, 207.

55. (a) Abbott, L.J. and Colina, C.M. (2011) *Macromolecules*, **44**, 4511–4519; (b) Wood, C.D., Tan, B., Trewin, A., Niu, H., Bradshaw, D., Rosseinsky, M.J., Khimyak, Y.Z., Campbell, N.L., Kirk, R., Stöckel, E., and Cooper, A.I. (2007) *Chem. Mater.*, **19**, 2034–2048.

56. Trewin, A., Willock, D.J., and Cooper, A.I. (2008) *J. Phys. Chem. C*, **112**, 20549–20559.

57. Larsen, G.S., Lin, P., Hart, K.E., and Colina, C.M. (2011) *Macromolecules*, **44**, 6944–6951.

58. (a) Hofmann, D., Fritz, L., Ulbrich, J., Schepers, C., and Böhning, M. (2000) *Macromol. Theory Simul.*, **9**, 293–327; (b) Hofmann, D., Heuchel, M., Yampolskii, Y., Khotimskii, V., and Shantarovich, V. (2002) *Macromolecules*, **35**, 2129–2140. (c) Heuchel, M., Fritsch, D., Budd, P.M., McKeown, N.B., and Hofmann, D. (2008) *J. Membr. Sci.*, **318**, 84–99; (d) Curco, D. and Aleman, C. (2003) *J. Phys. Chem.*, **119**, 2915–2922.

59. Lim, S.Y., Tsotsis, T.T., and Sahimi, M. (2003) *J. Phys. Chem.*, **119**, 496–504.

60. Abbott, L.J., McDermott, A.G., Del Regno, A., Taylor, R.G.D., Bezzu, C.G., Msayib, K.J., McKeown, N.B., Siperstein, F.R., Runt, J., and Colina, C.M. (2012) *J. Phys. Chem. B*, **117**, 355–364.

61. Del Regno, A. and Siperstein, F.R. (2013) *Microporous Mesoporous Mater.*, **176**, 55–63.

62. Jiang, S., Jelfs, K. E., Holden, D., Hasell, T., Chong, S. Y., Haranczyk, M., Trewin, A., Cooper, A. I. (2013) *J. Am. Chem. Soc.*, **135**, 17818.

14
Electrochemical Motors

Gabriel Loget and Alexander Kuhn

14.1
Inspiration from Biomotors

Fascinating autonomously moving microsystems are present in nature, which operate at dimensions where Brownian motion is considerable and viscosity effects dominate (at low Reynolds numbers). As examples at the microscale, one can cite eukaryote cells such as spermatozoa that propel themselves generating a mechanical wave with their flagella. Some prokaryotic entities such as the bacteria *Escherichia coli* (*E. coli*) or *Salmonella* (Figure 14.1a) use the corkscrew motion of their flagella to direct themselves at extremely high speeds, up to 60 bodylengths s^{-1} (in comparison, the speed of a cheetah, which is the fastest animal on earth, is around 25 bodylengths s^{-1}). At a smaller scale, within the cell, one can see the flagella motion being driven by the action of a remarkable rotating nanomotor, powered by the action of a proton flux [1, 2]. Other biomolecular nanomotors are active in cells, for example, kinesin and dynein, which travel along microtubule "tracks" in the cytoplasm to deliver vesicles, organelles, and other cargo at specific locations (Figure 14.1b) [3–5]. Most of these motors are powered by the spontaneous reaction of energy rich biomolecules, such as the hydrolysis of the biological fuel adenosine triphosphate (ATP). Inspired by the remarkable performances of these systems, and encouraged by the possibilities offered by nanotechnology [6], biologists, chemists, and physicists have developed in recent decades different strategies in order to try to compete with those biomotors [7–9].

One approach is using the biosystems as active parts directly. Several examples from the group of Whitesides illustrate this concept at the cellular level [12–14]. Controlling the morphology of filamentous *E. coli* by growing them in shaped polydimethylsiloxane (PDMS) microchambers results in specific motion [13]. Cargo transport and release was carried out using flagella-propelled unicellular algae, which were steered by phototaxis within a microchannel [14]. Bacteria attached to the surface of the particles can transfer their mechanical energy and propel them, as it has been demonstrated by Behkam *et al.* [15, 16]. At a smaller scale, a lot of effort has been carried out in order to induce the motion of nano- and microobjects by using biomolecular nanomotors [11, 17, 18]. In this context,

Discovering the Future of Molecular Sciences, First Edition. Edited by Bruno Pignataro.

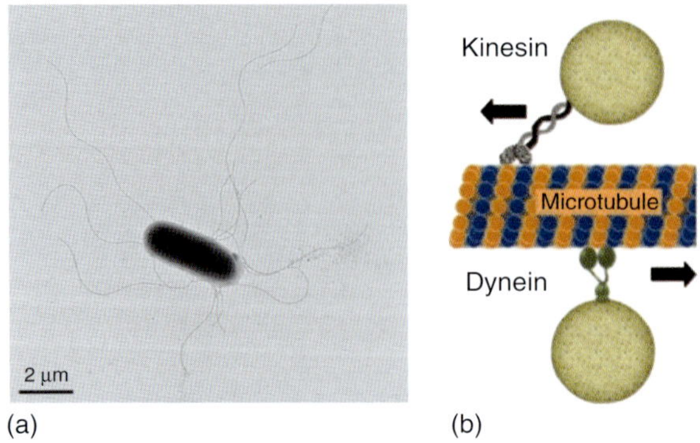

Figure 14.1 Examples of biomotors. (a) TEM micrograph of a *Salmonella* bacterium, reprinted from the OIST web site [10]. (b) Scheme showing the cargo transport by kinesin and dynein along a microtubule. Adapted and reproduced with permission from [11]. Reproduced with permission of John Wiley & Sons.

Montemagno used ATPase to rotate a biochemically attached nickel nanorod in the presence of ATP [19]. The effect of viscosity on the rotation of a microbead attached to ATPase has also been reported [20]. Vogel's group presented the use of kinesin-modified surfaces for directing the ATP-powered motion of microtubules within miniaturized systems [21–23]. This approach has recently been applied in biosensing [24, 25], cargo transport, and release [23, 26].

14.2 Chemical Motors

The above-mentioned biological systems are remarkably efficient, but they have a very limited lifetime because of their rapid degradation in a wide range of experimental conditions. This major drawback has led to many efforts for conceiving more stable synthetic motors with similar performances. Huge progress in the area of synthetic molecular motors [27, 28] using DNA or smaller molecules is currently made, but this topic exceeds the scope of this chapter. The most developed strategy is directly inspired by biological systems and is based on the use of a chemical fuel. The fuel induces the motion, by reacting at a precise position on the motor [29]. The two main mechanisms, self-electrophoresis and bubble propulsion, have been discussed in this context. After an interesting scientific debate [30], the first one has been widely adopted for explaining the motion of bimetallic nanorods (also referred to as nanosubmarines) [31] in the presence of hydrogen peroxide. These objects are usually synthetized by electroplating through alumina membranes containing cylindrical pores and are recovered after the dissolution of the membrane. They typically have a length in the micrometer range and are a few hundreds of nanometers wide (Figure 14.2b). Different metal combinations can be used, with one metal being the cathodic pole, where H_2O_2 is reduced, and the other one the

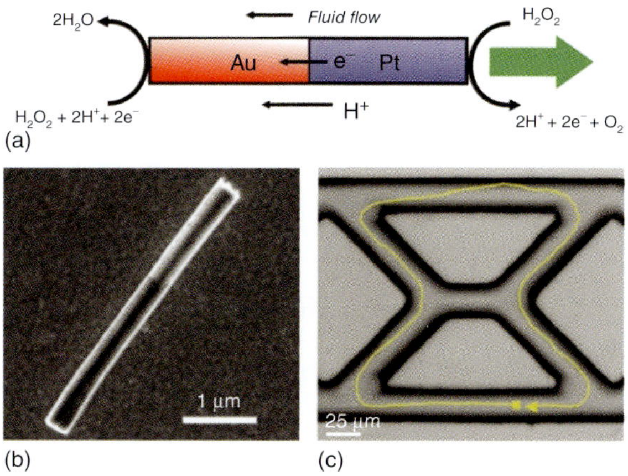

Figure 14.2 Self-electrophoretic swimmers. (a) Scheme showing the self-electrophoresis mechanism at a Pt/Au nanorod. (b) SEM image showing a Ni/Au nanorod. Adapted with permission from [30]. Copyright (2006) American Chemical Society. (c) Optical micrograph showing the motion track of a Pt-CNT/Au/Ni/Au nanomotor, which is steered around the central portion of a PDMS microstructure. Adapted with permission from [34]. Copyright (2008) American Chemical Society.

anodic pole, where H_2O_2 is oxidized. The respective locations of the spontaneous and simultaneous reactions can be deduced from the mixed potential value E_m of the shortcircuted two metals in the fuel solution, as demonstrated by Mallouk *et al.* [30]. Individually recorded current–voltage curves for each metal in a solution of H_2O_2 allow estimating the value of E_m, and deduce from the potential difference between E_m and the onset potential of the oxidation and reduction reaction which metal will host which reaction. As it is discussed in Section 14.5, under such conditions, the fuel transforms the bimetallic rod into a bipolar electrode (BE). It is worth mentioning that in this case, the bipolarity is induced by the chemical composition of the object and not by an external electric field. Classical rod compositions employ a good catalyst for H_2O_2 reduction, such as Pt [32] or Ni [33], as the cathodic pole and Au as the anodic pole. As illustrated in Figure 14.2a, the electron flow at the rod/solution interface is balanced by a proton flow in the electric double layer, inducing the motion in the opposite direction.

Many experiments, mainly reported by the groups of Wang and Ozin, have been performed with the aim to enhance the speed of these objects by materials engineering and tuning the medium composition. For example, the porosity of the Ni surface can be increased in order to boost the speed [35]. Strong speed enhancements have also been achieved by the incorporation of carbon nanotubes (CNTs) into the Pt matrix [36], the presence of Ag^+ ions [37] and hydrazine [36] in the mixture, and the use of an Au-Ag alloy instead of pure Au [38]. Using this improved system, speeds in the order of $\approx 150\,\mu m\,s^{-1}$, equivalent to ≈ 75 bodylength s^{-1}, have been reached [38]. The speed can be modulated by different stimuli [39], such as electrogenerated concentration gradients [40] and temperature [41]. Coupled with magnetically active parts, the nanorods can be steered using magnetic fields, as shown in Figure 14.2c [42]. Recently, several applications such as writing microstructures on surfaces [43], transporting cargos [44], delivering cargos in microfluidic channels [34, 45], and sensing [37, 46] have been carried out. Interestingly, these objects exhibit chemotaxis [47], that is, when exposed to a chemical gradient, they drive to the fuel rich area, similarly to biological entities. Rotational motion has been reported for Ni/Au rods with one part attached to a substrate [33], Au rods with a noncentered Pt catalyst [48], and a submillimeter gold gearlike structure with platinum strips in the tooth regions [49]. Using a self-electrophoresis mechanism, Sen *et al.* [50] recently reported the use of Pt/Cu rods powered by aqueous Br_2 or I_2. Mano *et al.* designed biochemical swimmers that operate with an analogous mechanism. Composed of a carbon fiber "body," carrying at both extremities glucose oxidase and bilirubin oxidase, respectively, they use biofuels such as glucose and O_2 [51]. Sen's group also reported the interesting concept of polymerization-powered motors [52]. In this system, a submicrometer spherical asymmetric particle (AP), half-coated with a Grubb's catalyst, is immersed into a solution containing the monomer fuel. The increase of the particle diffusion coefficient is attributed to the osmosis generated by the asymmetric polymerization [52].

The bubble propulsion mechanism was first introduced by Whitesides *et al.* [53]. They directed millimeter-sized objects, floating on a solution of H_2O_2, by means of the hydrodynamic flow generated by the bubble formation and released on a Pt surface during the fuel decomposition [53]. This mechanism, which is independent from the ionic strength of the medium, constitutes a very interesting alternative to the previously discussed self-electrophoresis [54]. This concept has been used at the micro- and submicroscale for moving asymmetric Pt/SiO_2 spheres [55], stogmatocytes (bowl-shaped polymers) loaded with Pt NPs [56], and silica beads tethered with a synthetic manganese catalase [57, 58]. An increased interest is currently focused on tubular microengines driven by bubble production (also referred to in the literature as microbots [59], microrockets [60], or microjets [61]). These objects are microsized tubes containing an inactive outer surface and a catalytically active inner surface. In the presence of the H_2O_2 fuel, O_2 bubbles nucleate and grow at the catalyst surface until they are expelled through the larger opening, inducing motion towards the opposite opening (Figure 14.3b).

Schmidt's group developed the use of rolled-up nanotechnology, based on photolithographic methods for the synthesis of tubular microengines [60, 63]. First, a prestressed inorganic nanomembrane is deposited on a photoresist on which other layers can be deposited (such as a ferromagnetic one), with the final one being the catalytically active layer. As shown in Figure 14.3a, the selective underetching of the sacrificial layer rolls up the membrane, creating the final microswimmer shape. Wang's group developed alternative strategies based on electrodeposition. The chemical etching of an Ag wire followed by the successive electrodeposition of metal layers can lead to a tubular microengine after Ag dissolution and dicing [64]. The successive electrodeposition of a conducting polymer, such as polypyrrole (PPy), poly(3,4-ethylenedioxythiophene) (PEDOT), or polyaniline (PAni), and the catalytic layer through the conical pores of a polycarbonate membrane can also

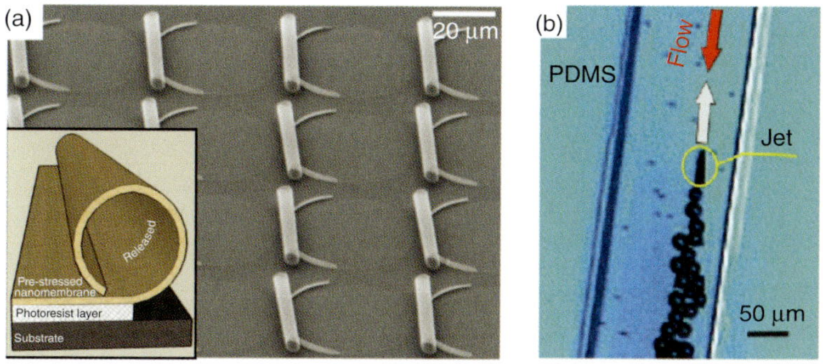

Figure 14.3 Bubble-propelled swimmers. (a) SEM image of an array of rolled-up nanomembranes. Inset: scheme illustrating the roll-up process of a nanomembrane into a tube on photoresist. Adapted with permission from [60]. (b) Optical micrograph showing a rolled-up microengine moving against the flow in a PDMS microchannel. Adapted with permission from [62]. Copyright (2011) American Chemical Society.

lead to tubular microengines after the removal of the template [65, 66]. The catalyst is usually Pt but can also be a layer of an enzyme such as catalase [67]. In this method, striking speeds of up to 1400 bodylengths s^{-1} have been recently reached at physiological temperature [66]. Coupled with a magnetic layer, they showed an impressive degree of motion control in lab-on-chip microchannels (Figure 14.3b) [62]. Their potential for bioanalytics has been demonstrated [46] in applications such as dynamic isolation of cells [59, 68], nucleic acids [69], proteins [70], and bacteria [71]. They have also been used for transport [62], assembly of particles [72], and microdrilling of biomaterials [73]. Recently, Zn-PAni tubular microengines have been propelled in strongly acidic media [74] and Al/Pd APs in alkaline solution [75].

14.3
Externally Powered Motion

The intrinsic problem in all these approaches is the requirement of a fuel molecule. Indeed, typical media need to be strongly acidic or contain H_2O_2, thus prohibiting their application in most biological systems. Therefore, alternative swimmers that can be used without fuel are attracting a lot of attention. In this context, the use of external electromagnetic fields seems to be an appealing approach. Some objects, namely surface walkers, require the use of a surface to break the spatial symmetry for generating a controlled motion. Doublets of paramagnetic beads [76], superparamagnetic chains [77], and ferromagnetic nanowires [78] have been propelled in rotating magnetic fields using this approach. An increasing interest is focused on biomimetic approaches. Bacteria's flagella motions have been mimicked using centimeter [79] and micrometer-sized (Figure 14.4) [80, 81] rigid helical propellers in rotating magnetic fields. These objects are usually obtained by multistep laser writing [82] or vapor deposition techniques [80]. Propellers

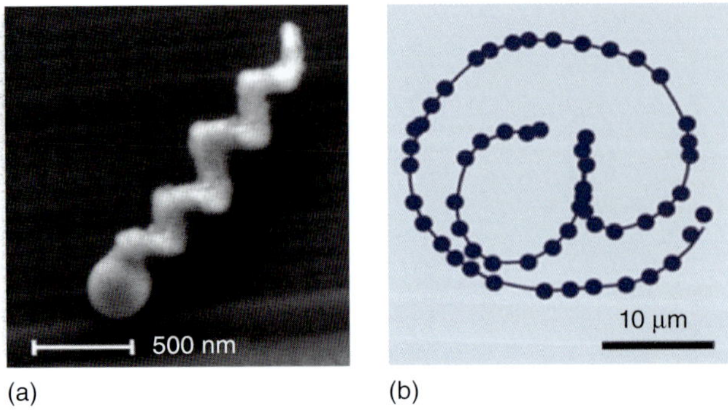

Figure 14.4 Magnetically propelled swimmers. (a) SEM image of a SiO_2/magnetic helical structure [80]. (b) Trajectory of an individual nanoswimmer in solution, navigating along a preprogramed track of the @-symbol. Adapted with permission from [80]. Copyright (2009) American Chemical Society.

having a flexible magnetic tail have also been proposed. Wang's group developed flexible metal motors that can be steered by a rotating field [83, 84]. Their synthesis was achieved by electrodeposition through alumina templates (similarly to the bimetallic nanorods discussed previously) and a subsequent partial dissolution of the silver segment, which leads to flexibility [83, 84]. Dreyfus *et al.* [85] reproduced the spermatozoa flagella's motion to drive a red cell attached to a chain of magnetic particles in an alternating magnetic field. Magnetic rigid helical motors have already been used for the transport and delivery of microparticles [82].

Electric field-driven motion is also a very promising area. Different electrokinetic phenomena are known to occur on particles when being exposed to an electric field, which can be used to induce motion [86]. In uniform DC fields, well-known electrophoresis makes charged particles migrate towards the electrode of opposite charge. In uniform AC fields, diodes can be propelled in a controlled way as first demonstrated by Velev *et al.* [87]. As depicted in Figure 14.5, the current rectification within the diode induces an electroosmotic ion flux in its electric double layer that propels the diode in the direction of either the cathode or the anode, depending on its surface charge. This phenomenon was used to drive translational (Figure 14.5) [87, 88] and rotational motion [87]. The interaction of the induced dipole of a

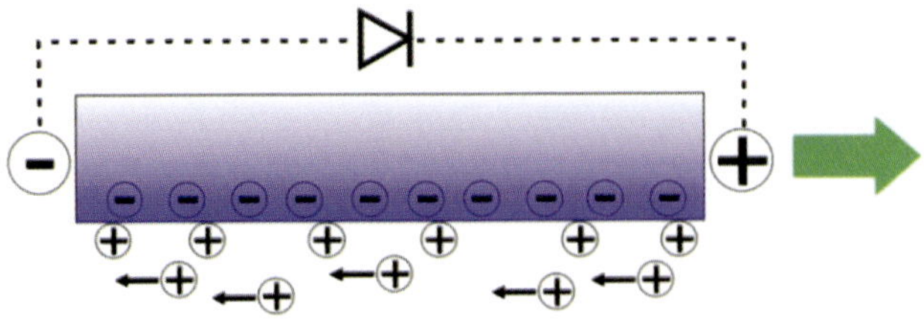

Figure 14.5 Scheme illustrating the origin of the motion of diodes in an AC electric field. The current is rectified within the diode, which induces an electroosmotic ion flux, propelling the diode.

particle with the gradient of a DC or AC field generates a dielectrophoretic force [86, 89]. The force intensity and direction depends on the particle size as well as the dielectric properties of the particle and the surrounding medium, which leads to the attraction or repulsion of the particle from the region with the highest electric field intensity [89]. The effect has been used for sorting nanotubes [90], cell separation [91], colloidal crystal assembly [92], and many other applications [89]. Electrowetting can also be used to move objects. Takei *et al.* [93] recently used this phenomenon to rotate a millimeter-sized capillary motor floating on a liquid drop.

14.4 Asymmetry for a Controlled Motion

Even though all the examples of moving objects presented so far are based on different mechanisms, they have in common the presence of asymmetry in one or the other form. Asymmetry is either incorporated in the structure of the swimmer, indeed bacteria (Figure 14.1a), magnetic swimmers (Figure 14.4), and microjets (Figure 14.3) are asymmetric in shape, or are asymmetric in composition as in the case of bimetallic nanorods (Figure 14.2) and diodes (Figure 14.5). Asymmetry can also be introduced by an electric field as, for example, in the case of dielectrophoresis. Asymmetry is crucial for the generation of controlled motion. In order to understand the importance of asymmetry for the controlled motion of an object isolated from the influence of other parameters, we propose the scheme shown in Figure 14.6. Therein we consider the model system of an isotropic particle that is submitted to the driving forces $\overrightarrow{F_{\text{drive}}}$ inducing the object's motion. $\overrightarrow{F_{\text{drive}}}$

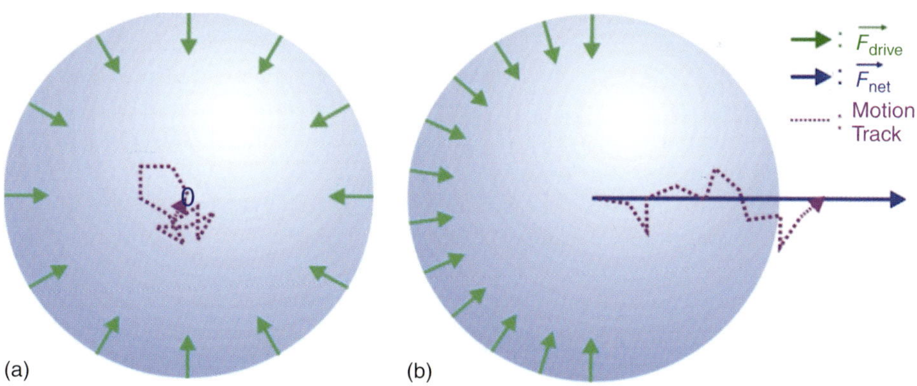

Figure 14.6 Asymmetry controls particle motion. (a) Scheme representing the net force $\overrightarrow{F_{\text{net}}}$ and the motion track of a particle, resulting from the integration of the driving forces $\overrightarrow{F_{\text{drive}}}$ being statistically distributed over the surface of an isotropic particle. (b) Scheme representing the net force $\overrightarrow{F_{\text{net}}}$ and the motion track of a particle, resulting from the integration of the driving forces $\overrightarrow{F_{\text{drive}}}$ being asymmetrically distributed over one hemisphere of an isotropic particle.

can be generated, for example, by the release of a bubble. In the situation shown in Figure 14.6a, the forces act at points that are distributed statistically over the particle surface, whereas in Figure 14.6b, the points of action are asymmetrically distributed over the particle by being localized only on one hemisphere. In each case, the object's motion is directed by $\overrightarrow{F_{net}}$, which results from the addition of all $\overrightarrow{F_{drive}}$ vectors. In Figure 14.6a, the vectors $\overrightarrow{F_{drive}}$ cancel each other out and no net motion results. In Figure 14.6b, in contrast, the asymmetric distribution of $\overrightarrow{F_{drive}}$ provokes that $\overrightarrow{F_{net}} > 0$ and a net motion is induced, perpendicular to the equatorial plane of asymmetry. In a real scenario, the simultaneous and homogeneous application of $\overrightarrow{F_{drive}}$ is less likely than a successive application in time, resulting in an erratic or Brownian motion (Figure 14.6a). In an analogous manner, oscillations around the major axis of motion can be provoked for the swimmer as shown in Figure 14.6b. The morphological or chemical asymmetry is usually introduced for micromotors by means of synthetic techniques, often multistep procedures that require complex equipments. In this context, the intrinsic asymmetric reactivity offered by the concept of bipolar electrochemistry, which is presented in the following section, is a very appealing approach for motion generation.

14.5 Bipolar Electrochemistry

The phenomenon of bipolar electrochemistry has been known for a long time and was originally used in industrial application for electrolysis [94] or batteries [95]. In the present decade, bipolar electrochemistry revealed attractive features for applications in the field of materials science and analysis [96–98]. The original term "bipolar electrode" is attributed to an object that, under certain conditions, exhibits different electrochemical behavior at its surface, on one side an oxidation, and on the other one a simultaneous reduction. BEs should not be mixed up with other objects that one can find in electrochemistry, such as bipolar plates (used for fuel distribution in fuel cells) or bipolar membranes (membranes composed by anion and cation exchange parts, generally used for electrodialysis). As in the case of the bimetallic nanorods discussed in Section 14.2, BEs can be objects exhibiting a chemical anisotropy, designed in order to promote oxidation and reduction reactions at the same time. In these examples, bipolar electrochemistry originates from the hybrid composition of the objects. It is important to note that, unlike these examples, this section and the following one is focused on objects that are, in most of the cases, chemically isotropic and exhibit a BE behavior only when exposed to an external electric field. This kind of BE can also be found in the literature under the name of "floating electrodes."

Let us consider a conducting object immersed in a solution and exposed to an external electric field (Figure 14.7). The total current flowing through the solution i_t will be divided into two fractions in the vicinity of the conducting object. A fraction of the current, which is called bypass current i_{bps}, will flow through the solution via

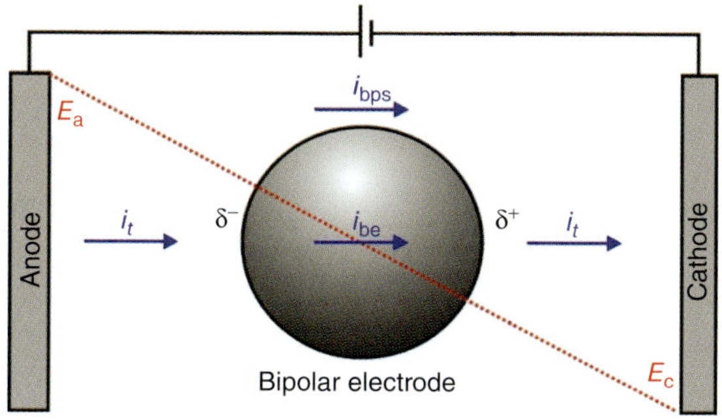

Figure 14.7 Scheme of a spherical bipolar electrode in solution exposed to an electric field.

the migration of charged species, whereas a current fraction i_{be} will flow through the conducting object via electronic conduction. Consequently, one can write the following equation:

$$i_t = i_{be} + i_{bps} \tag{14.1}$$

It follows that the ratio i_{be}/i_{bps} is correlated with the respective resistance of the object $R_{be,}$ and of the solution R_s (and the electron transfer resistance at the object/solution interface, which is neglected here). Therefore, working with a highly resistive solution and with a highly conductive object will result in minimizing i_{bps}, and almost all the current passing through the system will flow through the conducting object. With E_a and E_c being the potentials of the external feeder electrodes, d being the distance between the feeder electrodes and assuming that the electric field is linear, the value of the imposed electric field E can be calculated as:

$$E = (E_a - E_c)/d \tag{14.2}$$

As a consequence of the presence of the electric field, a polarization potential, which is a result of the difference of the solution potential value with respect to the conducting object arises. The polarization potential is maximal between the two extremities of the conducting object and its value, ΔE, is given by:

$$\Delta E = E\,L \tag{14.3}$$

with L being the characteristic length of the conducting object. It follows that, if ΔE is important enough, oxidation reactions can occur at the anodic pole of the conducting object, coupled with reduction reactions at the cathodic pole, in order to keep the electroneutrality in the object. Therefore, a field-generated asymmetric reactivity arises at the surface of the object, making it a BE. ΔE is the thermodynamic value that quantifies the driving force for the reactions at the extremities of the object. The BE can be any kind of conductive material with any characteristic dimension and geometry. However, as shown by Equation 14.3, a smaller object will require a higher external electric field in order to be polarized enough for inducing redox reactions.

The first example of bipolar microelectrodes goes back to 1986 when Fleischmann *et al.* [99] showed that electrochemical reactions such as oxygen and hydrogen evolution could be generated at bipolar Pt microelectrodes in a poorly conductive medium, when exposed to an electric field. Some publications discussed the fundamental aspects of bipolar electrochemistry in more detail [100, 101], but in this chapter, we focus more on practical applications of this concept for the design of micromotors.

14.6 Asymmetric Motors Synthesized by Bipolar Electrochemistry

As discussed previously, bipolar electrochemistry can be used to trigger electrochemical reactions. In principle, many electrochemical reactions can be generated at the extremities of BEs, among which we can cite the electrodeposition of metals. For example, as depicted in Figure 14.8a, for the asymmetric functionalization with gold at the cathodic pole of the substrate, using tetrachloroaurate, the following reaction has to be carried out:

$$[\mathrm{Au(III)Cl_4}]^- + 3e^- \rightarrow \mathrm{Au^0(s)} + 4\mathrm{Cl}^- \quad E^0_{\mathrm{AuCl_4^-/Au_0}} = 0.99\,\mathrm{V\ vs\ NHE} \quad (14.4)$$

where E^0 is the formal potential of the redox couple and NHE is the normal hydrogen electrode. In order to equilibrate the charge consumption, an oxidation, for example, of water, has to take place at the opposite extremity of the object:

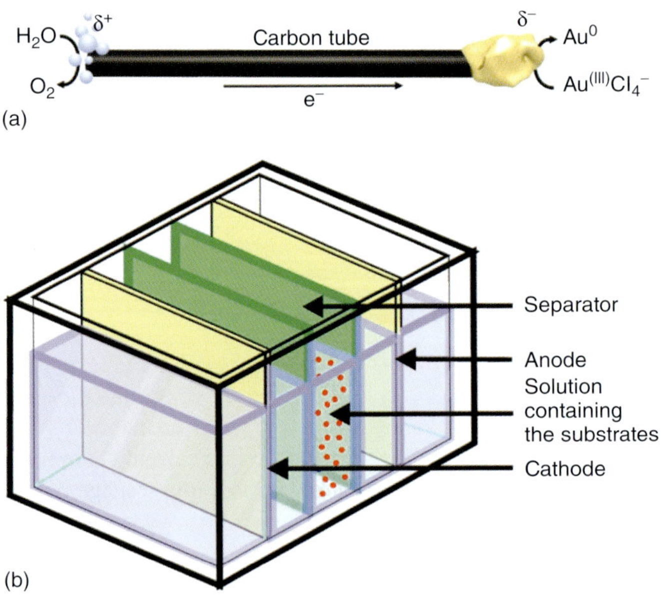

Figure 14.8 Bipolar electrodeposition. (a) Scheme of a carbon tube being modified by the bipolar electrodeposition of a gold cluster. (b) Scheme of a cell used for bipolar electrodeposition. Adapted and reproduced with permission from [102]. Reproduced with permission of John Wiley & Sons.

$$2H_2O(l) \rightarrow 4H^+(l) + O_2(g) + 4\,e^- \quad E^0_{O_2/H_2O} = 1.23\,V \text{ vs NHE} \tag{14.5}$$

It immediately follows that in this case the polarization has to generate a minimum potential difference of

$$\Delta V_{min} = |E^0_{AuCl_4^-} - E^0_{Au^0}| = 0.24\,V \tag{14.6}$$

Even though this potential range is easily accessible using a standard generator when dealing with macroscopic substrates, it becomes an intrinsic problem of the concept when dealing with micrometer- or nanometer-sized objects. In this case, the external electric field has to reach values of up to $MV\,m^{-1}$, conditions that seem to be incompatible with a normal industrial environment, especially when using aqueous solutions. Intrinsic side reactions accompanied by macroscopic bubble formation at both electrodes will disturb the orientation of the objects in the electric field. Bradley and coworkers could partly circumvent this problem by using organic solvents in order to enlarge the potential window of the electrolyte. In this way, it was possible to generate metal deposits in an asymmetric way on different objects on the micrometer scale [103, 104]. However, it was necessary to immobilize the objects on a surface in order to prevent them from rotating, which leads to a surface-confined process, as the ones reported in Sections 14.2 and 14.3 for synthesizing micromotors. Considering efficiency, a bulk method would be more attractive.

In this context, two methods were recently developed in order to perform the localized modification of conductive particles in the bulk using bipolar electrochemistry. The first one, namely, capillary-assisted bipolar electrodeposition (CABED), was based on using capillary electrophoresis technology. This setup allows applying high electric fields required for modifying small objects. A family of asymmetric metal-coated carbon microtubes (CMTs) [105] could be generated by combining metal deposition at the cathodic pole, with solvent oxidation at the anodic pole. This concept could be extended also to polymer–carbon–metal APs, and it is possible to use the technique at the nanoscale, as has been demonstrated for multiwall carbon nanotubes (MWCNTs) with a single gold NP at one end [106]. Even if this is a bulk technique, the accessible amount of APs is limited by the capillary volume. In order to circumvent this problem, another technology was developed that could allow generating larger quantities of APs. As shown in Figure 14.8b, the process uses a cell composed of one reaction compartment, in which the substrates and the reagents are located, separated from the two electrode compartments by two membranes. The feeding electrodes, which are immersed in a solvent containing no reactants or only supporting electrolyte, are connected to a high-voltage power supply. These separators are needed in order to limit problems that might occur when using strong electric fields (bubble formation, solvent evaporation, and convection following ohmic heating). The electrodes are separated from the reaction chamber containing the reactants and the objects that are to be modified, and their compartments are filled with cold solvent. Using this configuration, the electric field can be applied for relatively long periods of time. Additives can be employed in the reaction compartment; the use of a gelling agent, for example, helps to mechanically immobilize conductive beads during their modification, resulting in asymmetrically coated beads [102].

Bipolar electrodeposition is a very attractive approach for the synthesis of highly asymmetric structures. The particles that can be obtained are appealing objects for use as synthetic motors, as we discuss here. CMTs with a Ni patch at one extremity have been synthesized using the CABED process. These asymmetric structures, recovered after their modification, were suspended in water on a glass slide that was positioned in front of the objective of a transmission optical microscope. Using this setup, the motion of isolated Ni/CMTs was observed and recorded. The effect of an external magnetic field was studied and the presence of a permanent magnet in close proximity orientated the asymmetric object with the Ni "head" towards the magnet. A circular motion of the magnet around the microscope objective induced the object's rotation around its normal axis [107]. This is demonstrated by a series of optical micrographs in Figure 14.9, which show the controlled counterclockwise rotation of a Ni/CMT particle in the rotating magnetic field.

This proof-of-principle experiment shows that bipolar electrodeposition can be used to elaborate magnetic microswimmers. Owing to the fact that it is a straightforward process, bipolar electrodeposition is very competitive compared with other procedures employed to produce magnetic swimmers (vapor deposition or laser writing: see Section 14.3). The versatility offered by this technique is also a strong advantage. Indeed, a Ni patch was chosen, but several other metals could have been used.

After the discussion of the use of bipolar electrochemistry for generating magnetic motors, this section illustrates how it can be employed to generate chemically driven motors.

Platinum is a well-known catalyst for the decomposition of hydrogen peroxide:

$$H_2O_2(l) \xrightarrow{Pt} H_2O(l) + \frac{1}{2}O_2(g) \tag{14.7}$$

An evidence of its activity is given by the intense spontaneous production of O_2 bubbles at a Pt surface when being immersed into a H_2O_2 solution. Owing to this fact, Pt has been intensively used in order to power swimmers at different sizes by a bubble propulsion mechanism (Section 14.2). This propulsion mechanism is explored here using Pt/CMTs, which have been obtained through bipolar electrodeposition. Their ability to decompose hydrogen peroxide was tested first and the production of O_2 bubbles was observed immediately after the Pt/CMT hybrids were in contact with the reactant solution. The bubble generation can be tuned by the hydrogen peroxide content; for example, using a highly concentrated solution of 30 wt% H_2O_2 led to very intense bubble production.

The final topology of an electrodeposit on a carbon tube is a direct consequence of the orientation of the tube during the nucleation period, and it has been shown recently that, by playing with different parameters, the location of the deposit with respect to the tube, long axis can be tuned [108]. Figure 14.10a,b shows a schematic representation of the propulsion mechanisms, causing the motion of CMTs with a centered Pt deposit and a noncentered Pt deposit, respectively. At a CMT functionalized with a symmetric Pt particle at one extremity, the local

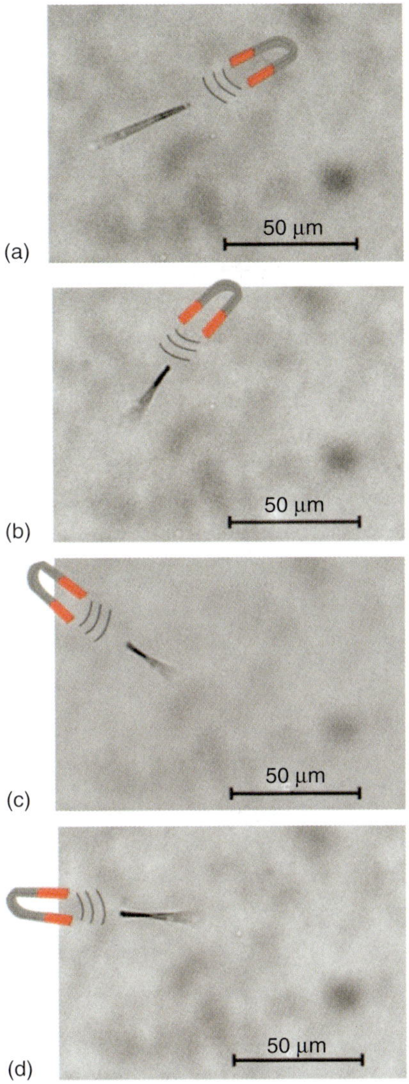

Figure 14.9 (a–d) Series of optical micrographs showing the manipulation of a nickel modified CMT, suspended in water, by applying a counterclockwise rotating magnetic field. Adapted and reproduced with permission from [107]. Reproduced with permission of Elsevier.

mechanical perturbations owing to the O_2 bubble formation and release generate a driving force $\overrightarrow{F_{net}}$ opposite to the direction of the translation motion [55]. In this ideal situation, a translation parallel to the CMT's longitudinal axis results. In the case of a CMT with a noncentered deposit, $\overrightarrow{F_{net}}$ is parallel to the CMT's normal axis, acting at its modified end and driving the rotation of the object, theoretically around its center of mass [48].

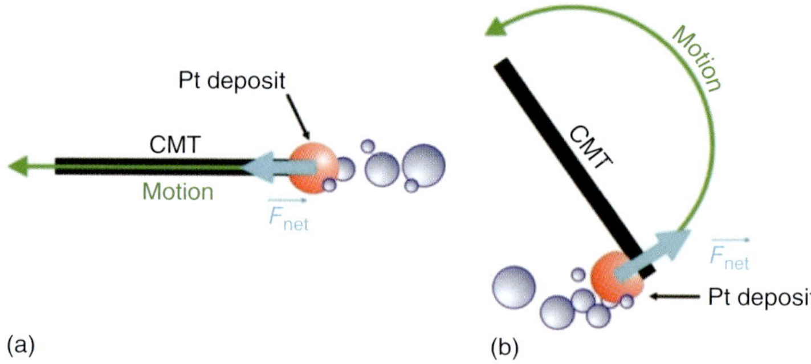

Figure 14.10 Schematic illustrations of the propulsion mechanism of Pt-modified carbon tubes. (a) A centered Pt deposit and (b) a noncentered Pt deposit.

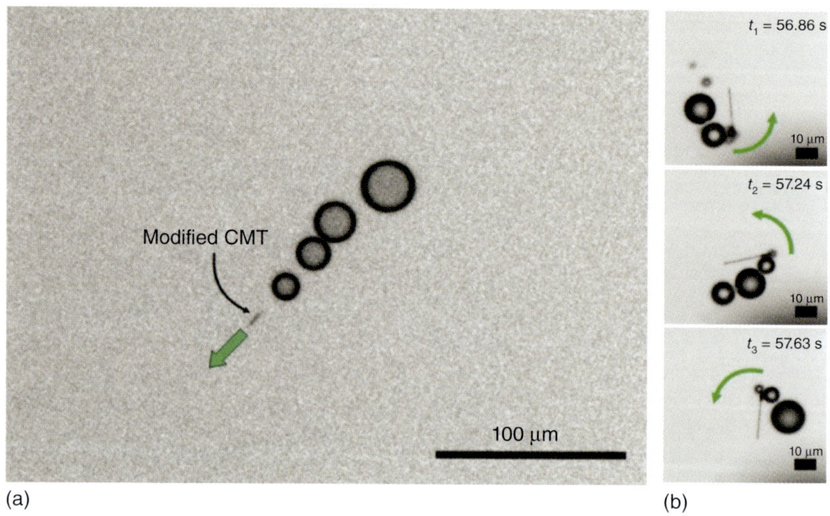

Figure 14.11 Motion of Pt-modified carbon microtubes. (a) Linear motion of a microswimmer leaving a bubble train behind, as observed under the optical microscope. (b) Series of optical micrographs of a counterclockwise rotating microswimmer. Adapted and reproduced with permission from [109]. Reproduced with permission of Elsevier.

Figure 14.11a shows an optical picture of a CMT during motion. The four dioxygen bubbles are formed successively from right to left, resulting in the propulsion of the carbon tube towards the bottom-left corner of the image. As mentioned earlier, this linear mode of motion can be explained by a CMT morphology with a centered Pt deposit and the related propulsion mechanism that is shown in Figure 14.10a. In addition to this linear mode of motion, rotation has also been observed. Figure 14.11b shows a selection of three optical images of a CMT performing a counterclockwise turn. These data are extracted from a large set of images in which the microhybrid is spinning several times. A time interval of

2.01 ± 0.14 s/turn has been estimated from 10 full turns, which allows calculating an angular speed $\omega = 29.9 \pm 2.1$ rpm corresponding to a frequency $f \approx 0.50$ Hz. The sequence of images in Figure 14.11b is in good agreement with the rotational motion of a CMT with a noncentered Pt deposit, depicted in Figure 14.10b. The CMT is rotating in the opposite direction with respect to the bubble evolution. The offset of the rotation axis with respect to the center of mass of the CMT/Pt object can be attributed to different parameters. An inhomogeneity of the spatial distribution and the bubble size can be caused by the deposit shape and its surface irregularities. Such phenomena can induce a certain shift of $\overrightarrow{F_{net}}$ with respect to the CMT normal axis, inducing an additional translational component to the rotation.

The majority of the literature processes used for the generation of chemical microswimmers are multistep procedures, whereas bipolar electrodeposition is straightforward and well adapted to generate such objects, especially for rotational motion where complex synthetic procedures have been employed usually in order to generate the appropriate particles [48, 49, 110].

14.7
Direct Use of Bipolar Electrochemistry for Motion Generation

As discussed previously in Section 14.4, the key concept for the propulsion of particles is asymmetry (Figure 14.6). The great majority of the techniques used for generating controlled motion of particles requires the use of intrinsically asymmetric swimmers. Because bipolar electrochemistry intrinsically provides a break of symmetry, which can be induced on any kind of conducting object, it is an appealing alternative to the existing propulsion mechanisms [111]. We present here its direct use for the propulsion of isotropic objects. Water splitting can be carried out at both extremities of a bipolar object. In an acidic aqueous solution, the following reactions take place (Figure 14.12):

$$2H^+(l) + 2e^- \rightarrow H_2(g) \quad E^0 = 0 \text{ V vs NHE} \tag{14.8}$$

$$2H_2O(l) \rightarrow O_2(g) + 4H^+(l) + 4e^- \quad E^0 = 1.23 \text{ V vs NHE} \tag{14.9}$$

At the cathodic and anodic pole of the object, gas bubbles are produced, which can be visualized under the optical microscope, as presented by Figure 14.12b for a 1 mm sized metal bead located in a 1.6 kV m^{-1} electric field. To maintain electroneutrality at the BE, the electron production and consumption must be equal at both sides. As a result, considering Equations 14.8 and 14.9, the produced H_2 volume (left side) is twice the produced O_2 volume (right side). In analogy with bubble-powered nanomotors (Section 14.2), this asymmetric bubble production is responsible for a directional motion. The obtained speeds were equal to $v = 20\ \mu m\ s^{-1}$, which corresponds to one bodylength every 14.2 s. As the produced O_2 bubbles hinder this motion, one can increase the speed by suppressing this reaction. The more favorable redox potential of the hydroquinone (HQ)/benzoquinone (BQ) redox couple, compared with the one of H_2O/O_2 (Equation 14.9), makes it an ideal

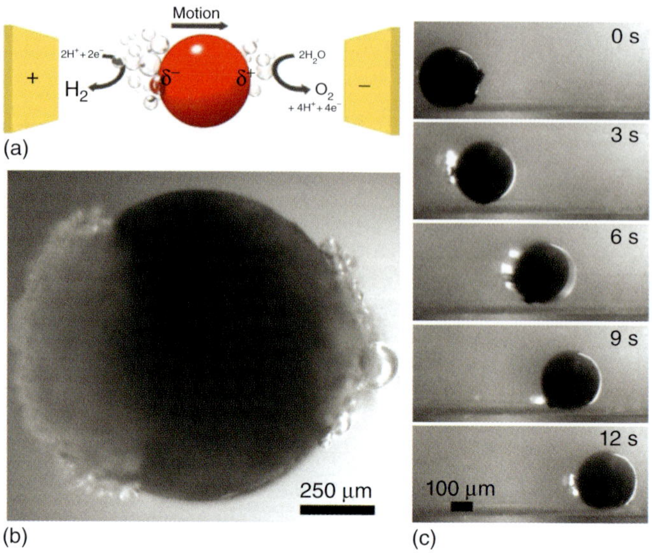

Figure 14.12 Translation motion induced by bubble production. (a) Scheme of water splitting, induced by bipolar electrochemistry. (b) Optical micrograph of a 1 mm stainless steel bead exposed to a 1.6 kV m^{-1} electric field in aqueous 24 mM H_2SO_4. The left part of the bead is the cathodic pole where H_2 bubbles are produced and the right part is the anodic pole where O_2 bubbles are produced. (c) Translational motion generated with a 275 μm GC sphere in a PDMS microchannel with a 4.3 kV m^{-1} electric field in an aqueous solution of 7 mM HCl and 14 mM HQ. Adapted and reproduced with permission from [111]. Reproduced with permission of Nature Publishing Group.

couple for suppressing the O_2 production. In an aqueous HQ solution, the following reactions happen at the cathodic and anodic poles of the bipolar object, respectively:

$$2H^+(l) + 2e^- \rightarrow H_2(g) \quad E^0 = 0\,V \text{ vs NHE} \tag{14.10}$$

$$HQ(l) \rightarrow BQ(g) + 2H^+(l) + 2e^- \quad E^0 = 0.7\,V \text{ vs NHE} \tag{14.11}$$

In this case, the bubble formation takes place only at the cathodic side of the object, leading to a higher speed (Figure 14.12c). Macroscopic BEs could be propelled over a distance of 1.5 cm with a speed $v = 512\,\mu m\,s^{-1}$ (about 1 bodylength every 2 s) in an electric field of 1.3 kV m^{-1}. The potential difference ΔV that is generated between the two sides of the bead is 1.3 V, a value that is above the thermodynamic value ΔV_{min} corresponding to the combination of both reactions. For micrometer-sized objects, as in the experiment reported in Figure 14.12c, a higher electric field is required to induce a ΔV of the same order of magnitude. An electric field of 4.3 kV m^{-1} can be applied, which corresponds to a ΔV of 1.2 V across the 275 μm-sized GC bead. On the basis of the data of Figure 14.12c, the bead velocity is $v = 62\,\mu m\,s^{-1}$, or 1 bodylength every 4.4 s. These experiments show that bipolar electrochemistry can directly induce a controlled motion of conductive objects by a bubble propulsion mechanism. The advantage of this mechanism, compared

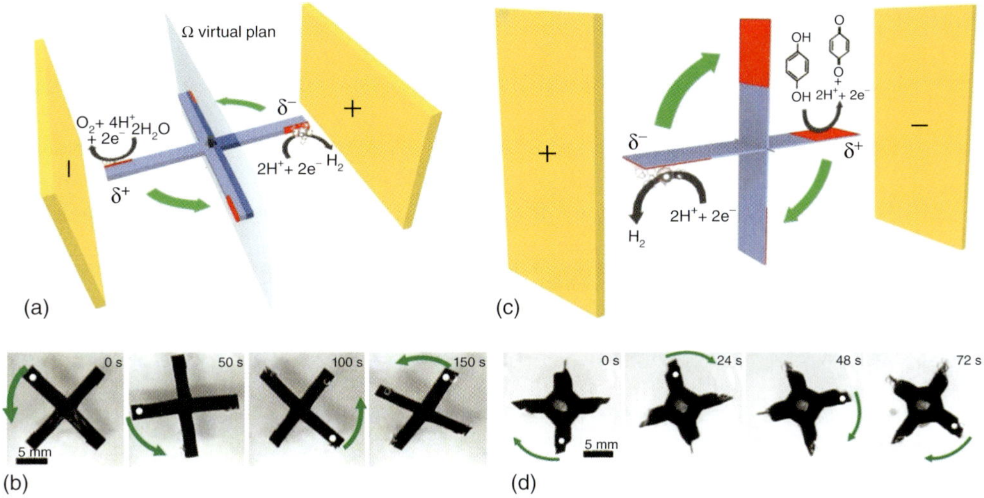

Figure 14.13 Bipolar rotors. (a) Scheme of a horizontal rotor powered by bipolar electrochemistry-induced water splitting; red areas represent the conductive parts exposed to the solution, whereas the blue areas have been shielded with an insulating polymer. (b) Series of photographs showing the rotational motion of a horizontal bipolar rotor in an electric field of 0.5 kV m^{-1} in 50 mM HCl. (c) Scheme of a vertical bipolar rotor powered by the reduction of protons coupled with HQ oxidation. (d) Series of photographs showing the rotational motion of a vertical bipolar rotor, in an electric field of 0.5 kV m^{-1} with 50 mM HCl and 100 mM HQ. Adapted and reproduced with permission from [111]. Reproduced with permission of Nature Publishing Group.

with most of the existing ones (Sections 14.2 and 14.3), resides in the fact that theoretically, every kind of conducting object can be propelled without the need of any synthetic procedure for breaking the symmetry of the system.

Besides being effective for inducing translations, the described concept can be adapted to generate rotational motion. Two kinds of bipolar electrochemical rotors have been designed, the first one rotating in a horizontal plane (Figure 14.13a) and the second one in a vertical plane (Figure 14.13c). To concentrate the bubble production on specific sites, the conducting rotor surface has been shielded with an insulating polymer (blue parts in Figure 14.13), except the extremities (red parts in Figure 14.13).

Figure 14.13b shows the rotation of a horizontal rotor powered by bipolar electrochemistry-induced water splitting. The water splitting reactions are induced at the extremities of the object by imposing an electric field $E = 0.5\,\mathrm{kV\,m^{-1}}$, corresponding to a ΔV of 8.5 V. An average angular speed $\omega = 1.9°\ \mathrm{s^{-1}}$ has been calculated, which correspond to 0.32 rpm.

A vertical rotor can also be designed and is a very good option to enhance the rotational speed, as this configuration takes advantage of the buoyancy of the bubbles generated below the rotor blade. In general, when gas is released from the object, and while the resulting bubble is still attached to the object, the expansion of the gas bubble in the liquid exerts a force on the object, which pushes the latter one in the opposite direction. In the particular case of the vertical rotor, the gas bubbles stick to the rotor blade and are located below the rotor blade because the rotor has been designed on purpose in that way. This means that in addition to the above-mentioned force, the buoyancy of the gas exerts a second type of force on the rotor blade (Figure 14.13c), leading in principle to a higher speed. However, if the bubbles adhere too strongly to the blade, they will slow down the motion once they are dragged downward (right side of the rotor). Therefore, one has to make sure that they will detach from the rotor when the blade is in the vertical position, essentially by the addition of small amounts of surfactant. The speed can be further optimized by suppressing the O_2 evolution again, which occurs at the opposite blade, by adding HQ, because generated O_2 bubbles might stick to the blade, and in this case the buoyancy slows down the rotation owing to the forces that are orientated in the wrong direction. A beneficial side effect of using HQ is the increase of the H_2 bubble production at the cathodic pole for the same external electric field, because a lower ΔV_{min} is required for the combination of these two redox reactions. This leads to an overall higher velocity. The motion is also more homogeneous in time than for the horizontal configuration, and the extrapolated average angular speed is 4.2° s^{-1} (0.70 rpm), more than two times higher compared with the previous speed. Although in this case the rotor is submitted to a sigmoid variation of the driving force, the angular speed is leveled out by the buoyancy effect specific to this setup.

These experiments illustrate that bipolar electrochemistry can be directly used for the rotation of objects. The angular speed can be tuned by changing different parameters such as the solution composition and the electric field strength. This kind of rotors could be used for producing additional mechanical energy in a setup where electric fields are used anyway, such as in electrolysis reactors.

The translation motions presented above were achieved in a horizontal plane. We now discuss the vertical ascension of BEs using a similar bubble propulsion

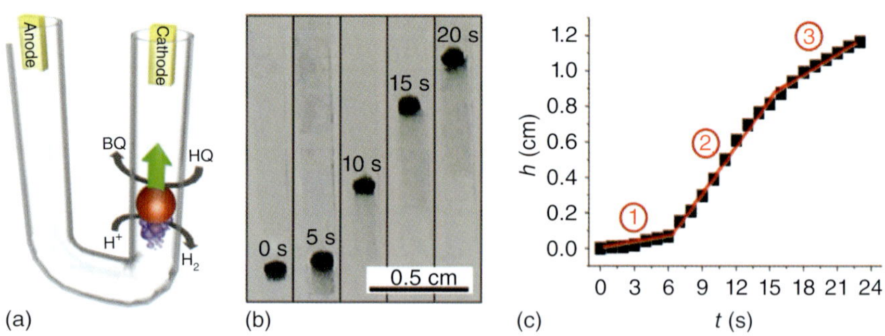

Figure 14.14 Levitation of a GC bead. (a) Scheme of the levitation mechanism for a conducting bead in a U-shaped cell. (b) Series of optical micrographs showing the rising of a carbon bead in a glass capillary. (c) Graph showing the height h evolution as a function of time t. Reproduced with permission of The Royal Society of Chemistry from Ref. [112].

mechanism. For these experiments, U-shaped glass capillaries (Figure 14.14a) were used as cells.

As before, the proton reduction (Equation 14.10) at the cathodic pole of the object, coupled with the HQ oxidation (Equation 14.11) at the anodic pole, induces the asymmetric production of bubbles at the object surface. If the external electrodes, localized at the capillary entrances, are polarized in such a way that the H_2 flow is generated underneath the bead, this bipolar mechanism can lead to the levitation of a conducting bead as illustrated in Figure 14.14a. GC has been chosen as the swimmer material because of its low density compared to many other conducting materials. This decreases the force needed to overcome gravity for an object with a given volume. Figure 14.14b shows the motion of a GC bead with a diameter around 1 mm in a glass capillary. The applied voltage being 270 V with a distance of 11.5 cm between the two feeder electrode, the overall electric field E is 23.5 V cm^{-1}. One can calculate that this corresponds to a polarization voltage ΔV of 2.3 V between the top and the bottom part of the bead. This value, which is more than three times higher than the theoretical value 0.7 V needed for inducing Equations 14.10 and 14.11 at both extremities of the object in standard conditions, shows that these two reactions should be highly favored. The resulting bubble production leading to the bead levitation can be clearly observed at the bottom of the bead in Figure 14.14b. Stopping the electric field directly induces a dropping of the bead, but a new ascension can be triggered by reapplying the electric field. One can easily imagine using this approach of BE levitation for stirring, cleaning, or unblocking microchannels.

On the basis of the observation that the swimmer motion can be controlled by changing the capillary shape, a Yo-Yo type motion could be generated on purpose using a conical capillary [112].

We demonstrate in the following section that the conductive bead can be used as a lifting motor moving through the capillary with a cargo attached to it. This cargo is a piece of polymer attached to the bead via a very thin polymer wire with a tiny drop of glue. The wire extremity is coated with black nail varnish in order to facilitate its visualization. The GC bead is the only conductive part. Figure 14.15 shows the cargo lifting with an electric field of $E = 23$ V cm^{-1}. The motion that drives the cargo over 2.7 cm is clearly due to the bubbles being generated only underneath the GC bead motor and not at the cargo or at the wire surface, because of their insulating character. The whole object moves with a maximum speed $v = 2$ mm s^{-1}. This proof-of-concept experiment makes clear that cargo lifting in fluid channels is possible using bipolar electrochemistry, and this phenomenon can be used as an alternative concept for moving particles in microfluidic systems. The towing speed of the cargo can be controlled by the applied external voltage, the length of the bipolar motor, and, most importantly, by the distance between the motor and the capillary walls.

Introducing such functionalities is of primary importance for potential applications of those moving systems, and it is therefore quite natural to look for other functions that could be carried out by such objects. In this context, the emission of light, simultaneously coupled with the motion, is an interesting issue to study,

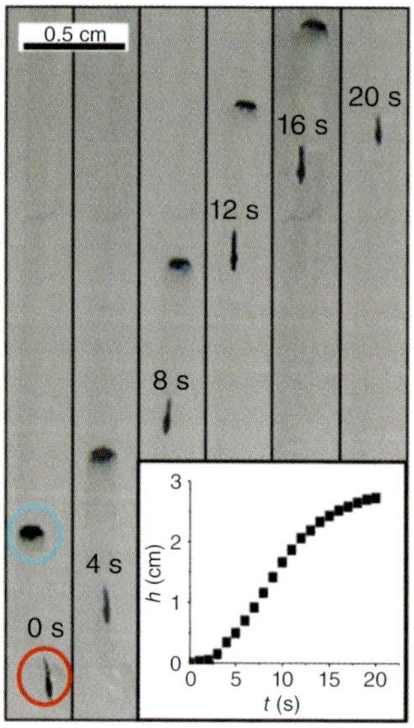

Figure 14.15 Series of optical micrographs showing the cargo lifter composed of the cargo (red circle) and the bipolar motor (blue circle). Inset: height position *h* of the cargo as a function of time *t*. Reproduced by permission of The Royal Society of Chemistry from Ref. [112].

because tracking moving objects in real time is a challenge. Therefore, intensive effort has been made to design particles with specific optical, in many cases fluorescent, properties. For example, the observation of most of the microswimmers presented in Sections 14.2 and 14.3 requires efficient microscopy setups [36] or a functionalization of the swimmers with a fluorophore [80] in order to monitor them while moving. In this context, it would be very helpful if the swimmer could act at the same time as an autonomous light source with the photon emission being intrinsically related to the motion. Here we present an original approach, where the moving object is simultaneously emitting light, produced by electrogenerated chemiluminescence (ECL).

In this section, we prove that bipolar electrochemistry is not only the asymmetric driving force for the particle propulsion but also responsible for light emission, thus combining in a synergetic way two redox reactions on the same object.

As in the previous sections, the motion is induced here by the produced H_2 bubbles owing to the reduction of water at the GC cathodic pole, which proceeds according to the following equation at neutral pH:

$$H_2O(l) \rightarrow \frac{1}{2} H_2(g) + OH^-(l) \tag{14.12}$$

Because of charge neutrality during bipolar electrochemistry, in the previous sections, water or a sacrificial reductant such as HQ was simultaneously oxidized at the anodic pole, generating oxidation products such as O_2 (Equation 14.9) or BQ (Equation 14.11). These anodic processes were not useful in the context of propulsion, and can therefore be replaced with electrochemical reactions leading to ECL emission. ECL is an electrochemical process that produces light. The ECL mechanism involves the following oxidation reactions:

$$TPrA(l) \rightarrow TPrA^{+}(l) + e^{-} \quad (14.13)$$

$$Ru(bpy)_3{}^{2+}(l) \rightarrow Ru(bpy)_3{}^{3+} + e^{-} \quad (14.14)$$

The generated reactive tripropylamine radical cation ($TPrA^{\bullet +}$) quickly undergoes a deprotonation reaction, forming a highly reactive reducing agent $TPrA^{\bullet}$. $TPrA^{\bullet}$ and $TPrA^{\bullet +}$ react with the ruthenium complex to generate the excited state $Ru(bpy)_3{}^{2+*}$ producing the ECL [113]. As depicted in Figure 14.16a, the asymmetric electroactivity induced by bipolar electrochemistry can induce the synergetic reduction of H_2O at the cathodic pole and oxidation of the ECL reagents at the anodic pole of the BE, which generates simultaneously motion and light emission of the bipolar swimmer.

The bipolar electrochemistry experiments were carried out with a levitation protocol presented before. The electrodes were inserted into the top part of the cell and were polarized in such a way that the H_2 flow is generated underneath the bead, as illustrated in Figure 14.16a. The electric field value was above the previously calculated theoretical threshold value for generating a synergetic levitation/ECL emission of the GC bead, as demonstrated by Figure 14.16b. The left picture of Figure 14.16b, obtained under ambient light, shows the GC bead in the capillary. The light was then turned off and the electric field applied. The ECL generated at the anodic pole of the BE, as shown in Figure 14.16b, was extremely bright and could be instantaneously observed with the naked eye. The initial time $t = 0$ s is defined as

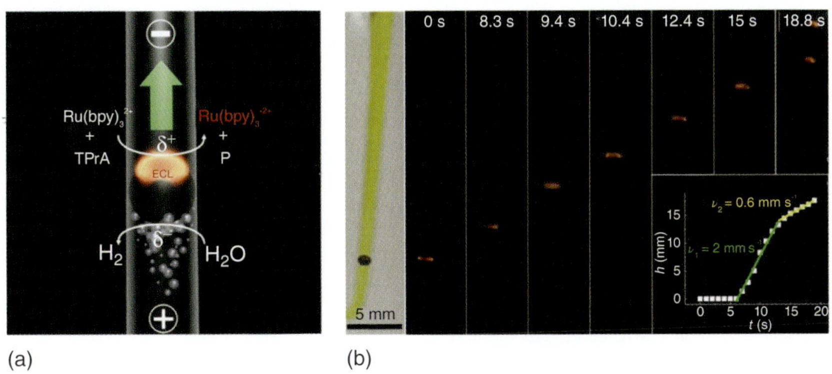

Figure 14.16 Light emission during bipolar levitation. (a) Scheme showing the synergetic reduction of H_2O at the cathodic pole and oxidation of the ECL reagents at the anodic pole, which induces simultaneous motion and light emission. P corresponds to a by-product of the tripropylamine radicals formed during the ECL process. (b) Series of optical pictures showing the levitation of a glassy carbon bead emitting ECL at different times. The bead position under white light is shown on the left picture, and the other images were taken in the dark. Inset: plot showing the height evolution h as a function of time t. Adapted and reproduced with permission from [114]. Reproduced with permission of John Wiley & Sons.

the time when the camera has been focused on the ECL signal, allowing its correct visualization. After 6 s, the levitation starts because of H_2 bubble accumulation underneath the bead. The height evolution and the levitation speed are comparable to the ones reported without ECL, showing that the ECL mechanism does not influence the bead propulsion. This work presented the first synergetic action of bipolar electrochemistry in terms of simultaneous propulsion and ECL generation, leading to the first example of a swimmer that is intrinsically coupled with chemically generated light. In the reported experiments, ECL provides a direct monitoring of the object motion, which is very useful when dealing with autonomous swimmers. In analogous experiments, other light emitting processes can be coupled with the motion, generating, for example, blue light during the propulsion [115].

All the objects described earlier were moving based on bubble propulsion triggered by bipolar electrochemistry. We now demonstrate that bipolar electrochemistry can also be used to propel macroscopic or microscopic metallic objects by a completely different mechanism, based on a dynamic self-regeneration process. This approach is the simultaneous deposition and dissolution of a material. The deposition occurring at the cathodic pole of the object is coupled with its dissolution at the opposite pole, resulting in a linear motion of the object (Figure 14.17).

In our experiments, metal objects composed of zinc dendrites were positioned within capillaries previously filled with an aqueous $ZnSO_4$ solution. When an external electric field is applied across the capillary, the following two reactions occur:

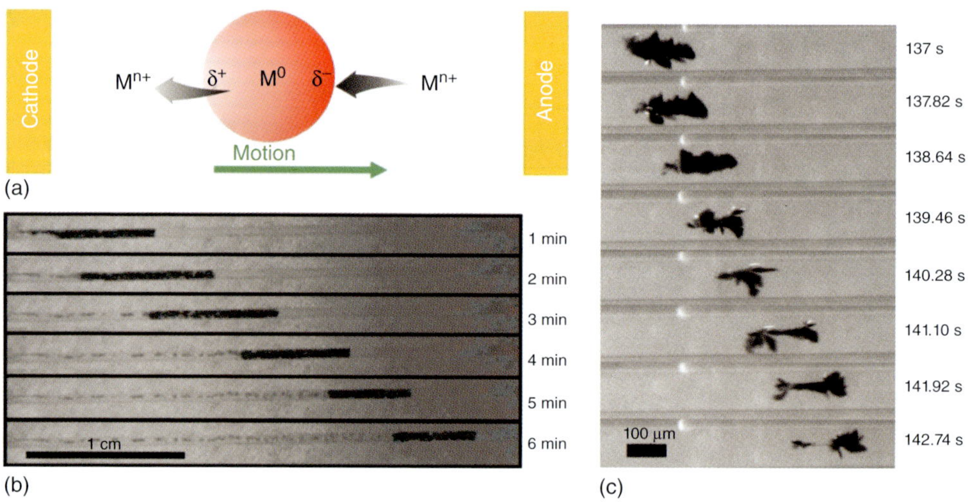

Figure 14.17 Dynamic bipolar self-regeneration. (a) Scheme of the principle. (b) Pictures of a zinc macroswimmer in a glass tube filled with a $ZnSO_4$ solution under the influence of an external electrical field, recorded at various times during the experiment. (c) Optical micrographs of a zinc dendrite in a glass capillary filled with a zinc sulfate solution at pH ≈ 5 under the influence of an external electric field at various times during the experiment. Reprinted with permission from Ref. [116]. Copyright (2010) American Chemical Society.

$$Zn^{2+}(l) + 2\,e^- \rightarrow Zn^0(s) \quad E^0 = -0.76\,V \text{ vs NHE} \tag{14.15}$$

at the cathodic pole of the zinc object and

$$Zn^0(s) \rightarrow Zn^{2+}(l) + 2\,e^- \quad E^0 = -0.76\,V \text{ vs NHE} \tag{14.16}$$

at the anodic pole. As only a single redox couple (Zn^{2+}/Zn^0) is involved in the bipolar process, ΔV_{min} is almost equal to zero (Section 14.4) and theoretically the application of low electric fields should be sufficient to trigger Equations 14.15 and 14.16 if one neglects eventually occurring overpotentials.

As a proof of principle, the experiments were carried out first at a macroscopic level. Under the chosen experimental conditions, zinc grew electrochemically with a dendritic morphology [117]. This property was used to generate freestanding zinc dendrites. Finally, an electric field with a value of 1.25 $kV\,m^{-1}$ was imposed. As discussed earlier, a smaller electric field could be imposed for triggering Equations 14.15 and 14.16, but the latter electric field value was set in order to reach sufficient speeds.

Figure 14.17b shows a series of optical micrographs recorded at different times during the experiment. The consequence of the object's dissolution at its anodic pole is a visible motion (from left to right) with a speed $v \approx 60\,\mu m\,s^{-1}$. This value represents an average speed, calculated using the distance covered by the object during the run. It seems hardly convertible into bodylength s^{-1}, owing to the inhomogeneous density of the object, leading to a continuous change of length during the bipolar regeneration. As expected, turning off the electric field stopped the motion. In these pictures, depending on the local pH conditions, the presence of a thin zinc oxide layer behind the moving object can be noticed. This suggests that the technique might be used for decorating surfaces with a conducting or semiconducting "ink," which could be engineered by adjusting the swimmer composition. Analogous experiments were performed at a microscopic scale. A zinc dendrite was isolated in a capillary with a diameter of 100 μm. The swimmer was then exposed to an external electric field of 7 $kV\,m^{-1}$, which resulted in a speed $v \approx 80\,\mu m\,s^{-1}$ (Figure 14.17c). The change of the swimmer morphology clearly shows that the motion was induced by the dynamic bipolar self-regeneration process and not by electrokinetic phenomena. Electroosmotic flow becomes relevant at the micrometer scale but is oriented in the opposite direction of the movement and therefore cannot explain the observed motion. It is verified that inert particles indeed move in the direction of the electroosmotic flow. The speed can be controlled by tuning the electric field value. It seems important to note that the object is constantly reconstructing itself during the run, rendering its atomic composition at the end different from the one in the beginning. This raises the philosophic question whether the object remains fundamentally the same when all its parts are replaced, very much comparable to what is discussed in the paradox of the ship of Theseus [118]. The notion of "motion" and "swimmer" can then also be critically regarded, and most likely these experiments should be considered more as the analog of a traveling physicochemical wave. The process can be generalized to other, nonnoble metals. In the future, dynamic bipolar self-regeneration could potentially be used for wireless localized deposition, fabrication of electrical microcontacts, or surface patterning.

14.8 Conclusion and Perspectives

Natural micro- and nanomotors are a great source of inspiration and numerous approaches based on chemical fueling or using external fields have been employed in order to propel particles in a controlled way. The chemical fueling methods have shown excellent results for steering particles that have characteristic dimensions from the millimeter [53] scale to the micrometer scale [30]. External fields can be used to control the motion of objects from the macroscale [93] down to the macromolecular level [119].

Asymmetry is a crucial parameter for the generation of controlled motion; therefore, the use of bipolar electrochemistry for the development of miniaturized motors is an interesting alternative in this context. The first strategy is using APs generated by bipolar electrodeposition and employing them directly as micromotors. In this way, hybrid carbon tube/nickel objects could be turned in rotating magnetic fields and carbon tube/platinum objects could be propelled following the oxygen generation occurring at the platinum surface when exposed to hydrogen peroxide solutions. Because bipolar electrodeposition is a single-step bulk technique, it is competitive with the other techniques usually employed for generating micro- and nanomotors, even with a complex design.

As bipolar electrochemistry intrinsically allows breaking symmetry, it can also be used as a direct trigger of motion for isotropic particles. On the basis of a bubble propulsion mechanism, it has been shown that water splitting reactions, induced at the reactive poles of a BE, can produce translational motion of millimeter and micrometer-sized conducting objects. By using this concept, the rotation of centimeter-sized objects can also be achieved. Furthermore, bubble accumulation under a swimmer can be used for the levitation of single BEs or for cargo lifting in fluid channels. It is also demonstrated that such a motion can be coupled with electrochemiluminescence at the anodic pole. Finally, a motion mechanism based on the simultaneous deposition/dissolution of a BE is developed. In this case, a metal object can be dynamically self-regenerated in a capillary filled with a metal salt solution.

So far, BEs having typical sizes from about 1 cm down to 100 μm have been steered. A first challenge concerning bubble propulsion and dynamic self-regeneration will be to decrease the microswimmer size (and, as a consequence, to use higher electric fields). This will allow testing these mechanisms at low Reynolds numbers, in the presence of significant Brownian motion and with competing electrokinetic phenomena. By engineering the swimmer shape, a three-dimensional motion based on these mechanisms is possible in principle. Furthermore, one can imagine an electrochemically triggered cargo release, induced by an electric field pulse or reversal. Beyond the academic interest, these mechanisms have to be tested for practical applications such as cargo delivery in lab-on-chip devices, or unclogging of fluid channels. Finally, the behavior of the metal swimmer during bipolar self-regeneration is also very interesting from an intellectual point of view, for example, with respect to the trajectory of the swimmer in different environments such as a microchannel with Y-junctions or in a labyrinth. All in all, the simplicity of the concept of bipolar electrochemistry and the fact that in principle any conductive particle can be set into motion makes it an attractive approach, complementing the already very rich tool case for controlled propulsion.

References

1. Berg, H.C. (2003) The rotary motor of bacterial flagella. *Annu. Rev. Biochem.*, **72** (1), 19–54.
2. Berg, H.C. (2008) Bacterial flagellar motor. *Curr. Biol.*, **18** (16), R689–R691.
3. Schliwa, M. and Woehlke, G. (2003) Molecular motors. *Nature*, **422** (6933), 759–765.
4. Lipowsky, R., Chai, Y., Klumpp, S., Liepelt, S., and Müller, M.J.I. (2006) Molecular motor traffic: from biological nanomachines to macroscopic transport. *Physica A*, **372** (1), 34–51.
5. Kon, T., Oyama, T., Shimo-Kon, R., Imamula, K., Shima, T., Sutoh, K. *et al* (2012) The 2.8a crystal structure of the dynein motor domain. *Nature*, **484** (7394), 345–350.
6. Feynman, R.P. (1960) There's plenty of room at the bottom. *Eng. Sci. (California Inst. Technol.)*, **23**, 22–36.
7. Wang, J. (2009) Can man-made nanomachines compete with nature biomotors? *ACS Nano*, **3** (1), 4.
8. Ozin, G.A., Manners, I., Fournier-Bidoz, S., and Arsenault, A. (2005) Dream nanomachines. *Adv. Mater.*, **17** (24), 3011–3018.
9. Ebbens, S.J. and Howse, J.R. (2010) In pursuit of propulsion at the nanoscale. *Soft Matter*, **6** (4), 726–738.
10. OIST *www.Oist.Jp/Photo/Salmonella-Bacterium-Flagella* (accessed 22 November 2013).
11. Hess, H., Bachand, G.D., and Vogel, V. (2004) Powering nanodevices with biomolecular motors. *Chem. Eur. J.*, **10** (9), 2110–2116.
12. Diluzio, W.R., Turner, L., Mayer, M., Garstecki, P., Weibel, D.B., Berg, H.C. *et al.* (2005) Escherichia coli swim on the right-hand side. *Nature*, **435** (7046), 1271–1274.
13. Takeuchi, S., Diluzio, W.R., Weibel, D.B., and Whitesides, G.M. (2005) Controlling the shape of filamentous cells of Escherichia coli. *Nano Lett.*, **5** (9), 1819–1823.
14. Weibel, D.B., Garstecki, P., Ryan, D., Diluzio, W.R., Mayer, M., Seto, J.E. *et al* (2005) Microoxen: microorganisms to move microscale loads. *Proc. Natl. Acad. Sci. U.S.A.*, **102** (34), 11963–11967.
15. Behkam, B. and Sitti, M. (2007) Bacterial flagella-based propulsion and on/off motion control of microscale objects. *Appl. Phys. Lett.*, **90** (2), 023902.
16. Behkam, B. and Sitti, M. (2008) Effect of quantity and configuration of attached bacteria on bacterial propulsion of microbeads. *Appl. Phys. Lett.*, **93** (22), 223901.
17. Van Den Heuvel, M.G.L. and Dekker, C. (2007) Motor proteins at work for nanotechnology. *Science*, **317** (5836), 333–336.
18. Goel, A. and Vogel, V. (2008) Harnessing biological motors to engineer systems for nanoscale transport and assembly. *Nat. Nanotechnol.*, **3** (8), 465–475.
19. Soong, R.K., Bachand, G.D., Neves, H.P., Olkhovets, A.G., Craighead, H.G., and Montemagno, C.D. (2000) Powering an inorganic nanodevice with

a biomolecular motor. *Science*, **290** (5496), 1555–1558.
20. Yamaguchi, S., Matsumoto, S., Ishizuka, K., Iko, Y., Tabata, K.V., Arata, H.F. *et al.* (2008) Thermally responsive supramolecular nanomeshes for on/off switching of the rotary motion of f1-ATPase at the single-molecule level. *Chem. Eur. J.*, **14** (6), 1891–1896.
21. Hess, H., Clemmens, J., Qin, D., Howard, J., and Vogel, V. (2001) Light-controlled molecular shuttles made from motor proteins carrying cargo on engineered surfaces. *Nano Lett.*, **1** (5), 235–239.
22. Clemmens, J., Hess, H., Doot, R., Matzke, C.M., Bachand, G.D., and Vogel, V. (2004) Motor-protein "roundabouts": microtubules moving on kinesin-coated tracks through engineered networks. *Lab Chip*, **4** (2), 83–86.
23. Schmidt, C. and Vogel, V. (2010) Molecular shuttles powered by motor proteins: Loading and unloading stations for nanocargo integrated into one device. *Lab Chip*, **10** (17), 2195–2198.
24. Fischer, T., Agarwal, A., and Hess, H. (2009) A smart dust biosensor powered by kinesin motors. *Nat. Nanotechnol.*, **4** (3), 162–166.
25. Rios, L. and Bachand, G.D. (2009) Multiplex transport and detection of cytokines using kinesin-driven molecular shuttles. *Lab Chip*, **9** (7), 1005–1010.
26. Hiyama, S., Moritani, Y., Gojo, R., Takeuchi, S., and Sutoh, K. (2010) Biomolecular-motor-based autonomous delivery of lipid vesicles as nano- or microscale reactors on a chip. *Lab Chip*, **10** (20), 2741–2748.
27. Von Delius, M. and Leigh, D.A. (2011) Walking molecules. *Chem. Soc. Rev.*, **40** (7), 3656–3676.
28. Sykes, E.C.H. (2012) Electric nanocar equipped with four-wheel drive gets taken for its first spin. *Angew. Chem. Int. Ed.*, **51** (18), 4277–4278.
29. Paxton, W.F., Sundararajan, S., Mallouk, T.E., and Sen, A. (2006) Chemical locomotion. *Angew. Chem. Int. Ed.*, **45** (33), 5420–5429.
30. Wang, Y., Hernandez, R.M., Bartlett, D.J., Bingham, J.M., Kline, T.R., Sen, A. *et al.* (2006) Bipolar electrochemical mechanism for the propulsion of catalytic nanomotors in hydrogen peroxide solutions. *Langmuir*, **22** (25), 10451–10456.
31. Pumera, M. (2010) Electrochemically powered self-propelled electrophoretic nanosubmarines. *Nanoscale*, **2** (9), 1643–1649.
32. Paxton, W.F., Kistler, K.C., Olmeda, C.C., Sen, A., St. Angelo, S.K., Cao, Y. *et al.* (2004) Catalytic nanomotors: autonomous movement of striped nanorods. *J. Am. Chem. Soc.*, **126** (41), 13424–13431.
33. Fournier-Bidoz, S., Arsenault, A.C., Manners, I., and Ozin, G.A. (2005) Synthetic self-propelled nanorotors. *Chem. Commun.*, (4), 441–443.
34. Burdick, J., Laocharoensuk, R., Wheat, P.M., Posner, J.D., and Wang, J. (2008) Synthetic nanomotors in microchannel networks: directional microchip motion and controlled manipulation of cargo. *J. Am. Chem. Soc.*, **130** (26), 8164–8165.
35. Zacharia, N.S., Sadeq, Z.S., and Ozin, G.A. (2009) Enhanced speed of bimetallic nanorod motors by surface roughening. *Chem. Commun.*, (39), 5856–5858.
36. Laocharoensuk, R., Burdick, J., and Wang, J. (2008) Carbon-nanotube-induced acceleration of catalytic nanomotors. *ACS Nano*, **2** (5), 1069–1075.
37. Kagan, D., Calvo-Marzal, P., Balasubramanian, S., Sattayasamitsathit, S., Manesh, K.M., Flechsig, G.-U. *et al.* (2009) Chemical sensing based on catalytic nanomotors: motion-based detection of trace silver. *J. Am. Chem. Soc.*, **131** (34), 12082.
38. Demirok, U.K., Laocharoensuk, R., Manesh, K.M., and Wang, J. (2008) Ultrafast catalytic alloy nanomotors. *Angew. Chem.*, **120** (48), 9489–9491.
39. Wang, J. and Manesh, K.M. (2010) Motion control at the nanoscale. *Small*, **6** (3), 338–345.
40. Calvo-Marzal, P., Manesh, K.M., Kagan, D., Balasubramanian, S.,

Cardona, M., Flechsig, G.-U. *et al.* (2009) Electrochemically-triggered motion of catalytic nanomotors. *Chem. Commun.*, (30), 4509–4511.
41. Balasubramanian, S., Kagan, D., Manesh, K.M., Calvo-Marzal, P., Flechsig, G.-U., and Wang, J. (2009) Thermal modulation of nanomotor movement. *Small*, **5** (13), 1569–1574.
42. Kline, T.R., Paxton, W.F., Mallouk, T.E., and Sen, A. (2005) Catalytic nanomotors: remote-controlled autonomous movement of striped metallic nanorods. *Angew. Chem. Int. Ed.*, **44** (5), 744–746.
43. Manesh, K.M., Balasubramanian, S., and Wang, J. (2010) Nanomotor-based 'writing' of surface microstructures. *Chem. Commun.*, **46**, 5704–5706.
44. Sundararajan, S., Lammert, P.E., Zudans, A.W., Crespi, V.H., and Sen, A. (2008) Catalytic motors for transport of colloidal cargo. *Nano Lett.*, **8** (5), 1271–1276.
45. Wang, J. (2012) Cargo-towing synthetic nanomachines: towards active transport in microchip devices. *Lab Chip*, **12**, 1944–1950.
46. Campuzano, S., Kagan, D., Orozco, J., and Wang, J. (2011) Motion-driven sensing and biosensing using electrochemically propelled nanomotors. *Analyst*, **136** (22), 4621–4630.
47. Hong, Y., Blackman, N.M.K., Kopp, N.D., Sen, A., and Velegol, D. (2007) Chemotaxis of nonbiological colloidal rods. *Phys. Rev. Lett.*, **99** (17), 178103.
48. Qin, L., Banholzer, M.J., Xu, X., Huang, L., and Mirkin, C.A. (2007) Rational design and synthesis of catalytically driven nanorotors. *J. Am. Chem. Soc.*, **129** (48), 14870.
49. Catchmark, J.M., Subramanian, S., and Sen, A. (2005) Directed rotational motion of microscale objects using interfacial tension gradients continually generated via catalytic reactions. *Small*, **1** (2), 202–206.
50. Liu, R. and Sen, A. (2011) Autonomous nanomotor based on copper–platinum segmented nanobattery. *J. Am. Chem. Soc.*, **133** (50), 20064–20067.
51. Mano, N. and Heller, A. (2005) Bioelectrochemical propulsion. *J. Am. Chem. Soc.*, **127** (33), 11574.
52. Pavlick, R.A., Sengupta, S., Mcfadden, T., Zhang, H., and Sen, A. (2011) A polymerization-powered motor. *Angew. Chem. Int. Ed.*, **50**, 9374–9377.
53. Ismagilov, R.F., Schwartz, A., Bowden, N., and Whitesides, G.M. (2002) Autonomous movement and self-assembly. *Angew. Chem.*, **114** (4), 674–676.
54. Huang, G., Wang, J., and Mei, Y. (2012) Material considerations and locomotive capability in catalytic tubular microengines. *J. Mater. Chem.*, **22** (14), 6519–6525.
55. Gibbs, J.G. and Zhao, Y.P. (2009) Autonomously motile catalytic nanomotors by bubble propulsion. *Appl. Phys. Lett.*, **94** (16), 163104.
56. Wilson, D.A., Nolte, R.J.M., and Van Hest, J.C.M. (2012) Autonomous movement of platinum-loaded stomatocytes. *Nat. Chem.*, **4** (4), 268–274.
57. Vicario, J., Eelkema, R., Browne, W.R., Meetsma, A., Crois, R.M.L., and Feringa, B.L. (2005) Catalytic molecular motors: fuelling autonomous movement by a surface bound synthetic manganese catalase. *Chem. Commun.*, (31), 3936–3938.
58. Pantarotto, D., Browne, W.R., and Feringa, B.L. (2008) Autonomous propulsion of carbon nanotubes powered by a multienzyme ensemble. *Chem. Commun.*, (13), 1533–1535.
59. Sanchez, S., Solovev, A.A., Schulze, S., and Schmidt, O.G. (2011) Controlled manipulation of multiple cells using catalytic microbots. *Chem. Commun.*, **47** (2), 698–700.
60. Mei, Y., Huang, G., Solovev, A.A., Ureña, E.B., Mönch, I., Ding, F. *et al.* (2008) Versatile approach for integrative and functionalized tubes by strain engineering of nanomembranes on polymers. *Adv. Mater.*, **20** (21), 4085–4090.
61. Sanchez, S., Ananth, A.N., Fomin, V.M., Viehrig, M., and Schmidt, O.G. (2011) Superfast motion of catalytic

microjet engines at physiological temperature. *J. Am. Chem. Soc.*, **133** (38), 14860–14863.
62. Sanchez, S., Solovev, A.A., Harazim, S.M., and Schmidt, O.G. (2011) Microbots swimming in the flowing streams of microfluidic channels. *J. Am. Chem. Soc.*, **133** (4), 701–703.
63. Mei, Y., Solovev, A.A., Sanchez, S., and Schmidt, O.G. (2011) Rolled-up nanotech on polymers: from basic perception to self-propelled catalytic microengines. *Chem. Soc. Rev.*, **40** (5), 2109–2119.
64. Manesh, K.M., Cardona, M., Yuan, R., Clark, M., Kagan, D., Balasubramanian, S. *et al.* (2010) Template-assisted fabrication of salt-independent catalytic tubular microengines. *ACS Nano*, **4** (4), 1799–1804.
65. Gao, W., Sattayasamitsathit, S., Orozco, J., and Wang, J. (2011) Highly efficient catalytic microengines: template electrosynthesis of polyaniline/platinum microtubes. *J. Am. Chem. Soc.*, **133** (31), 11862–11864.
66. Gao, W., Sattayasamitsathit, S., Uygun, A., Pei, A., Ponedal, A., and Wang, J. (2012) Polymer-based tubular microbots: role of composition and preparation. *Nanoscale*, **4** (7), 2447–2453.
67. Sanchez, S., Solovev, A.A., Mei, Y., and Schmidt, O.G. (2010) Dynamics of biocatalytic microengines mediated by variable friction control. *J. Am. Chem. Soc.*, **132** (38), 13144–13145.
68. Balasubramanian, S., Kagan, D., Jack Hu, C.-M., Campuzano, S., Lobo-Castañon, M.J., Lim, N. *et al.* (2011) Micromachine-enabled capture and isolation of cancer cells in complex media. *Angew. Chem. Int. Ed.*, **50** (18), 4161–4164.
69. Kagan, D., Campuzano, S., Balasubramanian, S., Kuralay, F., Flechsig, G.-U., and Wang, J. (2011) Functionalized micromachines for selective and rapid isolation of nucleic acid targets from complex samples. *Nano Lett.*, **11** (5), 2083–2087.
70. Orozco, J., Campuzano, S., Kagan, D., Zhou, M., Gao, W., and Wang, J. (2011) Dynamic isolation and unloading of target proteins by aptamer-modified microtransporters. *Anal. Chem.*, **83** (20), 7962–7969.
71. Campuzano, S., Orozco, J., Kagan, D., Guix, M., Gao, W., Sattayasamitsathit, S. *et al.* (2012) Bacterial isolation by lectin-modified microengines. *Nano Lett.*, **12** (1), 396–401.
72. Solovev, A.A., Sanchez, S., Pumera, M., Mei, Y.F., and Schmidt, O.G. (2010) Magnetic control of tubular catalytic microbots for the transport, assembly, and delivery of micro-objects. *Adv. Funct. Mater.*, **20** (15), 2430–2435.
73. Solovev, A.A., Xi, W., Gracias, D.H., Harazim, S.M., Deneke, C., Sanchez, S. *et al.* (2012) Self-propelled nanotools. *ACS Nano*, **6** (2), 1751–1756.
74. Gao, W., Uygun, A., and Wang, J. (2012) Hydrogen-bubble-propelled zinc-based microrockets in strongly acidic media. *J. Am. Chem. Soc.*, **134** (2), 897–900.
75. Gao, W., D'agostino, M., Garcia-Gradilla, V., Orozco, J., and Wang, J. (2013) Multi-fuel driven janus micromotors. *Small*, **9** (3), 467–471.
76. Tierno, P., Golestanian, R., Pagonabarraga, I., Sagueìs F. (2008) Magnetically actuated colloidal microswimmers. *J. Phys. Chem. B*, **112** (51), 16525.
77. Sing, C.E., Schmid, L., Schneider, M.F., Franke, T., and Alexander-Katz, A. (2010) Controlled surface-induced flows from the motion of self-assembled colloidal walkers. *Proc. Natl. Acad. Sci. U.S.A.*, **107** (2), 535–540.
78. Zhang, L., Petit, T., Lu, Y., Kratochvil, B.E., Peyer, K.E., Pei, R. *et al.* (2010) Controlled propulsion and cargo transport of rotating nickel nanowires near a patterned solid surface. *ACS Nano*, **4** (10), 6228–6234.
79. Ishiyama, K., Sendoh, M., Yamazaki, A., and Arai, K.I. (2001) Swimming micro-machine driven by magnetic torque. *Sens. Actuators, A: Phys.*, **91** (1–2), 141–144.
80. Ghosh, A. and Fischer, P. (2009) Controlled propulsion of artificial magnetic

nanostructured propellers. *Nano Lett.*, **9** (6), 2243–2245.
81. Zhang, L., Peyer, K.E., and Nelson, B.J. (2010) Artificial bacterial flagella for micromanipulation. *Lab Chip*, **10** (17), 2203–2215.
82. Tottori, S., Zhang, L., Qiu, F., Krawczyk, K.K., Franco-Obregón, A., and Nelson, B.J. (2012) Magnetic helical micromachines: fabrication, controlled swimming, and cargo transport. *Adv. Mater.*, **24** (6), 811–816.
83. Gao, W., Sattayasamitsathit, S., Manesh, K.M., Weihs, D., and Wang, J. (2010) Magnetically powered flexible metal nanowire motors. *J. Am. Chem. Soc.*, **132** (41), 14403–14405.
84. Pak, O.S., Gao, W., Wang, J., and Lauga, E. (2011) High-speed propulsion of flexible nanowire motors: theory and experiments. *Soft Matter*, **7** (18), 8169–8181.
85. Dreyfus, R., Baudry, J., Roper, M.L., Fermigier, M., Stone, H.A., and Bibette, J. (2005) Microscopic artificial swimmers. *Nature*, **437** (7060), 862–865.
86. Velev, O.D., Gangwal, S., and Petsev, D.N. (2009) Particle-localized ac and dc manipulation and electrokinetics. *Annu. Rep. Prog. Chem. Sect. C: Phys. Chem.*, **105**, 213–246.
87. Chang, S.T., Paunov, V.N., Petsev, D.N., and Velev, O.D. (2007) Remotely powered self-propelling particles and micropumps based on miniature diodes. *Nat. Mater.*, **6** (3), 235–240.
88. Calvo-Marzal, P., Sattayasamitsathit, S., Balasubramanian, S., Windmiller, J.R., Dao, C., and Wang, J. (2010) Propulsion of nanowire diodes. *Chem. Commun.*, **46** (10), 1623–1624.
89. Zhang, C., Khoshmanesh, K., Mitchell, A., and Kalantar-Zadeh, K. (2010) Dielectrophoresis for manipulation of micro/nano particles in microfluidic systems. *Anal. Bioanal. Chem.*, **396** (1), 401–420.
90. Krupke, R., Hennrich, F., Löhneysen, H.V., and Kappes, M.M. (2003) Separation of metallic from semiconducting single-walled carbon nanotubes. *Science*, **301** (5631), 344–347.
91. Pohl, H.A. and Hawk, I. (1966) Separation of living and dead cells by dielectrophoresis. *Science*, **152** (3722), 647–649.
92. Zhang, K.-Q. and Liu, X.Y. (2004) In situ observation of colloidal monolayer nucleation driven by an alternating electric field. *Nature*, **429** (6993), 739–743.
93. Takei, A., Matsumoto, K., and Shomoyama, I. (2010) Capillary motor driven by electrowetting. *Lab Chip*, **10** (14), 1781–1786.
94. Comninellis, C., Plattner, E., and Bolomey, P. (1991) Estimation of current bypass in a bipolar electrode stack from current-potential curves. *J. Appl. Electrochem.*, **21** (5), 415–418.
95. Saakes, M., Woortmeijer, R., and Schmal, D. (2005) Bipolar lead–acid battery for hybrid vehicles. *J. Power Sources*, **144** (2), 536–545.
96. Loget, G. and Kuhn, A. (2011) Shaping and exploring the micro- and nanoworld using bipolar electrochemistry. *Anal. Bioanal. Chem.*, **400** (6), 1691–1704.
97. Mavré, F., Anand, R.K., Laws, D.R., Chow, K.-F., Chang, B.-Y., Crooks, J.A. *et al.* (2010) Bipolar electrodes: a useful tool for concentration, separation, and detection of analytes in microelectrochemical systems. *Anal. Chem.*, **82** (21), 8766–8774.
98. Loget, G., Zigah, D., Bouffier, L., Sojic, N., and Kuhn, A. (2013) Bipolar electrochemistry: from materials science to motion and beyond. *Acc. Chem. Res.*, **46** (11), 2513–2523.
99. Fleischmann, M., Ghoroghchian, J., and Pons, S. (1985) Electrochemical behavior of dispersions of spherical ultramicroelectrodes. 1. Theoretical considerations. *J. Phys. Chem.*, **89** (25), 5530–5536.
100. Mavré, F., Chow, K.-F., Sheridan, E., Chang, B.-Y., Crooks, J.A., and Crooks, R.M. (2009) A theoretical and experimental framework for understanding electrogenerated chemiluminescence (ECL) emission at bipolar electrodes. *Anal. Chem.*, **81** (15), 6218–6225.
101. Loget, G. and Kuhn, A. (2013) Bipolar electrochemistry in the nanosciences,

in *Specialist Periodical Reports Electrochemistry* (eds R.G. Compton and J.D. Wadhawans), pp. 71–103.
102. Loget, G., Roche, J., and Kuhn, A. (2012) True bulk synthesis of janus objects by bipolar electrochemistry. *Adv. Mater.*, **24**, 5111–5116.
103. Bradley, J.-C. and Ma, Z. (1999) Contactless electrodeposition of palladium catalysts. *Angew. Chem. Int. Ed.*, **38** (11), 1663–1666.
104. Bradley, J.-C., Babu, S., and Ndungu, P. (2005) Contactless tip-selective electrodeposition of palladium onto carbon nanotubes and nanofibers. *Fullerenes Nanotubes Carbon Nanostruct.*, **13**, 227–237.
105. Loget, G., Lapeyre, V.R., Garrigue, P., Warakulwit, C., Limtrakul, J., Delville, M.-H. *et al.* (2011) Versatile procedure for synthesis of janus-type carbon tubes. *Chem. Mater.*, **23** (10), 2595–2599.
106. Warakulwit, C., Nguyen, T., Majimel, J., Delville, M.-H., Lapeyre, V., Garrigue, P. *et al.* (2008) Dissymmetric carbon nanotubes by bipolar electrochemistry. *Nano Lett.*, **8** (2), 500.
107. Loget, G., Larcade, G., Lapeyre, V., Garrigue, P., Warakulwit, C., Limtrakul, J. *et al.* (2010) Single point electrodeposition of nickel for the dissymmetric decoration of carbon tubes. *Electrochim. Acta*, **55** (27), 8116–8120.
108. Fattah, Z., Garrigue, P., Lapeyre, V., Kuhn, A., and Bouffier, L. (2012) Controlled orientation of asymmetric copper deposits on carbon microobjects by bipolar electrochemistry. *J. Phys. Chem. C*, **116** (41), 22021–22027.
109. Fattah, Z., Loget, G., Lapeyre, V., Garrigue, P., Warakulwit, C., Limtrakul, J. *et al.* (2011) Straightforward single-step generation of microswimmers by bipolar electrochemistry. *Electrochim. Acta*, **56** (28), 10562–10566.
110. Wang, Y., Fei, S.-T., Byun, Y.-M., Lammert, P.E., Crespi, V.H., Sen, A. *et al.* (2009) Dynamic interactions between fast microscale rotors. *J. Am. Chem. Soc.*, **131** (29), 9926.
111. Loget, G. and Kuhn, A. (2011) Electric field-induced chemical locomotion of conducting objects. *Nat. Commun.*, **2**, 535.
112. Loget, G. and Kuhn, A. (2012) Bipolar electrochemistry for cargo-lifting in fluid channels. *Lab Chip*, **12**, 1967–1971.
113. Miao, W., Choi, J.-P., and Bard, A.J. (2002) Electrogenerated chemiluminescence 69: the tris(2,2′-bipyridine)ruthenium(II), (Ru(bpy)32+)/tri-n-propylamine (TPrA) system revisited a new route involving $TPrA^{*+}$ cation radicals. *J. Am. Chem. Soc.*, **124** (48), 14478–14485.
114. Sentic, M., Loget, G., Manojlovic, D., Kuhn, A., and Sojic, N. (2012) Light-emitting electrochemical "swimmers". *Angew. Chem. Int. Ed.*, **51** (45), 11284–11288.
115. Bouffier, L., Zigah, D., Adam, C., Sentic, M., Fattah, Z., Manojlovic, D. *et al.* (2014) Lighting up redox propulsion with luminol electrogenerated chemiluminescence. *ChemElectroChem*, **1** (1), 95–98.
116. Loget, G. and Kuhn, A. (2010) Propulsion of microobjects by dynamic bipolar self-regeneration. *J. Am. Chem. Soc.*, **132** (45), 15918–15919.
117. Argoul, F. and Kuhn, A. (1993) Experimental demonstration of the origin of interfacial rhythmicity in electrodeposition of zinc dendrites. *J. Electroanal. Chem.*, **359** (1–2), 81–96.
118. Rea, M.C. (1995) The problem of material constitution. *Philos. Rev.*, **104** (4), 525–552.
119. Jiang, Y., Douglas, N.R., Conley, N.R., Miller, E.J., Frydman, J., and Moerner, W.E. (2011) Sensing cooperativity in ATP hydrolysis for single multisubunit enzymes in solution. *Proc. Natl. Acad. Sci. U.S.A.*, **108** (41), 16962–16967.

15
Azobenzene in Molecular and Supramolecular Devices and Machines

Massimo Baroncini and Giacomo Bergamini

15.1
Introduction

> But we are still blind ... blind and we don't have those tweezers we often dream of at night, the way a thirsty man dreams of springs, that would allow us to pick up a segment, hold it firm, and straight, and paste it in the right direction on the segment that has already been assembled. If we had those tweezers (and it's possible that, one day, we will), we would have managed to create some wonderful things. But for the present we don't have those tweezers, and when we come right down to it, we're bad riggers.
>
> *(The Wrench, Primo Levi)*

A current topic in supramolecular chemistry subject to intensive research and rapid expansion is the design of synthetic nanomachines that are able to carry out movements at the molecular and supramolecular scale triggered by external stimuli [1, 2]. Artificial nanomachines have been realized on the laboratory scale [3], and utilization of such systems to construct responsive materials [4] and surfaces [5], control catalytic processes [6], and develop test structures for information storage devices [7] and drug delivery [8] has been investigated.

In analogy to their macroscopic counterparts, molecular devices and machines need energy to operate and signal to communicate with the operator. Light provides an answer to this dual requirement. Indeed, a great number of molecular devices and machines are powered by light-induced processes, and light can also be useful to "read" the state of the system and thus to control and monitor its operation. In this regard, light energy possesses a number of advantages: (i) the amount of energy conferred to a chemical system by using photons can be carefully controlled by the wavelength and intensity of the exciting light, in relation to the absorption spectrum of the targeted species and (ii) such an energy can be transmitted to molecules without physically connecting them to the source (no "wiring" is necessary), the only requirement being the transparency of the matrix at the excitation wavelength. In this context, the readily induced and reversible

Discovering the Future of Molecular Sciences, First Edition. Edited by Bruno Pignataro.

trans–cis photoisomerization of the azo bond and the geometric changes that result when azo chromophores are incorporated into supramolecular systems and other materials allow systems incorporating azobenzenes to be used as photoswitches, effecting rapid and reversible control over a variety of chemical, mechanical, electronic, and optical properties. In this chapter, we report four examples in which azobenzene moieties are part of molecular and supramolecular architectures where photoisomerization controls molecular movements and nanoscale interactions.

15.2 Dendrimers

Dendrimers [9] are branched tree-like macromolecules that exhibit a defined structure and a high degree of order and complexity. Originally, the synthesis of such aesthetically pleasant macromolecules was really a challenge, but nowadays the development of the dendrimers chemistry has moved the interest in applications in such different fields as medicine, biology, and chemistry. By using appropriate synthetic strategies, it is possible to prepare dendrimers that contain select functional units in predetermined sites of their structure. It is thus possible to construct large nano-objects capable of performing complex functionalities that arise from the integration of the specific properties of the constituent moieties. Furthermore, in these supramolecular organized structures, dynamic three-dimensional cavities are present in which ions and small molecules can be hosted.

In this chapter, we report two examples in which the azobenzene units are placed in the core and in the periphery of the dendrimer structure. Dendrimers containing azobenzene groups [10, 11] in the core, branching points, or periphery can modify their structure and flexibility according to the isomerization state of the azobenzene units. In particular, structural changes in the peripheral units of a dendrimer can change the surface properties and cause rearrangements in the internal cavities. For all these reasons, dendrimers bearing azobenzene groups in the periphery could play the role of photoswitchable hosts.

15.2.1 Azobenzene at the Periphery

The first example is a fourth-generation dendrimer (**D**) of the poly(propylene amine) family functionalized with naphthyl and *trans*-azobenzene units. The investigated dendrimer (Scheme 15.1) comprises 30 tertiary amine units in the interior, and 32 naphthyl, and 32 *trans*-azobenzene units in the periphery. The coactor of the story is eosin Y (2′,4′,5′,7′-tetrabromofluorescein dianion, hereafter simply called eosin and indicated by **E**). We report, in particular, the studies of the uptake and release of **E** by the dendrimer **D** and the role of the photoisomerization process [12].

Dendrimer **D**(32t) (Scheme 15.1) is a fluorescent and photoreactive compound. The fluorescence of the naphthalene units is partially quenched by the tertiary amines (via electron transfer) as well as by the *trans*- and *cis*-azobenzene units

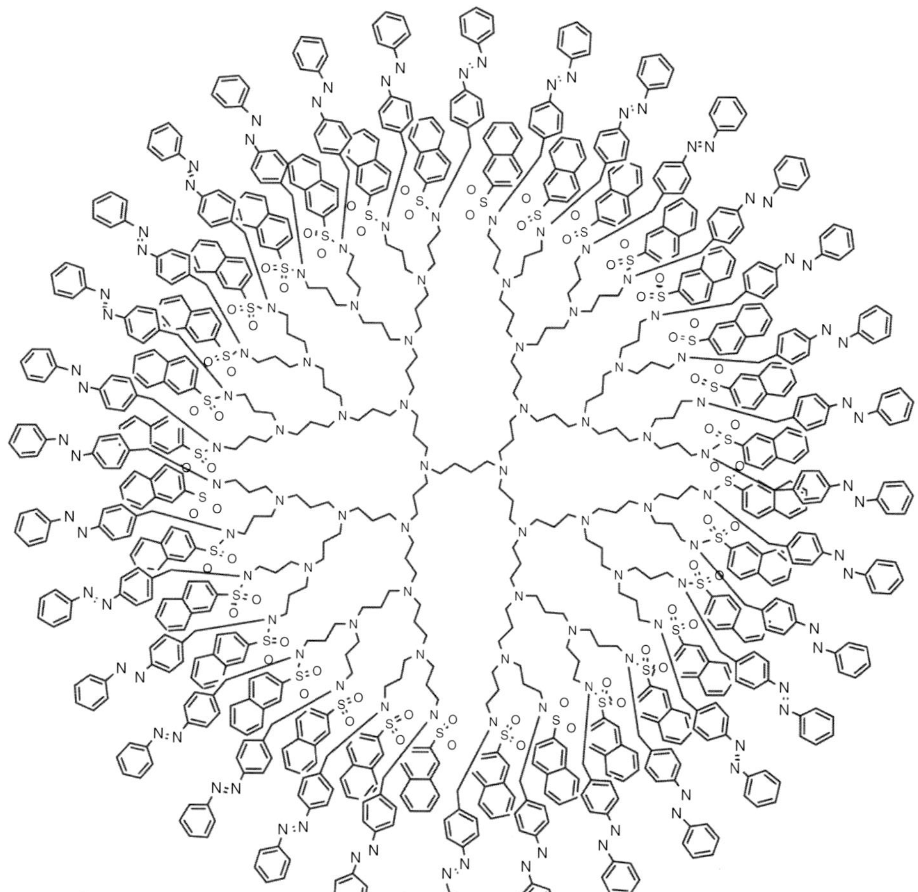

D(32t)

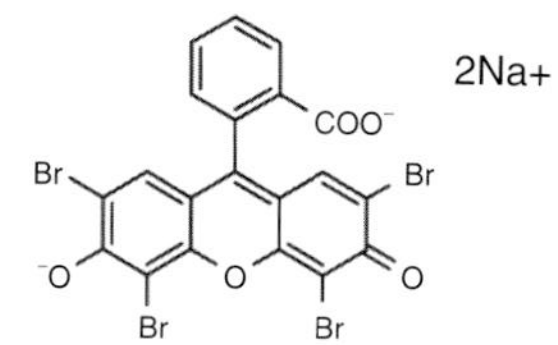

E

Scheme 15.1 Structure formulas of dendrimer **D** in its all-trans form **D**(32t) and eosin, **E**.

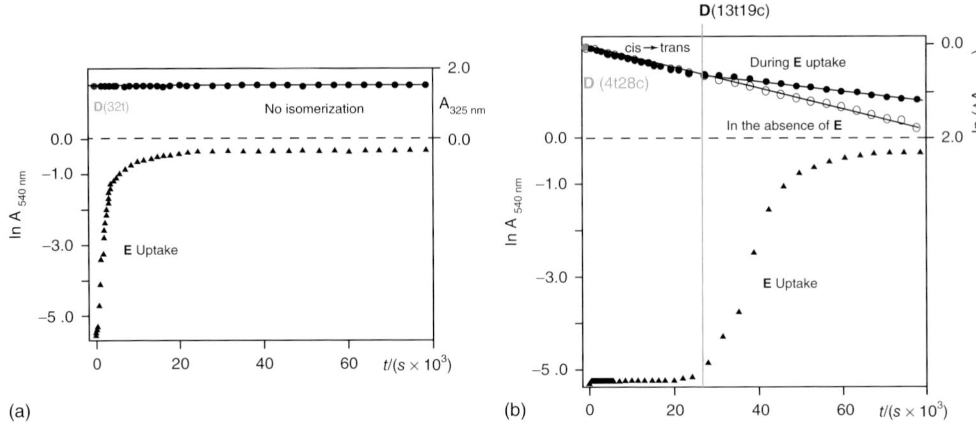

Figure 15.1 Kinetics of eosin uptake (full triangles, $A_{540\ nm}$) and cis → trans isomerization (full circles, $A_{325\ nm}$) for a dichloromethane solution of (a) **D**(32t) and (b) **D**(4t28c) in contact with a water solution of **E** at pH 10 at 298 K. For comparison purposes, cis → trans isomerization (open circles, panel b, $A_{325\ nm}$) for a dichloromethane solution of **D**(4t28c) not in contact with the water phase is reported.

(via energy transfer) [13]. In dichloromethane solution at 298 K on irradiation with 365 nm light, the all-*trans* dendrimer **D**(32t) is converted into species containing, as an average, 4 *trans*- and 28 *cis*-azobenzene units, **D**(4t28c).[1)]

It is well known [14, 15] that poly(propylene amine) dendrimers in dichloromethane solution can extract **E** from aqueous solution. The number of **E** extracted is strongly dependent on the generation number and pH of the aqueous phase, but is less sensitive to the nature of the units appended in the periphery. In the present case, dichloromethane solutions of dendrimer **D** are mixed and shaken with aqueous solutions of the disodium eosin salt at pH 7.0 (phosphate buffer). This pH value has been chosen because eosin is present only as its $\mathbf{E}^{2-}$ form and extraction is more efficient. Indeed, according to previous studies [14c], extraction efficiency is maximum between pH 5 and 7 and is practically negligible at pH 12. No extraction occurs when the dendrimers are not present in the dichloromethane phase. The number of eosin molecules extracted can be easily measured from the changes in the absorbance of the water phase at 540 nm, where **E** exhibits an intense absorption band.

To investigate the kinetics of eosin extraction as a function of azobenzene isomerization state, dichloromethane solutions of **D**(32t) and **D**(4t28c) were brought into contact with a water solution of **E** at pH 10. The water solution was buffered at pH 10 and not 7, as in the previous experiments, to slow down the extraction rate constant and make easier the detection.

As reported in Figure 15.1, substantial differences can be observed: in the case of **D**(32t), eosin uptake (triangles in Figure 15.1a) starts as soon as the two solutions

1) The absorption spectra indicate the overall number of *trans*- and *cis*-azobenzene moieties present in the experimental condition used.

are in contact, while for **D**(4t28c) a delay is observed (triangles in Figure 15.1b) and no eosin extraction takes place until a significant fraction of azobenzene units is converted from cis to trans, that is, when the **D**(13t19c) species is formed. Furthermore, eosin extraction is faster in the first case. It is noteworthy that the rate constant for the thermal cis → trans isomerization, monitored by absorbance changes at 325 nm, is the same, starting with a dichloromethane solution of **D**(4t28c) alone (empty circles in Figure 15.1b) or in contact with a water solution of **E** (full circles in Figure 15.1b) until no extraction of **E** takes place, while it is significantly slowed down as soon as eosin molecules are encapsulated in the dendrimers. Such a decrease of the cis → trans thermal isomerization rate is likely related to steric effects and suggests that the presence of the guest molecules modifies the dendrimer structure (see below).

In order to understand whether the isomeric form of the azobenzene units is related to the release of the guest, we have also compared the rate of eosin release in dichloromethane solution at 298 K from a **D**(32t) ⊃ 6**E** species (open triangles in Figure 15.2a) and a **D**(4t28c) ⊃ 6**E** (full triangles in Figure 15.2a), both of them prepared by the extraction of **E** from an aqueous solution at pH 7. We have found that **E** is released from the beginning following a first order process with the rate constant $1 \times 10^{-5}\,s^{-1}$. In an additional experiment, we have compared the rate of eosin release at 298 K from dichloromethane solutions of **D**(32t) ⊃ 6**E** and **D**(4t28c) ⊃ 6**E** in contact with an aqueous phase at pH 10. As shown in Figure 15.2b, eosin release occurs by a first-order process that is about four times faster in the case of **D**(32t) ⊃ 6**E** ($k = 1.33 \times 10^{-5}\,s^{-1}$ vs $3.09 \times 10^{-6}\,s^{-1}$). The results shown in Figure 15.2 demonstrate that the presence of *cis*-azobenzene units in the periphery of the dendrimer does reduce the rate of eosin release, as expected, because of the larger steric hindrance caused by the cis isomer compared with the trans isomer.

These experiments demonstrate that the isomerization state of the peripheral azobenzene units controls, to some degree, the *permeability* of the dendrimer cavities to **E** and, vice versa, eosin molecules hosted in the dendrimer cavities affect the velocity of thermal isomerization process of its peripheral azobenzene

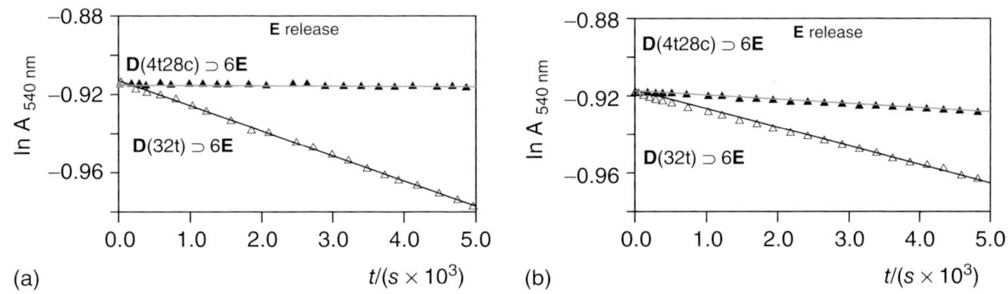

Figure 15.2 Kinetics of eosin release: (a) in dichloromethane solution at 313 K, from **D**(4t28c) ⊃ 6**E** (full circles) and **D**(32t) ⊃ 6**E** (open circles) and (b) in dichloromethane solution in contact with water at 298 K, from **D**(32t) ⊃ 6**E** (full triangles) and **D**(4t28c) ⊃ 6**E** (open triangles). Eosin release is measured in both cases from the decrease of absorbance at 540 nm in dichloromethane.

units. The results obtained suggest that a more extensive study of dendrimers with isomerizable azobenzene units in the periphery may lead to photocontrollable membranes and drug delivery systems.

15.2.2
Azobenzene at the Core

The second example of combination of dendrimer chemistry and photoisomerizable unit is a family of two dendrimers (Scheme 15.2) containing two coordinating units (1,4,8,11-tetraazacyclotetradecane, hereafter called cyclam) [16], linked by a

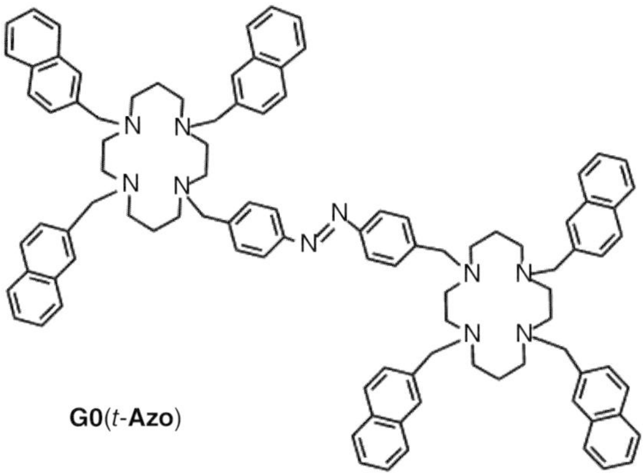

G0(*t*-**Azo**)

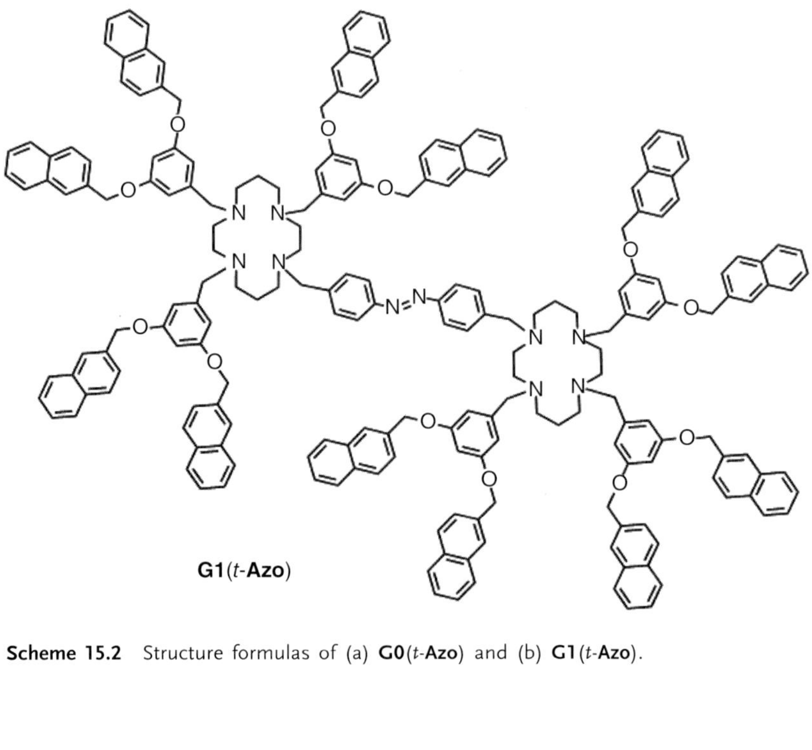

G1(*t*-**Azo**)

Scheme 15.2 Structure formulas of (a) **G0**(*t*-**Azo**) and (b) **G1**(*t*-**Azo**).

photoswitching azobenzene moiety, and functionalized at the periphery with 6 or 12 light-harvesting (naphthalene) chromophores in **G0**(*t*-**Azo**) and **G1**(*t*-**Azo**), respectively [17]. Because of their proximity, the various functional groups of the dendrimers interact: the azobenzene unit enables to control the distance between the two cyclam moieties on light stimulation. This leads to different coordination properties for the cis and trans isomers.

The absorption spectra of **G0**(*t*-**Azo**) and **G1**(*t*-**Azo**) in CH_3CN/CH_2Cl_2 1 : 1 (v/v) are similar to those obtained by the sum of the component spectra. They show the characteristic bands of both naphthalene at 275 nm and azobenzene at 336 ($\pi\pi^*$ transition) and 450 nm ($n\pi^*$ transition). The lowest energy excited state is localized on the azobenzene moiety. Regarding the emission properties, by excitation at 275 nm where most of the light is absorbed by the naphthalene chromophores, **G0**(*t*-**Azo**) and **G1**(*t*-**Azo**) exhibit a very weak luminescence. The former shows two bands at 335 and 470 nm, and the latter exhibits a maximum at 335 nm together with a shoulder at 400 nm. In analogy with the behavior previously reported for the two dendrimers containing only a cyclam core [18], we can assign the different emission bands to naphthyl localized excited states (λ_{max} = 335 nm), naphthyl excimers (λ_{max} about 390 nm), and naphthyl-amine exciplexes (λ_{max} = 470 nm) as a result of the interaction of an excited naphthalene with an amine group of the cyclam unit. The emission spectrum of **G1**(*t*-**Azo**) does not show the band at 470 nm: this result is likely due to the fact that energy transfer to the azobenzene core (see below) is faster than exciplex formation, whereas in **G0**(*t*-**Azo**), the close proximity of naphthalene and cyclam nitrogens enable exciplex formation in competition with energy transfer to the azobenzene core.

The photochemical experiments pointed out as on irradiation at 365 nm, where only azobenzene is absorbing light, the absorption spectra of **G0**(*t*-**Azo**) and **G1**(*t*-**Azo**) show a decrease of the $\pi\pi^*$ band at 336 nm and an increase of the $n\pi^*$ band of azobenzene. These spectral changes are characteristic of trans → cis isomerization. The *cis*-azobenzene species can be converted back to the trans isomer by irradiation at 436 nm in the $n\pi^*$ band.

In order to elucidate the interaction between naphthalene and azobenzene in the dendritic structures, photoisomerization of azobenzene has been investigated on selective excitation of the naphthalene at 275 nm, where more than 95% of the light is absorbed by the naphthalene for **G1**(*t*-Azo) and **G1**(*c*-Azo). The results show that azobenzene isomerization takes place, demonstrating that energy transfer from the naphthalene to the trans isomer of the azobenzene core occurs with an efficiency of 20 and 40% for **G0**(*t*-**Azo**) and **G1**(*t*-**Azo**), respectively.

On irradiation at 365 nm, the photochemical quantum yield $\Phi_{t\to c}$ and the percentage of trans at the photostationary state obtained are identical to those reported in the literature for a *trans*-azobenzene. Small differences are observed for **G1**(*t*-**Azo**): a lower value of $\Phi_{t\to c}$ on irradiation at 365 nm, and larger fraction of trans in the photostationary state is indicative of a higher stability of the trans isomer compared to the cis one in **G1**(*t*-**Azo**) with respect to **G0**(*t*-**Azo**).

It is well known that cyclam, one of the most extensively investigated ligands in coordination chemistry [19], is able to strongly bind a wide variety of metal ions.

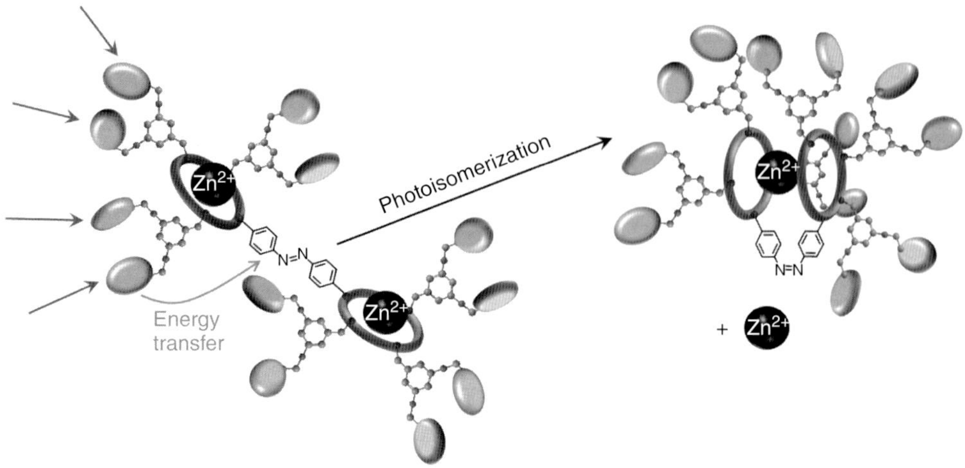

Figure 15.3 Schematic representation of the functions performed by **G1(Azo)**: different coordination ability by the cyclam moieties (solid circles) of **G1**(*t*-**Azo**) and **G1**(*c*-**Azo**), light harvesting by naphthalene units (gray ovals) and sensitized photoisomerization of the core azobenzene.

By monitoring the photophysical variations of the dendrimers on additions of a $M(CF_3SO_3)_2$ salt, it is possible to evaluate the stoichiometry and the association constants of the metal complex formed. In particular, the addition of Zn^{2+} to a solution of **G0**(*t*-**Azo**) causes a blue shift and an increase in the intensity of the $\pi\pi^*$ band of azobenzene together with a decrease of absorbance at 260 nm with isosbestic points at 268 and 342 nm. The plot of normalized absorption changes at 304 nm versus the amount of Zn(II) is almost linear and exhibits a plateau at about 2.3 equivalents of Zn(II) per dendrimer, demonstrating that up to two metal ions can be coordinated, that is, one per cyclam unit. By the global fitting of the absorption changes, an estimate of the association constants can be obtained: $K_1 = 7 \times 10^7\ M^{-1}$, $K_2 = 5 \times 10^6\ M^{-1}$ for 1 : 1 and 2 : 1 metal/dendrimer stoichiometry. In the 1 : 1 complex, only one of the two cyclam units is linked to Zn(II) because the *trans*-azobenzene unit keeps the two cyclam units quite far apart (Figure 15.3).[2] Titration with $Zn(CF_3SO_3)_2$ has also been performed on the solution obtained at the photostationary state after irradiation at 365 nm, where the ratio of cis and trans isomers is 95 : 5. The absorbance change at 304 nm reaches a plateau at about 1.4 equivalents of Zn(II) ions per dendrimer, instead of 2.3 equivalents observed in the case of the trans isomer. These results can be rationalized on the basis of a structural rearrangement that brings the two cyclam units much closer, so that one Zn(II) ion can be coordinated by both of them (Figure 15.3).[3] The coordination of the second Zn(II) ion is highly disfavored by the electrostatic repulsion because of the close proximity of two 2^+ ions. By fitting the absorption changes and taking into

2) HyperChem(TM) Professional 7.51 has been used to visualize the geometry of the 1 : 1 complex.
3) Since the coordination number of Zn(II) is lower than 8, the two cyclam moieties are forced to adopt a structure in which not all of the four N atoms are coordinated to a Zn(II) at the same time.

account the small percentage of trans isomer present at the photostationary state, the association constant is estimated as $K_1 = 1 \times 10^8\ M^{-1}$. This value is slightly higher than that found for the trans isomer, evidencing a positive effect of the second cyclam unit on the stability constant.

Qualitatively similar results are obtained in the case of **G1**(*t*-**Azo**) on titration with $Zn(CF_3SO_3)_2$: $K_1 = 7 \times 10^7\ M^{-1}$, $K_2 = 1 \times 10^5\ M^{-1}$. On titration with $Zn(CF_3SO_3)_2$ of the photostationary state of **G1**(*c*-**Azo**) obtained after the irradiation at 365 nm, and the absorbance variations at 304 nm are very similar to those reported for **G0**(*c*-**Azo**).

Photochemical experiments on the metal complexes of both dendrimers have been performed on addition of an excess of Zn(II) ions (6 equivalents per dendrimer). The presence of two metal ions per dendrimer in the trans isomer disfavors the trans → cis photoreaction, decreasing the value of $\Phi_{t \to c}$ on irradiation at 365 nm and the molar fraction of cis isomer at the photostationary state on irradiation at 365.

Regarding the isomerization sensitized by naphthalene excitation, a strong increase of the efficiency of energy transfer (η_{ET}) from naphthalene to azobenzene is observed with unitary values of η_{ET} for metal complexes of **G0**(*t*-**Azo**). This result is consistent with the fact that coordination of Zn(II) to the cyclam prevents exciplex formation.

15.3 Molecular Devices and Machines

The control of motion on the molecular scale is of fundamental importance for living organisms [20] and is one of the most fascinating challenges in nanoscience [21, 22]. Artificial molecular machines have been realized on the laboratory scale [2, 3], and utilization of such systems to construct responsive materials [4, 23] and surfaces [5], control catalytic processes [6], and develop test structures for information storage devices [7] and drug delivery [8] has been investigated. Nevertheless, the construction of synthetic nanoscale motors capable of showing directionally controlled linear or rotary movements still poses a considerable challenge to chemists [24]. Although a few examples of artificial molecular rotary motors [6, 25] and DNA-based linear motors [26] have been described, only one prototype of a fully synthetic linear motor molecule is available [27]. Moreover, the use of such systems to perform the tasks that natural molecular motors do [28] – particularly, active transport of substrates over long distances or across membranes – remains a very difficult endeavor, further complicated by the fact that most currently available synthetic molecular motors are based on sophisticated chemical structures and/or operation procedures [2, 3, 24].

In this context, the development of (supra)molecular systems that exhibit directionally controlled relative motions of their components, based on a minimalist design, and activated by convenient inputs, is of the highest importance. Particularly suitable species for the construction of linear molecular machines are pseudorotaxanes, supramolecular complexes minimally composed of an axle-like molecule surrounded by a macrocycle, and rotaxanes, structurally similar to pseudorotaxanes except for the presence of bulky groups at the extremities of the axle to prevent dethreading and render these systems kinetically inert [29, 30]. The operation of most pseudorotaxane and rotaxane machines is based on classical switching processes between thermodynamically stable states [3, 31]. It has become clear, however, that functional molecular motors will be realized only if the reaction rates between states can be controlled, thus enabling the implementation of ratchet-type mechanisms [32, 33]. Therefore, the ability of adjusting the threading–dethreading kinetics by modulating the corresponding energy barriers through external stimulation is an important goal. In this section, two examples of light-controlled molecular devices and machines are described. They present some small, although interesting, advancements in exploiting the peculiar properties of azobenzenes interaction with light to obtain useful functions.

15.3.1 Switching Rotaxane Character with Light

Here, we describe a self-assembling system that can be switched between thermodynamically stable (pseudorotaxane) and kinetically inert (rotaxane) forms by light irradiation. Compared with previous related studies, the present system is advantageous because (i) it can be controlled by light, (ii) it exhibits a clear-cut behavior, (iii) it is fully reversible, and (iv) it is structurally simple and easy to make.

The system is composed by the bis-azobenzylamine axle *E,E*-**1**H$^+$ and the dibenzo[24]crown-8 ether ring **DB24** (Scheme 15.3). The axle was synthesized in good yields by Mill's coupling of bis-aminodibenzylamine with *p*-methylnitrosobenzene in acetic acid, followed by anion exchange with NH_4PF_6. The ^1H NMR spectrum of a 1 : 1 mixture of *EE*-**1**H$^+$ and **DB24** in CD_3CN at room temperature shows sharp and well-resolved signals associated with the 1 : 1 complex [*EE*-**1**H ⊂ **DB24**]$^+$ and the free axle and ring components, as a result of slow chemical exchange. The stability constant of [*EE*-**1**H ⊂ **DB24**]$^+$ (K_{EE}) was

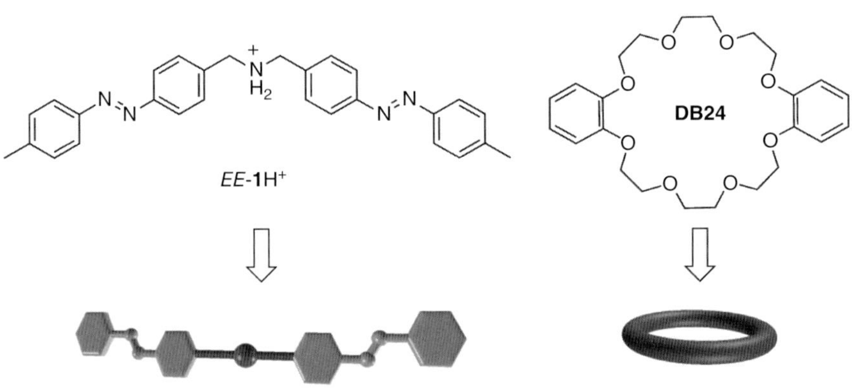

Scheme 15.3 Structure formulas of *EE*-**1**H·PF_6 and **DB24**.

obtained in different solvents at 298 K by single-point determination from ^{1}H NMR spectra. In acetonitrile, which is the solvent of choice for all the successive experiments, $K_{EE} = 800\ M^{-1}$. The equilibrium between the complex and its free components was established within the time necessary for mixing and recording the NMR spectrum. Hence, we can estimate a lower-limiting value for the threading rate constant $k(\text{in})_{EE}$ of about $5\ M^{-1}\ s^{-1}$.

Photoirradiation of *EE*-**1**H$^+$ at 365 nm afforded almost quantitative (>95%) isomerization to the *ZZ* isomer. The values determined for the $E \rightarrow Z$ photoisomerization quantum yield (0.095) and rate constant for $Z \rightarrow E$ dark isomerization ($1.4 \times 10^{-6}\ s^{-1}$) are typical of the azobenzene unit and are virtually unaffected by the presence of ring **DB24**. Thus, both end groups of 1H$^+$ – either alone or surrounded by **DB24** – can be effectively switched between the *E* and *Z* isomers.

Inspection of molecular models suggested that the *Z*-azobenzene unit could exert a stronger steric hindrance compared with *E*-azobenzene for the threading of **DB24** on **1**H$^+$. Indeed, no sign of complexation was observed by ^{1}H NMR soon after 1 : 1 mixing of **DB24** and *ZZ*-**1**H$^+$ in CD_3CN (Figure 15.4b). However, very slow changes, consistent with the formation of the [*ZZ*-**1**H ⊂ **DB24**]$^+$ complex, took place (Figure 15.5). The final spectrum, obtained 24 h after mixing, is identical to that shown in Figure 15.4a. A kinetic analysis of the NMR spectra afforded a rate constant $k(\text{in})_{ZZ} = 2.7 \times 10^{-3}\ M^{-1}\ s^{-1}$ while the stability constant, determined at equilibrium, resulted in $K_{ZZ} = 400\ M^{-1}$. On the basis of these data, we calculated a dethreading rate constant for the *ZZ* isomer, $k(\text{out})_{ZZ} = 6.8 \times 10^{-6}\ s^{-1}$. Therefore, the disassembly of the [*ZZ*-**1**H ⊂ **DB24**]$^+$ complex is three orders of magnitude

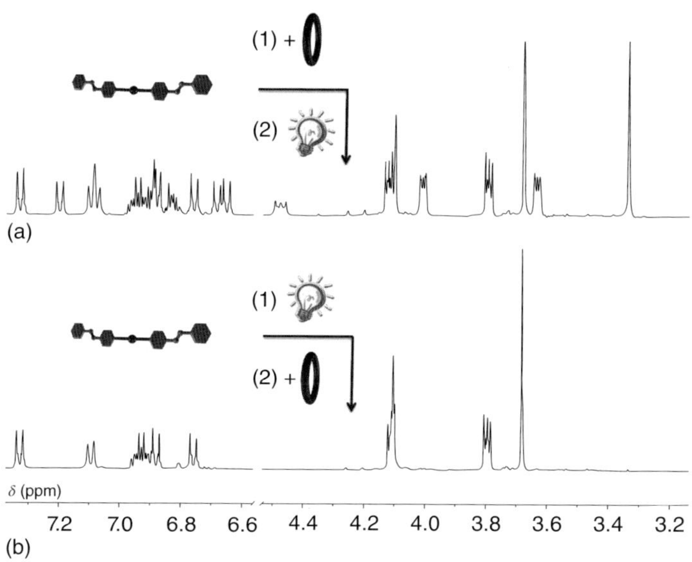

Figure 15.4 ^{1}H NMR spectra (CD_3CN, 298 K, 3 mM) of *EE*-**1**H$^+$ after: (a) addition of 1 equivalent of **DB24** and exhaustive irradiation at 365 nm and (b) exhaustive irradiation at 365 nm and addition of 1 equivalent of **DB24**.

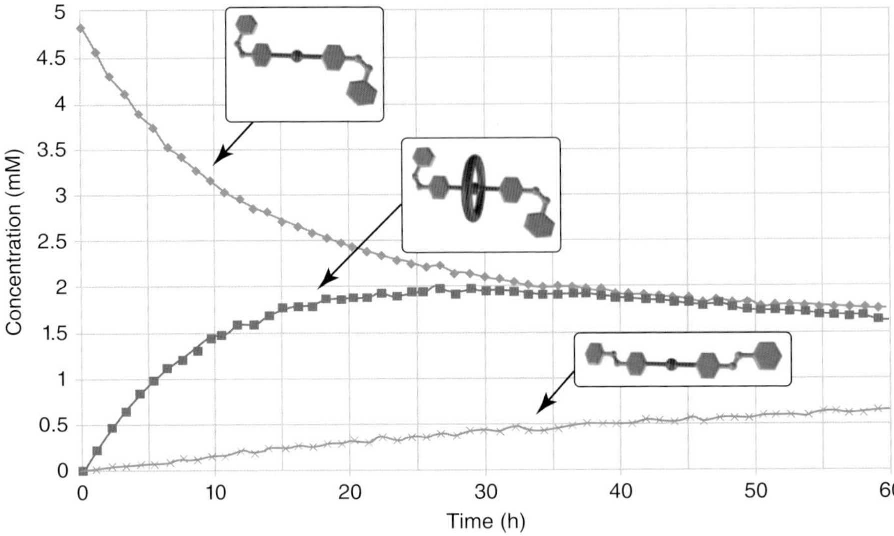

Figure 15.5 Concentration–time profiles, obtained from ^{1}H NMR spectroscopic data (CD_3CN, 298 K), of free (diamonds) and complexed (squares) *ZZ*-**1**H^+ on mixing *ZZ*-**1**H^+ (4.8 mM) and **DB24** (5.4 mM). The thermal regeneration of *EE*-**1**H^+ is also shown (stars). The solid lines represent the data fitting according to the model shown in Scheme 15.4.

slower than that of [*EE*-**1**H ⊂ **DB24**]$^+$ and takes place on the same timescale of the $Z \rightarrow E$ thermal isomerization. In other words, light irradiation transforms the [*EE*-**1**H ⊂ **DB24**]$^+$ pseudorotaxane into a rotaxane species, which remains kinetically interlocked for days (Scheme 15.4).

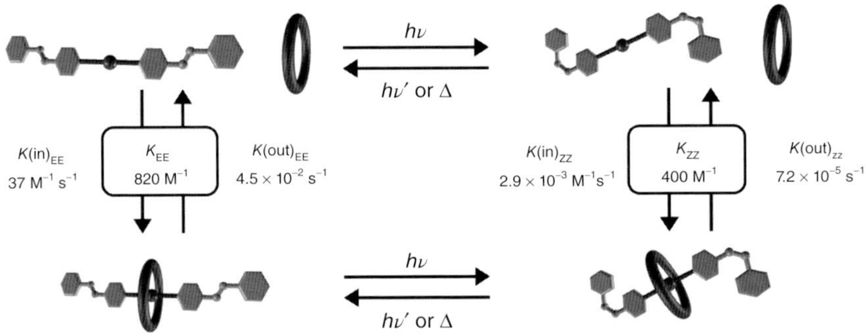

Scheme 15.4 Representation of the chemical equilibria and photochemical reactions involving components **1**H^+ and **DB24**.

This system is particularly appropriate for investigating if *Z*-azobenzene units can prevent dethreading when the intercomponent interactions between the ring and the axle are removed because such interactions can be virtually cancelled by deprotonating the ammonium center of the axle with a base, that is, using a stimulus orthogonal to that employed for switching the azobenzene end units. Interestingly, deprotonation of the ammonium center of [*ZZ*-**1**H ⊂ **DB24**]$^+$ (or [*EE*-**1**H ⊂ **DB24**]$^+$) with triethylamine resulted in a quantitative and fast dethreading. This observation indicates that the dethreading barrier is dramatically lowered by the thermodynamic destabilization of the threaded state. Such an effect is a consequence of the proximity of the ammonium center and azobenzene pseudo-stoppers. To attempt the dethreading of [*ZZ*-**1**H ⊂ **DB24**]$^+$ without weakening the intercomponent interactions, K^+ ions used as a competitive guest for crown ether **DB24** can be employed. Indeed, the addition of 2 equivalents of KPF_6 to the [*EE*-**1**H ⊂ **DB24**]$^+$ pseudorotaxane caused the immediate disassembly of the latter. Conversely, [*ZZ*-**1**H ⊂ **DB24**]$^+$ disassemble very slowly on addition of 2 equivalents of KPF_6. The kinetic analysis of the NMR spectra afforded a rate constant of $7 \times 10^{-6}\,s^{-1}$, which was in excellent agreement with the value calculated from the stability and threading rate constants (see above).

In summary, this system demonstrates that azobenzene units can be effectively employed as end groups in a molecular axle to implement photocontrol on the threading–dethreading rate with a crown ether ring. A system that exploits these results to assemble a nonsymmetric axle-ring system in which it is possible to control the threading direction by light irradiation is described in the next section.

15.3.2 Light-Controlled Unidirectional Transit of a Molecular Axle through a Macrocycle

Herein we describe the operation of a simple supramolecular assembly in which a molecular axle passes unidirectionally through the cavity of a molecular wheel in response to photochemical and chemical stimulation [34]. A system of this kind constitutes the first step toward the construction of an artificial molecular pump; it can also lead to the realization of molecular linear motors based on rotaxanes and rotary motors based on catenanes.

The working strategy of the system composed of a molecular wheel and a nonsymmetric molecular axle is shown in Figure 15.6. The axle molecule is made of three different functional units: (i) a passive pseudo-stopper (D), (ii) a central recognition site (S) for the wheel, and (iii) a bistable photoswitchable unit (P) at the other end. In acetonitrile solution, for kinetic reasons, the wheel threads through the axle exclusively from the side of the photoactive gate in its starting trans configuration (*E*) (Figure 15.6a), affording a pseudorotaxane in which the molecular wheel encircles the recognition site S. Light irradiation converts the *E*-P end group into the bulkier cis form (*Z*), a process that also destabilizes the supramolecular complex (Figure 15.6b). Therefore, a dethreading of the wheel is expected, which occurs by the slippage of the wheel through the D moiety of the axle (Figure 15.6c). The system is brought back to its initial state by photochemical or thermal conversion of the *Z*-P gate back to the *E* configuration (Figure 15.6d). Overall, the photoinduced directionally controlled transit of the axle through the wheel is obtained in accordance with a flashing energy ratchet mechanism [32, 33].

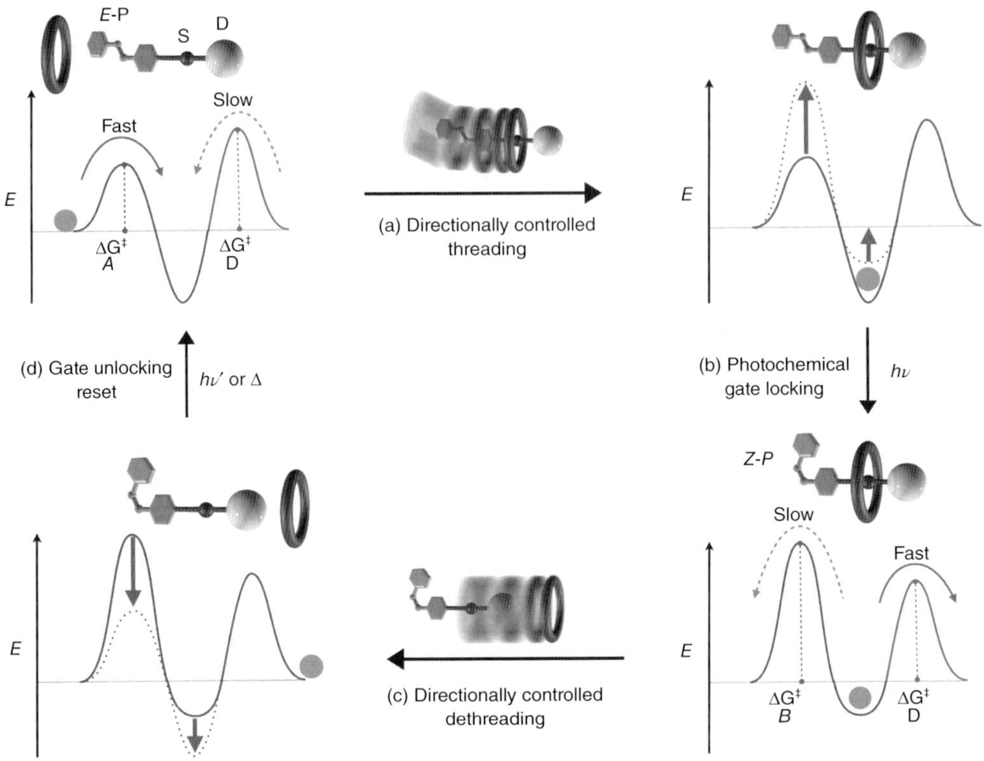

Figure 15.6 Strategy for the photoinduced unidirectional transit of a nonsymmetric axle through a molecular wheel. Simplified potential energy curves (free energy vs wheel-axle distance) for the states shown, describing the operation of the system in terms of a flashing ratchet mechanism.

Two basic requirements are needed for this strategy to work: (i) the kinetic barriers for the slippage of the wheel through the axle end groups [35] should follow the $\Delta G^{\#}(E\text{-P}) < \Delta G^{\#}(\text{D}) < \Delta G^{\#}(Z\text{-P})$ order and (ii) the wheel should form a more stable pseudorotaxane when the axle has the photoswitchable end group in its *E* configuration compared to the *Z* one. It is also important that the differences in the kinetic and stability constants are sufficiently large and that the photochemical interconversion of the P gate between its *E* and *Z* forms is fast, efficient, and reversible.

In the previous section [36], the formation of the supramolecular complexes (Figure 15.7) between the dibenzo[24]crown-8 wheel (**DB24**) and axle *EE*-**1**H$^+$, which is composed of a dialkylammonium recognition site with two azobenzene end units was described. The obtained results showed that the threading–dethreading rate constants are slowed down by four orders of magnitude when the *E*-azobenzene end units are photoisomerized to the *Z* form, practically transforming the complex from a pseudorotaxane into a rotaxane; moreover, the stability constant drops by a factor of 2 [37].

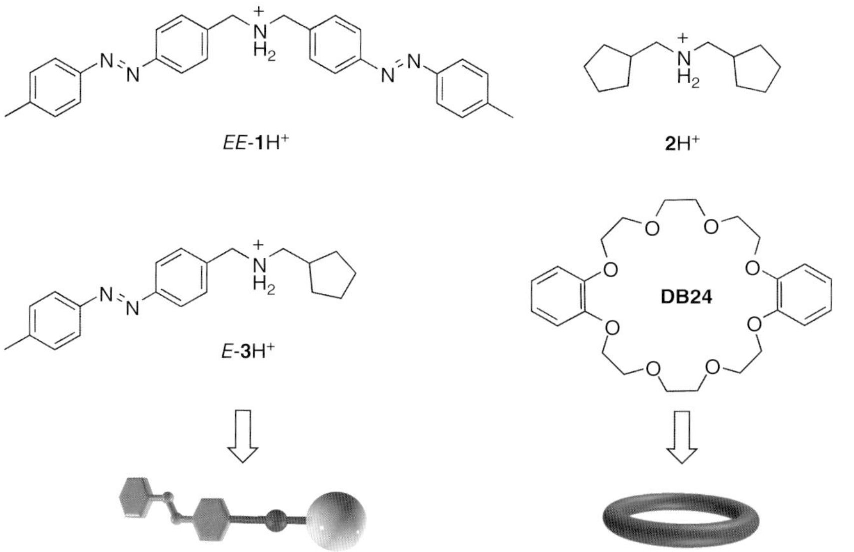

Figure 15.7 Structure formulas and cartoon representation of the examined axle and wheel components.

In a previous work by Stoddart *et al.* [38], it was reported that the bis(cyclopentylmethyl)ammonium ion **2**H$^+$ (Figure 15.7) is complexed by **DB24** to form a pseudorotaxane, with threading and dethreading rate constants that fall in between those observed for *EE*-**1**H$^+$ and *ZZ*-**1**H$^+$ with **DB24** [36]. Hence, the strategy shown in Figure 15.1 can be implemented with a nonsymmetric axle such as *E*-**3**H$^+$ (Figure 15.7).

^{1}H NMR spectroscopic titration experiments showed that in acetonitrile the wheel **DB24** threads through *E*-**3**H$^+$ exclusively from the *E*-azobenzene terminus. Irradiation of *E*-**3**H$^+$ with UV light affords *Z*-**3**H$^+$ in an almost quantitative way. The increased bulkiness of the azobenzene end group on photoisomerization forces *Z*-**3**H$^+$ to thread **DB24** through its methylcyclopentyl terminus. It is noteworthy that the $E \rightarrow Z$ photoisomerization of the azobenzene end group of **3**H$^+$ takes place efficiently also when it is surrounded by **DB24**. Therefore, it is possible to kinetically control the threading–dethreading side of **3**H$^+$ by photoadjusting the steric hindrance of its azobenzene end group.

In contrast with the results found for the [*EE*-**1**H ⊂ **DB24**]$^+$ and [*ZZ*-**1**H ⊂ **DB24**]$^+$ pseudorotaxanes, the stability constants of [*E*-**3**H ⊂ **DB24**]$^+$ and [*Z*-**3**H ⊂ **DB24**]$^+$ are identical within errors. Therefore, the dethreading of *Z*-**3**H$^+$ from the wheel cannot be caused by the same photochemical stimulus that triggers the azobenzene $E \rightarrow Z$ isomerization. Because deprotonation of the ammonium recognition site of [*Z*-**3**H]$^+$ with a base causes the fast dethreading from **DB24**, thereby neutralizing the stoppering ability of the *Z*-azobenzene unit, to promote the disassembly of the complexes, K^+ ions were used, which act as competitive guests for **DB24** [39]. The addition of 2 equivalents of KPF_6 causes the complete dethreading of both [*E*-**3**H

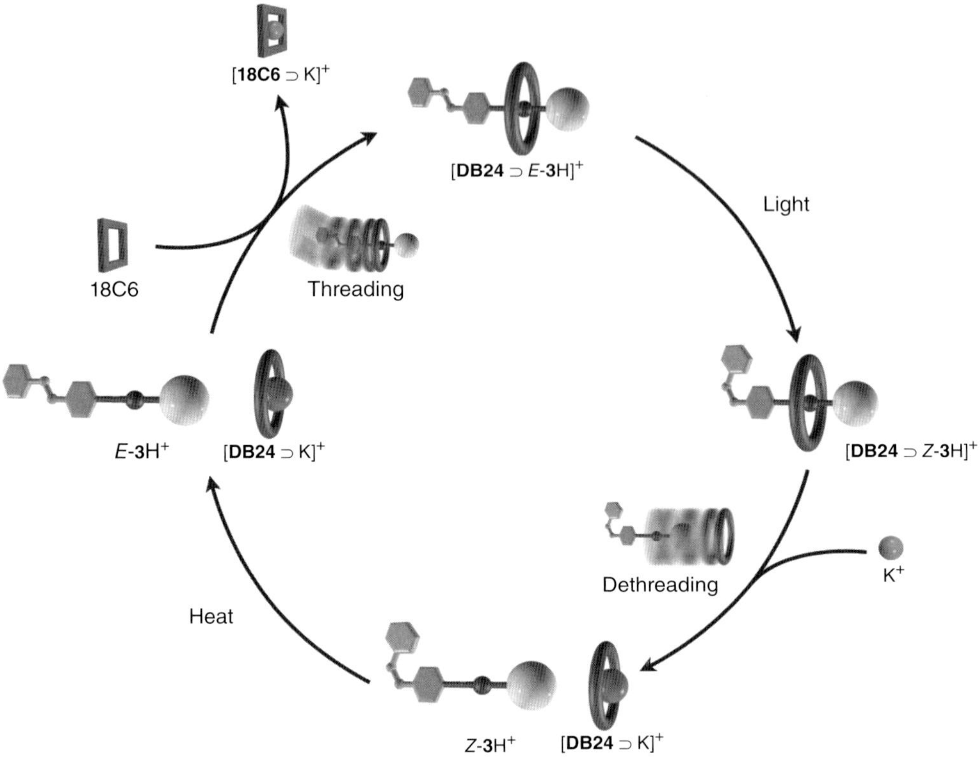

Figure 15.8 Representation of the photochemically and chemically controlled transit of **DB24** through **3**H$^+$.

⊂ **DB24**]$^+$ and [*Z*-**3**H ⊂ **DB24**]$^+$; however, while the K$^+$-induced disassembly of the former complex is fast, the latter one exhibits a dethreading half life of 51 min. This finding indicates that the chemically induced disassembly of *Z*-**3**H$^+$ and **DB24** takes place exclusively by the slippage of the wheel through the methylcyclopentyl unit of the axle.

The results of an experiment that illustrates the directional transit of the axle through the wheel are summarized in Figure 15.8: (i) *E*-**3**H$^+$ pierces **DB24** with its *E*-azobenzene side to form the [*E*-**3**H ⊂ **DB24**]$^+$ pseudorotaxane structure, which equilibrates fast with its free components; (ii) irradiation in the near UV converts quantitatively [*E*-**3**H ⊂ **DB24**]$^+$ into [*Z*-**3**H ⊂ **DB24**]$^+$, characterized by much slower assembly–disassembly kinetics; and (iii) the successive addition of K$^+$ ions promotes the dethreading of *Z*-**3**H$^+$ from **DB24** by the passage of the methylcyclopentyl moiety through the cavity of the wheel. It should be noted that equilibration of the [*Z*-**3**H ⊂ **DB24**]$^+$ complex with its separated components, which would cause the loss of the information on the threading direction of *E*-**3**H$^+$, is much slower than the time required for the activation of the dethreading stimulus (addition of K$^+$). Therefore, after the threading event, the system is "locked" by photoisomerization, and the successive addition of potassium ions causes dethreading in the same direction along which threading of E-**3**H$^+$ has initially occurred. The starting species E-**3**H$^+$ can be fully regenerated by thermal $Z \rightarrow E$ isomerization, and sequestration of K^+ by an excess of 18-crown-6 affords the reassembly of $[E$-**3**H $\subset$ **DB24**$]^+$ and the full reset of the system.

This supramolecular system, however, if it were to be incorporated in a compartmentalized structure (e.g., embedded in the membrane of a vesicle), could not be used to "pump" the molecular axle and generate a transmembrane chemical potential because the wheel component has two identical faces. Despite this deficiency, the described system is characterized by a minimalist design, facile synthesis, convenient switching, and reversibility: all these features constitute the essential requirements for real world applications.

15.4
Conclusion

The results described show that molecular and supramolecular systems capable of performing large-amplitude controlled mechanical movements on light stimulation can be obtained by careful incremental design strategies, the tools of modern synthetic chemistry, and the paradigms of supramolecular chemistry, together with inspiration from natural systems. Such achievements provide encouragement and confidence to devise future bold developments, namely (i) the design, construction, and characterization of more sophisticated artificial molecular motors and machines, showing more complex motions, better performances in terms of stability, speed, switching, and so forth and (ii) the use of such systems to perform molecular-level tasks such as uptake-release, transportation, catalysis, and macroscale mechanical movements. In the context of future developments, it is important to point out that the majority of the artificial molecular machines developed so far operate in solution, that is, in incoherent fashion, without the control of spatial positioning, and it seems reasonable that to find real technological applications, they have to be interfaced with the macroscopic world by ordering them in some way by which they can behave coherently and can be addressed in space. Apart from more or less futuristic applications, the study of motion at the molecular level and the extension of the concept of motor and machine to the nanoscale are fascinating topics for basic research.

References

1. Balzani, V., Credi, A., and Venturi, M. (2003) *Molecular Devices and Machines – A Journey into the Nano World*, Wiley-VCH Verlag GmbH, Weinheim.
2. Pease, A.R., Jeppesen, J.O., Stoddart, J.F., Luo, Y., Collier, C.P., and Heath, J.R. (2001) *Acc. Chem. Res.*, **34**, 433–444.
3. (a) Kay, E.R., Leigh, D.A., and Zerbetto, F. (2007) *Angew. Chem. Int. Ed.*, **46**, 72–191. (b) Balzani, V., Credi, A., and Venturi, M. (2008) *Molecular Devices and Machines – Concepts and Perspectives for the Nano World*, Wiley-VCH Verlag GmbH, Weinheim.

4. (a) van Delden, R.A., Koumura, N., Harada, N., and Feringa, B.L. (2002) *Proc. Natl. Acad. Sci. U.S.A.*, **99**, 4945–4949. (b) Pijper, D.L. and Feringa, B.L. (2007) *Angew. Chem. Int. Ed.*, **46**, 3693–3696. (c) Fang, L., Hmadeh, M., Wu, J., Olson, M.A., Spruell, J.M., Trabolsi, A., Yang, Y.-W., Elhabiri, M., Albrecht-Gary, A.M., and Stoddart, J.F. (2009) *J. Am. Chem. Soc.*, **131**, 7126–7134.
5. (a) Berna, J., Leigh, D.A., Lubomska, M., Mendoza, S.M., Perez, E.M., Rudolf, P., Teobaldi, G., and Zerbetto, F. (2005) *Nat. Mater.*, **4**, 704–710. (b) Harada, A., Kobayashi, R., Takashima, Y., Hashidzume, A., and Yamaguchi, H. (2011) *Nat. Chem.*, **3**, 34–37.
6. Wang, J.B. and Feringa, B.L. (2011) *Science*, **331**, 1429–1432.
7. Green, J.E., Choi, J.W., Boukai, A., Bunimovich, Y., Johnston-Halperin, E., Delonno, E., Luo, Y., Sheriff, B.A., Xu, K., Shin, Y.S., Tseng, H.-R., Stoddart, J.F., and Heath, J.R. (2007) *Nature*, **445**, 414–417.
8. Ambrogio, M.W., Thomas, C.R., Zhao, Y.-L., Zink, J.I., and Stoddart, J.F. (2011) *Acc. Chem. Res.*, **44**, 903–913.
9. (a) Vögtle, F., Richardt, G., and Werner, N. (2009) *Dendrimer Chemistry*, Wiley-VCH Verlag GmbH, Chichester. (b) Newkome, G.R. and Vögtle, F. (2001) *Dendrimers and Dendrons*, Wiley-VCH Verlag GmbH, Weinheim. (c)Fréchet, J.M.J. and Tomalia, D.A. (eds) (2001) *Dendrimers and Other Dendritic Polymers*, John Wiley & Sons, Ltd, Chichester.
10. Deloncle, R. and Caminade, A.-M. (2010) *J. Photochem. Photobiol., C*, **11**, 25–46.
11. (a) Park, S., Kadkin, O.N., Tae, J.-G., and Choi, M.-G. (2008) *Inorg. Chim. Acta*, **361**, 3063. (b) Liao, L.-X., Stellacci, F., and McGrath, D.V. (2004) *J. Am. Chem. Soc.*, **126**, 2181. (c) Grebel-Koehler, D., Liu, D., De Feyter, S., Enkelmann, V., Weil, T., Engels, C., Samyn, C., Müllen, K., and De Schryver, F.C. (2003) *Macromolecules*, **36**, 578. (d) Li, S. and McGrath, D.V. (2000) *J. Am. Chem. Soc.*, **122**, 6795. (e) Weener, J.-W. and Meijer, E.W. (2000) *Adv. Mater.*, **12**, 741.
12. Puntoriero, F., Ceroni, P., Balzani, V., Bergamini, G., and Vögtle, F. (2007) *J. Am. Chem. Soc.*, **129**, 10714.
13. Vögtle, F., Gorka, M., Hesse, R., Ceroni, P., Maestri, M., and Balzani, V. (2002) *Photochem. Photobiol. Sci.*, **1**, 45.
14. (a) Vicinelli, V., Bergamini, G., Ceroni, P., Balzani, V., Vögtle, F., and Lukin, O. (2007) *J. Phys. Chem. B*, **111**, 6620–6627. (b) Hahn, U., Gorka, M., Vögtle, F., Vicinelli, V., Ceroni, P., Maestri, M., and Balzani, V. (2002) *Angew. Chem. Int. Ed.*, **41**, 3595. (c) Balzani, V., Ceroni, P., Gestermann, S., Gorka, M., Kauffmann, C., and Vögtle, F. (2002) *Tetrahedron*, **58**, 629. (d) Balzani, V., Ceroni, P., Gestermann, S., Gorka, M., Kauffmann, C., Maestri, M., and Vögtle, F. (2000) *ChemPhysChem*, **4**, 224.
15. (a) Aumanen, J., Lehtovuori, V., Werner, N., Richardt, G., van Heyst, J., Voegtle, F., and Korppi-Tommola, J. (2006) *Chem. Phys. Lett.*, **433**, 75. (b) Teobaldi, G. and Zerbetto, F. (2003) *J. Am. Chem. Soc.*, **125**, 7388. (c) Köhn, F., Hofkens, J., Wiesler, U.-M., Cotlet, M., van der Auweraer, M., Müllen, K., and De Schryver, F.C. (2001) *Chem. Eur. J.*, **7**, 4126. (d) Baars, M.W.P.L., Kleppinger, R., Koch, M.H.J., Yeu, S.-L., and Meijer, E.W. (2000) *Angew. Chem. Int. Ed.*, **39**, 1285. (e) Sideratou, Z., Tsiourvas, D., and Paleos, C.M. (2000) *Langmuir*, **16**, 1766. (f) Pistolis, G., Malliaris, A., Tsiourvas, D., and Paleos, C.M. (1999) *Chem. Eur. J.*, **5**, 1440.
16. Bergamini, G., Marchi, E., and Ceroni, P. (2011) *Coord. Chem. Rev.*, **255**, 2458–2468.
17. Marchi, E., Baroncini, M., Bergamini, G., Heyst, J., Vögtle, F., and Ceroni, P. (2012) *J. Am. Chem. Soc.*, **134**, 15277.
18. Saudan, C., Balzani, V., Ceroni, P., Gorka, M., Maestri, M., Vicinelli, V., and Vögtle, F. (2003) *Tetrahedron*, **59**, 3845.
19. (a) Barefield, E.K. (2010) *Coord. Chem. Rev.*, **254**, 1607–1627. (b) Fabbrizzi, L., Licchelli, M., Mosca, L., and Poggi, A. (2010) *Coord. Chem. Rev.*, **254**, 1628–1636.
20. Mann, S. (2008) *Angew. Chem. Int. Ed.*, **47**, 5306–5320.

21. Jones, R.A.L. (2004) *Soft Machines, Nanotechnology and Life*, Oxford University Press, New York.
22. For recent outstanding examples see: (a) Ye, A., Kumar, A.S., Saha, S., Takami, T., Huang, T.J., Stoddart, J.F., and Weiss, P.S. (2010) *ACS Nano*, **4**, 3697–3701. (b) Lussis, P., Svaldo-Lanero, T., Bertocco, A., Fustin, C.-A., Leigh, D.A., and Duwez, A.-S. (2011) *Nat. Nanotechnol.*, **6**, 553–557. (c) Kudernac, T., Ruangsupapichat, N., Parschau, M., Macia, B., Katsonis, N., Harutyunyan, S.R., Ernst, K.-H., and Feringa, B.L. (2011) *Nature*, **479**, 208–211.
23. (a) Deng, H., Olson, M.A., Stoddart, J.F., and Yaghi, O.M. (2010) *Nat. Chem.*, **2**, 439–443. (b) Mercer, D.J., Vukotic, V.N.S., and Loeb, J. (2011) Linking. *Chem. Commun.*, **47**, 896–898.
24. (a) von Delius, M. and Leigh, D.A. (2011) *Chem. Soc. Rev.*, **40**, 3656–3676. (b) Coskun, A., Banaszak, M., Astumian, R.D., and Stoddart, J.F. (2012) *Chem. Soc. Rev.*, **41**, 19–30.
25. (a) Haberhauer, G. and Kallweit, C. (2010) *Angew. Chem. Int. Ed.*, **49**, 2418–2421. (b) Kelly, T.R., De Silva, H., and Silva, R.A. (1999) *Nature*, **401**, 150–152. (c) Hernández, J.V., Kay, E.R., and Leigh, D.A. (2004) *Science*, **306**, 1532–1537. (d) Canary, J.W. (2009) *Chem. Soc. Rev.*, **38**, 747–756.
26. (a) Sherman, W.B. and Seeman, N.C.A. (2004) *Nano Lett.*, **4**, 1203–1207. (b) Simmel, F.C. (2009) *ChemPhysChem*, **10**, 2593–2597. (c) Bath, J. and Turberfield, A.J. (2007) *Nat. Nanotechnol.*, **2**, 275–284.
27. von Delius, M., Geertsema, E.M., and Leigh, D.A. (2010) *Nat. Chem.*, **2**, 96–101.
28. Schliwa, M. (ed.) (2002) *Molecular Motors*, Wiley-VCH Verlag GmbH, Weinheim.
29. Sauvage, J.-P. and Dietrich-Buchecker, C. (eds) (1999) *Molecular Catenanes, Rotaxanes and Knots: A Journey Through the World of Molecular Topology*, Wiley-VCH Verlag GmbH, Weinheim.
30. Amabilino, D.B. and Stoddart, J.F. (1995) *Chem. Rev.*, **95**, 2725–2828.
31. Browne, W.R. and Feringa, B.L. (2006) *Nat. Nanotechnol.*, **1**, 25–35.
32. Astumian, R.D. (2007) *Phys. Chem. Chem. Phys.*, **9**, 5067–5083.
33. (a) Chatterjee, M.N., Kay, E.R., and Leigh, D.A. (2006) *J. Am. Chem. Soc.*, **128**, 4058–4073. (b) Serreli, V., Lee, C.-F., Kay, E.R., and Leigh, D.A. (2007) *Nature*, **445**, 523–527. (c) Alvarez-Pérez, M., Goldup, S.M., Leigh, D.A., and Slawin, A.M.Z. (2008) *J. Am. Chem. Soc.*, **130**, 1836–1838.
34. Baroncini, M., Silvi, S., Venturi, M., and Credi, A. (2012) *Angew. Chem. Int. Ed.*, **17**, 4223–4226.
35. Asakawa, M., Ashton, P.R., Ballardini, R., Balzani, V., Belohradsky, M., Gandolfi, M.T., Kocian, O., Prodi, L., Raymo, F.M., Stoddart, J.F., and Venturi, M. (1997) *J. Am. Chem. Soc.*, **119**, 302–310.
36. Baroncini, M., Silvi, S., Venturi, M., and Credi, A. (2010) *Chem. Eur. J.*, **16**, 11580–11587.
37. For other pseudorotaxane-type systems in which the threading-dethreading kinetics can be photocontrolled, see: (a) Hirose, K., Shiba, Y., Ishibashi, K., Doi, Y., and Tobe, Y. (2008) *Chem. Eur. J.*, **14**, 981–986. (b) Tokunaga, Y., Akasaka, K., Hashimoto, N., Yamanaka, S., Hisada, K., Shimomura, Y., and Kakuchi, S. (2009) *J. Org. Chem.*, **74**, 2374–2379. (c) Ogoshi, T., Yamafuji, D., Aoki, T., and Yamagishi, T. (2011) *J. Org. Chem.*, **76**, 9497–9503.
38. Ashton, P.R., Campbell, P.J., Chrystal, E.J.T., Glink, P.T., Menzer, S., Philp, D., Spencer, N., Stoddart, J.F., Tasker, P.A., and Williams, D.J. (1995) *Angew. Chem., Int. Ed. Engl.*, **34**, 1865–1869.
39. Takeda, Y., Kudo, Y., and Fujiwara, S. (1985) *Bull. Chem. Soc. Jpn.*, **58**, 1315–1316.

Index

Discovering the Future of Molecular Sciences, First Edition. Edited by Bruno Pignataro.